Sustainability of Engineered Rivers in Arid Lands

This interdisciplinary volume examines how ten arid or semi-arid river basins with thriving irrigated agriculture are doing now and how they may change between now and mid-century. The rivers studied are the Colorado, Euphrates-Tigris, Jucar, Limarí, Murray-Darling, Nile, Rio Grande, São Francisco, and Yellow. Engineered dams and distribution networks have brought large benefits to farmers and cities, but now these water systems face multiple challenges, above all climate change, reservoir siltation, and decreased water flows. Unchecked, they will result in reduced food production and endanger the economic livelihood of basin populations. The authors suggest how to respond to these challenges without loss of food production, drinking water, or environmental health. The analysis of the political, hydrological, and environmental conditions within each basin gives policymakers, engineers, and researchers interested in the water/sustainability nexus a better understanding of engineered rivers in arid lands.

Jurgen Schmandt is Professor Emeritus of Public Affairs at the University of Texas at Austin and Distinguished Fellow at the Houston Advanced Research Center, where he developed their sustainability program. He specializes in water policy and sustainable development. He previously worked at the German Academic Exchange Service, Organisation for Economic Co-operation and Development (OECD), and Harvard University. He used his work on water and drought in the Rio Grande basin and northeast Brazil to start the Sustainability of Engineered Rivers in Arid Lands (SERIDAS) project in 2013. He has published books and articles on technology and society, environmental regulation, climate change, water policy, and sustainable development.

Aysegül Kibaroglu is Professor of International Relations at MEF University, Istanbul. She was a visiting professor at the University of Texas at Austin. She is the author of *Building a Regime for the Waters of the Euphrates-Tigris River Basin* (2002, Brill) and co-editor of *Turkey's Water Policy: National Frameworks and International Cooperation* (2011, Springer). She has published in *International Negotiation*, *Water International*, *Water Policy*, *International Journal of Water Resources Development*, and *Global Governance*. She is a founding member of the Euphrates Tigris Initiative for Cooperation (ETIC).

Regina M. Buono is a nonresident scholar at the Center for Energy Studies at Rice University's Baker Institute for Public Policy and a doctoral candidate in public policy at the Lyndon B. Johnson School of Public Affairs at the University of Texas at Austin. She is a co-editor of *Regulating Water Security in Unconventional Gas and Oil* (2019, Springer) and has published in *Water International*, *Water Policy*, *Environmental Science and Policy*, and *Environmental Science and Technology*.

Sephra Thomas is a recent graduate from the University of Texas at Austin, where she completed her Master of Public Affairs and Master of Science in Environmental and Water Resources Engineering. Through her graduate career, she has edited and published a report on the sustainability of the Rio Grande / Rio Bravo, developed a MOOC for the NAIAD2020 project at IHE Delft, and researched the intersectionality of water management in Mexico City, Chicago, and the Nile Basin. She hopes to continue her career in water resources management in the intersection between policy and engineering.

Sustainability of Engineered Rivers in Arid Lands

Challenge and Response

Edited by

Jurgen Schmandt

University of Texas at Austin and Houston Advanced Research Center

Aysegül Kibaroglu

MEF University, Istanbul

Regina M. Buono

The University of Texas at Austin

Sephra Thomas

The University of Texas at Austin

CAMBRIDGE
UNIVERSITY PRESS

University Printing House, Cambridge CB2 8BS, United Kingdom

One Liberty Plaza, 20th Floor, New York, NY 10006, USA

477 Williamstown Road, Port Melbourne, VIC 3207, Australia

314–321, 3rd Floor, Plot 3, Splendor Forum, Jasola District Centre, New Delhi – 110025, India

103 Penang Road, #05–06/07, Visioncrest Commercial, Singapore 238467

Cambridge University Press is part of the University of Cambridge.

It furthers the University's mission by disseminating knowledge in the pursuit of
education, learning, and research at the highest international levels of excellence.

www.cambridge.org
Information on this title: www.cambridge.org/9781108417037
DOI: 10.1017/9781108261142

First published 2021

A catalogue record for this publication is available from the British Library.

Library of Congress Cataloging-in-Publication Data
Names: Schmandt, Jurgen, editor. | Kibaroğlu, Ayşegül, editor. | Buono, Regina M., editor. | Thomas, Sephra, editor.
Title: Sustainability of engineered rivers in arid lands : challenge and response / edited by Jurgen Schmandt, Houston Advanced Research Center, the University of Texas at Austin, Austin, United States ; Aysegül Kibaroglu, MEF University, Istanbul, Regina Buono, University of Texas, Austin, Sephra Thomas, University of Texas, Austin.
Description: Cambridge, UK ; New York, NY : Cambridge University Press, 2021. | Includes bibliographical references and index.
Identifiers: LCCN 2020046733 (print) | LCCN 2020046734 (ebook) | ISBN 9781108417037 (hardback) | ISBN 9781108261142 (epub)
Subjects: LCSH: Watershed management. | Arid regions. | Water-supply–Management. | Water resources development. | Dams–Management. | Sustainable development. | Irrigation.
Classification: LCC TC413 .S865 2021 (print) | LCC TC413 (ebook) | DDC 627/.12–dc23
LC record available at https://lccn.loc.gov/2020046733
LC ebook record available at https://lccn.loc.gov/2020046734

ISBN 978-1-108-41703-7 Hardback

We dedicate this volume to our mentors, who prepared us for the task of convening the international research team on the Sustainability of Engineered Rivers in Arid Lands, SERIDAS.

Jurgen Schmandt
Alexander King, British and international civil servant. Co-founder and president of the Club of Rome

and

George P. Mitchell, American businessman who built bridges between entrepreneurship and sustainable development

Aysegül Kibaroglu
Olcay Ünver, Professor of Practice, Ira A. Fulton Schools of Engineering, Polytechnic School, Environmental and Resource Management Program, Arizona State University, USA

and

Patricia K. Wouters, Director, Wuhan International Water Law Academy at the China Institute of Boundary and Ocean Studies (CIBOS) of Wuhan University, P. R. China.

Contents

Contributors

Jose Albiac
International Institute for Applied Systems Analysis (IIASA)-IA2, Laxenburg, Austria

John Berggren
Western Resource Advocates, Boulder, Colorado, United States

Regina M. Buono
Center for Energy Studies, Rice University's Baker Institute for Public Policy, Houston, Texas, United States; and Lyndon B. Johnson School of Public Affairs, the University of Texas at Austin, United States

Michael Cohen
Pacific Institute, Oakland, California, United States

Daniel Connell
Crawford School of Public Policy, Australian National University, Canberra, Australia

Encarna Esteban
Department of Economic Analysis, University of Zaragoza, Spain

Stephanie Glenn
Houston Advanced Research Center, Texas, United States

Taher Kahil
Water program, International Institute for Applied Systems Analysis (IIASA), Laxenburg, Austria.

Melvyn Kay
RTCS Ltd, Northamptonshire, United Kingdom

Douglas S. Kenney
Getches-Wilkinson Center for Natural Resources, University of Colorado Law School, Boulder, United States

Muhammad Khalifa
Institute for Technology and Resources Management in the Tropics and Subtropics (ITT), Technical University of Cologne, Germany

Aysegül Kibaroglu
Department of Political Science and International Relations, MEF University, Istanbul, Turkey

R. James Lester
Houston Advanced Research Center, Texas, United States

Antônio R. Magalhães
CGEE-Center for Strategic Studies and Management, Brasilia, Brazil

Eduardo Mansur
Food and Agriculture Organization, Rome, Italy

Eduardo Sávio P. R. Martins
Cearense Foundation for Meteorology and Water Resources, Fortaleza, Brazil

François Molle
IRD, University of Montpellier, France

Alexandra Nauditt
Institute for Technology and Resources Management in the Tropics and Subtropics (ITT), Technical University of Cologne, Germany

James E. Nickum
International Water Resources Association, Paris, France

Gerald R. North
Texas A&M University, College Station, United States

Lars Ribbe
Center for Natural Resources and Development (CNRD), Institute for Technology and Resources Management in the Tropics and Subtropics (ITT), Technical University of Cologne, Germany

Jurgen Schmandt
Houston Advanced Research Center, Texas, United States; and Lyndon B. Johnson School of Public Affairs, the University of Texas at Austin, United States

Jia Shaofeng
Center for Water Resources Research / Institute of Geographic Sciences and Natural Resources Research, Chinese Academy of Sciences

Justyna Sycz
Institute for Technology and Resources Management in the Tropics and Subtropics (ITT), Technical University of Cologne, Germany

Sephra Thomas
Lyndon B. Johnson School of Public Affairs and Cockrell School of Engineering, The University of Texas at Austin, United States

Olcay Ünver
UN-Water

George H. Ward
Center for Research on Water Resources, The University of Texas at Austin, United States

Part I Introduction

1 Introduction

Sustainability of Engineered Rivers in Arid Lands

Jurgen Schmandt

1.1 THE PROJECT IN BRIEF

In this volume we report on the goals, research methodology, findings, and policy recommendations of the multinational Sustainability of Engineered Rivers in Arid Lands (SERIDAS) project that we started in 2013. SERIDAS explores current and future water supply and demand in ten engineered river basins in arid lands, all equipped with multiple reservoirs that support irrigated agriculture, growing populations, and instream flow. The rivers included in the research project are representative of irrigation-intensive rivers in different parts of the world. According to the UN Food and Agriculture Organization, irrigation (from rivers and aquifers) accounts for 40 percent of global crop production (Alexandratos, 2013). How sustainable are engineered river basins in arid lands? To find the answer we study key challenges faced by the rivers and propose several response strategies.

In this introductory chapter we describe the organization of this volume; the various activities that preceded the start of SERIDAS; the selection and nature of rivers included in the project; the composition of our research team; the workshops we held in 2014, 2016, and 2017; as well as a few side events that were not formally part of SERIDAS but contributed to the research.

1.1.1 Challenge and Response

The challenge-and-response concept was central to Arnold Toynbee's study of human history (Toynbee, 1961). For example, a momentous challenge arose when the previously fertile grasslands of North Africa lost their reliable rains and the hunter-gatherer population was suddenly exposed to dangerous drought. Many of the elders predicted that rainfall would be normal again the next year. Everybody should stay. All perished. A second group moved southward, only to find even more extreme desiccation. They also perished. And still another group wandered east, where they found the Nile. These "communities that responded to the challenge of desiccation by changing their habitat and their way of life, and this rare double reaction was the dynamic act which created the Egyptiac and Sumeric civilizations out of some of the primitive societies of the vanishing Afrasian grasslands" (Toynbee, 1946, p. 70).

We find Toynbee's approach useful. It underlines several points that are central to our project. "Challenge" is caused by nature or society, or a combination of both. "Response" is neither automatic nor deterministic. "I believe that the outcome of a response to a challenge is not causally predetermined, is not necessarily uniform in all cases, and is therefore intrinsically unpredictable" (Toynbee, 1961, p. 257). Humans invent many, even conflicting, response strategies, and to successfully respond to a major challenge such as climate change may require "changing both habitat and way of life."

We use "challenge-and-response" to organize the chapters included in this volume. Following this introduction, Part II of the book, written by topical experts, examines the nature and significance of current challenges to engineered rivers in arid lands. Part III, written by river experts, presents a detailed analysis of past, current, and future conditions of the SERIDAS rivers. Part IV proposes management and policy changes – response strategies – that our team has identified; others may need to be developed. In Part V we summarize our findings and conclusions.

1.1.2 The Road to SERIDAS

I prepared for SERIDAS at different stages of my professional career. In the 1960s I worked at the Organization for Economic Cooperation and Development (OECD). At the time, the organization had just transformed itself from its original Marshall Plan mission – help with the reconstruction of Europe – to new policy goals; in particular, facilitating new ways of economic cooperation among member nations in an age increasingly dependent on advances in science and technology. My boss at the OECD was Alexander King, director of the science and education department. King was a Scotsman, chemist, diplomat, and a British and international civil servant. I was part of a team of young men and women trying to help member countries address policy issues related to science, technology, and education. King was also an outdoorsman with an interest in the environment – not at that time a policy priority. He hit it off with an Italian businessman – Aurelio Pecci – who shared King's thoughts about limits of natural resources and rapid population growth. Together they founded the Club of Rome. "We shared a vision of global dangers that could threaten mankind such as over-population, environmental degradation, worldwide poverty and misuse of technology" (King, 1984, p. 296). The term "sustainable development" was not used, but the new threats and possible remedies were defined, probably for the first time at the level of an international nongovernmental organization. The OECD, many years later, developed principles on water governance and sustainability that we use extensively in the SERIDAS project (Water International, 2018).

In the 1980s, now working at the University of Texas at Austin (UT Austin), I met George P. Mitchell – son of a Greek immigrant, student of geology and petroleum engineering, developer of an environmentally friendly city, founder of one of the largest independent oil and gas companies in the United States, successful developer of fracking natural gas, and – at the same time – ardent supporter of sustainable development. During my years working with Mitchell, I tried to understand how these different goals and achievements coexisted on his agenda (Schmandt, 2010): Natural gas is less harmful than coal and oil; population growth leads to environmental damage; and sustainable development seeks the optimal balance between economic and ecological well-being. In Mitchell's words,

> Sustainable societies are those that can reach and then sustain a decent quality of life for their citizens. To achieve sustainability in the world there must be a balance between ... environmental degradation, deforestation, desertification and food availability and other resources for the amount of people we have.
>
> (Mitchell, 1993)

As one of the first activities at the newly founded Houston Advanced Research Center (HARC), Mitchell sponsored the Woodlands Conference series that introduced the Club of Rome, Alex King, and their first book, *The Limits to Growth*, to the United States. He then funded selected HARC research projects, including the work of a small team that I directed. Going beyond HARC, he created the Cynthia and George Mitchell Foundation to support sustainable development projects in the United States. The government, he concluded, was not ready to use sustainability as a dominant guide for shaping the future. He educated himself on the role played by the National Academy of Sciences (NAS) in using scientific knowledge as a significant input for policy development. With a grant of $1 million he invited NAS to develop a multiyear program designed to better understand the linkages between economic development and humanity's global commons of atmosphere, land, and water (Schmandt, 2010). NAS added US$2 million of their own funds to start the Global Commons program. It was my privilege to work closely with this program for several years.

I worked for George Mitchell from 1984 to 2001 – part of the week in the Woodlands at HARC, where my group developed a research program on climate, water, and sustainability. During the other part of the week I worked on the same agenda, together with colleagues and graduate students, at the Lyndon Baines Johnson School of Public Affairs at UT Austin. This cooperative work led to numerous books and articles. I cite several, focused on water policy (Schmandt et al., 1988), climate change (Schmandt and Clarkson, 1992; Schmandt et al., 2011), and sustainable river management (Schmandt and Ward, 2000; Schmandt, 2006).

1.1.3 Test Study of the Rio Grande / Rio Bravo

Under EPA grant R824799, HARC convened a Mexican–US team for an interdisciplinary study of water supply and demand in the Lower Rio Grande, the 1,200 km river segment on the US–Mexico border that is replenished by two tributaries – the Conchos from Mexico (two-thirds of flow) and the Pecos from the United States (one-third). Amistad and Falcón reservoirs supply river water to the downstream impact area with thirty-one irrigation districts and multiple twin cities on both sides of the Rio Grande (Schmandt and Ward, 2000; Schmandt, 2002; Schmandt et al., 2013). Figure 1.1 presents a map of the study area.

As part of the study, we developed and tested a water budget model to measure water supply and demand in rivers with dominant reservoir operations (BRACERO). We then calculated supply and demand under alternative assumptions about future physical and social conditions in the basin: business as usual, drought of record, drought of record plus reduced inflow from the Conchos tributary, and sustainable development. We found

- Each decade Amistad and Falcón reservoirs lose 5 percent of storage to sedimentation.
- The sub-basin population will double between 2000 and 2030, reaching 4.9 million.
- Agriculture will lose part of its water allocation to cities.
- Most groundwater in this river segment is brackish and not available for alleviating drought conditions.
- In-stream flow will continue to decrease.
- Historically, the Conchos, a tributary from Mexico, provided two-thirds of the main stem flow in the Lower Rio Grande. Development in the Conchos basin will decrease this contribution.
- By 2030, the Lower Rio Grande will carry 30 percent less water than in the recent past.
- Agriculture will be able to cope with less water if farmers practice conservation, adopt less water-intensive irrigation

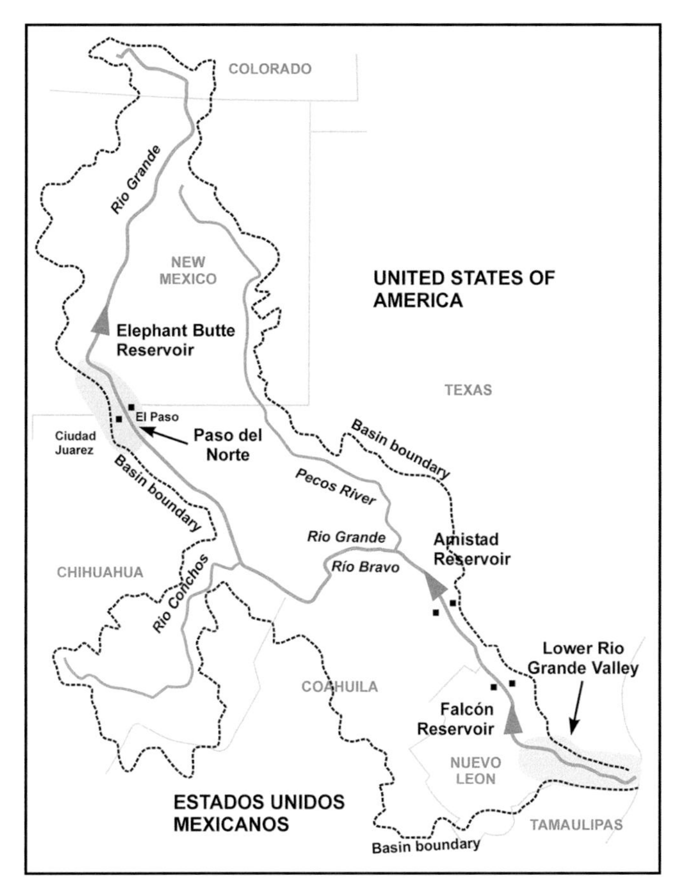

Figure 1.1 The Rio Grande / Rio Bravo region.

technology, and shift to less water demanding crops. This all-important finding is based on a case study of one of the thirty-one irrigation districts in the Lower Rio Grande sub-basin.

- The second most important adaptation strategy focuses on urban demand: Water conservation and repair of leaking distribution networks will allow cities to keep water use in check.

The results of the Rio Grande study gave us pause: Were they representative of engineered rivers in drylands elsewhere? What are the prospects worldwide? Discussing our Rio Grande results with colleagues around the world we found widespread interest in using and expanding the Rio Grande methodology. We started SERIDAS to find answers to these questions.

1.1.4 SERIDAS and ARIDAS

SERIDAS stands for Sustainability of Engineered Rivers in Arid Lands. We created the abbreviation SERIDAS by looking at Projeto Áridas as our model. Projeto Áridas studies drought conditions – both historical and current – in Brazil's northeastern state of Ceará (Ministry of Planning and Budget, 1995). The project was directed, over the course of many years, by Antônio Magalhães, a native of Ceará, an economist, a government and World Bank staffer, as well as one of the contributors to this volume. ARIDAS addresses key problems of northeast Brazil: over a hundred years of drought, widespread famine, poverty, and out-migration.

Having studied the drought-related problems of northeast Brazil, Antônio broadened the scope of his work to address problems in semi-arid drylands worldwide. In 1996, Jesse Ribot, Antônio Magalhães, and Stahis Panagides edited a book called *Climate Variability, Climate Change and Social Vulnerability in the Semi-Arid Tropics*. In 2011, Magalhães directed a particularly influential conference on climate, sustainability, and development in semi-arid regions. The conference led to a dryland call for action (Center for Strategic Studies and Management, 2011). By that time, I had worked with Antônio on several occasions in Fortaleza, the capital of Cearà and Brasilia. We continued our cooperation when he spent a year in Austin with me and a talented team of graduate students at UT Austin (Magalhães and Schmandt, 1998). The focus of ARIDAS on threats to reliable water supply for irrigated agriculture, drinking water, and the environment in arid regions guided us in the development of SERIDAS.

Thus, by the time SERIDAS got underway, we had developed a good understanding of key issues that we report on in this volume: multiple approaches to the management of water resources, the link between water supply and climate change, the importance of river reservoirs for irrigated agriculture in arid regions, the impact of population growth on water availability, and, above all, the need for sustainable development to guide economic growth, environmental protection, and food production. We did not have to start from scratch!

1.2 THE SERIDAS RIVERS

The SERIDAS team studies ten rivers from six continents: the Colorado and Rio Grande from North America, the São Francisco and Limarí from South America, the Nile from Africa, the Jucar from Europe, the Euphrates–Tigris and Yellow from Asia, and the Murray–Darling from Australia (Figure 1.2).

The rivers share several common features (Figure 1.3):

Nature: The main source of river water comes from upstream mountain snowpack or highland rains. Spring floods have brought sediment downstream for millions of years, creating fertile land ready for agriculture. The climate downstream is arid or semi-arid. With highly seasonal streamflow, agriculture is limited to spring and early summer.

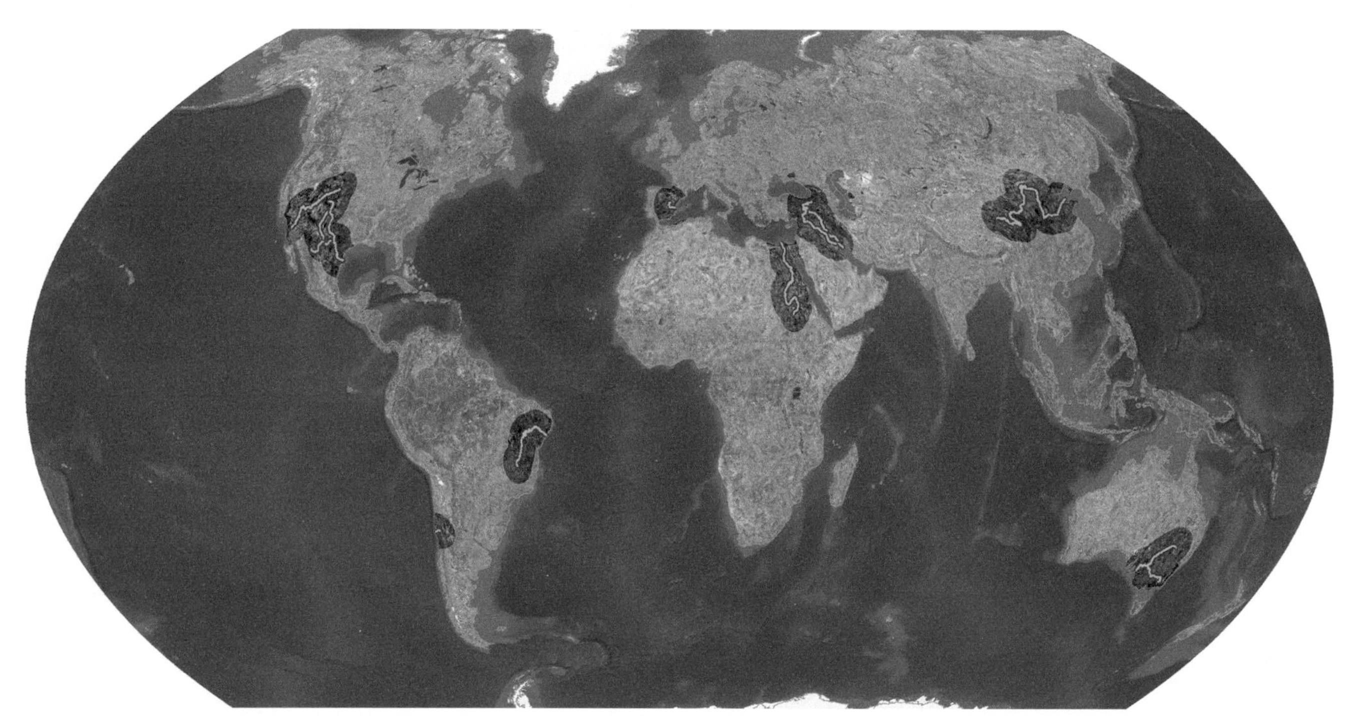

Figure 1.2 The SERIDAS rivers.
Figure by Houston Advanced Research Center (HARC)

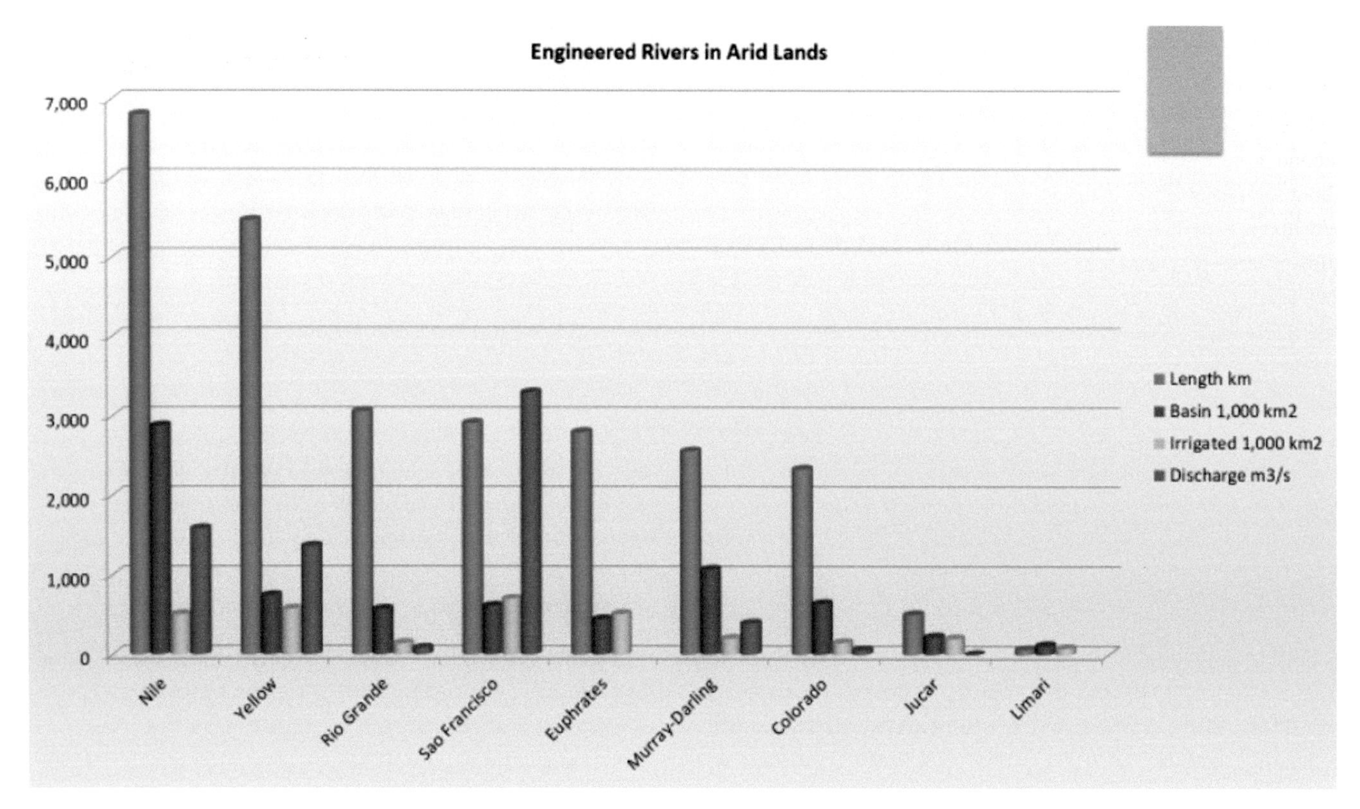

Figure 1.3 Physical characteristics of the SERIDAS rivers.
Figure by the Houston Advanced Research Center (HARC).

Engineering: Modern reservoirs – large dams built with steel-reinforced concrete – are designed to catch the spring flood and release water as needed by irrigated agriculture, cities, and instream flow throughout the year. The new technology dramatically increased agricultural productivity. Yet, reservoirs catch sediment. The result: Fertile soil no longer enlarges the agricultural area of downstream sub-basins. Instead, sediment slowly but surely reduces the storage volume of the reservoir. Removing sediment is technically possible but expensive. The first modern reservoir was built in 1902, when British engineers closed the Aswan Dam – now called the Aswan Low Dam – on the Nile. In 1916 the US Corps of Engineers closed Elephant Butte reservoir on the Rio Grande, an hour's drive north from El Paso, Texas, while simultaneously American civil engineers closed the Boquilla Dam on the Rio Conchos. Over the next several decades, all SERIDAS rivers became engineered rivers. In addition to reservoirs, often several on the same river, a network of distribution channels was built to bring river water to the farms and cities.

Social and ecological: Reservoirs control flooding, are the lifeblood of irrigated agriculture, provide water to industry and growing urban centers, are an important source of hydropower, regulate stream flow, and can be used to maintain environmental flow.

We assembled our team of river experts to project water supply and demand for the SERIDAS rivers in 2013. We also added a second group to the team – scholars who are knowledgeable about the aforementioned physical and social drivers of change. We hoped that the interaction between the two groups would allow river experts and topical experts to learn from each other as they researched new challenges and response strategies. To make this interaction work we organized three workshops, each supported by a grant from a major foundation.

1.2.1 Research Methodology: Austin Workshop 2014

The Cynthia and George Mitchell Foundation funded the first get together of the SERIDAS team, January 13–16, 2014, in Austin, Texas. The river experts presented an overview of current conditions and problems in each basin. The topical experts reviewed current understanding of river challenges. We then discussed how to assess future water supply and demand in response to climate change, reservoir sedimentation, changes in land use, and population growth.

We agreed that our work should be a contribution to the emerging field of sustainability science. Sustainability science aims at "understanding ... the interaction of global processes with the ecological and social characteristics of particular places and sectors" (Kates et al., 2001). A groundbreaking report by the National Academy of Sciences, *Our Common Journey* (National Research Council, 1999), defines additional characteristics of sustainability science. It must be integrative, participatory, and place-based. Policy recommendations should focus on a step-by-step approach, not a one-time blueprint. Our project uses the paradigms outlined by Robert Kates and the National Academy. We see global climate change and variability as key drivers of future conditions in river basins. Our team was formed with the express goal of being able to integrate natural and social science perspectives and methods. We work closely with stakeholders in the river basins when we ask them how more sustainable conditions can be reached in the future. We will present policy recommendations for responding to the challenges encountered in the river basins.

1.2.2 Mid-Term Review: Hanover Workshop 2015

The second SERIDAS workshop, supported by the Volkswagen Foundation, was held June 24–27, 2015, in Hanover, Germany. Team members presented drafts of river chapters and "challenge" chapters. Based on the presentations and discussion at the workshop we decided to use the following plan for projecting SERIDAS river basin futures. The plans were intended to guide the authors of the river chapters but also gave them discretion to make changes depending on availability of data and other circumstances.

1. Time frame: Each basin assessment will project changes for two timeframes: medium- and longer-term futures. The years 2040 and 2060 were mentioned. While comparability among basins is highly desirable, basin chapters use timeframes that can be addressed with available data.
2. River segments: The analysis would proceed by river segments, defined as the stretch of river from source to reservoir, reservoir to reservoir, or reservoir to mouth. Each river segment includes a hydrological and a socioeconomic (irrigated land and cities) component.
3. Change factors: Authors were to seek to quantify physical and social/economic changes due to climate variation and change, storage loss from reservoir sedimentation, groundwater/surface water connection, in-stream flow, land use (agricultural and urban), and population.
4. Future scenarios: We would use three scenarios, if possible, for each of the timeframe years. If data limitations made this impossible, authors would focus on 2040. All scenarios will combine the results of projected changes (#3 in this list) with (A) a business-as-usual scenario (current ways of using and managing water) and (B) a worst-drought-since-reservoir-closing scenario (reduced water supply).

5. Stakeholder input: Authors were encouraged to use surveys, site visits, and workshops to receive input on plans and options for using water more efficiently and sustainably. Input was desired from water users (irrigations districts, cities, industry, native populations), water management agencies, and nongovernmental organizations.
6. Integration of findings: Each basin analysis would summarize results by river segments and then by the entire river basin. Findings would be reported in four sections: water budget (quantitative analysis); water governance (qualitative analysis); environmental conservation; and policy advice on strengthening basin resilience and sustainability.

1.2.3 Policy Recommendations: Bellagio Workshop 2017

The workshop was supported by the Rockefeller Foundation, which invited us to meet at their conference site in Bellagio, Italy. After discussion we agreed that our policy recommendations would be based on these principles,

A reservoir-dominated river in arid lands is sustainable when five conditions are met,

1. Nature's water supply, averaged over the period of the most severe drought experienced in the historical record, delivers a dependable reservoir yield sufficient to meet human and ecological needs in the sub-basins created by engineered river structures.
2. To keep within the limits of the river's dependable yield, water managers and stakeholders jointly and proactively search for ways to use water more efficiently.
3. An ecologically prudent level of in-stream flow is maintained or restored.
4. Whenever observed or projected changes in the natural system or human actions modify river flow, the dependable yield of the reservoirs is recalculated and water management agencies, after consultation with governments and stakeholders, adjust rules for water allocations to match the new levels of dependable yield.
5. Individual reservoir impact assessments, including their dependable yield assessments, are aggregated into a basin-wide sustainability plan, which compares the results of reservoir assessments to existing water sharing agreements between upstream and downstream users as well as agricultural, urban, and industrial water right holders. If adjustments are necessary, new agreements are negotiated which should be based on equity considerations embodied in international law and the history of cooperation in the basin.

This new definition of sustainability of engineered rivers in arid lands guides the work presented in this volume. We amplify the five sustainability criteria in the chapters that follow.

1.3 SUPPORTING ACTIVITIES

In support of the SERIDAS project, teams of faculty and graduate students at the Lyndon B. Johnson School of Public Affairs conducted two studies. The first focused on two of our rivers: the Euphrates–Tigris and Rio Grande / Bravo (Kibaroglu and Schmandt, 2016). The second developed detailed policy recommendations for water management in the Paso del Norte sub-basin of the Rio Grande / Bravo (Schmandt and Stolp, 2018).

We also contributed to a joint workshop organized by the US and Mexican Academies (National Academies of Sciences, Engineering and Medicine, 2018). Findings of these activities will be referred to in several chapters of the volume.

REFERENCES

Alexandratos, N. and Bruinsma, J. (2013). *World Agriculture towards 2030/2050: The 2012 Revision*, ESA Working Paper No. 12-03, Rome: FAO, p. 151.

Center for Strategic Studies and Management (2011). *A Drylands Call for Action: Declaration of Fortaleza*, Brasilia, DF.

Kates, R. W. et al. (2001). *Sustainability Science*, John F. Kennedy School of Government, Harvard University, Faculty Research Working Paper RWP00–018.

Kibaroglu, A., Schmandt, J., and graduate students (2016). *Sustainability of Engineered Rivers in Arid Lands: Euphrates-Tigris and Rio Grande/Bravo*, University of Texas, LBJ School of Public Affairs. Policy Research Project Report 190.

King, A. (1984). The Launch of a Club. In P. Malaska and M. Vapaavuori, eds., *Club of Rome Dossiers 1965–1984*, Available at www.clubofrome.fi/index.php?id=14,43,0,0,1,0 pp. 55–59.

Magalhães, A. R., Schmandt, J., and graduate students (1998). *The Road to Sustainable Development: A Guide for Nongovernmental Organizations*, University of Texas, LBJ School of Public Affairs. Policy Research Project Report 120.

Ministry of Planning and Budget (1995). *A Strategy for Sustainable Development in Brazil's Northeast: Projeto Áridas*. Brasilia.

Mitchell, G. (1993). Comment in *The Woodlands Forum 10*, no. 2, p. 3.

National Academies of Sciences, Engineering and Medicine (2018). *Advancing Sustainability of U.S.–Mexico Transboundary Drylands: Proceedings of a Workshop*. Washington, DC: The National Academy Press.

National Research Council (1999). *Our Common Journey: A Transition toward Sustainability*. Washington, DC: National Academy Press.

Ribot, J. C., Magalhães, A. R., and Panagides, S. S. (1996). *Climate Variability, Climate Change and Social Vulnerability in the Semi-Arid Tropics*. Cambridge: Cambridge University Press.

Schmandt, J. (2002). Bi-national Water Issues in the Rio Grande/Rio Bravo Basin. *Water Policy*, 4(2), 137–155.

(2006). Bringing Sustainability Science to Water Basin Management. *Energy*, 31, 2014–2024.

(2010). *George P. Mitchell and the Idea of Sustainability*. College Station: Texas A&M University Press.

Schmandt, J. and Clarkson, J. (1992). *The Regions and Global Warming: Impacts and Response Strategies*. New York and Oxford: Oxford University Press.

Schmandt, J. and Ward, C. H. (2000). *Sustainable Development: The Challenge of Transition*. Cambridge: Cambridge University Press.

Schmandt, J., North, G. R., and Clarkson, J. (2011). *The Impact of Global Warming on Texas*, 2nd ed. Austin: University of Texas Press.

Schmandt, J., North, G. R., and Ward, G. (2013). How Sustainable Are Engineered Rivers in Arid Lands? *Journal of Sustainable Development of Energy, Water, and Environment*, 1(2), pp. 78–93.

Schmandt, J., Smerdon, E., and Clarkson, J. (1988). *State Water Policies*. New York: Praeger.

Schmandt, J., Stolp, C., and graduate students (2018). *Sustainable River Management on the U.S./Mexico Border: Recommendations for the Paso del Norte*, University of Texas, Austin, LBJ School of Public Affairs, Policy Research Project Report 202.

Toynbee, A. J. (1946). *A Study of History: Abridgement of Volumes I–VI*. London: Oxford University Press, p. 70.

(1961). *Reconsiderations*: Vol XII of *A Study of History*. London: Oxford University Press, pp. 254–262.

Water International (2018). Special Issue: *The OECD Principles on Water Governance*.

Part II Challenge

2 Global Climate Change and the Rivers

Gerald R. North

2.1 INTRODUCTION

Most readers of this book will already have some knowledge of the expected environmental changes in the twenty-first century due to the likely increase in greenhouse gases, primarily carbon dioxide. The purpose of this chapter is to present a review of the greenhouse effect and to give a status report on where the science stands today. Nearly all scientifically literate persons today agree with the proposition that the planet is warming and will continue to warm due to the increased concentrations of atmospheric carbon dioxide, but a brief recap of the evidence is useful for those whose concentration is not within the confines of the atmospheric sciences.

The key concept is the ongoing imbalance in the rate of energy received from the sun via absorbed light and the rate of emission of infrared radiation energy into space. When the two rates are equal, we are in balance and the climate will not be forced to change. The system is complicated and active areas of research continue into discerning all the ways the two streams of radiation can participate. But we know that if there is an imbalance, for example, more absorbed from sunlight than emitted back to space, we are likely to have a warming planet.

A simple example might be helpful. Consider weight loss or gain in an adult human. If a person ingests 2,500 calories/day and his/her weight remains steady, we can assume that his/her body is in a state of equilibrium, at least from an energy rate standpoint. Suppose the person increases food input to 3,000 calories/day. Then if the person's physical activity level does not change, we can expect the body weight to increase to a new, higher level after allowing suitable time for the new equilibrium to establish itself. We might even define "weight sensitivity" to be the change in body weight from one level to the new equilibrium weight in kilograms. Like the climate system, the human body is immensely complicated; however, the truth of the previous experiment seems obvious. Also, in analogy we can list a number of things that might change weight sensitivity. For example, additional food intake might induce more activity, thereby lowering sensitivity. On the other hand, more calories taken in might make the subject sleepy and induce positive feedback, increasing sensitivity. We could go on listing the many factors that could intervene, but if we average over many people and many external conditions, we are likely to find that weight sensitivity is positive, but with some wide error bars (uncertainty).

The case of diet and its effect on weight gain is clear. One big number, the energy budget, explains a lot. We know that, with all other things held constant, calorie input rate compared to the body's energy losses (perspiring, exhaling moist air, exercise) is key. It is inconceivable that eating fewer calories will lead to weight gain. I will come back to that other complex system the climate change situation later. In both situations one big consideration is key to understanding the problem: the conservation of energy.

2.1.1 Indicators of Past Climate Changes

Before launching into the theory of climate change, let us consider a few pieces of empirical information that lead us to see that climate is indeed changing and, from the magnitude of known mechanisms that drive global climate, can be estimated well enough to suggest that changes over the last century are unusual compared to changes over longer periods of time. Several independent research centers around the world have assembled data that can be aggregated to form a global average temperature. They are all in good agreement with the data shown in Figure 2.1 (taken from the very accessible website NASA/Goddard Institute of Space Studies). Simulations and statistical analyses make it clear that the trend is well outside the bounds of natural variability. The arrow I have imposed indicates an upward trend of about 1°C per century. Inspection of empirical data spanning several thousand years finds no century with a growth rate even remotely comparable. In fact, the current temperature is probably higher than it has been in over a millennium. This evidence is shown in Figure 2.2.

Why should this be the case? Here is a tip: The world underwent an unprecedented transformation in approximately 1781, the year of the patent for the invention of a continuously operating steam engine by Scotsman James Watt. Once this device was

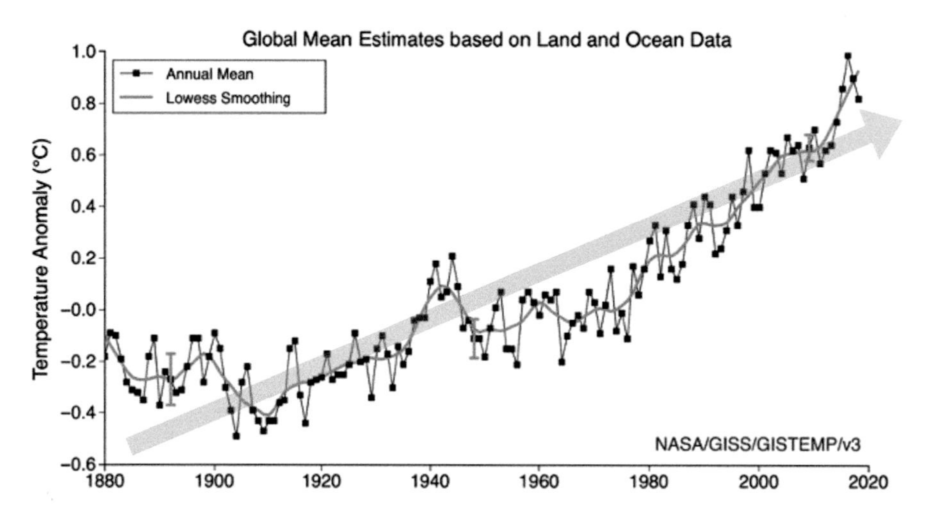

Figure 2.1 Last 200 years of global average surface temperatures. The arrow (added by the author) indicates the trend.
Figure by NASA/GISS (https://data.giss.nasa.gov/gistemp/graphs/), modified by the author in grayscale

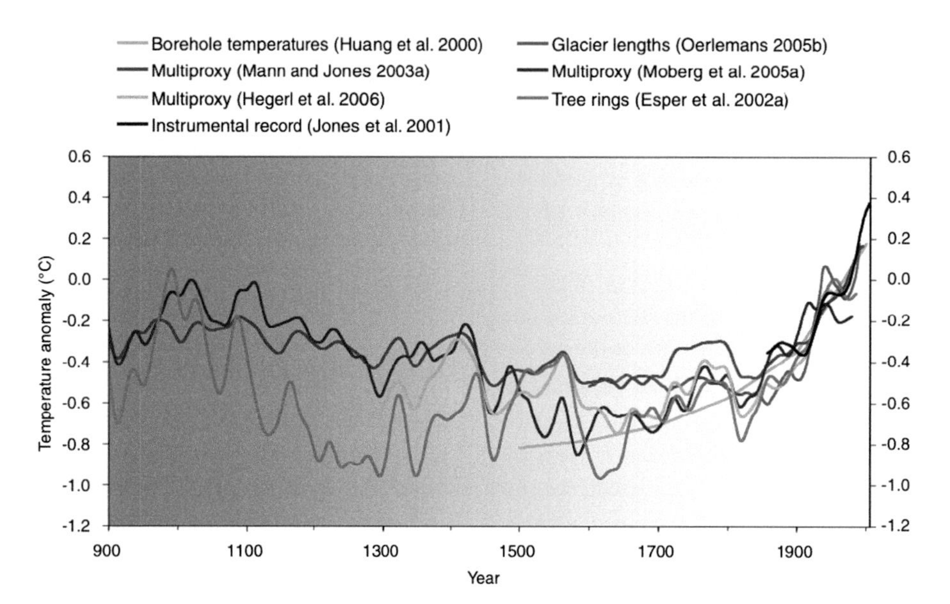

Figure 2.2 Last 1100 years of surface temperatures according to various proxy estimators. The background shading becomes darker the older the measurement, indicating that our certainty is murkier and murkier.
Source: Figure S-1 of Surface Temperature Reconstructions for the last 2000 Years. (National Research Council, 2006)

in play and water due to seepage could be drawn from coal mines, fossil fuels became available at low cost. At once, railroads could not only supply their own engines but steam-driven trains (and ships) could haul coal and other goods everywhere in bulk at low cost. Factories could produce goods with stationary steam engines. The Industrial Revolution was stoking up. Figure 2.3 shows how the beginnings of the exponential rise in CO_2 was coincident with this social event. Exponential growth of world population was enabled by the technological shift to massive rates of energy production. Figure 2.4 shows how global CO_2 concentration continues to increase exponentially along with these other driving factors.

2.1.2 Sea Level and Acidity

Further significant secular changes besides increases in temperature are occurring in the Earth System. Some of the changes are the result of global temperature increases. For example, there are increases in the occurrence of wildfires and heatwaves at local scales. But at the global scale we see sea level rising and at rates above those in the early records (from tide gauges and proxy data). Now we have satellite-borne instruments (radar altimeters) that can measure the rise more accurately. There are two reasons for the rise: (1) the heating of the oceans causes an expansion of the volume of water, and therefore depth, since shorelines are

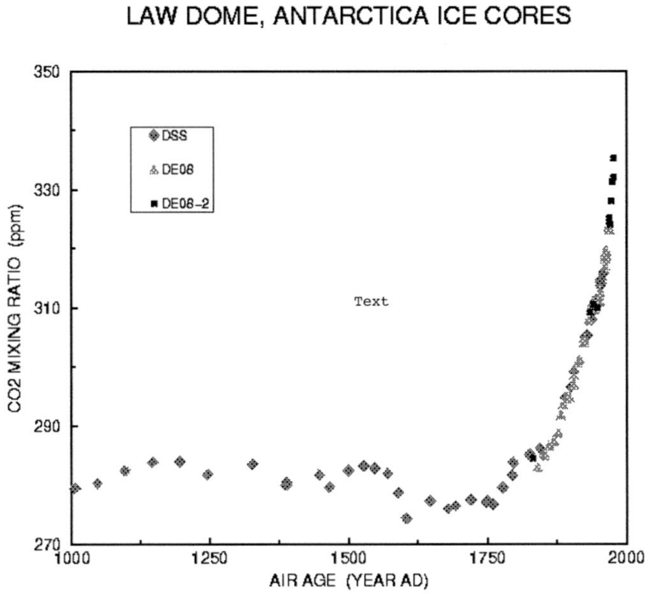

Figure 2.3 Long term CO_2 curve based on ice core data from Law Dome in Antarctica.

Figure adapted from Etheridge et al. (1996), made available on the Carbon Dioxide Information Analysis Center of the U.S. Department of Energy http://cdiac.ess-dive.lbl.gov/trends/co2/graphics/lawdome.gif

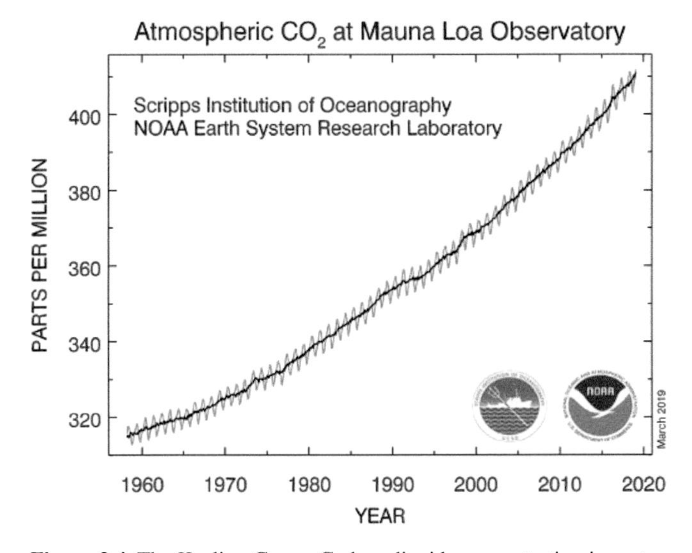

Figure 2.4 The Keeling Curve. Carbon dioxide concentration in parts per million background molecules in the atmosphere versus time as measured at Mauna Loa Observatory in Hawaii.

Figure modified to grayscale from the NOAA website www.esrl.noaa.gov/gmd/ccgg/trends/full.html

essentially fixed. This happens whether the heat absorbed is confined near the surface or is mixed throughout the sea's depths; and (2) the melting of glaciers takes water substance from land surfaces (receding mountain glaciers and the great ice sheets) – this happens even if the losses of ice are due to evaporation – and dumps it into the oceans as stream runoff or rain. The record of sea level increase is shown in Figure 2.5.

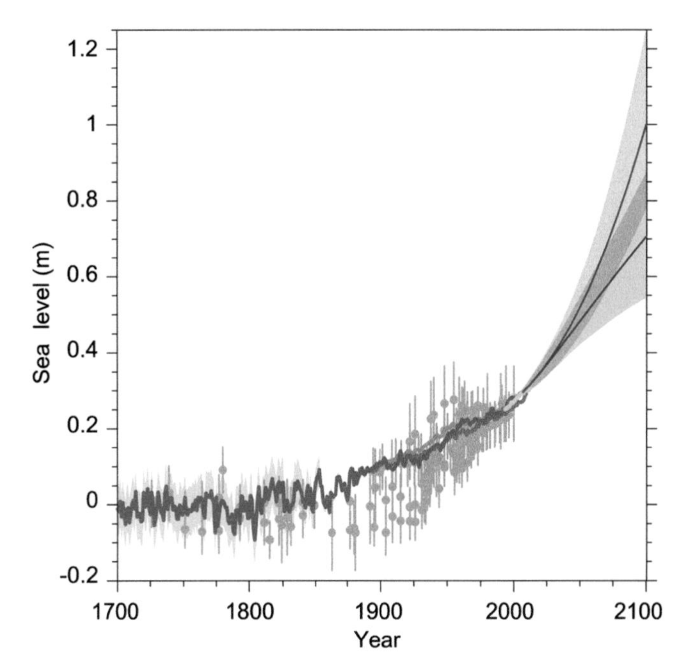

Figure 2.5 Past and projected sea level changes. Composites of paleoclimatic indicators (purple), tide gauge data central and ranges (blue, red, green), satellite altimeter (light blue). Estimates of future sea level based upon the CMIP5 suite of climate models for RCP2.6 and RCP8.5 (the latter is roughly "business as usual"). The shaded areas for the models indicate the range of uncertainty based upon the spread among models and the spread among emission scenarios.

Figure from IPCC (2013) TS TFE.2, figure 2

A second global phenomenon is the increasing acidity of open ocean waters. This is not due to warming but rather to the increasing partial pressure of CO_2 pressing on the ocean's surface. Such a trending partial pressure (proportional to concentration of CO_2) causes more of the gas to dissolve (as in your carbonated soft drinks) passing through a few chemical steps to form carbonic acid (H_2CO_3), a mild acid but enough over time to dissolve some calcium shells, in particular those in coral reefs. The acidity of the ocean water is quantified by the pH, where a value of 7 is neutral, 8 is slightly basic, and 6 slightly acidic. Ocean water is normally slightly basic with a pH of 8.10, but it has been steadily decreasing at about 0.03 per decade. Consistent with this lowering of the pH is the concentration of dissolved CO_2, which is increasing at a rate close to the rate of increase in the atmosphere. These and related data are collected at the Pacific Marine Environmental Laboratory (NOAA) available at www.pmel.noaa.gov/co2/file/Hawaii+Carbon+Dioxide+Time-Series.

2.1.3 Cryosphere Indicators

The cryosphere is the aggregate of the parts of the Earth System that are composed of solid water (ice) seasonally or throughout the year. Examples are sea ice, glaciers, ice sheets, snowpacks, and permafrost. All of these are changing to have a lower mass of water in solid form. Some familiar examples are from the area in the Arctic that is covered with sea ice (the kind that is not connected to land through, for example, a glacier's tongue extending into the sea). Sea ice largely still consists of ice formed from briny or salty waters, whereas ice on the land is generally formed from precipitation and is therefore fresh water. Note that sea level is not affected by melting sea ice that is floating but melting of land-based ice constitutes a net increase of sea water. Many factors are involved in the formation of sea ice in the Arctic, such as the oceanic and atmospheric wind currents that move floating ice around. However, the primary factor is increasing temperature. The problem is that it is getting warmer, leading to less ice cover. Figure 2.6 shows a remarkable decline in sea ice area in the Arctic. The geographical situation around Antarctica is different. Sea ice is not decreasing in the way it is in the north. Again, the current systems in adjacent oceans are the cause of resistance to declining ice coverage in the surrounding seas.

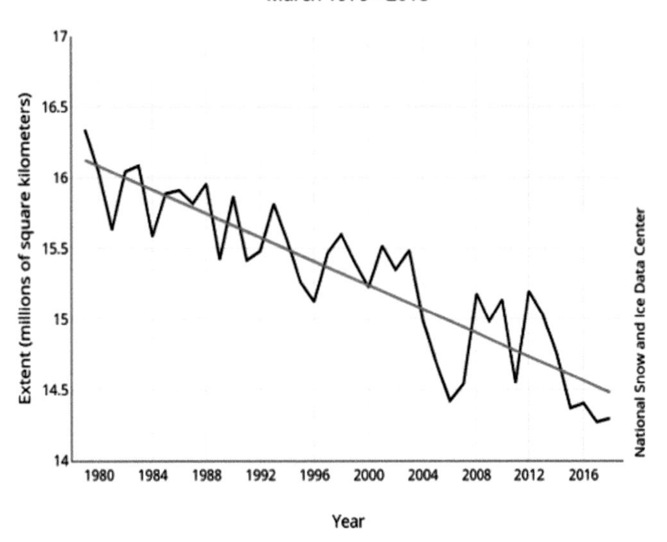

Figure 2.6 This graphic shows the declining area (millions of square kilometres) covered by sea ice in the Arctic Ocean. These measurements are collected and archived by the National Snow and Ice Data Center in Boulder, Colorado.
These and other data are found at http://nsidc.org/arcticseaicenews/

Ice of land-based polar regions is on the decline as well. Figure 2.7 shows the record of decline in ice mass on Greenland based on satellite gravity measurements (GRACE satellite system). Greenland alone has enough ice stored to raise sea level about 7.34 meters (24 feet) and Antarctica, 58.3 meters (189 feet). Of course, the complete melting of these ice sheets would take thousands of years. Nevertheless, calving of ice streams into the oceans is not yet an exact science, and the decay of an ice sheet is much faster than the building of one.

2.2 STATUS OF GLOBAL CLIMATE CHANGE RESEARCH

2.2.1 Planetary Energy Balance

A simplified discussion of how the global average energy budget works will suffice for our present purposes. As mentioned in the previous section, the annual and global average rate of absorbed incoming radiation (sunlight) and the same for (infrared) outgoing radiation emitted by the Earth provide the inputs. On average, sunlight absorbed is steady except for some fluctuations (due to aerosol loadings and cloud changes) that are of little consequence in this problem. On the other hand, emitted radiation to space can be reduced by several factors. If there were no gases in the atmosphere, outgoing radiation would simply be an increasing function of surface temperature. We can fill in the N_2,

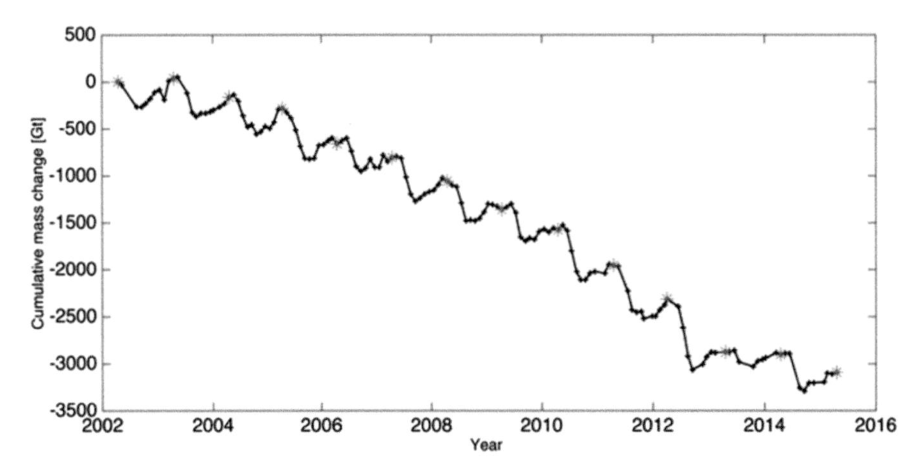

Figure 2.7 This graphic indicates the dramatic decline in mass of the ice in Greenland, as estimated from the GRACE satellites.
These and other data can be found at https://arctic.noaa.gov/Report-Card/Report-Card-2015/ArtMID/5037/ArticleID/219/Greenland-Ice-Sheet

O_2, and Argon (these comprise more than 99 percent of the molecules in the air) with the same dependence on surface temperature. Such a planet would be very cold, probably as much as 20°C colder than at present (probably leading to an iced-over planet). If a greenhouse gas (examples are CO_2, CH_4, and O_3 – in addition, water vapor, H_2O, also gets into the act; moreover, the greenhouse gases do not affect the passage of sunlight) is introduced even in trace amounts, some of the radiation upwelling from the surface will be absorbed at altitudes above the ground.

Greenhouse gases (except for water vapor, which is complicated because of condensation, etc.) have long lifetimes in the atmosphere and they mix vertically to an equilibrium "mixing ratio" with background gases. For CO_2 at present that is approximately four hundred and ten parts per million (i.e., 410 CO_2 molecules per 1 million background molecules, the others are even less) and because it is well mixed, the same ratio holds from bottom to top of the atmospheric column. Background molecule concentration falls off nearly exponentially as a function of elevation. Hence, if we double the concentration of CO_2 at ground level, after a few days it will double all the way up. Incidentally, temperature falls off nearly linearly as we go up at about 6°C/km up to about 10 km in middle latitudes and higher in the tropics.

CO_2 molecules absorb upwelling infrared radiation and emit (both up and down directions) infrared radiation. Emission into space occurs at a characteristic altitude that depends on the concentration of CO_2. Let us say that this happens at about 6 km above the surface. If we double the concentration, that characteristic altitude will rise (to where there the same number of absorbing CO_2 molecules reside above the new level). But it is colder up there than at the old emission level. A colder emitting "surface" means less radiation emitted to space than in the case where the emission level was lower and warmer.

If the system were in equilibrium before the doubling took place, we would find the same energy entering from sunlight, but cooling to space would decrease, leading to net warming. This imbalance will lead to warming of the entire column of air, including the surface, because of vertical mixing. Warming will continue until equilibrium is restored. This warming of the whole column of air brings the planet back into equilibrium.

Very reliable measurements and calculations suggest that if the air is dry when we double CO_2 levels, the temperature would rise by about 1.0°C, and it might take some years (because of the oceans – a deep pot of water – to heat up quickly) to re-establish equilibrium. But there is an additional issue: feedbacks. The easiest of these amplifiers to describe is water vapor feedback. Here goes! If the planet's surface warms a little, the water vapor concentration in the air will increase due to evaporation from the surface. Recall that water vapor will increase due to increased evaporation. It is complicated because of condensation in the air (clouds), but many experiments suggest that the effect of this feedback mechanism is such that it would cause warming to double to about 2.0°C. There are other feedbacks in the system, and they are less well understood. They include clouds, which affect both the incoming and outgoing radiation streams. It is not clear at this time whether cloud feedback is positive or negative since the two nearly cancel, but a survey of climate models suggests that the results of this and other feedbacks might bring total warming for doubling CO_2 to 3.0 ($\pm$1.5) °C.

Many studies have been conducted over the last few decades toward reducing this uncertainty. At this point this is about the best we can do. The value is definitely positive, and if we look at the spread of "climate sensitivities" across twenty or more different climate models from around the world this is also roughly the uncertainty we find. Figure 2.8 shows a schematic diagram of the flows of energy from the surface of the planet to outer space. The values in the figure come from many years of measurements.

2.2.2 General Circulation Models

The general approach to understanding the climate system is through models. These models are implemented on the world's fastest and most capacious (quick-access and huge storage capacity) computers. There is a general consensus that climate models can be sorted into a hierarchy of models, the simplest considering only the energy budget of the whole globe (as we have been describing), up to models that attempt to simulate the actual fluidic motions of the atmosphere and the oceans. Modeling the motions of these two media requires solving the equations of motion according to Newton's Second Law of Motion. The forces on an individual fluid element include gravity, horizontal and vertical (vector) components of pressure gradients, Coriolis Forces (due to the Earth's rotation), buoyancy, etc. In addition, mass must be conserved as a constraint, and that involves another equation. There is also an equation for pressure as it relates to density and temperature. This distribution throughout the atmosphere provides for buoyancy of grid-box-sized air or ocean parcels. Water vapor circulates in the atmosphere – it plays a role in density as well because energy is released as water changes phase. The analogue to water vapor in the ocean is salinity, and it – combined with temperature – determines the distribution of buoyancy in the oceans. Clouds and snow/ice cover contribute to the reflectivity of the Earth system to sunlight. Finally, the laws of thermodynamics determine the heating of the system. Heating occurs when there is an excess of absorbed or emitted radiation in a parcel of air/ocean. Changes of phase also transfer heat energy in an individual parcel of air wherein chemical or phase changes might occur. Finally, there are forces on parcels at interfaces: ground/atmosphere, seafloor/ocean, and air/ocean.

All of these can be expressed as partial differential equations involving independent variables consisting of spatial dimensions

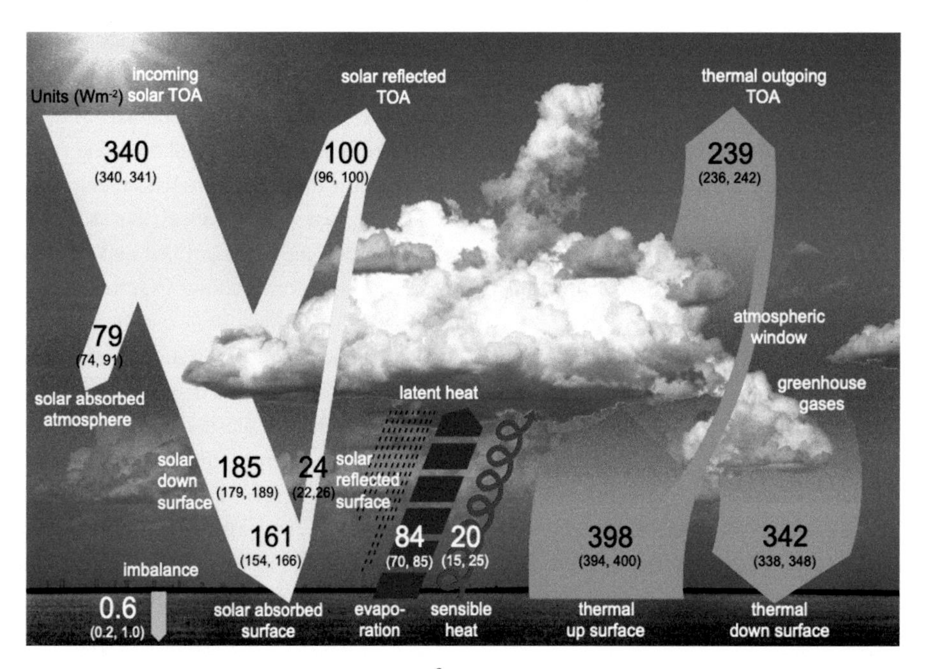

Figure 2.8 Schematic of balance of energy flux densities (Watts/meter^2) showing the different streams of energy into and out of the atmospheric system at the top and the flows between components within the system. Numbers in parentheses show the range with data from different studies. Modified to grayscale from figure 2.11 of IPCC (2013)

(latitude, longitude, elevation), as well as time. Internal boundaries are considered in ways that are subtle. The dependent variables are velocity of parcels, pressure, temperature, humidity, salinity, etc. There are several sticky issues and trade-offs in setting up such a system for solution on a computer. First and foremost is the sheer number of grid boxes that must be implemented in the numerical simulation. For example, a horizontal resolution of 20 by 20 km (divided into the surface area of the Earth, 500 million km^2) involves 1.25 million grid boxes. Multiply this with fifty layers in the vertical to obtain over 60 million boxes in the atmosphere. We are dealing with about ten variables, so we must evaluate all of them in every box at every time step moving forward. So, we now have at least 600 million arithmetic operations just to advance the solution one time-step, which involves about one minute of advance in the simulation. To get an hour of simulation time you need 36 billion accesses to the boxes to do a single arithmetic operation on all the variables. Our task is to do a simulation of a century (36 billion times 24 times 365 times 100 = 3.15 times 10^{16} calculations/century). With the newest computers and programming methods (especially parallel computing) it takes about five days of wall clock time to simulate a century of simulation time (R. Saravanan, private communication, April 5, 2018). Just a few years ago, computer speed was a bigger limit to the ability to gather statistics from model simulations.

The Intergovernmental Panel on Climate Change (IPCC) is a United Nations-supported consortium of over 100 governments that provide input to periodic assessments of the state of the art of climate research. There have been five such reports issued at 5- to 7-year intervals. The latest, Report Number Five (AR5) was released in 2013. Of special interest here is the latest news from climate model simulations. There is a total of roughly two dozen distinct climate models run by groups from nearly as many countries. There is some correlation among them because many of the individual components within the models (such as simulation of the fluid motions of the atmosphere or radiative transfer routines) are similar or imported from other groups. On the other hand, such important formulations as boundary treatments, and cloud physics and chemistry are quite different from one model to another. Even these small differences lead to different climate sensitivities and other minor deviations. When compared, the model sensitivities (to doubling CO_2 and waiting for equilibrium) show that the average across models continues to be about $3.0\pm1.5^{\circ}$C, a mean value and spread that have not changed much over the last four decades. The models continue to improve in the sense that we are gaining a better understanding of the simulations and their diverse results. The improvements are incremental from one report to the next, but confidence in the great General Circulation Models (GCMs) has continued to increase.

Figure 2.9 shows how as time evolves modelers have been able to add more and more physical components to simulation programs. This raises an interesting quandary for the community. Do we continue to improve the model physics so that we get the physics right in the atmosphere component before we begin coupling in other components of the climate system such as sea ice, hydrology, vegetation, etc.? The problem is that each new

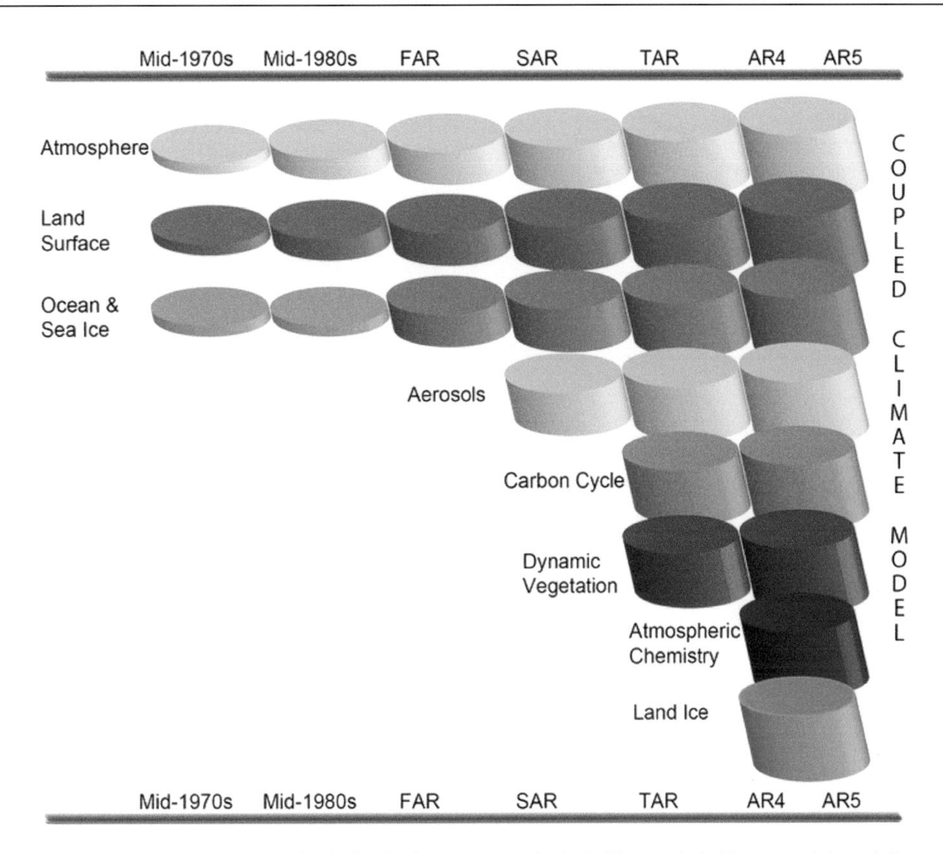

Figure 2.9 Changes in GCM components over time. The individual components included in coupled climate models and the growth of the number of components over time, as allowed by faster and more capacious computers.
Modified to grayscale from figure 1.13 of IPCC (2013)

feature takes up precious computer time and storage capacity. The competition is fierce since each public policy or impact community wants its favorite component included; for example, do we include an advanced package that couples an interactive soil moisture component to the atmospheric component? These are serious questions that the modeling community and its managers are grappling with. Both sides of this discussion are getting part of their way by extending the time period between releases of the IPCC Reports. This gives more time for improvements as well as fitting in new processes.

Of special interest to the audience of this book are the variables related to hydrology – or, especially, river flows across arid lands. These parameters include humidity, precipitation, evaporation, potential evaporation, and runoff. Humidity is easiest to predict since it is closely related to thermodynamics, especially surface temperature over the oceans. But over land it is also related to precipitation and local conditions. Precipitation is a singularly important dependent variable closely connected to the general circulation of the atmosphere. Precipitation almost always occurs where local air is rising. Precipitation is an order of magnitude more difficult to predict than temperature because the mechanisms that cause air to rise depend sensitively on the geographical dependence of instabilities (latitudinal thermal gradients, topography, tropical convection, etc.) in the atmosphere and these particular elements require high space/time resolution for accurate simulation. Experience in weather forecasting suggests that higher resolution (smaller grid boxes) does improve forecasts of precipitation, but they need only run for a few days while climate models need to run for decades and preferably include multiple realizations to estimate natural variability fluctuations, mainly from oceanic changes in sea surface temperature distributions.

Figure 2.10 is included to show the rapid increase in speed of calculations in supercomputers over time. A linear increase would indicate exponential growth. Note the slight tapering of the curves bending to slightly slower growth in recent years.

2.2.3 Oceanography and Climate

The climate system is dominated by the distribution of conditions at the ocean's surface. It is well known that warm spots at the ocean's surface can alter the circulation of the atmosphere, in terms of altering storm tracks in the middle latitudes. Such an alteration can determine the temperature and moisture budget for middle latitude regions, especially in winter. Some of the better-known cases include the distant influence of the El Niño cycle,

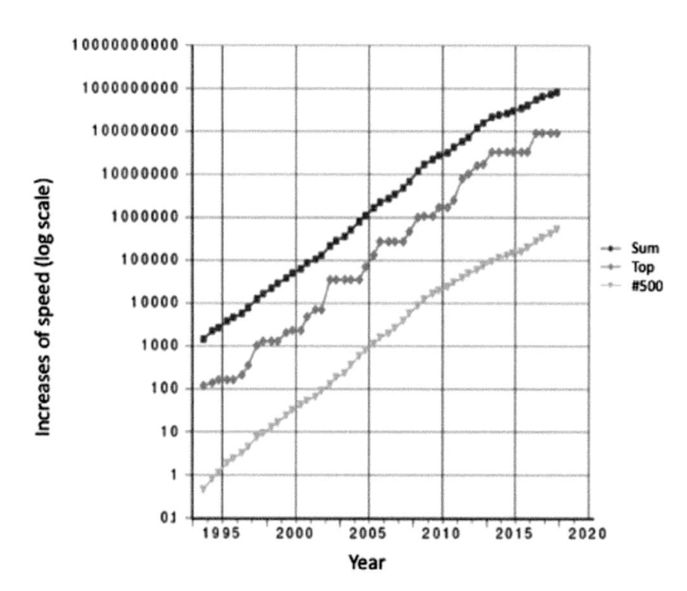

Figure 2.10 Graphic showing increases of speed (log scale) versus year based upon the fastest 500 devices. The curves indicate the number of 'floating point operations per second (FLOPs)' – these are arithmetic operations of numbers in scientific notation. A straight line on this graph indicates an exponential growth pattern. The upper curve is for the sum of the 500 fastest computers pooled together, the middle curve is the average speed of the top 500, and the lower curve is the slowest of the 500. Graphic created by the author using data from the TOP500 list. (www.top500.org)

which is dominant in the Equatorial Pacific. The influences are often great distances from the origin of the sea surface temperature anomaly (departure from long-term average).

The El Nino-Southern Oscillation (ENSO) cycle is a natural but slightly irregular cycle that affects the surface waters of the tropical Pacific. The cycle takes five to seven years to run its course. During the El Niño phase, convection moves from the East Indies toward the Date Line (roughly 180W), leaving drought conditions in the East Indies and Australia. The disturbance at the Date Line usually sets up a train of stationary waves that has influence all around the mid-latitudes in both hemispheres. During a La Niña (the opposite extreme from El Niño in the cycle) event, the coastal waters off Peru are also affected by anomalously cold conditions. Figure 2.11 shows patterns of influence for both extremes of the ENSO cycle in North America. ENSO is only the first of many such natural "oscillations" or "fluctuations" (those that are not so regularly repeating) of the ocean's surface, but it is probably the most prominent. Others include the North Atlantic Oscillation (NAO) and the Pacific Decadal Oscillation (PDO).

All these phenomena are properties of the nonlinear behavior of the coupled ocean-atmosphere system. Nonlinear systems like these are notoriously difficult to solve and understand. With new numerical techniques and the power of modern computers, we are now in a much better position to take them on. But even the detailed evolution of a single ENSO cycle cannot be said to be accurately predictable. The coupling of the two giant subsystems is mainly through the wind's stress driving upper level ocean currents, such as the Gulf Stream in the Atlantic. Of course, the Earth's rotation, expressed through the Coriolis Force, and constraints due to land/sea distribution and the bottom topography combine to guide ocean currents.

In addition, there are slower secular changes in the ocean's current patterns. Those related to density/buoyancy gradients are grouped under the name Thermohaline Circulation. They come about because of density differentials in ocean volume. Density can be affected by water temperature (colder water is denser) and by salinity (saltier water masses are denser). It is mainly through these circulations that water at or near the surface can sink to deeper parts of the water column. The world's oceans contain only a few of these "deep-water formation sites." Most of the ocean is stably stratified with warm (light density) layers lying above colder, saltier water. The major location of such down-welling is near the southern coast of Greenland, where the waters (fed by the Gulf Stream) are especially dense. There is no such place in the North Pacific. There are also some places along the coast of Antarctica where down-welling occurs. It is of interest to hydrologists in climate science to know that fresh water from river discharge into the ocean is a significant contributor to the salinity budget of the ocean.

The final contribution of the oceans to our expanding field of climate science is the delay of the response to energy imbalances of the system. If only the lowest 50–100 meters of well-mixed water just below the ocean surface are included in your model, the delay is short on climate scales (months to a year). For example, this is sufficient to simulate the seasonal cycle accurately where only the "well mixed layer" participates. But in the case of greenhouse warming, the delay can be decades. Just think, there are volumes of water (known as "old water") a few kilometres below the surface and in the middle of the Atlantic or Pacific that have not visited the surface in 800 years (radiocarbon dating). The actual delay might depend on such unpredictable quantities as the opening or closing of pathways for deepwater formation (due to nonlinear dynamics on long time scales in the oceans, some modeling studies have indicated slowdowns in thermohaline circulation).

2.3 FORCED CHANGE VERSUS NATURAL VARIABILITY OF CLIMATE

2.3.1 Climate "Forcings" and Feedbacks

Climate scientists have adopted the following two terms for types of climate change. A "forcing" is an imbalance at the top of the atmosphere of the incoming absorbed radiation and the outgoing (infrared) radiation. We can enumerate them as follows

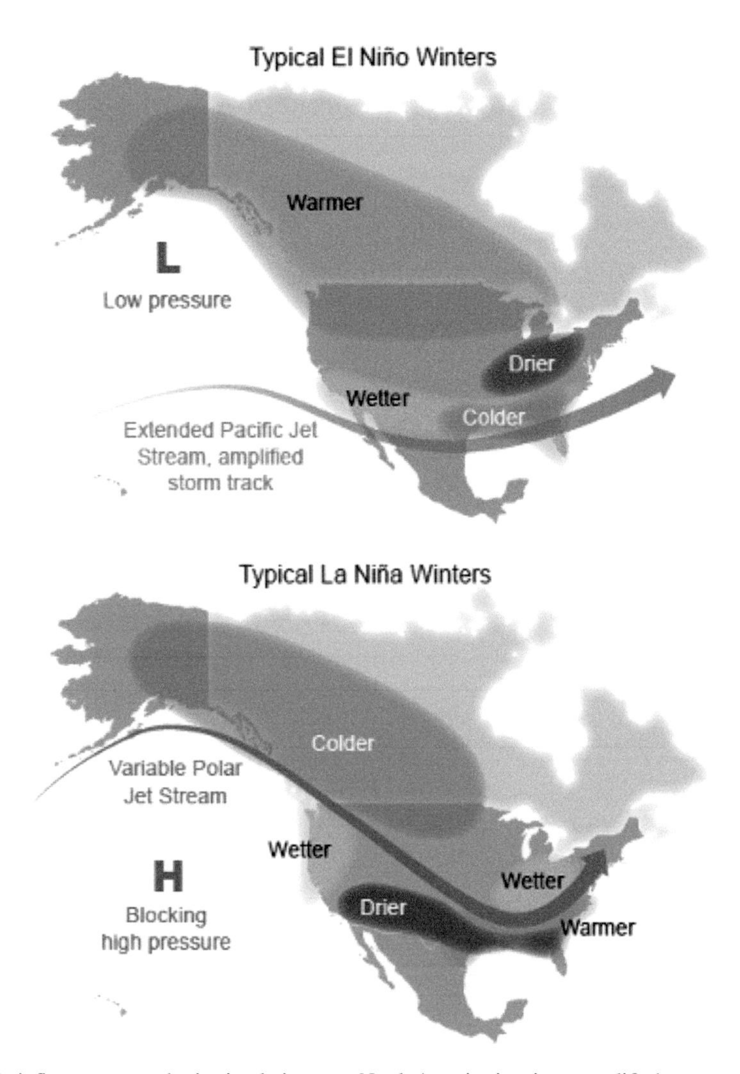

Figure 2.11 El Niño and La Niña influence atmospheric circulation over North America in winter, modified to grayscale from the US Climate Science Special Report, 2017, figure ES.7. The arrows indicate the average path of the Jet Stream for the two extremes of ENSO (El Niño-Southern Oscillation), upper: El Niño, lower La Niña. Most storms in winter tend to follow the arrows during these extreme events. Differently affected regions are distinguished by different shadings of grey.
Figure from the NOAA website, modified to grayscale

(a) A change in the concentration of greenhouse gases causes a warming
(b) Anthropogenic aerosol particles suspended in the air affect the incoming solar radiation, and most of these reflect some radiation back to space before it can warm the air below. However, some particles, such as soot, are black and can absorb sunlight and heat the air around them. Both kinds of aerosol particles come largely from human activities, although some are natural, such as wind-lofted dust from arid areas.
(c) Changes in solar energy output. Some scientists have argued that hypothetical changes could explain warming over the last century, but evidence appears to be slim.
(d) Some volcanic eruptions are massive enough and directed sufficiently vertically that they leave aerosol particles in the stratosphere. The stratosphere is very stable (very little convective turnover); hence, these particles can remain as a reflective shield for several years, after which stratospheric circulation combined with slow fallout removes them. These eruptions can easily be identified in the surface and atmospheric records. They tend to cool the planet almost instantly, followed by a slow recovery (several years) as the particles fall out and the oceans respond (usually, only the mixed layer is necessary for this response).

Natural variability describes the usually longer-term changes in the ocean at decadal or longer time scales. Climate models can demonstrate the existence of such phenomena in simulations, but they are not capable of identifying the timing of when that might

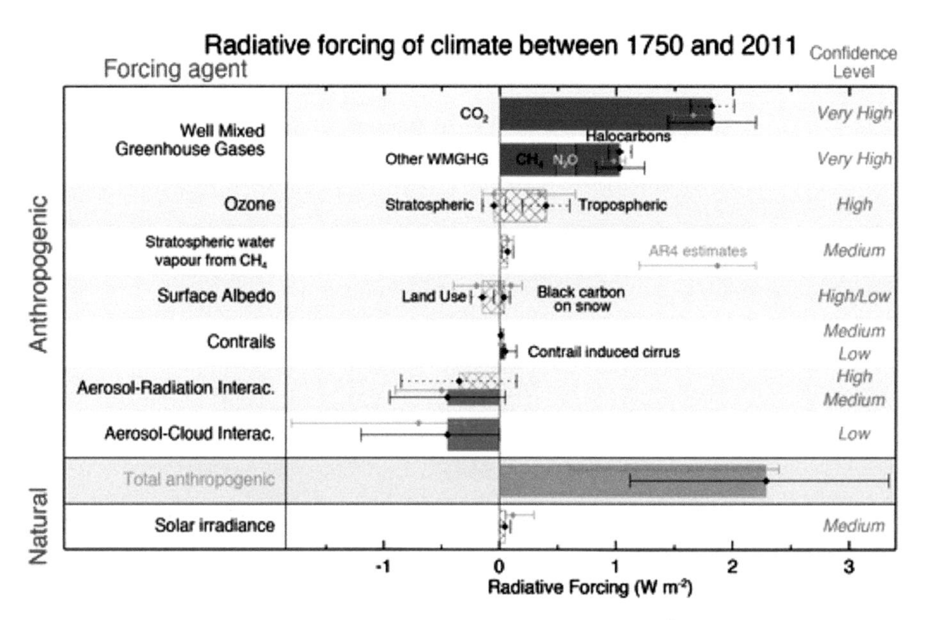

Figure 2.12 Radiative forcing due to different agents. The bar lengths are shown in Watts/meter^2. The red bar indicates direct forcing due to the contribution to the greenhouse effect of CO_2. The horizontal I-beams indicate uncertainty in each of the forcings. Note that the blue bars are coolants, but uncertainty is large.
Figure is modified from figure TS.6 of the IPCC (2013)

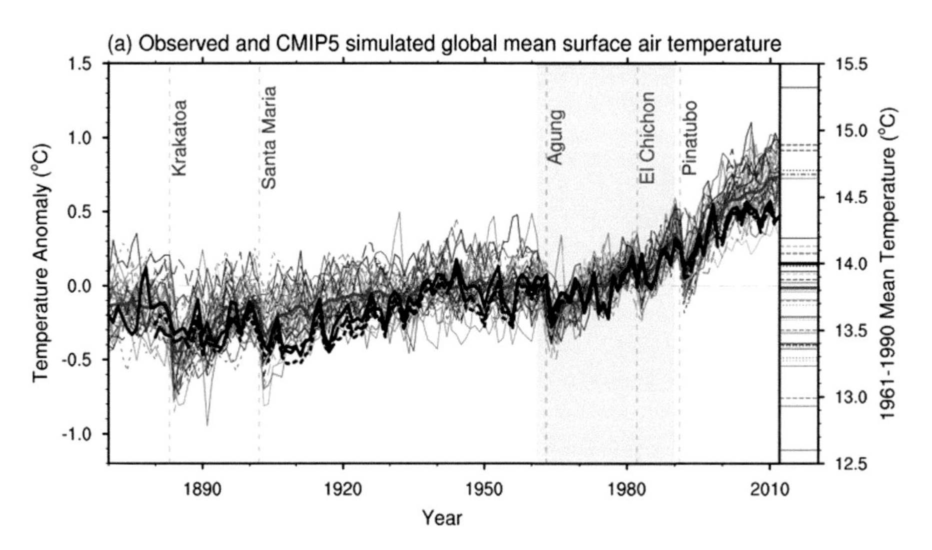

Figure 2.13 Illustration of an ensemble of runs from different models, modified from figure 9.8 of the IPCC (2013). The vertical dashed lines are labelled volcanic events that cause cooling because of tiny particles that remain in the stratosphere after eruption. The individual light shaded lines are simulations from different climate models from around the world. One of the lines is the ensemble mean of all the simulation ensemble members. The heavy black line is from observations. All model curves were adjusted to have zero deviation from the beginning. Numbers on the left ordinate are degrees C. On the right are shown unadjusted planetary mean temperatures of individual model runs. The data curve gets in line with the models after this graph is extended.

occur (as in a short-term weather forecast). This is one reason modelers like to run an ensemble of runs of the coupled ocean/atmosphere models. By this method these events occur randomly in time from one ensemble member to another. By looking at the statistics of individual realizations one can build a probability density distribution of such occurrences or simply look at the bundle of simulations – messy, but more revealing. The latter is the best way of looking at climate change from models. Figure 2.12 shows a bar diagram indicating the responses to different forcing agents based on current climate models. The "I beam" shaped horizontal lines indicate the range of uncertainty in the model estimates. Figure 2.13 shows realizations across many climate models simulating the last century of climate change. Figure 2.14 shows a 100-member ensemble of

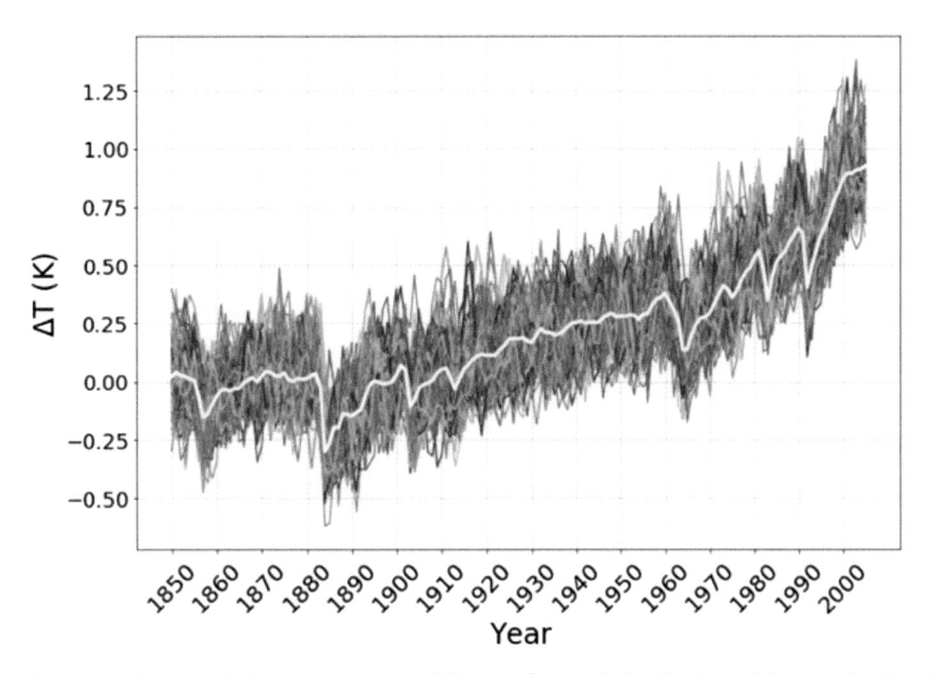

Figure 2.14 A 100-member ensemble of coupled ocean-atmosphere GCM runs from a single climate model (run at the Max Planck Institute in Hamburg). Each run has the same forcing from greenhouse gases, volcanic eruptions, and other forcings used in simulating the climate since 1850. The envelope of fluctuations about the mean (in white) shows the band of natural variability in a single model with sea surface temperatures computed within the model and interacting with the atmosphere. The downward kinks are due to imposed volcanic eruptions.
Source: Adams (2018). See also Dessler et al. (2018)

simulations from a single climate model (MPI-ESM1.1 model from the Max Planck Institute in Hamburg, see for example: Dessler et al. (2018).

2.4 ROBUST FINDINGS OF CLIMATE PROJECTIONS

2.4.1 Warming Everywhere

There are a few robust findings from the last few decades of study with climate models. The first and foremost result is that the globe is warming. This ubiquitous warming across the whole community of models appears in every one of the models that have been run in more than twenty countries. First, why do we care about the global scale? No one takes a picnic at the global scale. No one lives at the global scale. The reason this index is so important for climate scientists is that this is the scale at which the signal is most detectable. At smaller scales, such as a continent or country or even a county, the warming signal is about the same but the natural variability of weather and even sea surface temperatures masks the signal so much that it is nearly impossible to see if the data record is only a half century or less. Three factors help the global scale as an indicator: (1) In the data, averaging over the entire Earth's surface smooths out all of the local natural variability and even random measurement noise that is inevitable. For instance, often weather fluctuations on one side of the Northern Hemisphere are opposite to one 180° of longitude away. (2) In climate models, forcing by greenhouse gases increases and even the other forcings mentioned earlier are essentially global in extent. Intuition (and demonstrations with models) suggests that the response to forcing at the global scale will match that of the forcing itself. (3) At the global scale, most of the details of atmospheric circulation are lessened and thermodynamic considerations tend to dominate the larger the scale of the averaging.

The exact amount of global scale warming from one equilibrium state to another after a sudden change in forcing and waiting for the new equilibrium to be established (proportional to the climate sensitivity) differs from one model to another, as discussed earlier. The uncertainty of climate sensitivity can be as large as 50 percent. When we include time dependence, such as linearly increasing forcing due to carbon dioxide increases, there is a new difference among the models that is due to how a model treats the ocean. If we take this difference into account, the uncertainty becomes even larger. We can estimate that by the end of the century the global annual temperature will increase by about 3°C ± 1.5°C (or 5.4°F ± 2.7°F). This estimate is based on simulations of more than twenty climate models under moderate forcing from future emissions of CO_2. Most models show similar spatial patterns of warming so that a convenient index of the patterns consists of scaling according to the model's global sensitivity.

Robust findings are that as we go forward fifty or more years into the future, we can expect more warming in the north polar region and in the interiors of the large continents. Ocean surface temperatures will lag slightly (in the order of fifteen years) behind those over the continents. These features are clear in the observations and model simulations.

2.4.2 Hadley Cells

Atmospheric circulation patterns are remarkably stable although they rattle a bit. Most of the fluctuations appear in the middle latitudes, where instabilities due to thermal gradients lead to the convection that we in the middle latitudes known as passages of high- and low-pressure areas, associated fronts, etc. As you get closer to the Equator than thirty degrees of latitude, these instabilities and associated weather patterns disappear, and we are in what are called the subtropics. Except for the occasional tropical storm, the weather is mild, and the winds are steady due to the trade winds that blow along the surface toward the equator. As these lower atmospheric flows from the Northern and Southern Hemispheres converge, air is shoved upward. This band around the equator features rising air and subsequent heavy rain. Figure 2.15 is a map of annually averaged precipitation from a twelve-year record. The precipitation belt around the equator is clearly discernible. Also easily identified over land and ocean are the subtropical arid zones (white regions on the map).

Figure 2.16 shows a schematic of a cross section of rising air at the equatorial latitude, the horizontally moving surface layer (trade winds) that converges from both hemispheres. This cycle of winds comprising the subtropics is called the Hadley Circulation. For our purposes, the most important question is whether this huge feature is changing as we warm the planet. The answer to this question appears to be that the latitudinal extent of the Hadley Cell is expanding in both hemispheres. We are interested here because this means that the area known as the arid subtropics is expanding. A side effect of interest is that the middle-latitude storm belts will shift slightly toward their respective poles.

Figures 2.15 and 2.16 alert us that the schematic model of the Hadley Cell is idealized. Moreover, if we were to look at individual months of the TRMM data set (Available at: https://trmm.gsfc.nasa.gov/trmm_rain/Events/trmm_climatology_3B43.html) from which Figure 2.16 is derived, we would find that the ribbon of precipitation around the equator moves with the seasons, to reveal the monsoons and the associated seasonal rainy and dry seasons that occur in many regions, such as India, Australia, and Africa.

Most of the rivers that are the subject of this book cross these subtropical arid zones. These rivers originate from snowpacks in

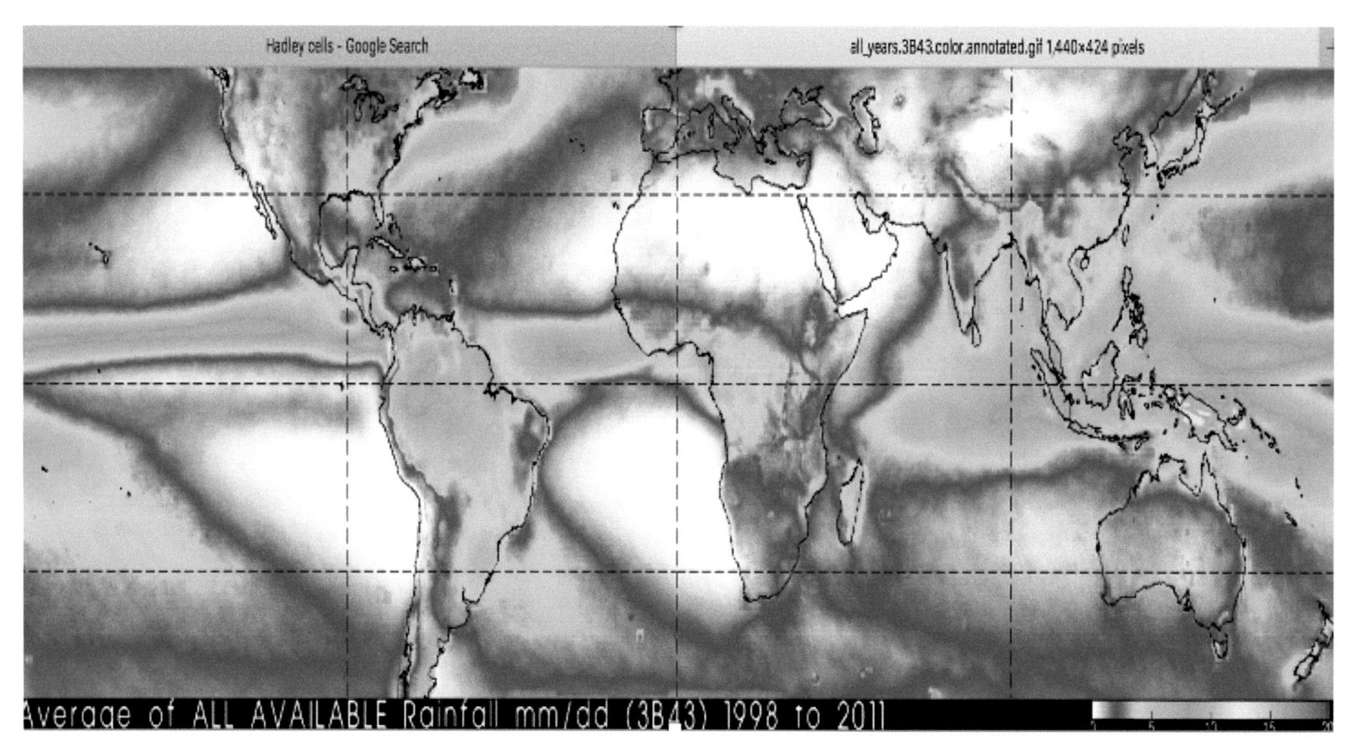

Figure 2.15 Rainfall climatology based on satellite measurements 1998–2011. White areas are dry, and darker shaded areas indicate more intense precipitation.
Source: Tropical Rainfall Measuring Mission (TRMM) website

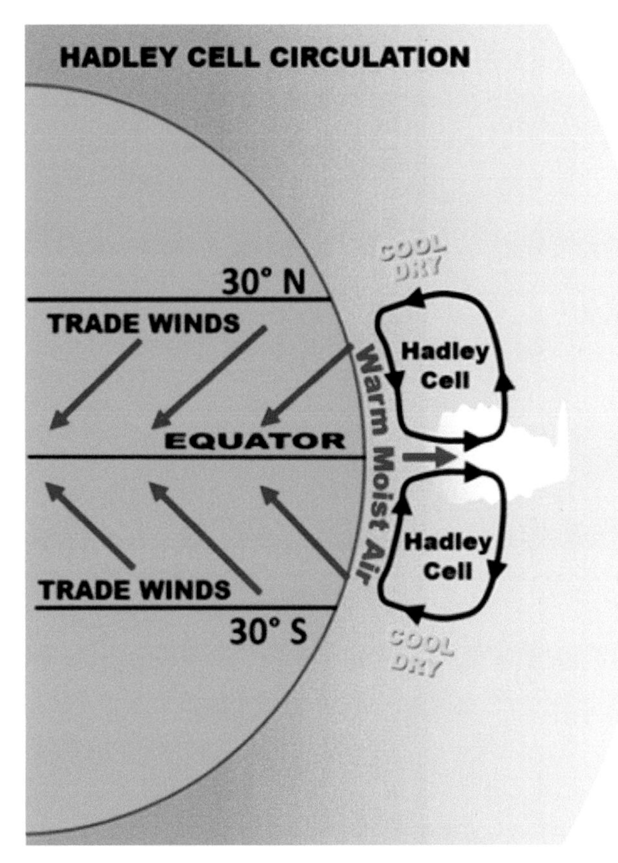

Figure 2.16 Schematic of Hadley Cells, which form a large-scale overturning circulation with low level convergence of moist air at the equator forcing vertical convection and precipitation. The lifted air (now dry) sinks over the subtropics and returns in a shallow flow (Trade Winds) toward the equator. The shallow return path over the subtropics is generally free of precipitation. Figure is not to scale; the thickness of the atmosphere is greatly exaggerated herein.
Illustration by Doug Robbins, modified from the work of Tinka Sloss, New Media Studio, reprinted with permission from http://dougrobbins.blogspot.com/
.

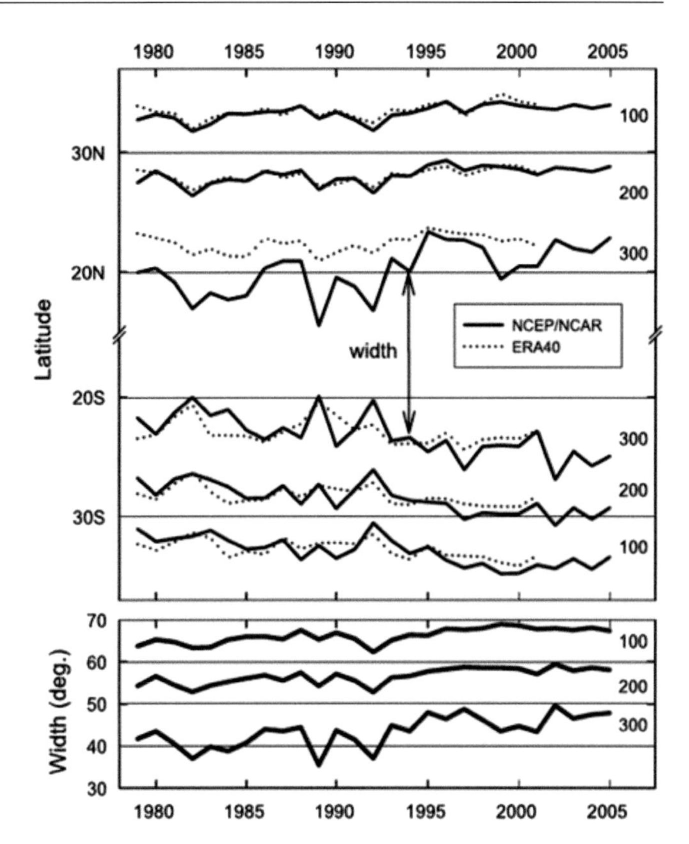

Figure 2.17 Estimates using different methods of the width of the Hadley Cell indicating that over the period 1978 to 2006 the trend is positive.
Modified to greyscale from Seidel et al. (2007), doi:10.1038/ngeo.2007.38

high elevations in middle latitudes (or in some cases such as the Nile the origin is in the high elevation tropics). The Yellow River in China is an exception, staying in the middle latitudes where conditions might differ from the other rivers in our study.

Numerous studies have indicated that the Hadley Cell is expanding and that it will continue this trend if global warming continues. Qiang Fu and his colleagues have presented both original and accumulated evidence from both modeling and data sources that firmly indicate that the Hadley Cells in each hemisphere are expanding (Seidel et al., 2007). Of course, as they expand the rainy belts recede toward their respective poles. When this happens, lands that now experience regular precipitation from the passage of high- and low-pressure areas (particularly in their winter seasons) may not receive as much or any rainfall from these sources as the end of the century draws nearer. In the semi-arid zones, warming will be occurring while precipitation may be lessening.

There is now evidence from real data as well as model runs that indicate that the Hadley Cells in each hemispheres are widening. The models include the forcing of carbon dioxide increases. Figure 2.17 shows some studies using different indicators that this is happening.

The seasonal swing of precipitation is global in its effects. Figure 2.18 shows data from a twelve year climatology of data from the TRMM satellite. The swing is evident in the monsoon regions of Africa and India. Other more subtle but important seasonal changes can be seen in monthly data that is now available at the Global Precipitation Mission (GPM) website where TRMM and GPM data are merged for a longer data set.

2.4.3 Precipitation and Evaporation: Aridity

Precipitation is fundamental to the hydrological cycle. We can get a feel for the seasonality of precipitation from the TRMM data referred to earlier.

Evaporation from the Earth's surface is measured by a couple of indices that I will discuss. The amount of water per m^2 per unit of time issued from the soil is the evaporation rate. Water is also emitted into the atmosphere from the leaves of plants.

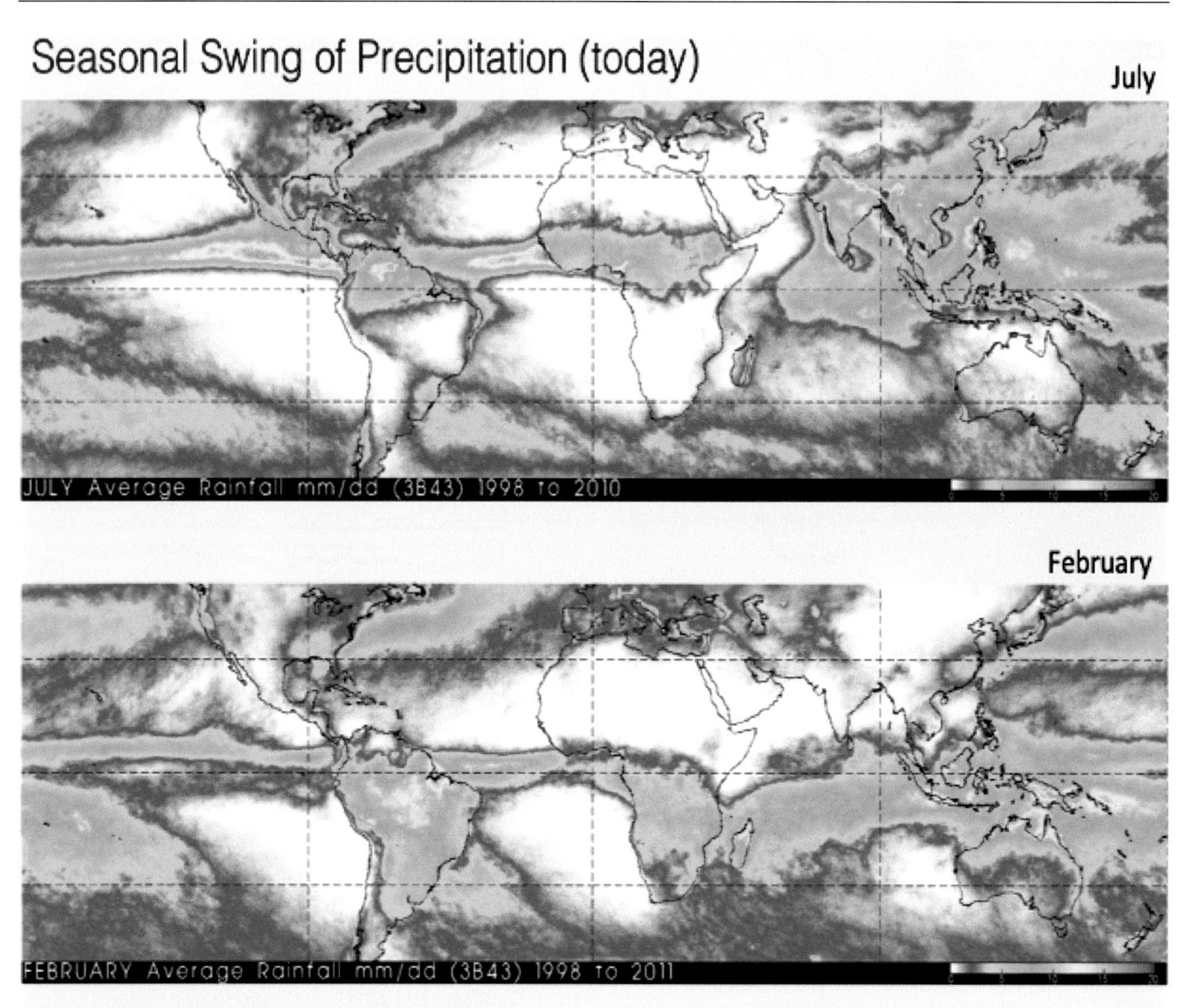

Figure 2.18 The seasonal swing of precipitation, based on 12 years of data from the Tropical Rainfall Measuring Mission (TRMM). Individual months (in color) available at the TRMM website

That water is brought up to openings on the leaf surface through stems, branches, and trunks from below the soil surface via the roots of the plant. This source of moisture released into the atmosphere is called transpiration. The sum of evaporation and transpiration is called evapotranspiration. These quantities are the actual releases of moisture into the air, and in practice they can be measured at a given time or averaged over days, months, and years.

Another indicator of interest is potential evapotranspiration (PET). It is a measure of the environment's demand for evapotranspiration under certain conditions, for example, relative humidity and wind. PET is dependent on external circumstances that could in principle be computed in a numerical weather or climate model. Strictly speaking, PET refers to standard conditions (grass cover shading the soil, moist soil, etc.), but it is commonly used in general conditions. Reporting stations in the field use standard conditions. Actual evapotranspiration occurs (theoretically) when adequate moisture in the soil is available.

One measure of the moisture conditions in a climatic region over land is the aridity index (AI), which is defined as the precipitation rate divided by the PET. Since AI can be computed from climate model simulations, we can estimate how this index changes as we project forward in time in model simulations. First, consider the change projected for precipitation. Figure 2.20 shows that the dry lands expand. While precipitation increases over some of the land, PET increases even more, causing the dryer regions to expand in area. Figure 2.21 shows further evidence focussed on the American great plains area.

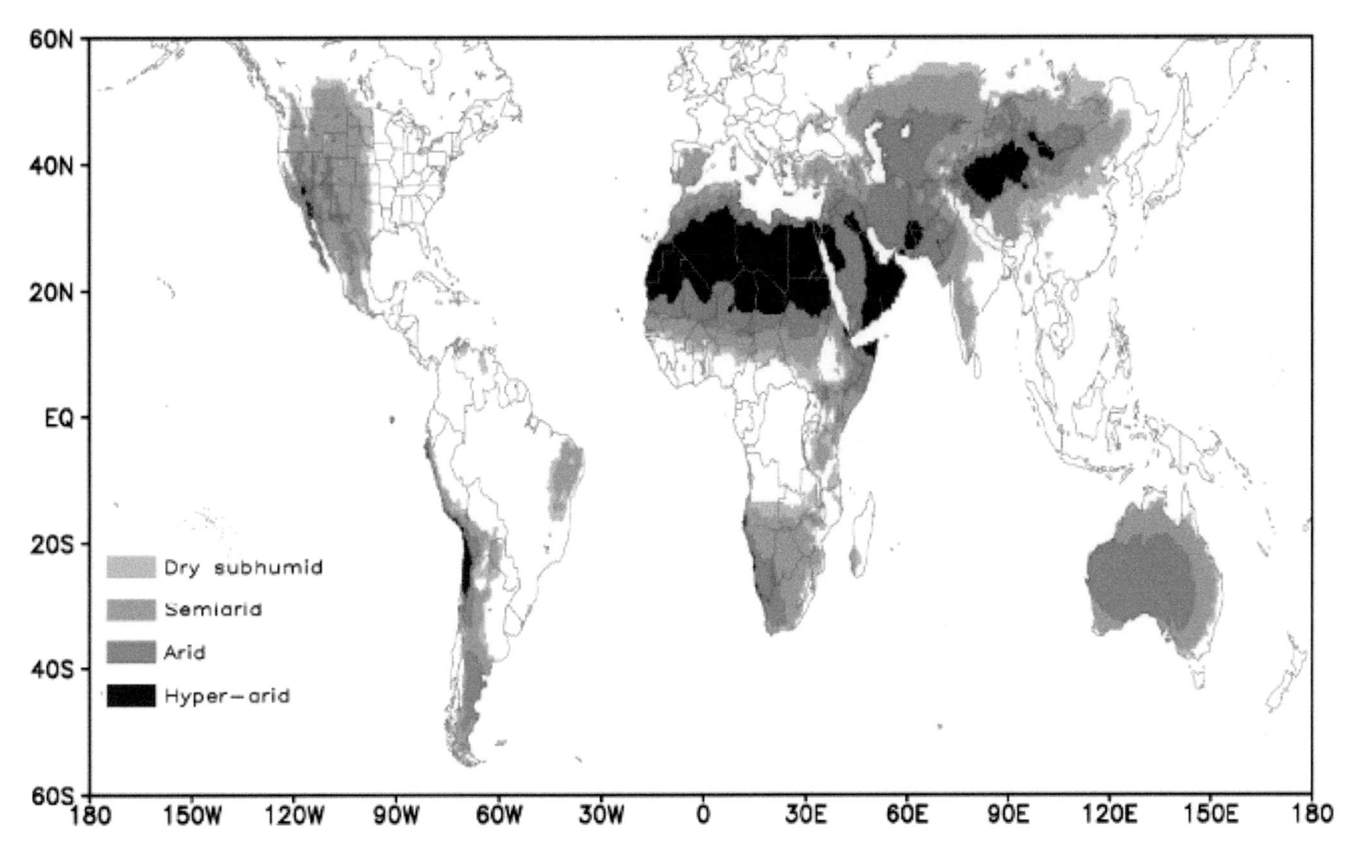

Figure 2.19 Present climatology of dry lands, based on the annually averaged aridity index (AI: precipitation/PET). Dry sub-humid lands are considered to have $0.05 < AI < 0.65$ (lightest gray).
Source: Feng and Fu (2013)

2.4.4 Precipitation Changes

The denominator of the aridity index is precipitation. Current precipitation is shown in Figures 2.16 and 2.19. Figure 2.22 (from the AR5 of the IPCC (2013)) shows changes in seasonal rainfall from the beginning of the twenty-first century to the last (20-year average). The patterns shown for the "business as usual" emission path in Figure 2.23 reinforce the idea that the subtropic Hadley cells (see Figures 2.16 and 2.19) are widening in both hemispheres. Note that many of the drying areas are the very ones considered in this volume. These are mostly semi-arid lands crossed by the rivers in question: the Colorado, Rio Grande, Sao Francisco, Nile, Murray-Darling, etc. Only the Yellow River and the Nile River (in summer) are spared reduced precipitation.

2.4.5 Steady Evaporation Increases

Throughout most regions of the planet, the rate of evaporation is likely to increase. The increase is mainly due to increased surface temperatures. These are shown for annual averages in Figure 2.24. Some river basins such as the Rio Grande see little change in evaporation, since there is very little soil water there to evaporate. This is true in some other areas of interest as well, such as Southern and Mediterranean Africa, the Tigris/ Euphrates, Murray- Darling (actually, all of Australia), and Spain.

2.4.6 Runoff

Runoff (precipitation minus evaporation) is the source of water for tributaries and eventually the large rivers of the Sustainability of Engineered Rivers in Arid Lands initiative (SERIDAS) basins. Figure 2.24 shows the results of 33 climate model simulations following the RCP 8.5 ("business as usual") emission scenario. We find the expected pattern of less runoff in the arid subtropics. Hard hit in this parameter is the western coast of Chile. Somewhat less affected are Central America, parts of tropical Brazil, Southern Africa, and most of southern Europe and Turkey. While it is not clear how robust the African seasonal simulation of precipitation is, there seems to be an increase in runoff at the headwaters of the Nile. According to these simulations, the Middle East and the American Southwest will see less runoff, but not as severely as in some places.

Before leaving the section on hydrology, it is important to recall the great uncertainty in predicting precipitation statistics

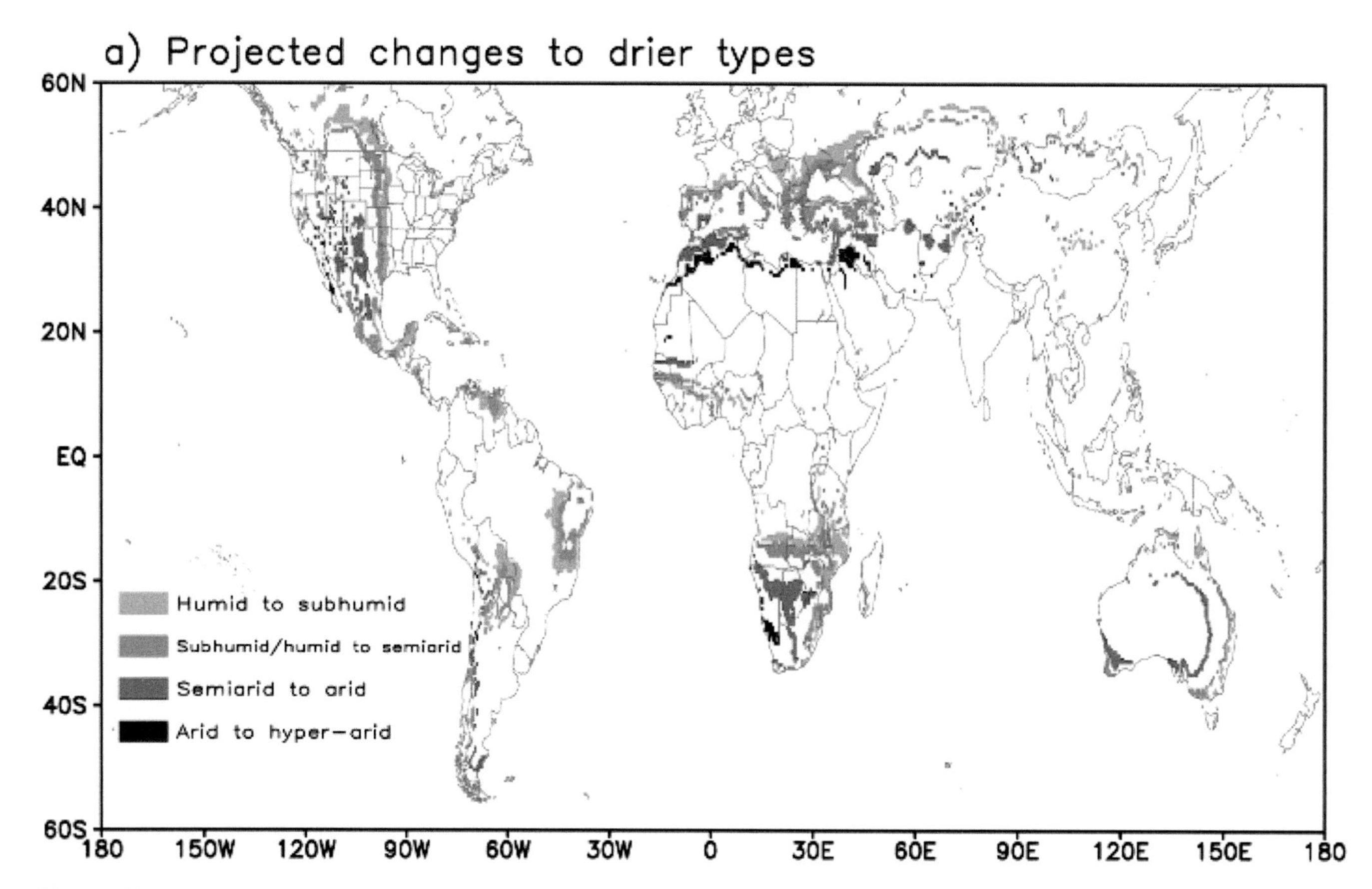

Figure 2.20 Changes in the Aridity Index from present to an average taken in the last decades of the century.
Figure modified from Feng and Fu (2013)

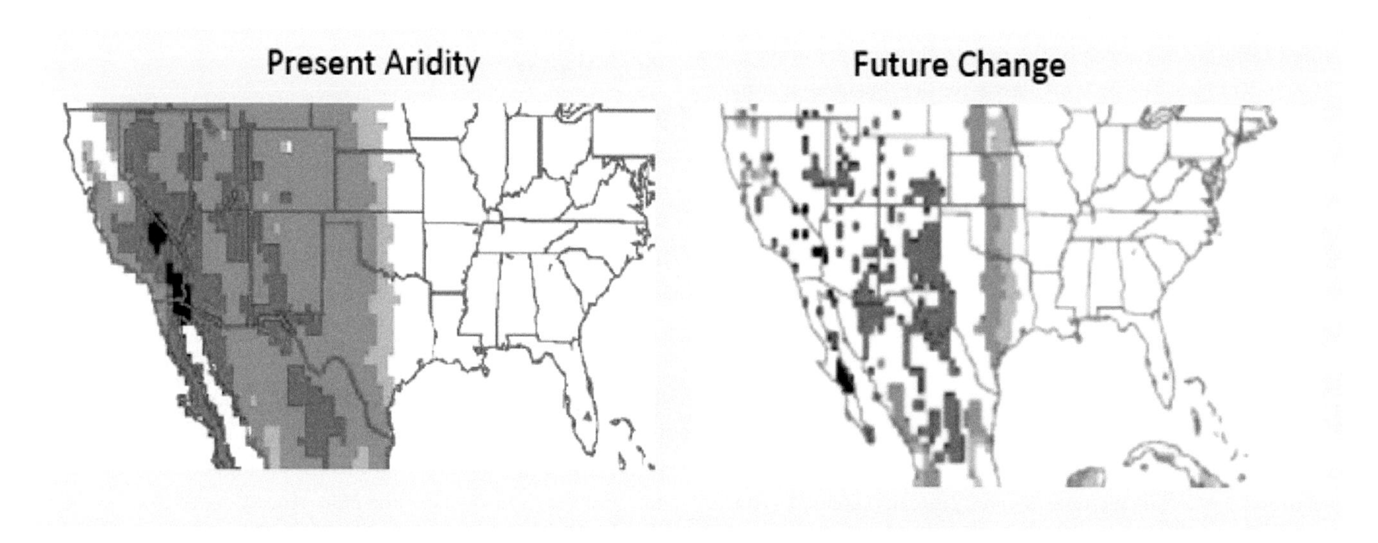

Figure 2.21 Left panel, the present aridity index. Right panel, the difference from that at the end of the century, assuming a "business as usual" emissions scenario. Dryer areas expand.
Figure created by the author using select portions of figures from Feng and Fu (2013)

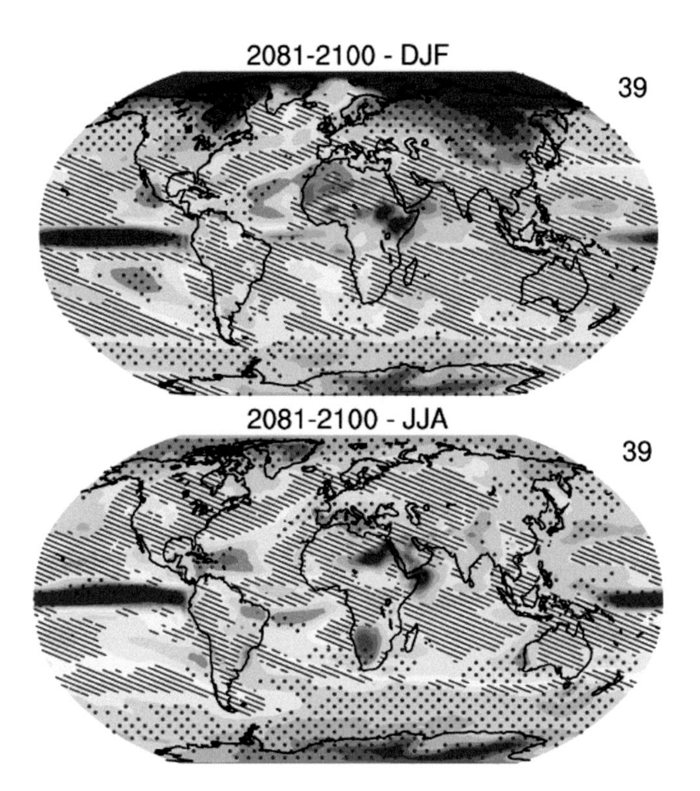

Figure 2.22 Change in precipitation (%) from the beginning to the end of the twenty-first century for two opposite seasons. Changes are for an average of 39 model simulations using the RCP 8.5 emission path (essentially "business as usual"). Hatching indicates regions where the multi-model mean is less than one standard deviation of internal variability. Stippling indicates regions where the multi-model mean is greater than two standard deviations of internal variability and where 90 percent of models agree on the sign of change.
Modified from figure12.22, IPCC (2013)

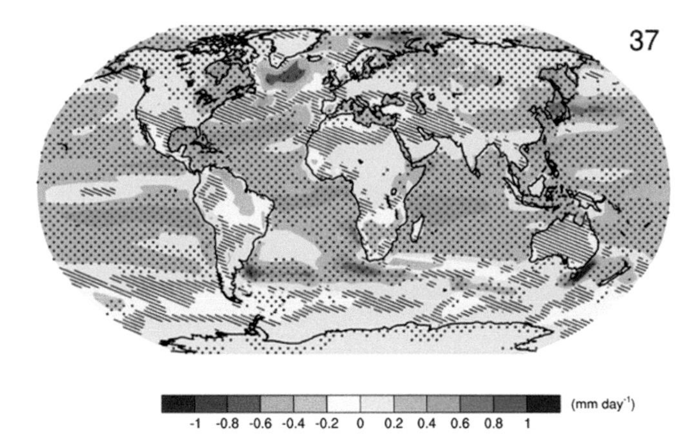

Figure 2.23 Evaporation changes over the twenty-first century from 37 climate model simulations.
Modified from figure 12.25, IPCC (2013)

Changes in Runoff to 2100

33

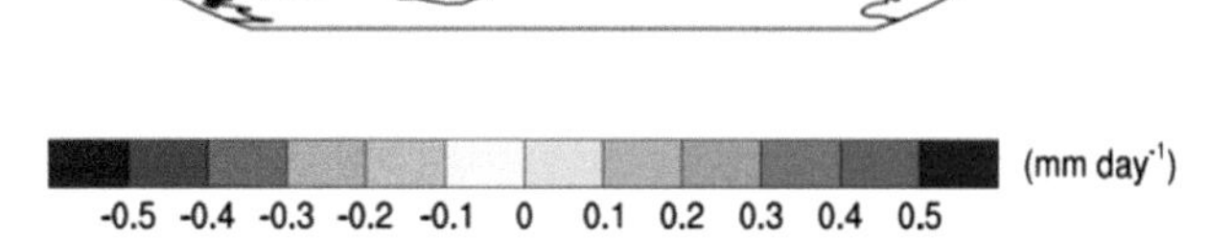

Figure 2.24 Runoff change (precipitation minus evaporation) over the twenty-first century as simulated by 33 climate model runs for the RCP 8.5 ("business as usual") emissions scenario.
Modified from figure 12.24, IPCC (2013)

with general circulation models. Recall that unlike the surface thermal patterns, which are mainly controlled by thermodynamics, precipitation is mostly determined by lifting parcels of moist air until they expand, cool, and, if they reach saturation, produce cloud droplets and ice particles that if sufficiently large produce precipitation. These are quite a few steps to line up. Lifting is caused by several possible mechanisms, (1) orographic, i.e., airflow pushed up a mountainside by large-scale circulation; (2) forced convergence of moist air at the surface such as the trade winds near the equator; and (3) instabilities in middle latitudes due to the north–south gradient of temperature. While GCMs exhibit all of these features, they are very delicate and one must consider that some of the processes are of much smaller scale than the computer box size for numerical simulation (currently about 20 km, say the size of a typical county in the United States). From one GCM simulation to another there will be large differences in simulation results, because the empirical parameters used at the "sub-grid" scales cannot yet be modeled with great confidence. What are robust are the general qualitative features of precipitation and the pattern of the Hadley Cells' expansion, on which most of the models agree.

Once we have GCM-simulated results for precipitation, we have next to model the surface characteristics such as potential evapotranspiration (PET) and evaporation. Using these quantities, we can estimate such quantities as runoff and, eventually, river flow. The estimation of future hydrological parameters requires a computer model, which is by no means simple. Problems include the variability and spatial heterogeneity of soil and vegetative cover. How do these quantities react to the simulated atmosphere above (winds, near surface temperature, humidity, sunlight, infrared emission, and absorption)? This stack of blocks is likely to be a bit wobbly by the time we climb to the top. Pooling data from many model simulations and looking at the statistics probably gives us some idea of what lies in the future, but we must remember that beyond the uncertainties listed here there is the most important driver of all, future emission scenarios.

2.4.7 Natural Fluctuations in Precipitation Patterns

There is one other consideration in precipitation patterns that must be discussed. That is the internal oscillations, of which the El Nino Southern Oscillation (ENSO) is probably the largest, but there are many other such month-to-decadal scale fluctuations that seem to be driven mostly by ocean surface characteristics such as temperature. The patterns of sea surface temperature change over these periods and are governed by nonlinear dynamics of flows at the interface between air and sea. ENSO is the most studied and perhaps best understood. It affects air flows and stationary waves in the air that stretch from the tropical Pacific across North America, but there are "connections" in both hemispheres and at all longitudes. The phases of the ENSO phenomena can lead to drought or flood conditions in many areas. More information about these effects can be found on NOAA websites.

The point here is that these phenomena are ever present in the global system and they tend to contribute significantly to natural variability in precipitation patterns. Coupled Ocean Atmosphere GCMs do exhibit these behaviors, but it must be stressed that much work lies ahead in improving our ability to model them with the precision we desire.

2.5 CONCLUSIONS

This chapter has attempted to present the method of climate modeling to estimate future conditions in the Earth-hydrological system. We are interested in patterns of rainfall and surface hydrological parameters such as potential evapotranspiration, actual evaporation, and ultimately runoff into selected river basins. Most of the river basins in the SERIDAS group run over currently arid subtropical lands after originating in middle latitudes, often at high elevations featuring melt water from snowpacks.

We know enough about climate models to tell that the planet will be warmer by century's end by about 3°C ± 1.5°C. The polar regions and the interiors of the larger landmasses will warm slightly more, with ocean surfaces lagging by a few decades. Another important robust finding is that the mid-latitude storm belts will recede gradually toward their respective poles. This process will cause the Hadley Cells to widen. Consequently, the arid subtropical zones will expand. Warming in the cloud-free subtropics will lead to more warming and therefore more evaporation (if there is any soil water there to evaporate).

Of course, climate change is only one of many factors that humans will inflict on the subtropics. These include population growth, demanding more potable water and water for other uses to be drawn from rivers. The shortage of fresh water in the subtropics will be one of the most important challenges of our coming time.

REFERENCES

Adams, B. (2018). Understanding the Impact of Internal Variability on Our Estimates of the Transient Climate Response. Master's Thesis Report. College Station: Texas A&M University.

Etheridge, D. M. et al. (1996). Natural and Anthropogenic Changes in Atmospheric CO_2 over the Last 1000 Years from Air in Antarctica Ice and Firn. *Journal of Geophysical Research*, 101(D2), pp. 4115–4128.

Feng, S. and Fu, Q. (2013). Expansion of Global Drylands under a Warming Climate. *Atmospheric Chemistry and Physics*, 13 (9), pp. 10081–10094. https://doi.org/10.5194/acp-13-10081-2013

Hartmann, D. L. (2014). *Global Physical Climatology*, 2nd ed., San Diego, CA: Academic Press.

IPCC (2013). *AR5 Climate Change 2013: The Physical Science Basis*. Cambridge and New York: Intergovernmental Panel on Climate Change.

Johanson, C. M. and Fu, Q. (2009). Hadley Cell Widening: Model Simulations versus Observations. *Climate*, 22, pp. 2713–2725.

National Research Council (2006). *Surface Temperature Reconstructions for the Last 2,000 Years*. Washington, DC: National Academies Press. https://doi.org/10.17226/11676

Seidel, D. J. and Randel, W. J. (2007). Recent Widening of the Tropical Belt: Evidence from Tropopause Observations. *Journal of Geophysical Research*, 112, D20113. https://doi.org/10.1029/2007JD008861

Seidel, D. J., Fu, Q., Randel, W. J., and Reichler, T. J. (2007). Widening of the Tropical Belt in a Changing Climate. *Nature Geoscience*, 1, 21–24.

USGCRP (2017). *Climate Science Special Report: Fourth National Climate Assessment*. Volume I. Washington, DC: U.S. Global Change Research Program, p. 470. https://doi.org/10.7930/J0J964J6

3 Reservoirs

Design, Functions, Challenges

George H. Ward

3.1 THE NEED FOR RESERVOIRS

In most places on the terrestrial earth, most of the time, precipitation does not occur. A complex combination of meteorological processes is required to create precipitation, and this combination arises sparsely in space and separated in time. For example, precipitation is frequently associated with singularities in the structure of the atmosphere such as fronts (abrupt changes in the properties of the atmosphere) and vortices (concentrations of rotation), which form, move across the landscape, and disappear, controlled by dynamics operating on a range of space-time scales. Despite the sporadic occurrence of precipitation, widespread access to water by all living things is due to an important feature of the hydrological cycle: the ubiquitous opportunity for storage. Water is stored in surface depressions, in layers of the soil, in the network of stream channels, and in rock formations (to say nothing of organic matter in the ecosystem, especially vegetation). The retention in the various types of hydrological storage ranges from minutes to more than millennia, at least twelve orders of magnitude.

In regions where the variability of precipitation is particularly extreme, humans have augmented the storage available in nature by constructing dams – barriers to the flow of water – in streambeds to create impoundments. This practice dates back thousands of years in some places (e.g., Schuyler, 1909; Smith, 1976; Billington et al., 2005; Oestigaard, 2019), but within the past two centuries these structures and their capabilities for management of water have achieved new levels of sophistication (Smith, 1976). Fundamental to the function of the dam is the reservoir of stored water behind it. Broadly, this function is either detention or retention, i.e., whether the storage is temporary or indefinite. The most common objective of detention is flood control, in which the high flows of the river are temporarily trapped and released downstream at a lower flow rate to reduce or prevent flooding. Another objective is intra-annual water supply in regions with a strong annual seasonality in runoff, storing water during the wet season for use during the dry season. A common objective of retention is inter-annual water supply, in which water is stored during a period of surfeit in precipitation to be used as needed in a period of drought. The most important reservoir objectives in social and economic development in the past two centuries include flood control, hydroelectric generation, municipal and industrial water supply, and irrigation water supply. Other objectives include navigation, recreation, sediment control, steam-electric cooling, and environmental flows.

For most of the history of dam construction, reservoirs were designed to be single purpose, dedicated to one basic function: irrigation, or hydroelectric power, or flood control, etc. The operation of a flood-control dam is to keep the reservoir empty, in preparation for a flood event. A water-supply project, in contrast, is kept full, i.e., at "conservation level," to the extent possible. Only in the past century have multipurpose projects come into widespread use. The oldest multipurpose project this writer could locate in a (superficial) literature search is the Low Aswan Dam on the Nile, completed by the British in 1902 (and whose dam height was successively raised in subsequent projects, e.g., Oestigaard, 2019). In the current database of the International Commission on Large Dams (ICOLD), 27 percent of the classified dam projects are multipurpose. In this database (which are large dams registered by the countries that are members of ICOLD), the most frequent function of single-purpose projects is irrigation, which is also the most common function of multipurpose projects. Operation of a multipurpose project is much more complex than a single-purpose project because functions may have conflicting strategies, flood-control versus conservation, for example. The usual strategy is to divide the multipurpose reservoir into vertical zones or layers, known as "pools," each of which is dedicated to a specific function, such as hydroelectric power, flood-control, and water-supply, and operated like independent single-purpose reservoirs (Wurbs, 1992). These may be augmented by pools for buffer storage, flood surcharge, dead (or inactive) storage, and recreation.

3.2 RESERVOIR DESIGN AND PERFORMANCE

3.2.1 Reservoir Water-Balance Modeling

Design of a reservoir is based on a water-budget analysis of the contemplated project. In the engineered river basins of concern in this volume, the water-supply function is of uppermost importance, for which the key performance parameter is yield. The design objective could be to determine the yield from a particular reservoir configuration, or to determine the necessary capacity to achieve a desired yield. Detailed site information is required, which includes geology and soils data, accurate topography of the dam location and the projected impoundment, and data on local hydrometeorology. Most important is a long record of streamflow for the river(s) to be impounded. Decades of data are generally desirable in order to encompass the range of variation in streamflow. This is especially important for the arid to semi-arid basins of the Sustainability of Engineered Rivers in Arid Lands initiative (SERIDAS) rivers, since high variability is a nearly universal characteristic of such hydroclimates (Slatyer and Mabbutt, 1964; Pilgrim et al., 1988; Graf, 1988; Dettinger and Diaz, 2000; Loik et al., 2004; Cudennec et al., 2007; Petrie et al., 2014; Zhang et al., 2014). Underlying the use of historical streamflow data for reservoir design is one of the principal protocols of natural science, to "study the past, if you would divine the future," e.g., McCarroll (2015).[1] More specifically, a record of measurement of sufficient length will exhibit the full range of hydrological conditions for which the reservoir is to be designed. This is not so much a protocol, however, as a postulate, as will be seen.

Streamflow is the main component of the total inflow to the reservoir, which also includes any tributary flows that would be impounded, runoff from the peripheral watershed of the reservoir (not included in the streamflow gauges), and the net of precipitation over evaporation at the reservoir surface (typically a negative number in arid and semi-arid environments). Designating the mean of this inflow over the reporting time interval $\Delta t_i \equiv t_i - t_{i-1}$ as $\overline{Q_i}$, the volume of water flowing into the reservoir during the reporting interval is the product $\Delta t_i\ \overline{Q_i}$. Therefore, the time history of volume $V(t)$ of water contained in a reservoir of capacity C is

$$V_i \equiv V(t_i) = max\left\{ min\left\{ V_{i-1} + \sum_{i=1}^{n} \Delta t_i(\overline{Q_i} - Q_w), C \right\}, 0 \right\} \tag{1}$$

where the values of $V(t)$ are given at discrete points in time $t = t_0, t_1, t_2, \ldots, t_N$, Q_w is a withdrawal rate (in volume of water per unit time, assumed constant), and the time history is begun at time t_0 with a specified volume $V(0)$. The most common time interval for conservation reservoir design is a month or a year. Neither of these intervals is constant (witness the number of days in a month), hence the need for the subscript on Δt_i and its appearance within the sum of (1). Some version of the mathematical model (1) is employed in the process of reservoir design, as well as in the analysis of performance of existing reservoirs, usually with more complex specification of how the reservoir will be operated, details on the hydraulic equipment of the dam, and features of the spillways and release structures (e.g., Beard, 1966).

Equation (1) is a water budget for a simple reservoir, with uncontrolled spills when its capacity is exceeded. Inspection of this equation will disclose that whenever the sum in (1) exceeds the capacity C, V_i is reset to C, i.e., the surplus water is "spilled" over the dam. When the sum falls negative, V_i is set to 0, i.e., the reservoir is dry. The occurrence in time of these extrema of reservoir contents is a useful indicator of how well the reservoir will meet its conservation objective. The key is the sign of the $(\overline{Q_i} - Q_w)$ term in the summand of (1). The time period over which V_i is computed may be divided into intervals in which $(\overline{Q_i} - Q_w)$ is negative or nonnegative. Within each nonnegative interval, i.e., in which $(\overline{Q_i} - Q_w) \geq 0$, V_i may acquire the value C. If it does so, then the first point at which $V_i = C$ and the starting point of the next interval in which $(\overline{Q_i} - Q_w) < 0$ define a time interval over which $V_i = C$. This is called a *full interval* and the bounding times are *full points*. Engineers sometimes refer to this as a wastage interval, because the contents in excess of C are considered to be "lost" downstream. Similarly, within each interval in which $(\overline{Q_i} - Q_w) < 0$, V_i may fall to zero. If so, then the first point at which $V_i = 0$ and the starting point of the next interval in which $(\overline{Q_i} - Q_w) \geq 0$ define an interval over which $V(t) = 0$, known as a *fail interval*, because the reservoir fails to meet the "demand" Q_w, with bounding times the *fail points*. A time interval bounded by a full point where $V_i = C$ and a fail point where $V_i = 0$ with no intervening extrema is a *depletion* or *drawdown period*. An interval bounded by a fail point and a full point with no intervening extrema is a *filling* or *storage period*. Examples are shown in Figure 3.1 (in which the hypothetical reservoir volume is scaled to unity). In the present context, it is the conservation function of a reservoir that is of primary interest, for which the occurrence of a pattern of low streamflows, or drought, is paramount. During such a pattern, we look for drawdown periods terminating in fail intervals.

3.2.2 Dependable Yield

The conceptual model of the reservoir water budget presented in Equation (1) of the previous section provides an opportunity to explore an important parameter quantifying the water supply achievable from a reservoir. For a specific time history of reservoir contents $V(t)$, the draft Q_w that eliminates all fail intervals so that there remain only discrete fail points is the *dependable yield* Q_d (also called the *firm yield*, *safe yield*, *assured yield*, and, in

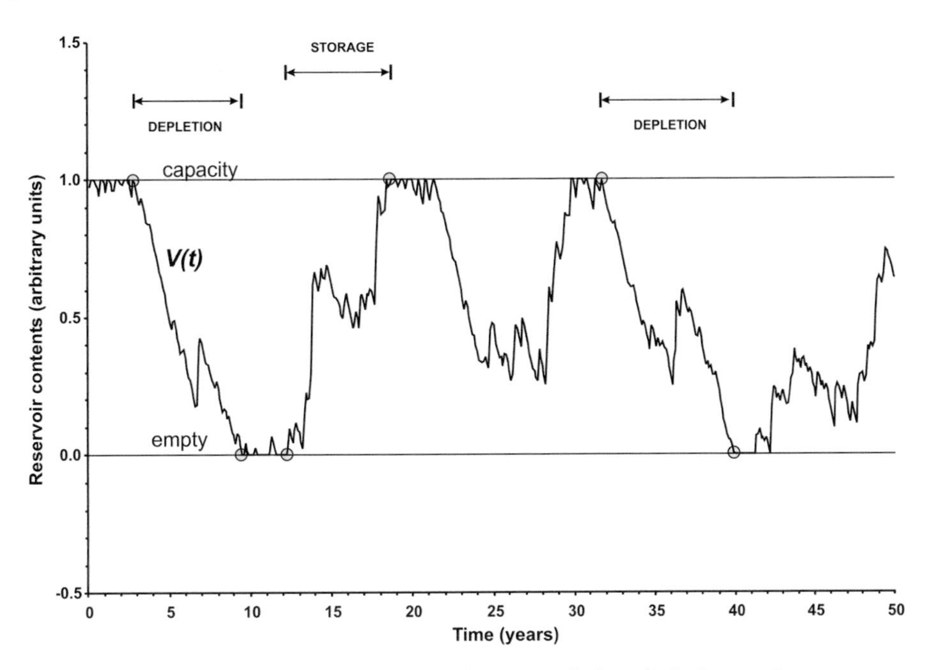

Figure 3.1 Depletion and storage periods in the simulated time history of contents of a hypothetical reservoir.

recent years, *sustainable yield*). While theoretically there can be multiple fail points under a dependable-yield draft, in realistic cases this is exceptional, and for practical application there is exactly one. The practical definition of dependable yield is therefore the constant draft for which the value of zero contents occurs exactly once in the reservoir volume time history $V(t)$. The depletion period culminating in that fail point is the *critical drawdown period.*

Dependable yield is determined iteratively using a model like (1), with the fixed input of the time history of inflow into the reservoir. Successive values of withdrawal rate Q_w are input to the model, and the resulting time series of volume $V(t)$ examined for the occurrence of failure intervals. The method is suggested by the model results of Figure 3.2, though of course more than three iterations are necessary to converge to the value of dependable yield, in this case $Q_d = 0.814\ \overline{Q}$. (The draft is sometimes given as a fraction of the period-of-record, POR, mean *flow* $\overline{Q}$. Draft as a percentage of the POR mean is referred to as *development* or *degree of regulation.*) The dependable yield can never exceed this POR mean flow. It will equal the POR mean only in the degenerate case of a constant inflow over the period of record. Thus, the case of a draft of $\overline{Q}$ necessarily results in fail intervals, evident in Figure 3.2. Successively reducing this fraction, we find that 90.7 percent of the POR mean still exhibits fail intervals, though considerably less than the 100 percent case. Finally, 81.4 percent achieves exactly one fail point, at about 29 years, and is therefore the dependable yield.

An alternative, and in fact the most common, definition of dependable yield is that draft which can be maintained constantly without failure throughout the time history of reservoir storage. This is equivalent to the definition given previously. To clarify, in the time history of storage with a withdrawal equal to the dependable yield, the occurrence of zero is an instantaneous value. Before, as the contents are decreasing to zero, the withdrawal of Q_d continues. Just as the storage zeroes, inflow increases, the storage increases, and the withdrawal is uninterrupted. There is no failure to deliver Q_d. It should also be noted that dependable yield is a transitional value. Any larger values of draft, however small, result in failure intervals and shortfalls in delivery. For any smaller values of draft, there is always water remaining in the reservoir.

Dependable yield is an old concept, arising from a basic problem in the use of any water resource (stream or river, aquifer, lake, or reservoir), viz. to determine the guaranteed yield from that resource. For management of the water resource, this is valuable information. The dependable yield can be allocated among users of the water resource. So long as the drought of record, which determines the dependable yield, is the worst possible drought and the draft does not exceed the firm yield, water demands will always be met.

Fundamental to the analysis are the properties of the cumulative of inflow $Q(t)$, which appears as the summation in (1), known as the "flow-mass curve" (e.g., Chow, 1964), or simply the "mass curve." The earliest reporting of a reservoir yield analysis using mass-curve methodology is attributed to Rippl (1883). The method employs a volume versus time plot on which are constructed straight-line segments, paralleling hypothetical draft flows, tangent to "humps" or "valleys" in the mass-curve (e.g., Chow, 1964; Fiering, 1967; Fredrich, 1975; McMahon and Mein, 1978, 1986; Mays and Tung, 1992). The appearance of

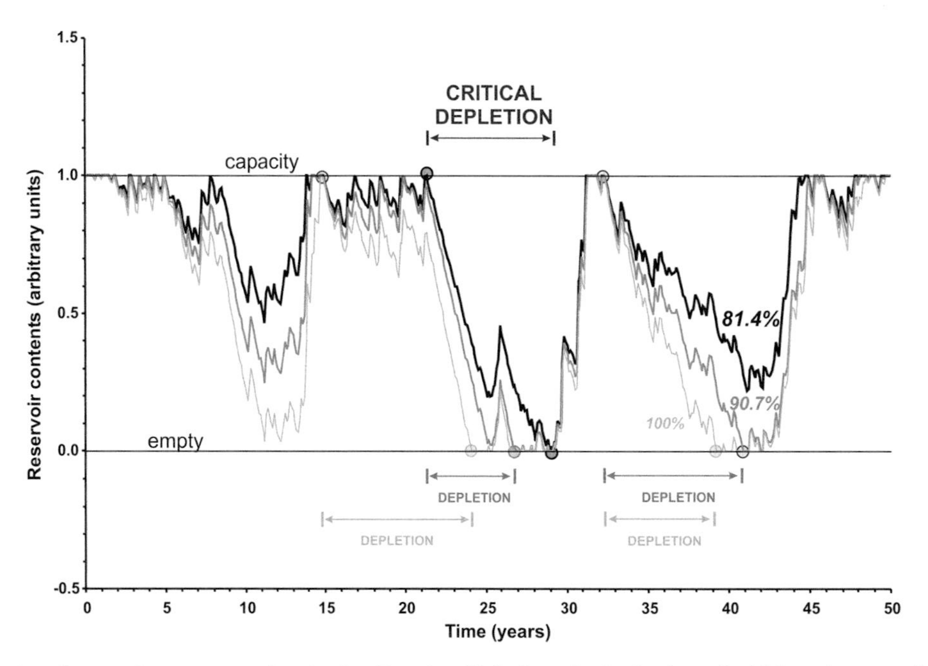

Figure 3.2 Time histories of reservoir contents produced using Equation (1) for hypothesized values of withdrawal, converging to the dependable yield, expressed as a percentage of the POR average inflow.

prominent humps and valleys in the cumulative is derived from the high autocorrelation of streamflow time series.

The example of Figure 3.2 is a hypothetical reservoir with a synthetic inflow. As an example for a real reservoir, Equation (1) was applied to Lake Corpus Christi on the Nueces River in south Texas, which is the principal water supply for Corpus Christi and neighboring communities. Figure 3.3 is the 1945–1960 segment of an application of Equation (1) for 1915–1960, nearly a half century of data. This segment encompasses the Drought of the Fifties, the drought of record for much of the south-central United States and northern Mexico. The firm-yield flow for the simulation through 1960 is 9.430 m^3/s with the critical depletion ending in 1957.

This would have been the drought of record for the reservoir, like most of the surrounding region, except that in the 1960s the Nueces basin was subjected to a shorter but more intense drought, shown in Figure 3.4. The critical depletion period during this drought terminates in late 1964, with a corresponding dependable yield of 7.425 m^3/s, about 80 percent of the value from the Drought of the Fifties. In fact, the drought of the 1960s has been the drought of record for water-supply planning in the Nueces basin for decades. And then came the twenty-first century.

Starting in 2007, the basin was subjected to a drought even more severe than those of the twentieth century, see Figure 3.5. The dependable yield for this drought proves to be 5.913 m^3/s from the critical depletion period terminating in early 2012. This is 80 percent of the yield from the drought of the 1960s and 63 percent of the yield from the Drought of the Fifties. Clearly,

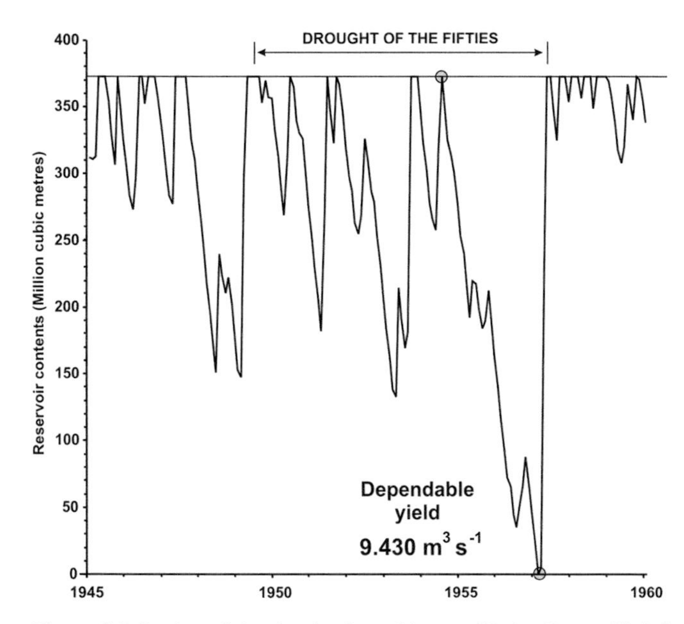

Figure 3.3 Portion of simulated volume history of Lake Corpus Christi for the Drought of the 1950s, showing dependable yield of 9.430 $m^3 s^{-1}$.

this is a new drought of record for the basin. At this writing, this drought is still underway.

Dependable yield is based on the most severe historical drought. Its use as a planning or management parameter therefore makes the implicit assumption that this historical event is the worst possible. This can be a decided advantage in courting approval for reservoir projects, as the engineer can point to an indisputable, historic drought as the justification for conservation and management. Indeed, the use of firm yield in both design and

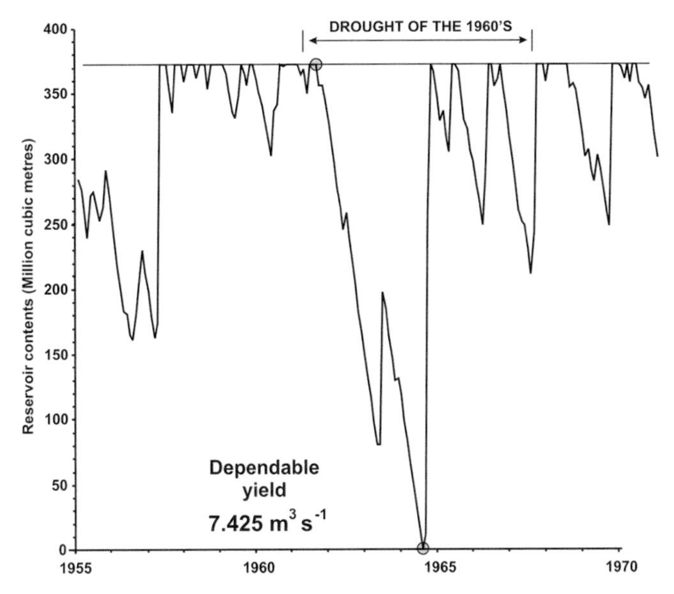

Figure 3.4 Portion of simulated volume history of Lake Corpus Christi for the drought of the 1960s, showing dependable yield of 7.425 m^3s^{-1}.

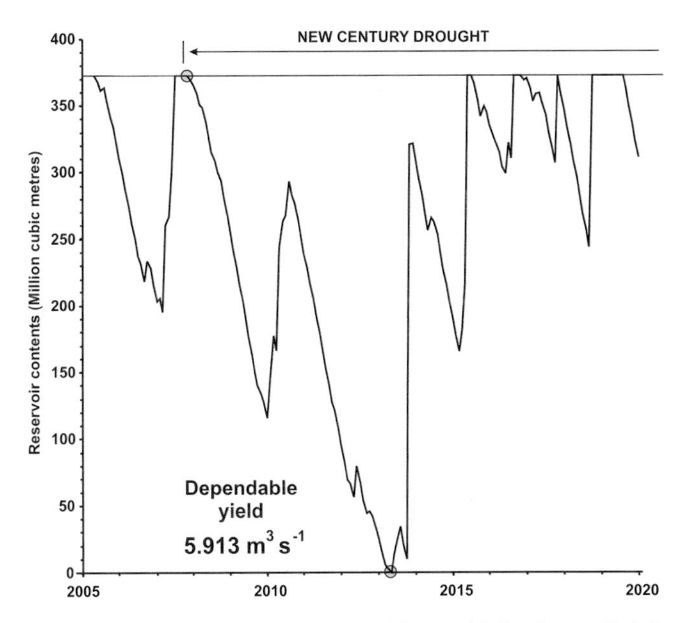

Figure 3.5 Portion of simulated volume history of Lake Corpus Christi for the drought starting in 2007, showing dependable yield of 5.913 m^3s^{-1}.

water-use allocation remains popular today (e.g., Werick and Whipple, 1994; Leib and Stiles, 1998; Zarriello, 2002; Bureau of Reclamation, 2008; Schnabel Engineering, 2008, 2011; New Jersey Geological and Water Survey, 2011; Maryland Department of the Environment, 2013; Laskey, 2017).

The example of Lake Corpus Christi illustrates that this assumption may be the great weakness of dependable yield as an operational parameter. From a scientific standpoint, given our pitifully short instrumental records of streamflow, the assumption is without foundation, even without invoking climate change. In the Nueces basin, three record droughts occurred successively within 70 years, representing a total reduction in the estimated available water-supply of about 40 percent. (It is actually even worse than this. The streamflow record starts in 1915, during a severe drought in the basin that would continue until 1919, even worse in this region than the Dust Bowl of the 1930s. The dependable yield for this drought was 9.469 m^3/s, slightly larger than that of the Drought of the Fifties, with the critical depletion terminating in March 1918. This would make four successive droughts of record for the Nueces in just over a century.)

This does not mean, however, that dependable yield is without value as a hydrometeorological analysis parameter. On the contrary, it is a parameter that compactly quantifies long-term water supply in drought-prone basins, is directly comparable to the separate and total water demands in a basin, and can be used to estimate shortfalls. It is also readily adaptable to future scenarios of altered watersheds, upstream diversions, and changed climate.

3.3 WATER QUALITY

While the primary emphasis of the chapters in this volume is on water quantity, particularly under conditions of limited water supply relative to the various demands on the resource, the quality of the water is of equal importance. A bountiful supply of water of inadequate quality creates the same critical shortcoming in meeting the needs of the river basin as inadequate quantity. Therefore, while space does not allow quality to be addressed here in the same detail as quantity, it is necessary to occasionally summarize issues concerned with quality.

Concerning water quality, it must be observed that a reservoir is a different kind of watercourse than the river or its tributaries. There are profound contrasts between the two types of watercourses. Most salient is that a reservoir is a lake, albeit artificial. (There are exceptions, of course. Lake Victoria became a reservoir when its natural outlet was replaced by the Owens Falls Dam in 1954, yet it is certainly not artificial.) The breadth, depth, and volume of the reservoir are larger by orders of magnitude than the river, which means that currents arising from the inflow of the river are typically much smaller than those created by, say, wind stress on the surface of the lake. With its large surface area and long fetches, the reservoir is subject to substantial wind waves and internal circulations. Most importantly, the reservoir may develop density stratification. The driving force for this is the receipt of solar radiation at the surface, which penetrates the lake, heating the near-surface waters. This heated water, measured by water temperature, is mixed vertically and downward by ambient turbulence, derived mainly from wind-driven waves and

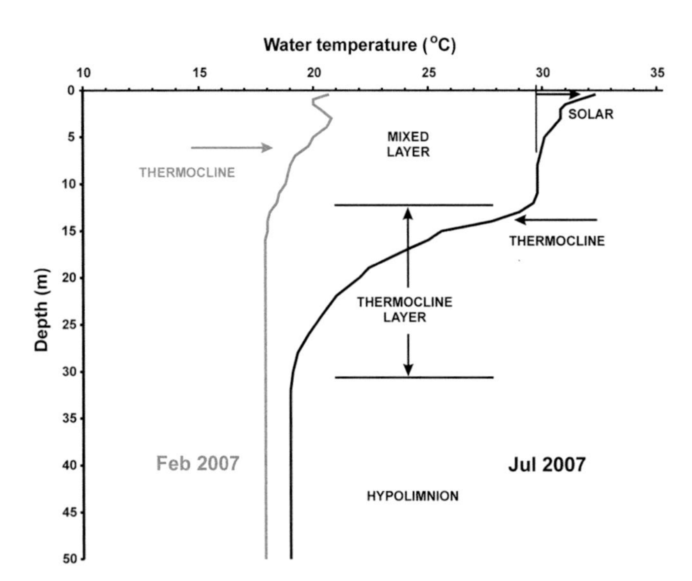

Figure 3.6 Temperature profiles from the reservoir of High Aswan Dam (Lake Nasser) in cool and hot seasons of 2007, plotted from field measurements of Abd Ellah (2009).

currents. Temperature is a key biological parameter. It is also important to the dynamics of water movement because it is the main determinant of the water density structure.

Example profiles of temperature during the cool and warm seasons (February and July, resp.) of 2007 in Lake Nasser, the reservoir of the High Aswan Dam, in the northern region near the dam are shown in Figure 3.6, based on data of Abd Ellah (2009). A pronounced decline in temperature with depth during the warm season is evident. There are several important layers defined by the behavior of this decline. (The reader is cautioned that there are conflicts in terminology among limnologists, dynamists, and oceanographers.) The depth at which the *rate* of decline is maximum is the *thermocline*, indicated in Figure 3.6. The thermocline divides the water column into two layers, the *epilimnion* from the surface to the thermocline, and the *hypolimnion* from the thermocline to the bottom. Both the epilimnion and hypolimnion are conceived ideally to be isothermal water volumes except in the layer of temperature roll-off containing the thermocline, referred to as the *thermocline layer*, which includes parts of the epilimnion and hypolimnion. The quasi-isothermal layer from the surface to the top of the thermocline layer is the *mixed layer*, see Figure 3.6. (In limnology, this is the epilimnion, the thermocline layer is the *metalimnion*, and the hypolimnion is the isothermal layer beneath the metalimnion.) There is occasionally a layer at the surface where the water is slightly warmer than the isothermal profile of the mixed layer. This results from short-term insolation under calm conditions and is sometimes referred to as a "radiation thermocline" or "solar thermocline." This is present in the July profile of Figure 3.6.

The importance of the thermocline is that it is a level of maximum increase of water density with depth. Its delineation of the transition from epilimnion to hypolimnion is not simply a convention of definition, but is a real physical barrier between the two layers. This is the level most resistive to vertical movement of water. In other words, the thermocline is the surface of highest gravitational stability. A parcel perturbed away from its position at the thermocline will find itself surrounded by water of different density, so that gravity will force the parcel back to the thermocline, irrespective of whether it is displaced upward or downward. In contrast, an isothermal layer has no variation in density in the vertical, so parcels of water may be moved up or down in the water column without encountering any resistance of buoyancy.

The formation and depth of the thermocline are governed by the net heat inputs at the surface and the downward turbulent transport of this heat against the density gradient. In most stratified reservoirs, this follows a seasonal cycle, diagrammed schematically in the time-depth cross section of Figure 3.7. The influx of heat continues through the warm season, increasing the temperature of the epilimnion and the stability of the thermocline. Maximum heat accumulation is attained by the end of the warm season. Then as air temperature cools, the epilimnion begins to lose heat and the vertical temperature gradient decreases, until turbulence is able to overcome the thermocline stability and mix through a greater depth, perhaps the entire water column, by late in the cool season. With the onset of the next warm season, the heat influx turns positive and the development of stratification begins again with a deepening thermocline. The temperatures depicted in Figure 3.7 are typical of tropical and subtropical lakes. Temperate lakes are cooler. (The winter dissolution of stratification is sometimes known as "overturn," e.g., Abu-Zeid, 1987. Technically, Figure 3.7 depicts vertical mixing. Overturn is a phenomenon of temperate or subarctic lakes whose temperatures fall to 4°C, the maximum density of water.) No scale is given for depth in Figure 3.7, because the depth and thickness of the layers of stratification are strongly influenced by the dimensions and layout of the reservoir and by its regional climatology. Depths of the thermocline in large reservoirs typically range 10–60 m.

The hydrometeorology accompanying the seasonal variation in heat flux governs the dynamics of the development and dissipation of thermal stratification. Reservoirs within the influence of synoptic disturbances in the westerlies will experience major storms during the equinoctial seasons, which enhance vertical mixing. Some rivers exhibit seasonal or intra-annual flooding, which may produce sufficient inflow to completely replace the reservoir water, thereby eliminating its stratification. This is often the case for Lake Nasser, with the Nile floods in August through October.

Dissolved oxygen (DO) time-depth contours are also plotted in Figure 3.7. DO is equally as important as temperature as a water-

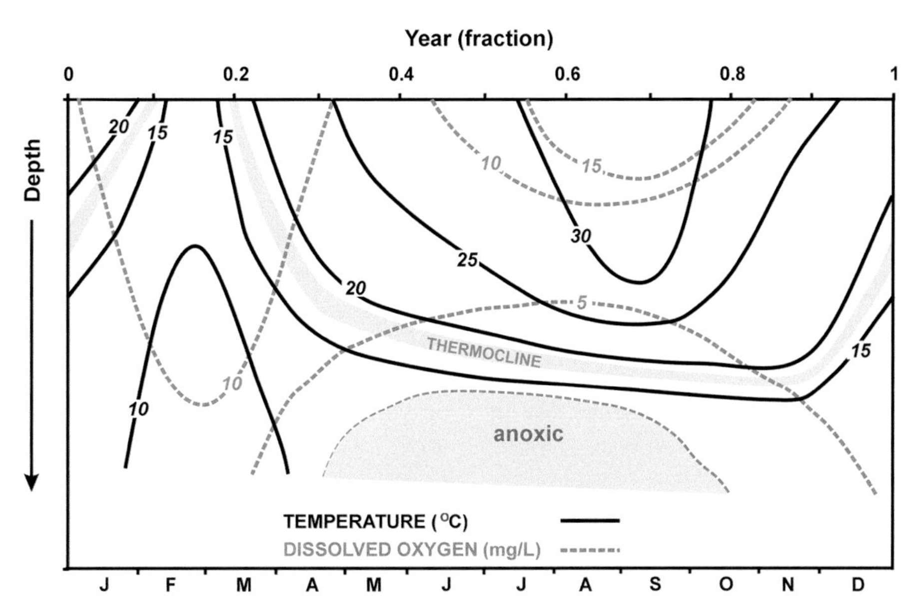

Figure 3.7 Schematic time-depth cross section of reservoir temperature and dissolved oxygen structure in subtropics of the northern hemisphere (see text).

quality indicator. It also has value in conjunction with temperature as a hydrodynamic tracer. Oxygen is injected into the water column at the surface by physical aeration and within the near-surface layer (specifically, the euphotic zone) as a byproduct of algal photosynthesis. It is removed from the epilimnion by respiration of plankton and macrofauna, and from the entire water column by bacterial respiration. While the latter occurs throughout the water column, it is concentrated in the bottom sediments and near-bottom waters where sinking organic matter accumulates. Generally, the reservoir produces a denser and more diverse community of phytoplankton than the river. Primary production increases with insolation and water temperature, so there is a strong seasonal variation, as indicated by the surface maximum in DO in the late boreal summer of Figure 3.7. The extent to which high-DO waters at the surface penetrate into the waters below the level of photosynthesis is a measure of vertical turbulence. In deeper lakes, the maximum decline in DO (the "oxycline") is located just above the thermocline, and the DO in the hypolimnion is virtually zero due to the inability of the aerated water to penetrate the gravitational stability of the thermocline. Abu-Zeid (1987) reports depths in Lake Nasser to the anoxic layer of about 10 m in May–July (consistent with Figure 3.6) and 40–50 m in October, after the flood season.

The fact that the reservoir is a large body of generally quieter water than the river it impounds means that it traps a significant amount of the river-borne sediment. This is addressed in more detail in Section 3.4, but in the present context several water-quality implications can be identified. These are related to the budget of the macronutrients nitrogen (N) and phosphorus (P) in reservoir waters. Generally, fine sediments (silts and clays) act as "carriers" of N and P, either through sorption of inorganic nutrients to sediment particles or because the fine sediments include organic matter incorporating these nutrients. Much of the nutrient influx to the lake results in high concentrations of N and P sequestered in the deposited sediments. This sequestration is temporary, however, because the nutrients are reintroduced into the water column in biologically assimilable forms, referred to as "internal" loads.

This flux depends on mobilization – the conversion of the sequestered nutrient to a water-soluble form – and on transport from the bed sediment to the water column. Biochemical mobilization is effected by bacterial decomposition of organic nitrogen and phosphorus, producing, respectively, ammonia, predominantly NH_4^+, and phosphates. The bacteria responsible are anaerobes, which can thrive only in an anoxic environment. Physicochemical mobilization of phosphorus includes desorption and dissolution, notably associated with reduction of iron and the release of ferric-bound phosphorus (e.g., Mortimer, 1941, 1942; Boström et al., 1988). Transport of dissolved nutrients between the sediment pore water and overlying lake water is a diffusive flux.

Under warm-weather stratification with an anoxic hypolimnion, as indicated for May through October in Figure 3.7, the stability of the thermocline isolates the hypolimnion from the turbulence of the epilimnion so the hypolimnion is stagnant. Hydromechanically, the hypolimnetic water column is laminar, and transport processes are molecular. During the rest of the year, oxic waters reach the sediments, creating an oxidizing upper layer in the sediments overlying the reducing anaerobic zone, which has the net effects of oxidizing some of the ammonia and precipitating much of the phosphates. Under either scenario, there is a continuous flux of these inorganic nutrients into the water column, ultimately enhancing production in the epilimnion. (For details, consult Berner, 1980; Boudreau, 1997; DiToro, 2001.)

During the period of stratification, these nutrients are confined in the hypolimnion due to the stability of the thermocline, except when strong storms rupture the thermocline and mix hypolimnetic water to the surface. In large lakes, the epilimnion may be moved from one end of the lake to another by wind stress, upwelling the hypolimnion at the opposite end (e.g., Lorke et al., 2003). Otherwise, these nutrients do not reach the epilimnion until the seasonal elimination of stratification. For those sediments that are deposited on the periphery of the lake in relatively shallow water, there is another access to the nutrients by the ecosystem: macrophytes may root in the bed sediments for direct uptake.

The accumulation of nutrients in the reservoir sediments, and their mineralization and release to its waters represent an energy supply to the base of the ecosystem, its autotrophs. An oversupply of nutrients can engender an excess of primary production, characterized by excessive algal blooms and vegetation growth, referred to as eutrophication.

3.4 RESERVOIR CHALLENGES

3.4.1 Sedimentation

A universal statement about reservoirs is the precept (borrowing from the syllogism exemplar):

All reservoirs are mortal

by which is meant that any reservoir has a finite lifetime due to the accumulation of sediment behind its dam. Banerji and Lal (1974) are blunt, "The destiny of a reservoir is to silt up." Smith (1976) opens his history, "As will emerge time and again throughout the story [of the history of dams], dams always act as traps for silt." Mukherjee et al. (2013) state, "All the reservoirs in the world receiving water from natural rivers are bound to be silted up by these rivers."

Rates of reservoir siltation[2] vary considerably from country to country, within river basins, and from reservoir to reservoir, depending on surficial geology, watershed land use, and hydrometeorology, to say nothing of upstream dams. The data of Mahmood (1987) indicate a world-wide average of 40–50 functional years of reservoirs remaining. Kondolf and Farahani (2018) determined that global loss of storage to siltation exceeded new storage development around the year 2000. That is, about 2000 the net global storage peaked and began to decline. For storage per capita, the peak occurred around 1980, and has declined since. Estimates of the global annual capacity loss due to siltation range 0.5–1.5 percent, and the cumulative capacity loss over the lifetime of the reservoirs around 2000 ranges 10–15 percent, though by individual subcontinents 3–45 percent (Mahmood, 1987; Sumi and Hirose, 2009). Wang and Hu (2009) report that 2.3 percent of total reservoir capacity in China is lost annually, and twenty large reservoirs in China have lost 66 percent of their capacity.

While regional factors may trump climate for a specific reservoir, there seems to be a proclivity for increased siltation in arid and semi-arid lands (Israelsen, 1943; Graf, 1988; Thornes, 1994). For example, in Spain only about 12 percent of reservoirs have silting problems but about half of these have lost more than 50 percent of their storage capacity (Avendaño-Salas et al, 2000). The reservoir of the Valdeinfierno Dam on Rio Guadalentin filled to the brim with sediment in about 50 years (Smith, 1976). In Texas, the Austin Dam on the Colorado River (which bears no relation to the river of Chapter 13) lost over 40 percent of its capacity to siltation in its first four years of operation (Smith, 1976). Chanson (1998) reports some spectacular incidents of reservoir siltation in Australia of reservoirs filled within 25 years.

In modern practice, as part of the design process, the engineer may estimate the rate of sediment accumulation to confirm that siltation will not compromise reservoir function, for which some design operational period is assumed, and may reserve a portion of the reservoir volume for sediment accumulation, consistent with this anticipated lifetime of the structure (Kondolf and Farahani, 2018). Only the largest projects warrant the challenge and expense of sophisticated sediment modeling. Sometimes, especially for dams constructed prior to around 1960, the problem of siltation has been simply ignored in the hope that it will not prove crucial, at least in the near future. In describing the value of designing a hypothetical reservoir with capacity considerably greater than the average annual inflow, Schuyler (1909) concludes that the annual loss in volume to siltation would be less than 1 percent, whereupon, "the period of usefulness of the works would be vastly increased, and the consideration of the problem of silt disposal would be left for future generations to solve." Nearly a century later, Chanson (1998) observes, "The issue [siltation] was ignored too long and it is only recently that researchers and engineers have acknowledged the matter."

Accurate prediction of sediment accumulation in a reservoir is an extraordinarily difficult problem. This derives from four facts. (1) Sediment brought into the reservoir is governed by the mechanics of sediment transport. This is a complex subject that addresses how movement of a fluid may mobilize and carry masses three- (for water) to a thousand-times (for air) denser, involving topics such as turbulence and flow-induced instabilities. (2) Even though there are equations, both empirical and quasi-theoretical, for many aspects of sediment transport, the data necessary for their application is frequently lacking. (3)

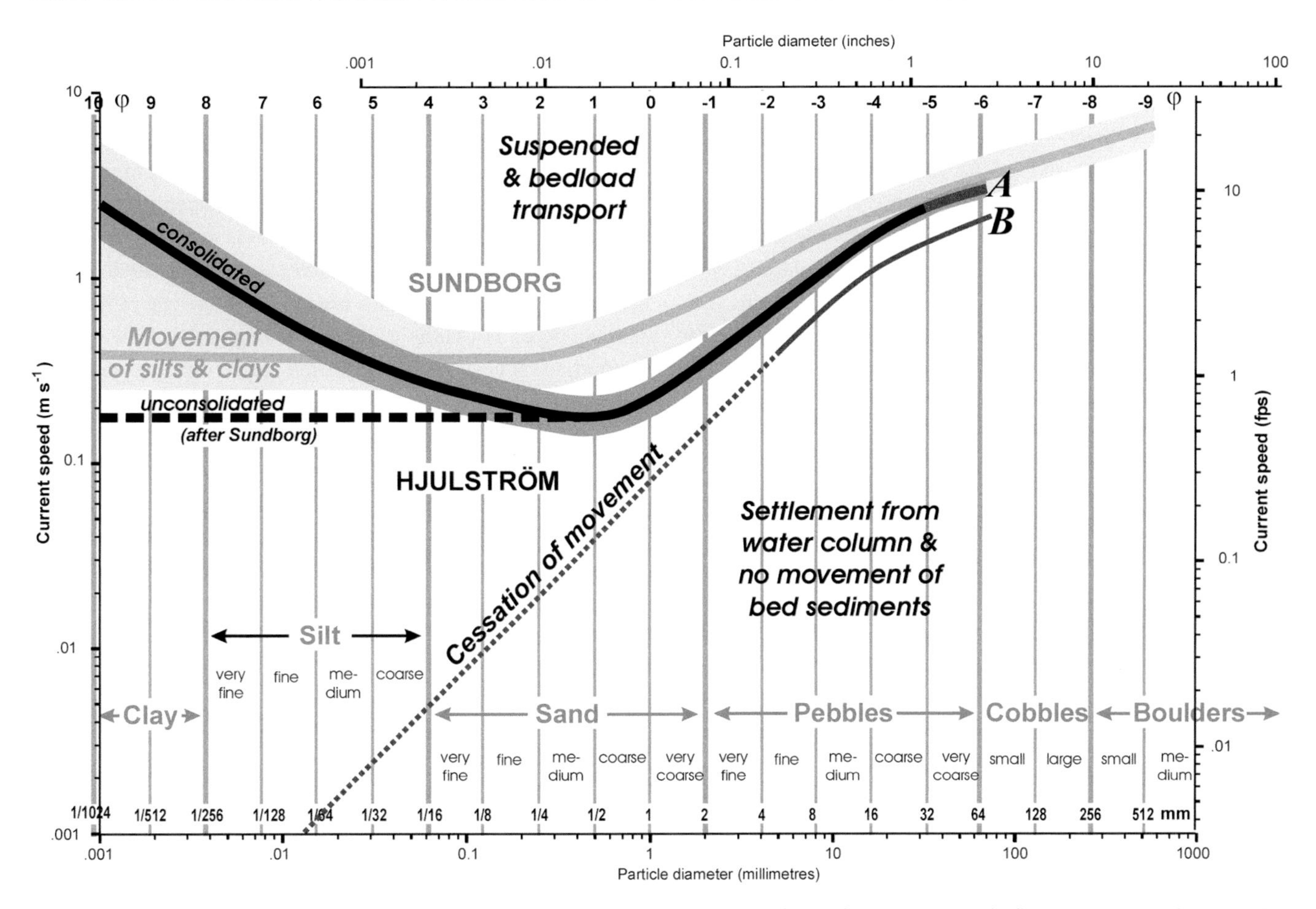

Figure 3.8 Hjulström–Sundborg diagram with Wentworth–Krumbein grade scale, showing sediment movement in flowing water as a function of particle size and vertical-mean current (see text).

Sediment transport does not operate in isolation but involves land use, soils, geomorphology, and hydrometeorology for its complete treatment. (4) A highly variable proportion of the sediments are carried downstream through the dam by flowing water, depending on details of flow variability and the morphology of the reservoir (and including flushing as a management action). All of the approximations and uncertainties attending the quantification (or neglect) of each of these accumulates, like silt in the reservoir.

Figure 3.8 graphically summarizes much of the physics underlying erosion and deposition by water. The lower abscissa is "diameter" in millimeters on a logarithmic scale, divided into classes by powers of two (which is also logarithmic). These define the grade categories formalized by Wentworth (1922), shown in grey font. This grade classification was later supplemented by the *φ-scale*, suggested by Krumbein (1934), see also Krumbein and Aberdeen (1937), where $\varphi = -\log_2 d$, given at the top of the figure, d denoting the particle size in mm.

Superposed on the particle-size scales of Figure 3.8 are two sets of curves that qualitatively display the behavior of the different grain sizes in water of mean current speed given by the ordinate of the graph. The earliest of these is due to Hjulström (1935, 1939), a broad convex-downward curve (which he labeled "A" as shown in Figure 3.8) above a thinner, partially dashed curve (labeled "B"), together forming the famous Hjulström diagram for sediment in a flowing stream. Curve "A" represents incipient motion of particles on the riverbed under the sustained current speed of the ordinate. The lower curve "B" represents the current speed at which particles already carried in suspension begin to fall out of the flow. Hjulström states that the curve is generally based on the work of Schaffernak but without citation (perhaps Schaffernak, 1922), and that "B" depicts competency of the flow as defined by Gilbert (1914).

The region between the curves "A" and "B" is transitional, in which only some particles are in motion, or are exhibiting different modes of motion (sliding, rolling, saltation and suspension). Perhaps unfortunately, Hjulström's (1939) designations of the three regions of the graph demarcated by these two curves were "erosion" (above "A"), "transportation" (between the curves), and "deposition" (below "B"). Erosion and deposition properly refer to the result of a deficit or surfeit in sediment budget of the bed, resp., whereas the diagram is intended to depict the ability of the current to move particles of sediment.

The particle size exhibited by a rock fragment is determined by its mineral composition, notably its crystalline structure and resistance to chemical and mechanical disintegration, and its history of exposure to weathering and erosive processes (e.g., Friedman et al., 1992). The majority of sediment size classes are derived from silica-based minerals, mainly quartz, feldspars, and clay minerals. The finer particles in a watercourse, *viz.* silts and clays, generally contain a higher proportion of layer silicate (phyllosilicate) minerals. These particles are electrochemically charged on their surface and edges, creating a net negative charge on the particle (Holtz et al., 2010). In water with a concentration of free ions, this net charge is decreased, and the electrostatic repulsion reduced so that molecular attraction becomes dominant. The net effect is that these particles tend to be cohesive and form flocs (Huang et al., 2006).

In the Hjulström diagram, there is a pronounced minimum in the fine sands of current needed to suspend and transport sediment: smaller and larger diameters require a higher current speed for resuspension. Though Hjulström discussed the effect of cohesiveness in creating resistance to movement, it was the insight of Sundborg (1956) that the range of the quantity of cohesive sediments would lead to a spread in the curve. He showed this in a similar diagram inferred from his studies of the Klarälven River, the second set of curves in Figure 3.8. The lower boundary, corresponding to an absence of cohesive silts and clays ("unconsolidated"), has been added to the Hjulström curve. The diagram serves to display generally the currents necessary to mobilize sediments in moving water, and the effects of grain size and cohesiveness.

Once a dam is constructed, model availability notwithstanding, field observation is the most reliable method for determining actual sedimentation rates. Siltation is frequently monitored by repeated bathymetric surveys over time. The methodology has been greatly improved by modern fathometer technology, global positioning satellites, and geographical information systems. This type of data can be used to determine historical and present siltation rates for an existing reservoir and provide an empirical basis for extrapolation to the future.

Bathymetric surveys (and many numerical models) are based on an underlying hypothesis that all siltation occurs within the open-water regions of the reservoir. For many lakes, significant siltation also occurs in the riverine approaches. These are the backwater reaches subject to inundation by the downstream impoundment. The modest reduction in flow velocity due to the widening cross section is sufficient to actuate settling of the coarser suspended particles (Figure 3.8), which form deltas. These deltas provide broad substrates for additional deposition, through which river flow is confined to narrow distributaries. Deltaic systems are further subjected to sediment reworking by later flood events, except where vegetation has rooted in the nutrient-rich sediments, reinforcing the deposits against erosion.

The process of reservoir siltation can be mitigated by only three broad strategies, (1) retain the sediment upstream from the reservoir, (2) divert sediment around the reservoir or route it through the reservoir, to avoid deposition, or (3) remove the deposited sediment from the reservoir (mainly by dredging or flushing). Any of the standard strategies for any one of these, if feasible at all, is usually expensive to implement. However, as circumstances have become desperate, these strategies are finding increasing application, and a modest body of engineering methods has been developed (Morris and Fan, 1998).

Upstream reservoir development frequently reduces the siltation threat to projects lower in the basin, by the simple (and usually inadvertent) sharing of the riverine sediment load. Ten years after deliberate impoundment began on Elephant Butte on the Rio Grande, a sedimentation survey was performed that determined the reservoir had lost 6.7 percent of its capacity (Israelsen, 1943), which would imply that the reservoir would be completely silted in about 150 years. More recent surveys (through 2007) indicate a considerable reduction in siltation, with complete siltation in about 335 years, more than doubling the projected life of the reservoir. While the early siltation surveys may entail considerable uncertainty, there is little doubt that the siltation problem in Elephant Butte has benefited from the considerable reservoir construction upstream in the watershed. A similar reduction in siltation has been observed in Lake Amistad due to reservoir construction on the Rio Conchos (see Chapter 14).

Perhaps the most extreme example of strategy (1) is the construction of a separate reservoir for the explicit purpose of sediment (and flood) control, thereby protecting the structures downstream. The sediment-control reservoir can be said to be sacrificial. This is not a modern strategy. For example, a dam at Ponte Alto on the Fersina River (Italy) has existed for this purpose since the sixteenth century, being re-built or heightened several times as it filled with silt and debris (Smith, 1976). A modern example is the sediment and flood control project of the US Corps of Engineers on the upper Rio Grande, consisting of Cochiti, Galisteo, Jemez Canyon, and Abiquiu (see Chapter 14).

3.4.2 Storage and Conveyance Inefficiencies

The objective of the water-supply reservoir is, obviously, water supply. It is therefore ironic that the distribution system for delivering water is subject to substantial losses, and that the retention of water in the reservoir itself entails major consumption. The delivery system is the network of canals by which water is transported from storage to the user. These typically are earthen ditches especially subject to infiltration, leakage, and evaporation. For irrigation, the additional transport to the target plants in the field is often inefficient as well, much water being wasted as runoff, infiltration into unplanted soils, or discharge onto nontarget vegetation ("weeds"). Taken together, these wastages in the delivery and debouchment of water are conveyance losses (though some researchers separate conveyance and application losses, e.g., Jägermeyr et al., 2015). Chapter 16 addresses in detail the dominating role of irrigation in basin water management.

The range of conveyance loss is large and highly dependent on the construction and soils of the canal and plumbing of the irrigation. Israelson (1943) offers an educated guess of 15 percent to more than 60 percent in the United States; Jägermeyr et al. (2015) judged greater than 50 percent globally. Measurement is difficult and uncertain, and tends to be granular, concentrated on a specific canal or field.

Jägermeyr et al. (2015) implemented a different approach, augmenting a numerical global land-surface water-budget model with deterministic process models for each of the above wastages (except canal leakage). The model was given inputs for soils, vegetation, irrigation areas, crops, canal systems, and types of irrigation water application, together with subregional data from the Food and Agriculture Organization's AQUASTAT data base, MIRCA2000 land-use data set, and a data compilation of 1900–2005 global irrigated land. The model simulation employed a 0.5° × 0.5° grid with a 1901–2009 execution period and was driven with the Climate Research Unit high-resolution gridded monthly climatology and the Global Precipitation Climatology Centre gridded monthly precipitation data files.

Results from the simulation were summarized as 1980–2009 averages. These are optimistic in two senses, that irrigation water is assumed to come from "local, renewable" sources, e.g., reservoirs and groundwater replenished by recharge, not from, say, fossil groundwater, and that all fugitive outflows from field irrigation (canal seepage, runoff, percolation, soil storage) are assumed to be recoverable, except for that portion of the flow that reaches the ocean. The baseline simulation shows 51 percent of water diverted for irrigation to be "consumed," that is, returned directly to the atmosphere, of which 26 percent is "beneficially consumed," that is, transpired by the target crop. The former is the usual definition of irrigation efficiency (see Chapter 16). Jägermeyr et al. (2015) focus on the latter as *beneficial irrigation efficiency* (the product of conveyance efficiency and field application efficiency). Areas in which rice irrigation is extensive, such as southeast Asia, are found to be highly inefficient (<30 percent), while Europe and North America exhibit efficiencies much higher than the global mean, due mainly to efficient field application (sprinkler and drip systems). Global averages of beneficial efficiency by field application are field 29 percent, sprinkler 51 percent, and drip 70 percent.

Evaporation from the surface of a reservoir is a significant loss of water, especially in arid and semi-arid climates. For Lake Nasser, long-term evaporation is about 10 percent of the reservoir volume (from salt-budget data reported by Abu-Zeid, 1987). For Elephant Butte, annual evaporation is about 3 metres, and in volume is about twice the annual usage of the City of Albuquerque.

The evaporation rate from a free water surface is given by the Dalton equation

$$E = 1.26\,x\,10^{-3} W\,(e_s - e_a) \tag{2}$$

for E = evaporation in m/d, W = wind speed in m/s at 2 m height, e_s and e_a = vapor pressure at the water surface and the atmosphere (2 m height), resp., see, e.g., Hess (1959), Brutsaert (1982), and Dingman (2002). The volume evaporated from the reservoir per day, then, is EA in m^3/d, for A the reservoir surface area in m^2. Vapor pressure may be computed from the saturation vapor pressure, which is a function only of temperature, given by the Clausius–Clayperon equation (Hess, 1959; Wallace and Hobbs, 2006)

$$e(T) = 6.11\ exp\left\{5427\left(\frac{1}{273} - \frac{1}{T+273}\right)\right\}$$

where $e(T)$ denotes the saturation vapor pressure in hPa (mb) for T in °C. Assuming that vapor pressure at the water surface is saturated at surface water temperature T_s, then (2) becomes $E = 1.26 \times 10^{-3}\ W\ [e(T_s) - r\ e(T_a)]$ for r and T_a the relative humidity (fraction between 0 and 1) and air temperature, resp., at 2 m. While wind speed, air temperature, and relative humidity are standard climatological data, water temperature at the surface may not be readily available, in which case an estimate can be had by taking $T_s = T_a$, so that

$$E = 1.26 \times 10^{-3} W\,(1 - r)\,e(T_a) \tag{3}$$

There are no good remedies available for dealing with evaporation losses from reservoirs. One strategy is to store the water where there is limited exposure. For a small municipality, water tanks may suffice. For the volumes of water entailed in irrigation supply, or water supply for a major city, this is impractical. Aquifer storage and retrieval is employed in a few locations where the economics and logistics permit, evidently with some success. There is some experimentation underway with spreading chemical monolayers on the reservoir to prohibit or reduce exchange with the atmosphere. This strategy has its own problems. The chemical employed must be nontoxic and biodegradable. It must also be recognized that any monolayer that acts as a barrier to evaporation will also interfere with transfer of gases across the surface, such as oxygen and nitrogen, with deleterious consequences for the aquatic ecosystem.

3.4.3 Impacts on River

Not only does the reservoir replace a segment of the river but it also influences the unimpounded reaches of the river both upstream and downstream. While most of the impacts are downstream, upstream the reservoir creates a zone of backwater, i.e., quasi-permanent inundation, frequently with the morphology of a delta. This region can exhibit eutrophication due to trapping and accumulation of nutrients (see Section 3.3), with associated degradation in water quality, and negatively impacted ecosystem.

Downstream, there are several categories of effects of the reservoir. The first is alteration in the river hydrograph, with increased periods of low (or zero) flow and diminishment of pulses. Alterations in the river hydrograph occur as soon as deliberate impoundment is begun. For retention reservoirs, if there is no consideration of riverine impacts, impoundment will entail an extended period of zero flow while the reservoir pool is filled. The need for preservation of at least some of the features of the time pattern of river flow, referred to as "environmental flows," is addressed by Dr. Glenn and Dr. Lester in Chapter 6.

The second category is stimulated channel erosion and deepening due to sediment "starvation," a consequence of reservoir siltation. Unless the reservoir is equipped with a sluice gallery, or some sort of high-flow sediment by-pass channel, the impacts on the downstream channel are largely unavoidable, but generally take place over a period of years. This affords some opportunity for compensatory modifications of human structures such as bridges, diversions, and drainageways, and for adaptation of the ecosystem to the changing channel morphology.

The third is degraded water quality in the river because of releases from the reservoir. This can occur if the reservoir becomes eutrophic, and the release contains harmful algae, such as blue-greens (which are, in fact, bacteria). Deep reservoirs may have releases from the hypolimnion, with colder temperatures and reduced oxygen content compared to the riverine environment. This has become a significant problem in Australia, where low-level releases from most reservoirs, especially large, deep reservoirs, are creating a form of "thermal pollution," in this case pollution with unusually cold water, see Sherman (2000), Lugg and Copeland (2014), and Weber et al. (2017). Thermal alterations are also reported on the Yangtze River due to a series of dams culminating in Three Gorges. These impacts are more complex, however, because the temperature impacts are highly variable, both increases and decreases, with dam and with season, a result of the depth of the reservoirs and the extent of stratification, and are cumulative as one proceeds downstream (He et al., 2020).

3.4.4 Demands and Conflicts

While the motivation for construction of a reservoir, i.e., its design objective, forms the basis of its intended function and operation, the fact is, once built, a reservoir is simply a big tank to hold water. The social, political, and hydroclimatic environments surrounding that structure can be expected to evolve in time. Associated with this evolution, the requirements for how water is to be managed, and the role that the reservoir will play, can be expected to change as well. In some of the river basins addressed in this study, there have already been exceptions to the initial operating rules, sometimes temporary in response to a short-term crisis, sometimes permanent as dictated by changing needs. For example, the initial purpose of La Boquilla on the Rio Conchos (Rio Grande basin) was power generation for mining operations, but a century later most of its capacity is dedicated to irrigation and municipal water supply.

Several potential trends have been identified in the chapters of this book, including population increases, agricultural needs and the security of food supply, and changing climate. These are dealt with in detail in their respective chapters. Here we simply identify the challenges and their particular importance to reservoirs. Alteration of the operation of a reservoir can be expected to be protested, perhaps, by conflicting interests in the society served by that resource. When a water supply reservoir is full, demands for withdrawals typically exceed those allocations based on the ever-present possibility of the beginning of a critical drawdown. As populations increase and more of the watershed below the dam is developed, the needs for both water supply and flood protection increase, but with different distributions of risk among the population served.

As a further example, the indication for future climates in arid and semi-arid climates is increased variability (see the presentation by Professor North in Chapter 2), so that there are prospects for droughts of increased frequency and severity, and also exposure to floods of greater magnitudes. Whether the reservoir will manage for either entails increased risk for the other. This is particularly true for those reservoirs with both flood control and conservation pools, because any shifting of the boundary between the pools represents a shifting of the relative risks, yet public demands for such changes are usually influenced by the most recent event of flood or drought.

It is clear that greater sophistication in the specification and implementation of reservoir operation will be demanded in the future, not only in the technical dimensions of reservoir function but also in the ability of its management to assess, fairly allocate, and negotiate its service to its society.

REFERENCES

Abd Ellah, R. A. (2009). Thermal Stratification in Lake Nasser, Egypt, Using Field Measurements. *World Applied Sciences Journal*, 6(4), pp. 546–549.

Abu-Zeid, M. (1987). Environmental Impact Assessment for the Aswan High Dam. In A. K. Biswas and S. B. C. Agarwal, eds., *Environmental Impact Assessment for Developing Countries*. Amsterdam: Elsevier Ltd., pp. 168–190.

Avendaño Salas, C., Sanz Montero, M., and Cobo Rayan, R. (2000). *State of the Art of Reservoir Sedimentation Management in Spain*. In International Workshop and Symposium on Reservoir Sedimentation Management, pp. 27–34, Toyama: Water Resources Environment Technology Center-Japan (WEC).

Banerji, S. and Lal, V. (1974). Silting of Reservoirs: Indian Data and the Needed Direction of Efforts. *Proceedings of the Indian National Science Academy*, Section A40(5–6), pp. 356–365.

Beard, L. R. (1966). *Methods for Determination of Safe Yield and Compensation Water from Storage Reservoirs*. TP-3, Hydrologic Engineering Center, Davis, CA: US Army Corps of Engineers.

Berner, R. A. (1980). *Early Diagenesis: A Theoretical Approach*. Princeton, NJ: Princeton University Press.

Billington, D., Jackson, D., and Melosi, M. (2005). *The History of Large Federal Dams: Planning, Design, and Construction in the Era of Big Dams*. Denver, CO: Bureau of Reclamation, Department of the Interior.

Boström, B., Andersen, J., Fleischer, S., and Jansson, M. (1988). Exchange of Phosphorus across the Sediment–Water Interface. *Hydrobiologia*, 170, pp. 229–244.

Boudreau, B. P. (1997). *Diagenetic Models and Their Implementation: Modelling Transport and Reactions in Aquatic Sediments*. Berlin: Springer-Verlag.

Brutsaert, W. (1982). *Evaporation into the Atmosphere*. Dordrecht: D. Reidel Publishing Company.

Bureau of Reclamation (2008). *Water Supply and Yield Study*. Mid-Pacific Region, Sacramento, CA: Bureau of Reclamation.

Chanson, H. (1998). Extreme Reservoir Sedimentation in Australia: A Review. *International Journal of Sediment Research*, 13(3), pp. 55–63.

Chow, V. T. (1964). Runoff. In V. Chow, ed., *Handbook of Applied Hydrology*. New York: McGraw-Hill Book Company, pp. 14–54.

Cudennec, C., Leduc, C., and Koutsoyiannis, D. (2007). Dryland Hydrology in Mediterranean Regions: A Review. *Hydrological Sciences Journal*, 52(6), pp. 1077–1087.

Dettinger, M. and Diaz, H. (2000). Global Characteristics of Stream Flow Seasonality and Variability. *Journal of Hydrometeorology*, 1, pp. 289–310.

Dingman, S. L. (2002). *Physical Hydrology*, 2nd ed. Upper Saddle River, NJ: Prentice-Hall, Inc.

DiToro, D. M. (2001). *Sediment Flux Modeling*. New York: John Wiley & Sons.

Fiering, M. B. (1967). *Streamflow Synthesis*. Cambridge, MA: Harvard University Press.

Fredrich, A. J. (1975). Reservoir Yield. Vol. VIII of *Hydrologic Engineering Methods for Water Resources Development*. Davis, CA: Hydrologic Engineering Center, US Army Corps of Engineers.

Friedman, G., Sanders, J., and Kipaska-Merkel, D. (1992). *Principles of Sedimentary Deposits*. New York: Macmillan Publishing Company.

Gilbert, G. K. (1914). The Transportation of Debris by Running Water. Professional Paper 86, U.S. Geological Survey. Washington, DC: Government Printing Office.

Graf, W. L. (1988). *Fluvial Processes in Dryland Rivers*. Berlin: Springer-Verlag.

He, T., Deng, Y., Tuo, Y., Yang, Y., and Liang, N. (2020). Impact of the Dam Construction on the Downstream Thermal Conditions of the Yangtze River. *International Journal of Environmental Research and Public Health*, 17, p. 2973. https://doi.org/10.3390/ijerph17082973

Hess, S. L. (1959). *Introduction to Theoretical Meteorology*. New York: Holt, Rinehart & Winston.

Hjulström, F. (1935). The Morphological Activity of Rivers as Illustrated by the River Fyris. *Bulletin of the Geological Institution of the University of Upsala*, 25, pp. 221–527.

(1939). Transportation of Detritus by Moving Water. In P. D. Trask, ed., *Recent Marine Sediments, a Symposium*. London: Thomas Murby & Co, pp. 5–31.

Holtz, R., Kovacs, W., and Sheahan, T. (2010). *An Introduction to Geotechnical Engineering*, 2nd ed. Upper Saddle River, NJ: Prentice Hall.

Huang, J., Hilldale, R., and Greimann, B. (2006). Cohesive Sediment Transport. In C. T. Yang, ed., *Erosion, and Sediment Manual*. Denver, CO: Bureau of Reclamation, U.S. Department of the Interior.

Israelsen, O. W. (1943). *The Foundation of Permanent Agriculture in Arid Regions*. Paper 51, Utah State Univ. Faculty Honor Lectures. Available at: http://digitalcommons.usu.edu/honor_lectures/51

Jägermeyr, J., Gerten, D., Heinke, S. et al. (2015). Water Savings Potentials of Irrigation Systems: Global Simulation of Processes and Linkages. *Hydrology and Earth System Sciences*, 19, pp. 3073–3091.

Kondolf, G. and Farahani, A. (2018). Sustainably Managing Reservoir Storage; Ancient Roots of a Modern Challenge. *Water*, 10, p. 117. https://doi.org/10.3390/w10020117

Krumbein, W. C. (1934). Size Frequency Distributions of Sediments. *Journal of Sedimentary Research*, 4(2), pp. 65–77.

Krumbein, W. C. and Aberdeen, E. J. (1937). The Sediments of Barataria Bay (Louisiana). *Journal of Sedimentary Research*, 7(1), pp. 3–17.

Laskey, F. A. (2017). *Report on 2016 Water Use Trends and Drought Status*. Water Policy & Oversight Committee Report, Boston: Massachusetts Water Resources Authority.

Leib, D. and Stiles, T. (1998). Yield Estimates for Surface-Water Sources. In M. Sophocleous, ed., *Perspectives on Sustainable Development of Water Resources in Kansas*. Lawrence: Kansas Geological Survey, pp. 158–169.

Loik, M., Breshears, W., Lauenroth, W., and Belnap, J. (2004). A Multi-Scale Perspective of Water Pulses in Dryland Ecosystems: Climatology and Ecohydrology of the Western USA. *Oecologia*, 141, pp. 260–281.

Lorke, A., Müller, B., Maerki, M., and Wüest, A. (2003). Breathing Sediments: The Control of Diffusive Transport across the Sediment–Water Interface by Periodic Boundary-Layer Turbulence. *Limnology and Oceanography*, 48(6), pp. 2077–2085.

Lugg, A. and Copeland, C. (2014). Review of Cold-Water Pollution in the Murray-Darling Basin and the Impacts on Fish Communities. *Ecological Management and Restoration*, 15(1), pp. 71–79.

Mahmood, K. (1987). *Reservoir Sedimentation: Impact, Extent, and Mitigation*. Tech. Pap. No. 71, Washington, DC: World Bank.

Maryland Department of the Environment (2013). *Guidance for Preparing Water Supply Capacity Management Plans*. Baltimore: Maryland Department of the Environment.

Mays, L. and Tung, Y. (1992). *Hydrosystems Engineering and Management*. New York: McGraw-Hill Book Company.

McCarroll, D. (2015). "Study the Past if You Would Divine the Future": A Retrospective on Measuring and Understanding Quaternary Climate Change. *Journal of Quaternary Science*, 30(2), pp. 154–187.

McMahon, T. and Mein, R. (1978). *Reservoir Capacity and Yield*. Amsterdam: Elsevier Scientific.

(1986). *River and Reservoir Yield*. Littleton, CO: Water Resources Publications.

Morris, G. and Fan, J. (1998). *Reservoir Sedimentation Handbook: Design and Management of Dams, Reservoirs, and Watersheds for Sustainable Use*. New York: McGraw-Hill Book Co.

Mortimer, C. H. (1941). The Exchange of Dissolved Substances between Mud and Water in Lakes, Part I. *Journal of Ecology*, 29, pp. 280–329.

(1942). The Exchange of Dissolved Substances between Mud and Water in Lakes, Part II. *Journal of Ecology*, 30, pp. 147–201.

Mukherjee, B., Das, S., and Mazumdar, A. (2013). Mathematical Analysis for the Loss of Future Storage Capacity at Maithon Reservoir, India. *Asian Research Publishing Network (ARPN) Journal of Engineering and Applied Sciences*, 8(10), pp. 841–845.

New Jersey Geological and Water Survey (2011). *Estimating the Safe Yield of Surface Water Supply Reservoir Systems*. Trenton, NJ: Guidance manual, Department of Environmental Protection.

Oestigaard, T. (2019). The First Aswan Dam in Egypt: A Useful Pyramid? In T. Ooestigaard, A. Beyene, and H. Ögmundardóttir, eds., *From Aswan to Stiegler's Gorge: Small Stories about Large Dams*, Current African Issues No. 66. Uppsala: The Nordic Africa Institute, pp. 23–39.

Petrie, M., Collins, S., Gutzler, D., and Moore, D. (2014). Regional Trends and Local Variability in Monsoon Precipitation in the Northern Chihuahuan Desert, USA. *Journal of Arid Environments*, 103, pp. 63–70.

Pilgrim, D., Chapman, T., and Doran, D. (1988). Problems of Rainfall-Runoff Modelling in Arid and Semiarid Regions. *Hydrological Sciences Journal*, 33(4), pp. 379–400.

Rippl, W. (1883). *The Capacity of Storage-Reservoirs for Water-Supply*. Minutes of Proceedings of the Institution of Civil Engineers 71, pp. 270–278.

Schaffernak, F. (1922). *Neue Grundlagen für die Berechnung der Geschiebeführung in Flußläufen*. Vienna and Leipzig: Franz Deutike.

Schnabel Engineering (2008). *Water Supply Assessment for Talking Rock Creek Dam No. 13*. Project 07170030, Alpharetta, GA: Schnabel Engineering.

(2011). *Safe Yield Analysis, Glades Reservoir–Cedar Creek Reservoir, Hall County Georgia*. Project 10217007, Alpharetta, GA: Schnabel Engineering.

Schuyler, J. D. (1909). *Reservoirs for Irrigation, Water-Power and Domestic Water-Supply*, 2nd ed. New York: John Wiley & Sons.

Sherman, B. (2000). *Scoping Options for Mitigating Cold Water Discharges from Dams*. Consultancy Report 00/21, Canberra: CSIRO Land and Water.

Slatyer, R. and Mabbutt, J. (1964). Hydrology of Arid and Semiarid Regions. In V. Chow, ed., *Handbook of Applied Hydrology*. New York: McGraw-Hill Book Company, pp. 24–46.

Smith, N. (1976). *A History of Dams*. Secaucus, NJ: Citadel Press.

Soothill, W. E. (1910). *The Analects of Confucius*. Yokohama: Fukuin Printing Company, Ltd.

Sumi, T. and Hirose, T. (2009). Accumulation of Sediment in Reservoirs. In Y. Takahashi, ed., *Water Storage, Transportation, and Distribution*. Encyclopedia of Water Sciences, Engineering and Technology Resources, of the Encyclopedia of Life Support Systems (EOLSS), developed under the auspices of the UNESCO, Paris: Eolss Publishers. Available at www.eolss.net

Sundborg, Å. (1956). The River Klarälven: A Study of Fluvial Processes. *Geografiska Annaler*, 38(2), pp. 125–237.

Thornes, J. B. (1994). Catchment and Channel Hydrology. In A. Abrahams and A. Parsons, eds., *Geomorphology of Desert Environments*. Cambridge: Cambridge University Press, pp. 257–287.

Wallace, J. M. and Hobbs, P. V. (2006). *Atmospheric Science*, 2nd ed. Burlington, MA: Academic Press.

Wang, Z. and Hu, C. (2009). Strategies for Managing Reservoir Sedimentation. *International Journal of Sediment Research*, 24, pp. 369–384.

Weber, M., Rinke, K., Hipsey, M., and Boehrer, B. (2017). Optimizing Withdrawal from Drinking Water Reservoirs to Reduce Downstream Temperature Pollution and Reservoir Hypoxia. *Journal of Environmental Management*, 197, pp. 96–105.

Wentworth, C. K. (1922). A Scale of Grade and Class Terms for Clastic Sediments. *Journal of Geology*, 30(5), pp. 377–392.

Werick, W. and Whipple, W. (1994). *Managing Water for Drought*. Rep. 94-NDS-8, Institute for Water Resources, Alexandria, VA: US Army Corps of Engineers.

Wurbs, R. A. (1992). *Military Hydrology: Report 21, Regulation of Streamflow by Dams and Associated Modeling Capabilities*. Misc. Pap. EL-79-6, Vicksburg, MS: US Army Engineer Waterways Experiment.

Zarriello, P. J. (2002). *Simulation of Reservoir Storage and Firm Yields of Three Surface-Water Supplies, Ipswich River Basin, Massachusetts*. Water-Resources Investigations Report 02-4278, Denver, CO: US Geological Survey.

Zhang, Q., Peng, J., Singh, V. P., Li, J., and Chen, Y. D. (2014). Spatio-Temporal Variations of Precipitation in Arid and Semiarid Regions of China: The Yellow River Basin as a Case Study. *Global and Planetary Change*, 114, pp. 38–49.

Notes

1 Widely attributed to Confucius, though this writer has been unable to verify the source. The closest approach, perhaps, is, "The past [is] the mirror of the future" (The Analects Book II, XXIII, Soothill, 1910).

2 The nouns "silt" and "sediment" are synonymous, unless the specific grain-size range is indicated for "silt."

4 Depletion of Groundwater

The Surface–Groundwater Connection

Stephanie Glenn

4.1 GROUNDWATER

High quality water supplies are critical for drinking water and for municipal and irrigation activities. Managing surface and groundwater quality and quantity effectively is becoming more and more important as population continues to grow and the threat of drought in arid lands persists. An integrated approach to river basin management that includes the analysis of surface and groundwater interactions is crucial in prioritizing resources for future water demands. Currently it is estimated that groundwater provides drinking water to at least 50 percent of the global population. In terms of groundwater usage, it is estimated that groundwater accounts for 43 percent of all the water used for irrigation (FAO, 2010).

Increasing water demands means more reliance on both surface and groundwater in the future. A recent study that used the results of global water demand and hydrological models paired with a dynamically simulated water use detailed the results of the researchers' global overview of water use. The results showed that globally, water withdrawal increased by more than 60 percent from 1979 to 2010. The study also indicated that from 1990 to 2010 global groundwater reliance increased (Wada et al., 2014).

Arid and semi-arid lands represent over 30 percent of the Earth's land surface and are home to over 20 percent of world population (United Nations, 2018). Freshwater is a limiting resource on these lands, and withdrawal of groundwater substantially exceeds recharge. Groundwater over-pumping and the resulting aquifer depletion is a common problem around the world. Over-pumping of the aquifer can result in irreversible consequences, such as saltwater intrusion, land subsidence, and sinkholes (USGS, 2000). To manage groundwater sustainably, the effects of groundwater pumping must be quantified, which requires identifying interactions between the surface and groundwater, as well as between the different layers of the aquifer system itself.

Figure 4.1 shows groundwater extraction in Sustainability of Engineered Rivers in Arid Lands initiative (SERIDAS) countries and territories (SERIDAS basin areas with data available shown – note not all SERIDAS basins had data available) (Margat and van der Gun, 2013). Population trends continue to increase, and as is shown here, arid regions rely heavily on groundwater as part of their total water allocation. Some countries, such as the United States and Mexico, have groundwater abstraction rates that are large compared to other populous regions. The impacts of climate change will lead to substantial changes in groundwater recharge. Changes to surface flow, timing, and intensity of precipitation, as well as increasing temperatures in arid lands leading to increased evaporation rates will all put pressure on water resources. Studies show that it is expected that climate change will result in arid and semi-arid regions becoming more dependent on groundwater (Van der Gun, 2012). In addition, while the exact impacts of climate change on groundwater in terms of recharge may be difficult to estimate in all areas, increased pressure on water resources overall will mean more demand on groundwater resources. This will be particularly severe in arid and semi-arid regions, where climate change predictions include increased frequency and intensity of droughts (Taylor et al., 2012).

Water management in the face of climate change includes planning for changing precipitation and snowmelt patterns, with effects such as groundwater depletion due to sea level rise, recharge by salt water (intrusion), and anthropogenic responses to droughts such as increasing numbers of reservoirs, which can lead to changing recharge patterns. During times of drought, communities often rely more heavily on groundwater resources, which can result in increased pumping and strain on water supply. Increased pumping can cause water levels to drop, which results not only in decreased groundwater availability but also deterioration of groundwater quality. Drought will place more stress on aquifer systems, increasing the need to know how decreasing water levels will impact groundwater quality at varying geospatial scales. Assessing the effects of increased pumping and decreased recharge rates due to droughts and climate variability on groundwater will be critical for sustainable future management.

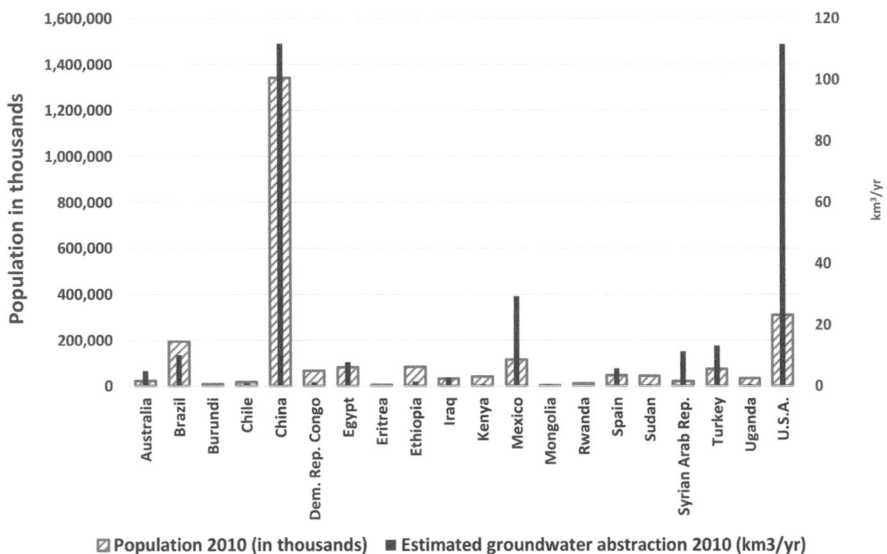

Figure 4.1 Groundwater abstraction and population in select SERIDAS river basins.
Figure by Houston Advanced Research Center (HARC) using data from Margat and van der Gun (2013) and UN World Population Prospects (2011)

4.1.1 Groundwater Use and Recharge

Recharge is the rate at which water percolates down to the aquifer. Recharge varies widely among aquifers as well as within an individual aquifer, and rates can vary widely spatially as well as temporally. Rates of recharge can be impacted by several factors, including precipitation changes, altered pumping regimes, development on recharge zones (increased impervious zones), and changes to soil and lens structures.

Groundwater-tracer techniques are considered the most reliable estimators of recharge. Certain types of physical techniques such as surface water and unsaturated and saturated zone data provide estimates of potential recharge. Modeling can provide estimates from published data such as precipitation, vegetation, soils, and aquifer zones. For example, in the state of Texas, the Gulf Coast Aquifer recharge rate ranges from 0.0004 inches (.0101 mm) to 2 inches (50.8 mm) per year, while in the High Plains Aquifer recharge rates range from 0.004 inches (.0101 mm) to 11 inches (280 mm) per year.

Engineered rivers and creation of reservoirs alter flow regimes, especially overbank flooding, which impacts groundwater recharge and, ultimately, aquifer storage. In addition, the creation of reservoirs changes rates of evaporation. Reservoirs have larger surface areas than rivers and facilitate faster evaporation rates of water molecules at the air–water interface. This increases the rate of water loss to evaporation, which impacts the total water budget for the system.

4.1.2 Groundwater Management Case Study: Texas

Often, surface water and groundwater are managed as two different entities. River basin management often does not include impacts on groundwater; the same holds true for groundwater management. Texas water law is an excellent example of this fact. In Texas, surface water is owned by the state of Texas and managed under the doctrine of prior appropriation ("first in time, first in right"). There are thorty-three Texas river authorities and similar surface water entities. Water rights include principles such as "run of the river" and prioritizing withdrawal permits.

Groundwater law in Texas is governed by the rule of capture, which states that groundwater pumping is an unrestricted right (see Texas aquifers in Figure 4.2). The Texas Supreme Court adopted the rule of capture in 1904. Whoever pumps the groundwater owns it, and pumping cannot be restricted even if others are harmed. The rule of capture protects the liability of anyone who pumps groundwater. Groundwater is a major source of water in Texas, providing 60 percent of the 16.1 million acre-feet of water used in the state; of this, 61 percent of groundwater used in Texas is for agriculture.

Surface and groundwater resources are treated differently under the legal and policy framework in Texas. In general, surface water is considered a public resource and is allocated under a permit system by the Texas Commission on Environmental Quality. Groundwater has not been regulated in the past; resources are utilized under the common law principle

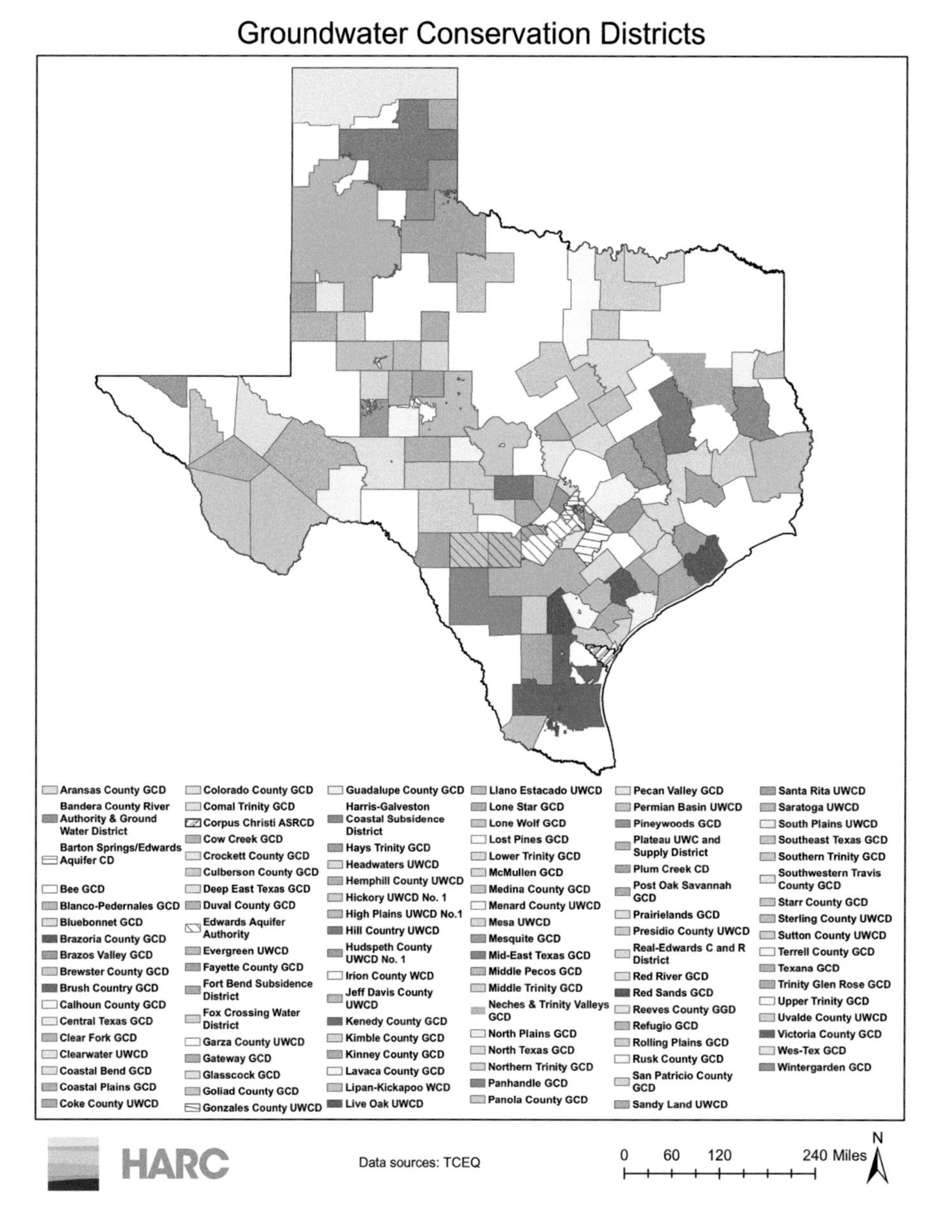

Figure 4.2 Major aquifers in Texas.
Figure by HARC using data from the Texas Water Development Board (2018)

of the rule of capture. This rule has been modified in areas where public need is greater, e.g., in the Edwards Aquifer area, where excessive pumping of groundwater would kill endangered species. In most of Texas, groundwater is considered private property associated with ownership of the surface. This has been modified in those parts of the state where a groundwater district has been established to manage those resources in the interest of the state (Figure 4.3). Groundwater districts are a recognition of the public interest in water held in aquifers.

Groundwater conservation districts (GCDs) provide limited regulation for groundwater management. A few aquifers are strictly regulated, such as the Edwards Aquifer, where the rule of capture has been replaced. There are ninety-eight GCDs covering 56 percent of the state's surface area, which includes 89 percent of the developed groundwater. Many GCDs are single county districts, even though the aquifers are all multiple-county.

When water supply is challenged by drought or excess demand, managers often look for new sources of water, such as additional wells or new reservoirs. It can be difficult to develop new reservoirs – they are extremely costly to build and can take years to complete. In addition, reservoirs require a significant amount of land and their impact on affected ecosystems must be taken into consideration. Building reservoirs in the arid climate of West Texas is problematic due to rapid evaporation and loss of water in the reservoir.

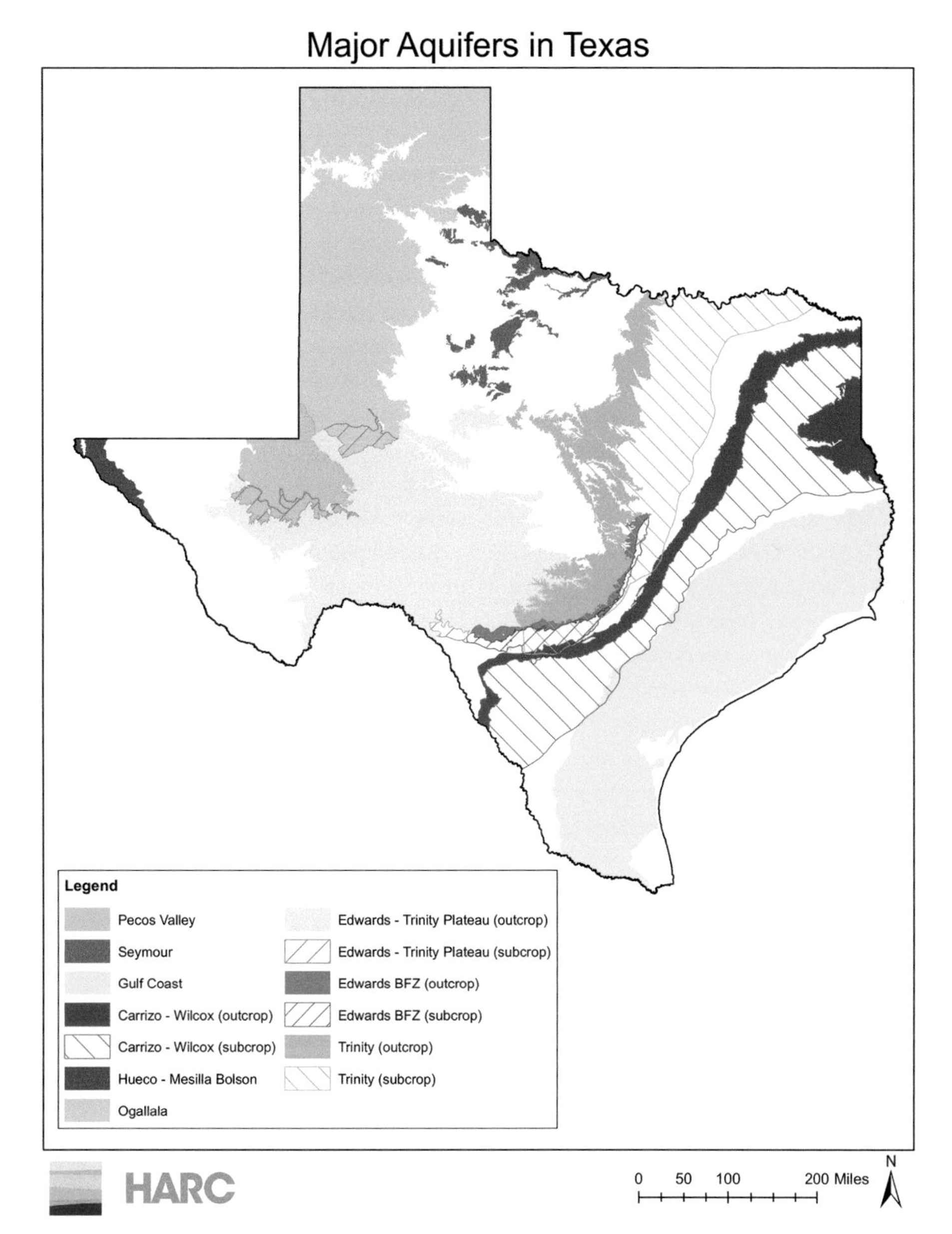

Figure 4.3 Texas groundwater conservation districts.
Figure by HARC using data from the Texas Commission on Environmental Quality

Because of the difficulty and expense of developing new reservoirs, extracting groundwater can be an attractive option. Groundwater is a crucial source of freshwater for human activities, for as world populations continue to increase, so does reliance on groundwater for a freshwater supply. Texas is among the top five states responsible for the largest groundwater withdrawals in the United States. In the 2017 Texas State Water Plan (TWDB, 2018), surface water is more than two-thirds of the existing water supply for municipal, manufacturing, steam-electric, and mining water needs. Groundwater, on the other hand, is more than three-quarters of the existing water supply for irrigation and livestock water. Groundwater supplies are projected to decrease 24 percent, from 7.2 million acre-feet per year in 2020 to 5.4 million in 2070 (TWDB, 2018). The 2017 State Water Plan discusses at length the strategies necessary to successfully manage the state's surface and groundwater considering decreasing supplies and increasing demand.

4.1.3 Engineered Alterations and Groundwater

Engineering rivers often means increases in impervious surfaces and resulting changes to recharge. A 2009 report by the National Resource Defense Council, American Rivers, and Smart Growth America (Otto et al., 2009) explained a detailed analysis determining groundwater recharge "losses" (the amount of water not infiltrated into an aquifer due to impervious surfaces) in the major American metro areas. While not the worst on the list, Dallas was calculated at 6.2 billion to 14.4 billion gallons and Houston was calculated at 12.8 billion to 29.8 billion gallons potentially not infiltrated per year. With planning on the part of engineered basins, Texas can lower these losses.

More resource planning needs to focus on the connections between land use, groundwater, and surface water, specifically when analyzing engineered river basins. Watershed stewardship should optimize benefits to both the surface water and groundwater. Rivers created by groundwater (such as the San Marcos, Comal, and Devils rivers) and springs have a circular connection to groundwater, with springs feeding rivers and rivers feeding aquifers. As an example of how important groundwater can be to a river basin, the Devils River in Texas is one of two major tributaries to the Rio Grande. Downstream from its headwaters, a series of springs from the Edwards-Trinity (Plateau) aquifer provide up to 80 percent of the river's baseflow (Texas Parks and Wildlife, 2002). In some parts of the Rio Grande, the contribution of groundwater helps to overcome water quality problems presented due to low inflows; some segments north of the Amistad Reservoir have impaired water quality due to increasingly high salinity. However, groundwater contributions help to mitigate this problem as river flows out of Big Bend National Park in Texas (Bennet et al., 2012).

Innovative ways of thinking are needed for water stewardship in arid lands with high evaporation rates. In the 2012 Texas State Water Plan, major reservoirs were the most expensive items. But consider that in the 2011 drought, the evaporation loss from Lakes Travis and Buchanan was greater than the water used by the city of Austin during the same time. Instead of reservoirs, different types of water storage such as water security in the form of stored aquifer water (ASR: Aquifer Storage and Recovery – see later discussion) should be explored.

4.1.4 Surface and Groundwater Interactions

The circular connection of groundwater and surface water is an integral part of river basin management. River basin withdrawals can deplete groundwater or pumpage of groundwater can deplete basins. Pollution of surface water can cause degradation of groundwater; degraded groundwater quality can in turn impact surface water. Effective water stewardship requires management that factors in the entire river basin – not analyzing groundwater and surface water as separate entities, but rather analyzing them together as a system. There is a need to understand quantitative linkages between groundwater and surface water.

Surface water is exposed to the atmosphere in rivers, streams, and reservoirs but groundwater is covered by soil or rocks. Some streams and aquifers are directly connected through springs and recharge structures. In limestone, it is possible for rivers to flow underground through cave systems. There is a continuum of distinction between surface and groundwater. At one end is the water in aquifers that has been sequestered for thousands of years and can only be brought to the surface by drilling wells, e.g., the Ogallala Aquifer in Texas. Then there are aquifers that can be recharged slowly from isolated surface recharge structures. Closest to surface water are those aquifers that can be rapidly recharged and connect directly to surface waters through springs. Many shallow aquifers in Texas are recharged annually from surface precipitation and their waters return to the surface in springs to feed the flow of streams and rivers. Some shallow coastal aquifers recharge rapidly and flow into the bays and lagoons along the coast, thus returning to surface water.

Karst aquifers are a type of aquifer that is particularly sensitive to the surface and groundwater connection. Karst aquifers are known as such for their limestone and dolomite rock formations.

Springs, caves, and sinkholes form when carbon dioxide-enriched water dissolves the rock. Groundwater in karst aquifers provides about 25 percent of the United States' groundwater drinking supply. The large fissures formed during dissolution ensure that groundwater is quickly recharged; however, this also means karst aquifers are sensitive to contamination. Karst habitats contain unique environments that promote a high degree of niche habitats (a species may only live in one sinkhole, spring, or cave). Karst formations such as caves, sinkholes, springs, and seeps provide habitat for hundreds of sensitive species, including many that are threatened and endangered.

Decline in aquifer water levels impacts surface water bodies. Groundwater over-pumping has been shown to lower water levels in lakes, wetlands, and springs, reduce stream flow in river systems, and reduce spring flow (Katz et al., 1997). Wetlands find it hard to adapt to changing water levels and are sensitive to reduced water supply (Mitsch and Gosselink, 1993). The effects of groundwater pumping on surface water depends on a variety of factors. One factor is the distance between the pumping well (s) and the surface water body; the closer the well and the greater the rate of pumping, the larger the impact. Another factor is the rate and duration of pumping – how long a well is pumped and at what rate makes a big difference in the intensity of impact. Geologic and hydrologic characteristics of the aquifer is another factor; aquifers with Karst limestone features have more porous features and often greater interconnectivity between surface and groundwater. Unconfined aquifers are directly connected to the surface, and the groundwater levels are dependent on relatively constant recharge. Confined aquifers are saturated layers of pervious (porous) rock bounded above and below by largely impervious rock. Another factor is the vertical permeability and thickness of the confining beds of aquifers, and the degree of interconnection (Stewart, 1968).

4.1.5 Water Supply

Water is an essential resource for human survival and an essential ingredient of many human activities, especially agriculture. Population growth and economic activity have resulted in concomitant pressure on the water supply. Many arid regions have surface water resources that are inadequate for certain types of economic activity, and many of these regions have supplemented surface water resources with large amounts of groundwater. Even in the wetter regions, there are concerns about obtaining sufficient water resources for the future. In many areas, both surface and underground sources are declining or are over-committed and are insufficient to support future population or economic growth.

Water supply challenges pose a variety of problems. Allocation and user conflicts are problems in many arid regions. Interbasin transfers have been proposed or are in place to move allocated water between resources. Interaquifer transfers are another method of allocation transfer that move groundwater resources across the boundaries of groundwater regions. Groundwater removal for large water transfer projects can initiate conflicts between the many users of the same aquifer. A common conflict for water supply is between municipal users and agricultural users. In most areas, this conflict is exacerbated by inefficiencies by both user groups. Large quantities of water are used on inefficiently irrigating crops and lawns with excessive water needs. A recent study shows show groundwater is overexploited in many large aquifers that are critical to agriculture, especially in Asia and North America (Gleeson et al., 2012). Figure 4.4 shows usage by sector (Irrigation, Domestic Use, Industry) for groundwater extraction in SERIDAS countries and territories (SERIDAS basin areas with data available shown – note not all SERIDAS basins had data available) (Margat and van der Gun, 2013). The vast majority of SERIDAS basins use the groundwater extracted for irrigation purposes.

While water managers have an obligation to provide water to meet citizens' needs, they also have an obligation to protect the environment. Removal of all the available water or failure to return enough could lead to a dry riverbed. This has happened to the Rio Grande in Texas and could happen to other rivers. At some level of reduced flow, the ecosystems in that watershed, both aquatic and terrestrial, suffer damage (see Chapter 6 on Environmental Flows). Groundwater in arid rivers with connectivity between surface and groundwater also suffers from a dry riverbed, as do the ecosystems it supports – sometimes sensitive ecosystems that support niche habitat species. In addition, groundwater resources are often critical to important wetland ecosystems. Aquatic ecosystems provide free services to the human population that may be difficult to replace.

Water marketing is receiving a great deal of interest; water is a natural resource that is currently marketed in many ways and the market could be expanded by interbasin and interaquifer transfers. The conflict that could arise is between the small rural landowner or community dependent on traditional groundwater

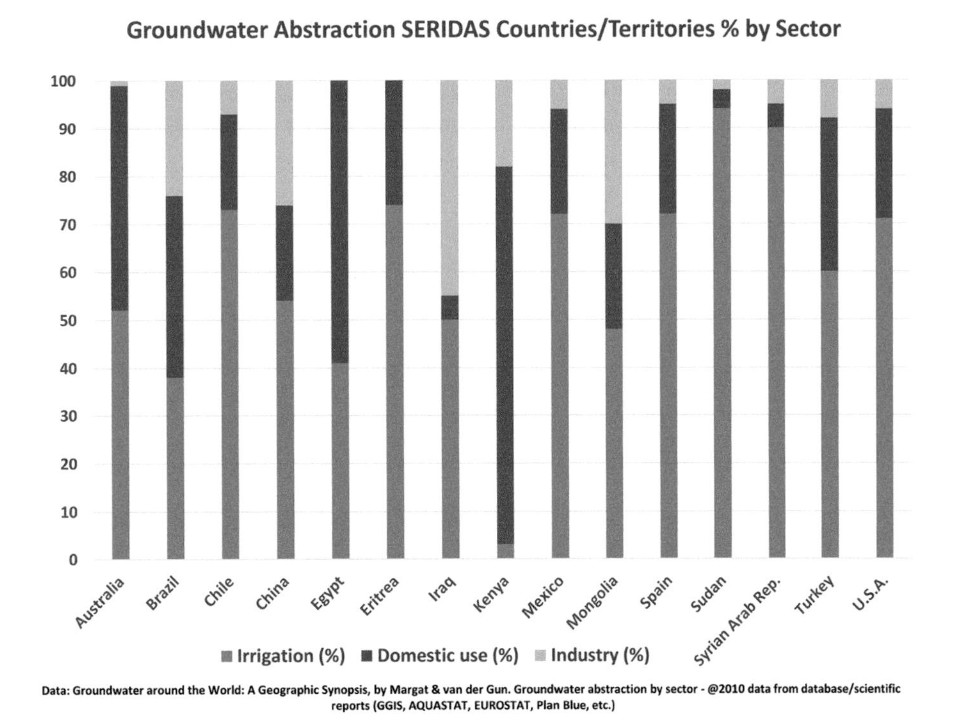

Figure 4.4 Groundwater abstraction usage by sector.
Figure by HARC using data from Margat and van der Gun (2013)

wells or surface reservoirs and large municipal or industrial users with greater willingness to pay for the resource. Water marketing in arid areas could also exacerbate the problem of groundwater over-pumping. Subsidence is always a matter of concern with groundwater over-pumping; and once the aquifers are compacted, the effect is irreversible. Water marketing groups have been targeting brackish aquifers to pump, treat, and bottle the area. Models should be developed, and the available allocation determined, so that over-pumping does not occur.

As water utilization intensifies, problems associated with surface and groundwater quality are also increasing. Groundwater over-pumping impacts can include lowering of aquifers, lowering the amount of water available for surface runoff, and deterioration of water quality (Winter, Harvey et al., 1998). One example of water quality deterioration is the potential for saltwater intrusion. Saltwater intrusion is the movement of saline water into the freshwater of aquifers. The process occurs naturally over many years in most coastal aquifers. However, over-pumping of coastal aquifers makes saltwater intrusion occur at a much faster rate – sometimes to such an extent the freshwater storage in an aquifer is entirely replaced by the saltwater. This process can be exacerbated by sea-level rise.

Water issues are normally classified according to water quality and water quantity or supply. Water quality standards are set depending on the use of the water; for example, drinking water must be of higher quality than water used for aquatic recreation or support of aquatic life. Water quality may be reduced by pollution from human activities or natural contaminants. Water supply is determined by analyzing the amount of safely available water in rivers and streams or underground aquifers. Suitability for various uses often depends on the concentrations of criteria pollutants in the water, and the concentration of those pollutants is directly dependent on the quantity of water in which they occur. Water quality and quantity are closely related; decreasing the volume of water in a river or aquifer can lower the water quality.

4.2 FUTURE THINKING FOR SUSTAINABILITY MANAGEMENT OF GROUNDWATER

Managing groundwater and surface water as a single resource does add layers of complication.

In the SERIDAS Rio Grande basin, five major aquifers – Bolson, Edwards-Trinity, Edwards, Carrizo-Wilcox, and Gulf Coast – are found just in the Texas portion of the basin. Quantifying interactions on all levels is critical. For example, if surface water is withdrawn to irrigate a field, some of the water will be lost due to evaporation and use by crops, while some may percolate to the groundwater. However, this kind of circular water budget determination will be necessary for future management decisions.

Sustainably managing groundwater in the SERIDAS river basins will involve a multi-pronged approach, which will vary considerably dependent on available financing, data, technology, and both community and political support. An important first step is establishing a water budget that incorporates both groundwater and surface water and projects water availability based on climate change. Of course, a great deal of this is dependent on the level of data available and the capability of the managing entity to calculate desired future conditions with this dataset.

Climate change and the resulting changes, as known, in precipitation, temperature, and evaporation patterns will possibly change the percentage of groundwater and surface water usage. The water budget should plan for future scenarios that incorporate climate change impacts as well as how those impacts will affect the circular relationship between groundwater and surface water. In addition, extreme event scenarios must by analyzed; climate change will likely bring more rainfall during the wet season and less rainfall during the dry season to these river basins. Further, drought events in arid areas are predicted to increase in frequency and severity. Figure 4.5 shows the SERIDAS river basins, with current global groundwater basins and areas of low rainfall. Many of these arid basins are predicted to get even less rainfall with climate change impacts, stressing resources that are now in short supply.

The impacts of climate change and increases in population will result in increased need for water supply; this means increased reliance on both groundwater and surface water (specific to basin). Future management scenarios will need to take this into account. Another factor that is important for managing sustainably is determining to what extent existing dams change recharge, and if aging infrastructure or increased sedimentation will impact that in the future. If the river basin has groundwater desalinization projects in practice or scheduled for the future, their impact on surface or groundwater quality should be incorporated into the water budget process.

4.2.1 Increasing Need for Water Supply and Innovative Technology

Less than one percent of the total water on our planet is currently considered suitable for most human uses. Much of it is unsuitable due to high concentrations of salts. New technology that would convert water from unsuitable to suitable for various human uses would increase supply. Care needs to be taken about unintended

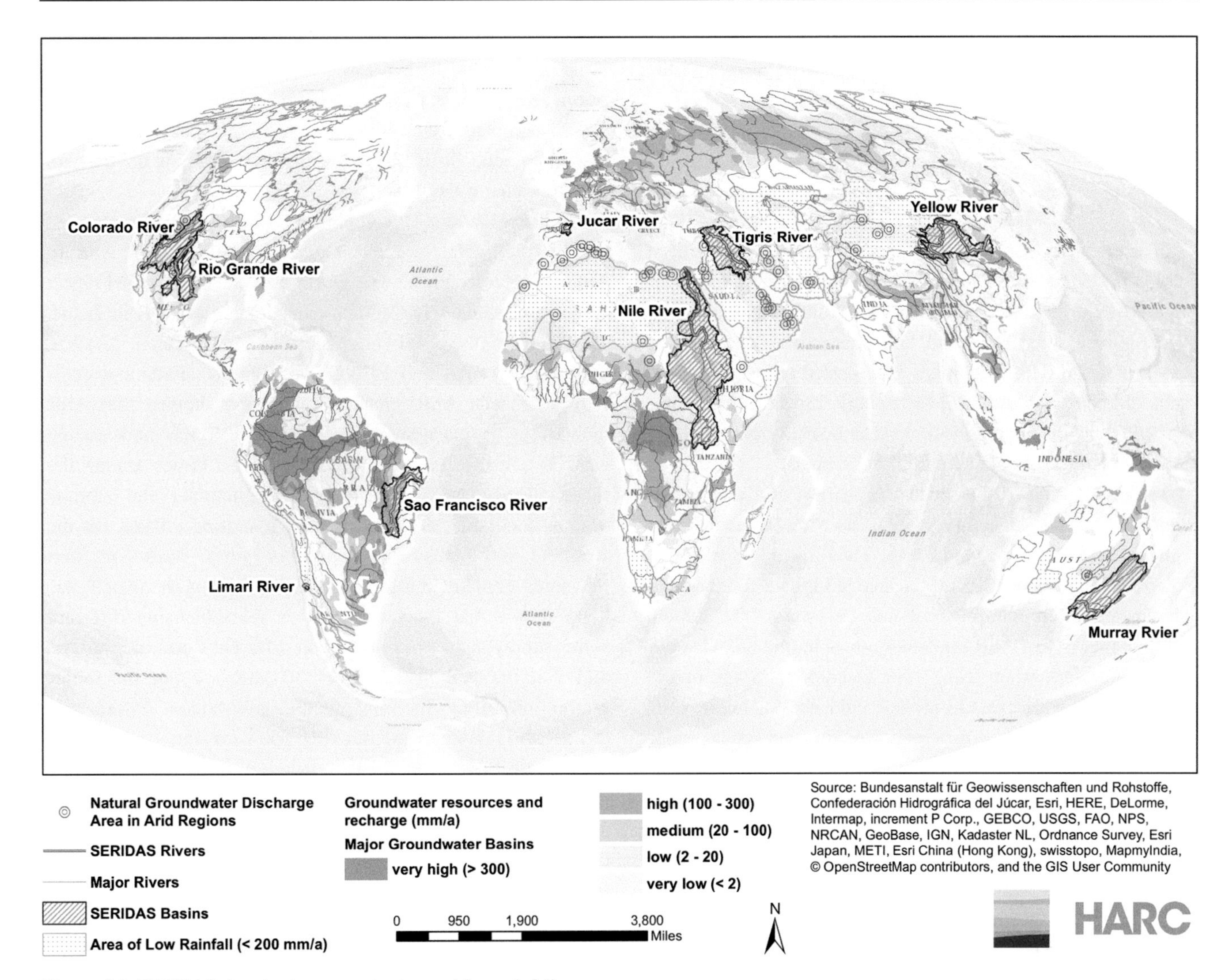

Figure 4.5 SERIDAS river basins, groundwater, and low rainfall areas.
Figure by HARC using data from Bundesanstalt für Geowissenschaften und Rohstoffe, Confederación del Júcar, Esri, HERE, DeLorme, Intermap, increment P Corp., GEBCO, USGS, FAO, NPS, NRCAN, GeoBase, IGN, KadasterNL, Ordnance Survey, Esri Japan, METI, Esri China (Hong Kong), swisstopo, MapmyIndia, © OpenStreetMap contributors, and the GIS User Community

consequences of new technology, though. Desalination, for example, generates large quantities of salt that must be disposed of. Similarly, seeding clouds to produce rain, as takes place in China, can increase water supply at the expense of somewhere else.

There are other opportunities for technology in the management of groundwater. Aquifer recharge can be manipulated, and more research is needed on technologies for measuring, acquiring, and processing water from aquifers.

Storage above ground in surface water reservoirs is problematic because of water losses due to evaporation, transpiration, and siltation. An alternative that is underutilized in many areas is called Aquifer Storage and Recovery (ASR). ASR involves the storage of water in a suitable aquifer and stored as groundwater. Water is pumped (or other means of artificial recharge are used) into the aquifer through an ASR well when water is available, and water is recovered from the same well or different wells in the same aquifer when the water is needed.

The largest operational ASR site to date is at Las Vegas, Nevada, with about thirty wells and total recovery capacity of 112,014-acre feet/year (100 mgd) (CERP, 2001). The state of Texas currently has three successful examples of ASR, supporting the water needs of the cities of San Antonio, El Paso, and Kerrville. The current capacities of the three Texas ASR systems are San Antonio Water System at 60 mgd, Kerrville at 2.65 mgd, and El Paso Water Utilities (EPWU) at 10 mgd (Malcolm Pirnie Inc. et al., 2011). The San Antonio Water System Twin Oaks ASR facility is the third largest ASR wellfield in the United States, behind Las Vegas and Calleguas Metro Water District in California (68 mgd) (Malcolm Pirnie Inc. et al., 2011).

4.2.2 Water Supply: ASR as a Water Storage Alternative

ASR has been successful in other states and nations and has been in use in the United States for over 40 years. The recovered water is used for a variety of purposes, including drinking water, irrigation, agriculture, industrial, and other purposes. Some projects have succeeded while other have posted failures; it is the results of both that move ASR technology forward. When planning ASR implementation, using lessons learned from past successes and failures should be part of the process.

Brown et al. (2006) discussed lessons learned from a review of fifty ASR projects, including projects from the United States, England, Australia, India, and Africa. They found that ASR can be a cost-effective method of water storage for both confined and unconfined aquifers with brackish and freshwater. The authors reviewed a variety of ASR implementations to find out what did and did not work when using ASR technology. Some of the major findings, which should provide guidance for future ASR projects everywhere, are that well clogging can be a major problem, and the use of regular back flushing should be considered at the beginning. The authors also found that various water quality issues can be problematic depending on the site; discussed the importance of well hydraulics and modeling or analyzing well interference effects (both on the ASR system and nearby users); and stressed the importance of well-designed monitoring equipment.

The state of Florida has been successful in adopting ASR to meet its water needs, incorporating the technology as part of its water supply program for several years now. The state has more than fifty constructed or proposed ASR facilities. The Comprehensive Everglades Restoration Plan (CERP) relies heavily on an ASR component, and the proposed plan would mean implementation of the largest scale ASR facility in the United States (CERP, 2001). The CERP study on issues associated with ASR includes water quality, regional flow pattern changes (hydraulics), mercury bioaccumulation, hydrogeology uncertainties, rock fracturing, and recovery rates. Several pilot ASR studies are in place to study these issues and make recommendations for the final large-scale ASR project.

Other lessons learned come from an ASR program in the Antelope Valley of Southern California (Rydman, 2012). Groundwater in the area had been pumped to such an extent that there was significant land subsidence, making the aquifer a good candidate for ASR implementation. Some of the lessons learned from this project are similar to those discussed earlier: the importance of monitoring equipment, concern over flow patterns (hydraulics), water quality, and well clogging (stressing the importance of not letting a well sit idle for too long, backflow, freeze protection, and video inspections). The study also emphasized the importance of working with staff members in the water district of concern to ensure everyone understands the concept and the value of the project.

The Texas Water Development Board (TWDB) contracted in 2010 with Malcolm Pirnie for an assessment of ASR in Texas (Malcolm Pirnie Inc. et al., 2011). This detailed report discussed extensively the then-current state of ASR in Texas. The authors discussed the ASR systems in Texas (SAWS, City of Kerrville, and EPWU) and the successes and challenges of each ASR system. The report showed that ASR technology is feasible and adaptable in Texas and can be used as a viable means for storing water. Issues among the three current ASR systems include defining the storage volume available, defining optimal operating criteria for the wellfield, and Texas water regulatory issues. The report from this study concluded that the "existing ASR systems in Texas have shown that the technology is feasible using different water supply sources as well as in different types of aquifers, and that the technical aspects of ASR are not the major factors inhibiting its implementation"(Malcom Pirnie Inc. et al., 2011). Rather, the sticking point for adoption of ASR technology is often lack of knowledge about ASR and how it could be implemented, as well as a perceived lack of ability to protect the stored water. ASR could provide a less costly and more readily available source of water storage for Texas.

Managing surface and groundwater quality and quantity effectively is becoming more and more important globally as populations continue to grow and the threat of drought persists in arid and semi-arid regions. Utilizing technologies such as ASR will provide alternative sources of readily available water without having to develop new reservoirs, especially in arid regions with high evaporation rates.

4.3 CONCLUSIONS

There are many questions that need to be explored to better understand the groundwater and surface water connection in engineered rivers in arid lands. Managers need to determine what portion of the overall water budget is groundwater, both in terms of availability and usage. For some river basins, there is not enough data to answer this question quantitatively. That should be noted, as well as what would be required to gather the necessary data. A qualitative discussion would also be useful in areas where data are lacking. In addition, how climate change and the resulting changes in precipitation, temperature, and evaporation patterns impact the percentage of groundwater and surface water usage will be vital knowledge for decision making.

Various factors will be important in this determination. Planning efforts should assess if there will be increased reliance on groundwater or surface water in the future. If so, will this threaten quality, and what steps could be taken to mitigate the impacts? Other factors will be the movement of water in and out of river basins and whether engineered canals and reservoirs change recharge. If groundwater desalinization projects are in place or planned, how will they impact surface or groundwater quality? Policies for each basin and separate governing entities within the basin should be factored into the planning efforts. For example, Texas regulates groundwater and surface water separately, but New Mexico recognizes them legally as a connected entity. Management plans in each place will necessarily reflect that.

Good groundwater management is tailored to the resources and needs of communities, while planning for reliable supplies regardless of high impact events such as drought. Sustainable groundwater management, which meets all the stakeholder needs now and in the future, is essential to a reliable and resilient water system. Growing population and the impacts of climate change will impact water supplies globally. Planning for groundwater and surface water as one resource will be critical to determine availability, circular impacts, and needs assessment. In regions with arid lands, thoughtful and evidence-based planning for future water resources will be vital to ensure sustainability.

REFERENCES

Bennett, J., Brauch, B., and Urbanczyk, K. M. (2012). *Estimating Ground Water Contribution from the Edwards Trinity Plateau Aquifer to the Big Bend Reach of the Rio Grande, Texas.* South-Central Section – 46th Annual Meeting.

Brown, C. J., Hatfield, K. et al. (2006). *Lessons Learned from a Review of 50 ASR Projects from the United States, England, Australia, India and Africa.* Universities Council on Water Resources (UCOWR) and National Institutes of Water Resources (NIWR). Annual Conference: Increasing Freshwater Supplies, Santa Fe, New Mexico.

CERP (2001). *Comprehensive Everglades Restoration Plan Aquifer Storage and Recovery Program. South Florida Water Management District and the United States Army Corps of Engineers.*

FAO (2010). *The Wealth of Waste: The Economics of Wastewater Use in Agriculture.* Rome: FAO.

Gleeson, T., Wada, Y., Bierkens, M. F. P., and van Bee, L. P. H. (2012). Water Balance of Global Aquifers Revealed by Groundwater Footprint. *Nature*, 488, pp. 197–200. https://doi.org/10.1038/nature11295

Katz, B. G., DeHan, R. S. et al. (1997). Interaction between Ground Water and Surface Water in the Suwannee River Basin, FL. *Journal of the American Water Resources Association*, 33(6), pp. 1237–1254.

Malcolm Pirnie, Inc.; L. Aquifer Storage and Recovery Systems et al. (2011). *An Assessment of Aquifer Storage and Recovery in Texas.*

Margat, J. and Van der Gun, J.(2013). *Data Request from the Authors and the Publisher of Groundwater around the World: A Geographic Synopsis.* Leiden: CRC Press/Balkema of the Taylor, and Francis Group.

Mitsch, W. J. and Gosselink, J. G. (1993). *Wetlands.* New York: Van Nostrand Reinhold.

NGWA (2016). *Facts About Global Groundwater Usage.* Available at: www.ngwa.org/Fundamentals/Documents/global-groundwater-use-fact-sheet.pdf

Otto, B. et al. of American Rivers; Lovaas, D. et al. of Natural Resources Defense Council, and Bailey, J. of Smart Growth America (2009). *Paving Our Way to Water Shortages: How Sprawl Aggravates the Effects of Drought.*

Rydman, D. (2012). Lessons Learned from an Aquifer Storage and Recovery Program. *Journal of the American Water Works Association*, 104(9).

Stewart, J. W. (1968). *Hydrological Effect of Pumping from the Floridan Aquifer in Northwest Hillsborough, Northwest Pinellas and Southwest Pasco Counties, Florida.* United States Geological Survey Report, Tallahassee, Florida.

Taylor, R. G., Scanlon, B., Döll, P. et al. (2012). Ground Water and Climate Change. *Nature Climate Change.* https://doi.org/10.1038/nclimate1744

Texas Parks and Wildlife (2002). *Ecologically Significant River and Stream Segments of Region J (Plateau).*

Texas Water Development Board (TWDB) (2018). *2017 State Water Plan: Water for Texas.* Available at www.twdb.texas.gov/waterplanning/swp/2017/

UN DESA (2011). *World Population Prospects: The 2010 Revision, Highlights, and Advance Tables.* Working paper No ES/P/WP. 220. New York: United Nations, Department of Economic and Social Affairs, Population Division.

United Nations (2018). *Desertification Decade.* Available at www.un.org/en/events/desertification_decade/whynow.shtml

USGS (2000). *Land Subsidence in the United States.* Available at water.usgs.gov/ogw/pubs/fs00165/

Van der Gun, J. et al. (2012). *United Nations World Water Assessment Programme, Groundwater and Global Change: Trends, Opportunities and Challenges.* UNESCO Publishing.

Wada, Y., Wisser, D., and Bierkens, M. F. P. (2014). Global Modeling of Withdrawal, Allocation and Consumptive Use of Surface Water and Groundwater Resources. *Earth System Dynamics*, 5, pp. 15–40. https://doi.org/10.5194/esd-5-15-2014

WHYMAP (2018). *World-Wide Hydrogeological Mapping and Assessment Programme*. Data sources for WHYMAP groundwater resources data include BGR (Bundesanstalt für Geowissenschaften und Rohstoffe). Available at www.whymap.org/whymap/EN/Home/whymap_node.html

Winter, T. C., Harvey, J. W. et al.(1998). *Ground Water and Surface Water: A Single Resource*. Denver, CO: USGS: US Geological Survey Circular 1139.

5 Endangered Food Security

Olcay Ünver, Eduardo Mansur, and Melvyn Kay

5.1 INTRODUCTION

Water scarcity is one of the fundamental challenges to sustainable development. This is not just a physical issue; it is also caused by institutional, economic and infrastructure-related constraints and is linked to pressures that emanate from population growth and mobility, socioeconomic development, dietary changes and climate change.

Water management is now everyone's responsibility, and particularly those involved in agriculture. Agriculture accounts for almost 70 per cent of global freshwater withdrawals (for irrigation, livestock and aquaculture) and has multiple implications for sustainable food and agricultural systems (FAO, 2017c). In arid regions, such as those identified in the Sustainability of Engineered Rivers in Arid Lands initiative (SERIDAS), agriculture and irrigation can account for withdrawals as high as 85 per cent of available water resources. By 2050, if society continues to pursue the current 'business as usual' model, global water demand could exceed supply by over 40 per cent, which would put 45 per cent of the global GDP, 52 per cent of the world's population and 40 per cent of grain production (WWAP, 2016) at risk. Groundwater resources are already being over-exploited. This resource accounts for one third of water withdrawals for irrigation and countries that rely on groundwater are already operating beyond the threshold of physical sustainability, with the demand for water exceeding renewable supplies (Wada et al., 2012).

Since 2012, the World Economic Forum (WEF, 2020) has put water at the top of the agenda as one of the greatest risks facing the world. Water scarcity is also a major concern within the United Nations 2030 Sustainable Development Agenda. The Agenda calls for sustainable development that goes beyond the divide of 'developed' and 'developing' countries; it is the collective responsibility of all, and solutions require fundamental changes in the way societies produce, protect and consume. It recognises the interdependencies across the various development sectors, it discourages fragmented or 'silo' approaches to resource management, which has been the past tradition, and encourages integration and cooperation as the most effective means of making best use of limited natural, financial and social capital.

Water flows through all seventeen Sustainable Development Goals (SDGs). SDG 6 – known as the 'Water Goal' – focuses attention on sustainable management of water for all. In 2018, the UN Deputy Secretary General described SDG 6 as the 'docking station' for all the SDGs and the 2030 Agenda (SIWI, 2018). SDG6 recognises the limited nature of renewable water resources, the inefficiencies of traditional 'silo' approaches to managing water and the need for Integrated Water Resources Management (IWRM) for people, industry, agriculture and the environment. In many countries, planning for IWRM is well advanced, but implementation is still in its infancy. There is no 'one size fits all' solution and countries must seek their own unique way of meeting their water challenges based on local physical, social and economic circumstances.

Agriculture consumes large quantities of water. It has a reputation for inefficient water use and is often seen as one of the major causes of water scarcity. However, there are two sides to every coin. On the positive side, although water demand for agriculture continues to increase, its share of global withdrawals has decreased over the past two decades. Significant water savings can be made in agriculture (FAO, 2017c), as it plays an important role in bringing renewable water supplies into balance with increasing demand, particularly in the world's large, arid river basins (SERIDAS, 2021).

Across the world, water for agriculture faces two major challenges. The first challenge is integrating agriculture into IWRM. Even though agriculture in dry regions can consume up to 85 per cent of available freshwater, in most countries, IWRM has traditionally focused on public water supply (Shah, 2016). Changing this culture to include water for agriculture, as well as water for the environment, will take time. The second challenge is coordinating water needs within agriculture. Agriculture is a highly fragmented industry and is largely organised around commodities, such as plants, dairy, meat and fish, rather than natural resources. Research, extension and marketing support for farmers also tend to have a commodity focus. In the past, when resources were more plentiful and demands were fewer, a commodity focus made sense; fragmentation did not create too many problems and offered a sensible way of managing specialised services and support for farmers. But as demand and competition for resources have increased, so have the problems, and stakeholders are recognising the need for more coordinated efforts within the agricultural sector to avoid potentially inefficient use of limited natural resources.

Assessing just how much water agriculture needs and the potential for savings is fraught with difficulties. Unlike domestic water demand, which is narrowly focused on available water resources and predicted population growth, agricultural water planners face a complex and interconnected mix of drivers and pressures, including national food policies, population growth, lifestyle changes, dietary preferences that are transforming food and agricultural systems, and climate change. Adding to this complex mix and equally important is the role that land and soils play in producing food. From a farming perspective, water, land and soil resources are inseparable and symbiotic. In 2004, Hodgson commented that the lack of harmonisation and coordination of water, land and soil resources can adversely impact rural and urban lives, the economy and the environment, especially where natural resources are already under intense pressure (Hodgson, 2004).

These factors together not only make it extremely difficult to assess demand but also encourage a 'silo' approach to resource planning and management among government departments. Resources are no longer plentiful and the prospects of coping with resource scarcity, particularly water, challenge usual approaches to resource management.

This, of course, is not a new challenge. Integration has been vigorously promoted at an international level for several decades and more recently in the UN 2030 Sustainable Development Agenda. But there is little evidence of integration within agriculture and among natural resources planners and managers on the ground.

A review of progress on SDG 6 in 2018 stated that '[p]utting Integrated Water Resources Management (IWRM) into practice will be the most comprehensive step that countries make towards achieving SDG 6' (United Nations, 2018). In this paper we view agriculture through a 'water lens' to argue that a significant step towards coping with water scarcity in agriculture will require increasing cooperation among agriculture's sub-sectors and better coordinated planning and management of limited water, land and soil resources on which agriculture depends.

5.2 AGRICULTURE THROUGH A WATER 'LENS'

Over the past century, rising demand for food and fibre has been met with expanded use of natural resources, water, land, and soils through increased productivity. Agricultural production, as a result, has almost tripled over the past 50 years. Global population increased from 3 to 7 billion between 1961 and 2011, yet the estimated global average per capita food supply also increased by 30 per cent. Consumptive water use increased as well, but estimates indicated that more food was produced with less water – 'more crop per drop' (Lundqvist and Unver, 2018). However, food demand is expected to increase further and soon will outstrip the available resources if current production and consumption practices continue.

In areas where renewable water resources are already fully used, water scarcity is becoming a bigger constraint to agricultural production than the availability of good agricultural land. Although only one-third of the global land area (or 4.8 billion hectares) is used for agricultural production or forests, the more accessible land is already cultivated with little room for expansion. Thus, most of the increases in production have come from a shift from extensive to intensive farming systems. The downside is the increased pressure on water, land and soil resources, which in turn have degraded the very resources on which sustainable production depends (FAO, 2011a). Expanding and intensifying production in irrigated and rainfed agriculture, livestock production and aquaculture will bring new environmental externalities that will impact water quality and quantity.

Globally, it is estimated that there are more than 570 million farms; some 475 million are smaller than 2 ha, and more than 500 million are family farms (Lowder et al., 2016). Climate change will affect every aspect of food production in most regions, particularly the low- and middle-income countries, where millions of smallholder farmers depend on agriculture and are vulnerable to food insecurity. It will be most impactful through extreme floods and droughts, which are predicted to become more frequent and more severe. In sub-Saharan Africa, estimates indicate that by 2080 land areas experiencing severe climate or soil constraints will increase from 35 million hectares to 61 million hectares (9–20 per cent of the region's arable land) (GWP, 2015). This is further exacerbated by increasingly erratic rainfall and the shift in small streams from perennial to intermittent flow.

Planning for climate change is like planning for sustainable agricultural and water and land management practices. Adopting climate change adaptation initiatives with disaster management initiatives that focus on floods and droughts will be crucial, as currently these initiatives tend to operate within their own silos.

Some respite from the drive for sustainability has come from the OECD review on global food demands (OECD-FAO, 2018). They report that global pressures on food production are changing and expect food demand to slow down up to 2027, mainly driven by slowing demand in China. However, the pressures to produce 'more crop per drop' are likely to remain because the demand for limited water resources will continue to increase across all water-using sectors, not just agriculture.

5.2.1 Agriculture's Sub-sectors

The principal sub-sectors within agriculture – arable and livestock farming, and aquaculture – have traditionally been planned, managed and administered as separate entities. However, they are inextricably linked through their dependence on the same water, land, and soil resources, though not always at the same time and in the same place. Some activities are independent, some are complementary and others create conflict. When resources are scarce, understanding these interrelationships and coordinated action will be vital in resolving conflict and maximizing benefits.

Rainfed farming is the world's most common farming system, particularly practised by the many millions of smallholder farmers. In 2007, a major water study assessed global water use in agriculture and found 80 per cent was 'green' water[1] (Falkenmark and Rockström, 2004), used for rainfed arable and livestock farming to produce 60 per cent of the world's food and fibre (IWMI, 2007). In areas of high and reliable rainfall, like northern Europe, crop yields are good, and production is reliable. But in areas of low, erratic and unreliable rainfall, such as the drier regions of Africa and south Asia where many of the poor and disadvantaged live, crop yields are low and uncertain. As a result, yields are far below their potential in many low-income countries. Conservation agriculture is promoted to encourage farmers to make better use of rainfall by integrating soil and water management practices, improving rainfall infiltration and harvesting water to reduce water losses, improve yields and raise the overall water productivity of rainfed systems to produce 'more crop per drop' (FAO, 2017a).

Irrigation farming – both total and supplementing rainfall – can provide opportunities to increase food production and food security, and improve the lives of millions of people. Irrigation accounts for the remaining 20 per cent of agriculture's global freshwater water use, which is 'blue' water abstracted from rivers and groundwater. Abstraction varies among regions depending on climate and the prominence of irrigated farming in the economy. Withdrawals for irrigation in Africa and Asia account for more than 80 per cent of total water withdrawals; in Europe, withdrawals are just over 20 per cent (FAO, 2016).

Since 1900, the global irrigated area has grown from 40 million ha to over 340 million ha. Only 20 per cent of the world's cultivated land is irrigated, yet it produces some 40 per cent of the world's food and fibre needs. Irrigated farming already provides more than half the domestic food in China, India, Indonesia, and Pakistan, which account for almost half the irrigated land area. Sub-Saharan Africa is more likely to expand irrigated area to increase food production, whereas Asia will tend to increase cropping intensities on existing farms.

Irrigation brings together blue water, land and soils in a highly structured manner: physically, technically and institutionally. Unlike rainfed farming, blue water has a high opportunity cost and competition when it is scarce. When effectively managed, it can also help protect aquatic environments and increase water security. But if poorly managed, it can do irreparable harm to cultivable land, soils and freshwater ecosystems. Too often, irrigation is poorly managed and has a reputation for low-value, wasteful and inefficient use of water. The global average agricultural water use efficiency is estimated to be 55 per cent, with national figures ranging from 40 to 65 per cent as measured by crop water use divided by water withdrawals (Hoogeveen et al., 2015). The implication is that much of the water withdrawn never reaches the crops due to conveyance losses such as leakage, evaporation and misuse. There are already growing concerns about the over-exploitation of groundwater for irrigation (Wada et al., 2012). Poor irrigation practices have impacts beyond wasting water, including increasing soil salinity, waterlogging, soil erosion, and increasing pollution through the misuse and leaching of agri-chemicals. While these problems have technical solutions, many of them persist because of weak and poorly integrated services that support irrigated farming.

Livestock farming requires water for producing animal feed, drinking and cleaning. It specializes in the production of dairy products, eggs, wool, hides, and other goods as well as meat. Despite growing concerns over serious environmental pollution problems and the impacts of climate change, for many people livestock farming is a vitally important sector both socially and economically. Globally, livestock farming employs over 1.3 billion people and generates livelihoods for over 1 billion of the world's poorest people (FAO, 2006). It contributes 40 per cent of the global value of agricultural output. It is one of the most dynamic parts of the agricultural economy, and is driven by population growth, rising affluence and urbanisation (WWAP, 2012).

Livestock farming has grown rapidly since the turn of this century as people have become wealthier and diets have shifted in many countries from plant-based diets to meat and dairy products. The demand for water has also grown to meet the growing demand for animal feed and fodder crops; for drinking, cooling, and cleaning on the farm; and for processing products along the value chain.

Zimmer and Renault (2003) estimated that globally some 2,000–3,000 km^3 of freshwater was required to grow animal feed and fodder, around 40 per cent of total water use in agriculture. However, much of this was 'green water' on grasslands and feed and fodder crops grown in water-abundant areas. In 2012, a (rough) estimate suggested that 13 per cent of blue water was consumed to irrigate feed, fodder and pasture, which could potentially be used for other purposes (WWAP, 2012). Water withdrawals for livestock will inevitably have grown over the past decade in line with the rapid increase in pig and poultry production, much of which is highly dependent on irrigated feed stuff. Added to this was a global estimate of 16 km^3 for animal drinking water and 6.5 km^3 for animal servicing requirements, both modest in comparison to the irrigation requirement (FAO, 2006). Concerns continue to grow about livestock production and the use of limited blue water resources and pollution caused by waste from farms, abattoirs, and processing plants discharged into water courses untreated into water bodies, thus reducing the availability of freshwater.

Data on freshwater aquaculture and its relationship with agriculture are sparse, though FAO (2018) estimates that globally some 11.6 million tonnes were harvested from inland capture fisheries. Most freshwater fish are harvested from natural aquatic ecosystems, such as lakes and rivers, but harvesting fish in tanks and irrigation systems is a long-established practice dating back many centuries, mainly in China. Over the last century, production has grown, particularly in Asia and to some extent in Africa. It is an important source of food and income in many developing countries and food-insecure areas. Most fish are harvested and consumed in-country.

Inland capture fisheries production is often overlooked, mainly because of lack of awareness among governments of the contribution they make and the ecosystems that support them. Data collection is not a priority, particularly in low-income countries. The activity is dispersed and not usually associated with intensive yields, and so it is vulnerable to impacts from more influential development sectors that compete for freshwater, like agriculture and energy. Fish do not consume water (rather they 'borrow' it) but development in other sectors can reduce the quantity and quality of water available for production (FAO, 2018). Water scarcity, competition from other users and environmental degradation all have negative impacts on aquaculture production. A diversification of global food production that includes aquaculture offers enhanced resilience, but its promise will not be realised if government policies fail to provide incentives for resource-use efficiency, equity and environmental protection (FAO, 2017c).

5.3 INTEGRATING WATER, LAND, AND SOILS

Sustainable development across agriculture's sub-sectors must acknowledge that natural resources – water, suitable land and soils – for food production are limited and must also support vital ecosystem services. Development can only be sustainable if it works within those constraints over time and across locations and sectors (Weitz et al., 2014). In many countries, the pressure to produce more food and energy crops while maintaining the natural aquatic environment are so great that resources are being degraded (FAO, 2011a).

Every land-use decision has a water footprint and vice versa, yet land and water are usually managed as if they are separate and distinct resources. One exception, of course, is among farmers. For them, the connections and the need for coordination are self-evident and part of their everyday life. In a dry climate, land without water is of little use, as is access to water without land. Securing access to land can open opportunities to secure access to water and enable farmers to invest with confidence in management practices and technologies that enable them to improve their livelihoods and use limited resources wisely.

There are historic reasons for treating water and land separately based on Ancient Roman law when land and water rights were first established as separate entities. This tradition, at least from a legal perspective, continues to this day in many countries (Hodgson, 2004), as the legal profession sees no reason to integrate water and land laws. However, the view from those who manage limited natural resources is rather different.

Knowledge about the physical interactions between water, land and soil is extensive and well known. Ways in which land and soils are managed influence the quantity and quality of water runoff into water courses and, in turn, water quantity and quality affect how the land and soils are used for agriculture. The interactions are numerous. Changes in plant cover can change water consumption and hence runoff into rivers and aquifers (FAO, 2002). Soil compaction can reduce infiltration, encourage runoff and increase peak flows in water courses. Excessive irrigation or rainfall can cause soil erosion, leach fertilisers and other farm chemicals, and pollute water sources. Irrigation can also change soil structure, causing soils to slake,[2] reduce infiltration and increase salinity in soils and groundwater. The overall impact is to undermine sustainable food productivity and production.

At a landscape level, changing land-use can have lasting and irreversible impacts on groundwater abstraction, which many smallholder farmers have come to rely on for irrigation as well as cities and towns for water supply. Many cities are expanding rapidly, with consequent increases in domestic water demand that conflict with rural communities that rely on groundwater for cropping and aquatic environmental requirements (Shah, 2014). Large-scale land acquisitions, such as those taking place across sub-Saharan Africa, are as much about acquiring water, often unintentionally, as they are about acquiring land (Woodhouse, 2012). Virtual water trading, or the transfer of water from 'water-rich' to 'water-poor' countries embedded in food imports, follows the same unintentional consequences. This is influenced as much by an abundance of land as well as water. Assessing food security based only on water can seriously distort food trade decision-making (GWP, 2015). Although much is known about water, land and soil resources, little is known about how we might benefit from managing them in a more coordinated manner and, importantly, the costs and benefits of doing so. There has been some progress within water, land and soil communities to recognise the need for integration but little progress on the ground.

Over the past 25 years, there has been strong movement across the water and water-using sectors to integrate as a means of making best use of limited water resources. Known as Integrated Water Resources Management (IWRM), the Global Water Partnership was established to develop and promote this concept (GWP, 2000). This includes water use in agriculture, and some argue that IWRM should be changed to ILWRM to reflect the importance and inter-connectedness of land and water management. However, concerns about water scarcity have tended to dominate international development rather than land and soils resources. The IWRM concept has gained significant momentum and is now enshrined in SDG 6 within the UN 2030 Development Agenda.

Land managers have also recognised the need for integration through their desire for Sustainable Land Management (SLM). SLM is defined as 'the use of land resources, including soils, water, animals and plants, for the production of goods to meet changing human needs, while simultaneously ensuring the long-term productive potential of these resources and the maintenance of their environmental functions' (United Nations, 1992). More recently management community EcoAgriculture Partners (2012) advocated a 'whole landscape approach' or Integrated Landscape Management (ILM). This recognised the important contribution that farmers make to managing water, land, and soils, and influencing the multifunctional nature of the landscape. It also recognised that farmers cannot do this alone and institutional support was needed to provide all the benefits that society now expects from natural resources. The report suggested 'the new reality was one of shared dependency on limited resources' and 'seeking to address the challenges of food production, ecosystem management, and rural development by reaching across traditional sectoral boundaries to find partnerships that solve what are clearly inter-connected problems'.

Those who champion soil resources have also been active at international and national levels, though rather in isolation, establishing a Global Soil Partnership (GSP) in 2011 to promote better understanding of soils, and establishing an Intergovernmental Technical Panel on Soils (ITPS) that provides technical and scientific guidance and complements the work of the Intergovernmental Panel on Climate Change (IPCC) and other bodies. Soils are often seen as implicit within land management and although they are clearly an essential element for sustainable development, they have not captured international attention in quite the way that water and land have. On World Soil Day in 2014, the FAO Director General described soils as the 'nearly forgotten resource'. In 2015, FAO described soils as the core of land resources and the foundation of agricultural development and ecological sustainability. If water was the 'life blood' of agriculture, then soils were the 'body' providing valuable services, such as producing food, regulating climate and safeguarding ecosystem services and biodiversity (FAO, 2015).

In 2015, a revised edition of the World Soil Charter was issued, incorporating topics related to environment, climate change and urban sprawl, as the primary normative policy document to promote sustainable soil management at all levels (FAO, 2015). Guidelines were then published in 2017 to support policy decision-making on sustainable soil management at all levels (FAO, 2017d). This may all seem remote from the realities of local soils management, but the greater concerns are the consequences of soil degradation and loss of production that go beyond local and national boundaries and can interrupt international food supply chains.

5.4 EVIDENCE OF HOLISTIC APPROACHES

Resource scarcity, be it water, land or soil, is now beginning to drive changes in the way we manage resources in the future. The laudable efforts of various theme- or sector-based organisations recognise the threat, usually through their own particular perspective, and are now showing signs of moving from acknowledging the need to implementation, all driven by the same desire for sustainable development. Examples include integration within the water and water-using sectors that spill over into all aspects of development, including food security, poverty alleviation, economic growth and aquatic ecosystem conservation. Others approach sustainable development through a focus on conserving the landscape (e.g., source-to-sea, ridge-to-reef, integrated landscape management), developing eco-system-based agriculture (e.g., agroecology) and nature-based solutions, all of which include essential elements of water management. All these approaches align with the climate change agenda. The following are some encouraging examples of a holistic approach.

5.4.1 An IWRM Approach

In 2012, UN-Water surveyed 134 nations across the world to assess progress within the water sector towards IWRM. Some 83 per cent had embarked on reforms to improve the enabling environment for IWRM, 65 per cent had developed IWRM plans, and 34 per cent reported they were at an advanced stage of putting plans into practice. Continued progress was reported in a review of progress with SDG 6, including IWRM in 2018 (United Nations, 2018), though there are some concerns over how progress is being measured using qualitative questionnaires. IWRM initiatives, by their nature, are catchment based, often large in scale, and mostly involve government agencies, as they are usually the natural custodians of national water resources, including their planning and management.

Although globally the emphasis is still on planning rather than implementation, there are some countries and river basins where IWRM is being put into practice. Brazil is a country that offers an example of an ongoing IWRM approach to water management. Although Brazil's rivers account for nearly 18 per cent of the world's total, there is persistent water scarcity in the main São Francisco River basin in what is the poorest part of the country. The São Francisco is 2,863 km long, drains over 600,000 km^2, and provides water for over 13 million people. Management of this extraordinarily complex basin is undertaken by the Sao Francisco River Basin Committee, which has been actively pursuing an IWRM approach to planning and managing limited water resources since 2000. The committee seeks to reconcile the water demands of irrigated agriculture; power generation (dam construction and reservoir operation); navigation; domestic water supply to a predominantly urban population; dilution of urban, industrial and mining-related effluents; and ecosystem maintenance (Braga and Lotufo, 2008). Over the past 20 years, both physical and institutional structures have continued to develop with a water-centric focus that demonstrates progress as well as the complexities of managing large and diverse but interconnected water resources.

A second example is Chile, a country that has adopted IWRM as an essential element of economic, social and environmental development despite never having had an explicit IWRM policy (Lenton and Muller, 2009). The country comprises some 200 river basins that drain this long, narrow country westward from the Andes into the Pacific Ocean. Water resources are characterised by extraordinary heterogeneity ranging from 500 m^3 per person per year in the arid north to more than 10,000 m^3 in the wetter south. Since the late 1970s, Chile has pursued an IWRM approach to achieving the '3 Es' (GWP, 2000) – maintaining macro-economic equilibrium, strengthening the role of markets in efficiently allocating resources and opening up the economy to world markets while exploring products for which the country had an advantage – almost all of which use water in their production process. These objectives led to increased economic efficiency that in turn increased water use, highlighting a need to better address social equity and environmental stability concerns. Rather than calls for immediate integration of all water-related planning and management, Chile adopted a long-term sequential and adaptive approach to water resources and economic development that has served the country well over the past 50 years.

Shah (2014) reported on several successes in Italy, Sweden, Switzerland, the United States, the Caribbean and among countries in Central and Eastern Europe. Shah suggested that progress relied on a gradual nuanced approach rather than forcing the pace of change, which can be counterproductive. He also observed that countries at different stages of development have different needs and capabilities, and these must be reflected in the way IWRM is approached.

The water sector is often criticised for focusing on IWRM as being too water-centric. The nexus approach that links water, food and energy – although similar in many ways to IWRM – has enabled the water sector to integrate more fully with those who use water as well as those within the sector. The key difference is that IWRM begins with water when considering the relationships among water, energy and food, whereas the nexus thinking begins with an interrelated system and then notes the two-way relationships between water, food or energy and the other resources.

5.4.2 A Landscape Approach

The Kagera river basin, shared by Burundi, Rwanda, the United Republic of Tanzania and Uganda is an example of a landscape approach to sustainable development. Key to sustainability in this case was maintaining the Kagera River flow regime, which contributes to the water levels of Lake Victoria and the outflow to the Nile River, while maintaining the riverine wetlands, water and pasture quality, and associated livelihoods of some 20 million (mostly rural) people.

The Kagera project adopted a basin-wide integrated landscape approach to restore degraded lands; sequester carbon and adapt to climate change; conserve agro-biodiversity and sustainable use; and increase agricultural production, contribute to food security, sustain rural livelihoods and protect the basin's international waters (FAO, 2017b). This holistic perspective was made operational through coordinated efforts across sectors and national boundaries within the Kagera basin.

Since the project began in 2010, the results have demonstrated the importance of developing knowledge and capacity among land users to understand the many facets that threaten their household livelihoods on family farms and to take coordinated action. Land users learned to organise themselves to respond to changing social, economic and environmental trends, including pressures on land resources and the impacts of land degradation. They needed to understand the impacts of land fragmentation and overexploitation as well as the increasing threat from weather variability and climate change. The project found that educating land users about the crucial role that soil restoration and water conservation play in food security, climate change adaptation and mitigation, and essential ecosystem services on which they depend were all vital elements in alleviating poverty and sustaining development. Given the right tools and incentives, farmers have few problems in understanding the connections and the need for coordination, as they are self-evident and part of their everyday life.

5.4.3 An Agro-Ecosystems Approach

Yahara Pride Farms close to the city of Madison, Wisconsin, USA is an example of an agro-systems approach to water management and how engagement from stakeholders from many sectors outside agriculture is an essential ingredient. Madison had serious water quality problems caused by excess phosphorous affecting municipal water supplies. Phosphorous is an essential element of plant and animal growth. However, only small amounts are needed, and too much is harmful – particularly in surface water bodies where it can increase biological productivity and accelerate eutrophication, a natural ageing process in lakes and rivers. The solution was to encourage local farmers to adapt their agronomic practices to keep phosphorous on the land and not in the city's waterways (FAO, 2011b).

Yahara Pride Farms (YPF) is a farmer-led, not-for-profit organisation, working with the Clean Lakes Alliance, to improve soil and water quality. YPF was created by a group of farmers and local business leaders to improve and protect the land and waterways around the city. YPF runs a cost-share programme that provides products and services that keeps phosphorus on fields so that crops can use it. A key partner is the Madison Metropolitan Sewerage District, which funds farm practices that prevent the need for more burdensome capital investments in water treatment. Cost analysis has shown that it is cheaper to fund conservation than water treatment, and both urban and rural communities benefit from the initiative. The changes began in 2010, and in 2017 revised soil and water conservation practices were being used on nearly 25 per cent of the farmland in the catchment area. This is seen as a long-term project and YPF believes in rewarding longevity to ensure that farmers maintain conservation practices.

5.4.4 Nature-Based Solutions (NBS) and Climate Change

In 2016, the International Union for Conservation of Nature (IUCN) described nature-based solutions (NBS) as a relatively 'young' concept still in the process of being framed (Cohen-Shacham, 2016). IUCN promotes collaborative initiatives that have many similarities with other sustainable development approaches but with a focus on ecosystem restoration, protection and management, and using green and natural infrastructure that can also bring benefits to society and biodiversity.

An example is the Namibian Government's commitment to work with development stakeholders to deliver its Nationally Determined Contribution (NDC) under the Paris Agreement on climate change mitigation and adaptation. Namibia is highly vulnerable to climate impacts; half the population relies on subsistence agriculture, and water scarcity is a serious threat to the people's welfare and the national economy. To reduce this vulnerability, Namibia sought a diversity of solutions to improve water security, prevent desertification and increase resilience to flooding. To achieve these goals, Namibia has adopted the NDC Partnership's integrated planning process to strengthen coordination, resource mobilisation and transparency for implementing the NDC.

Namibia released its NDC Partnership Plan in September 2018 (Ramalho, 2019), which identified priority areas, including nature-based solutions for agriculture, forestry and other land-uses through actions such as preventing desertification and mitigating the risks to wildlife through constructing fire breaks and restoring rangeland. The programme is led by Namibia's Ministry of Environment and Tourism, which coordinated the NDC Partnership and is supported by a biodiversity economy programme supported by Germany, France and the EU. Through this initiative, policies and projects are being implemented to secure and diversify livelihoods in local communities that depend on natural resources, thus improving human wellbeing while conserving nature. Given Namibia's vulnerability to climate impacts, but also its unique climate and geology, the country is pursuing ambitious and innovative systems and solutions, while keeping nature conservation in its core pursuits. It is early days at the time of writing this chapter, but collaboration is at the heart of this development initiative.

5.4.5 Regional Projects

There is a plethora of examples where broad socioeconomic frameworks aiming to improve livelihoods factor a strong land and water development component into planning and management. Examples include the Tennessee Valley Project in the United States, perhaps the oldest comprehensive river basin management agency, born in the crisis of the Great Depression in 1933 (Miller and Reidinger, 1998); the Southeastern Anatolia Project of Turkey, which impacts both the transboundary Tigris and Euphrates rivers (Unver, 2007; Kibaroglu, 2002); and multiple projects and programmes in India (Biswas et al. 2009; Shah, 2013).

5.5 CONCLUSIONS

Agriculture is a major water consumer with a reputation for inefficient water use and is often seen as one of the major contributors to water scarcity. Increasingly, it is now being seen as a sector where significant savings can be made. In this paper we viewed agriculture through a 'water lens' to argue that the most significant step towards coping with water scarcity in agriculture will come from increasing cooperation among agriculture's main sub-sectors of arable and livestock farming, and aquaculture, and better coordinated planning and management of limited water, land and soil resources.

Agriculture is a highly fragmented industry that has traditionally focused on producing commodities such as plants, dairy, meat and fish, rather than concerns over natural resources, and it has largely ignored the inseparable and symbiotic nature of water, land and soil resources on which the sector it totally dependent. Increasing demands for water and the potential for savings in the agricultural sector suggest that it is timely to seek ways of making better use of limited natural resources by coordinating use among arable and livestock farming and aquaculture.

Over the past 20 years or so there has been much laudable rhetoric in the global development community and among the organisations that champion sustainable use of water, land and soils expressing the need for holistic approaches to resource planning and management. Evidence suggests that many are now finding ways of moving from rhetoric to action using their own particular perspective to reach the common desired goal of sustainable development. Examples of holistic approaches focus not on specific resources but on wider objectives such as conserving the landscape, developing ecosystem-based agriculture, and nature-based solutions, all of which include essential elements for coping with water scarcity. Additionally, such broad collaborative efforts provide opportunities to bring the UN Sustainable Development Goals and climate change objectives together to maximise overall benefits and resource efficiencies and tap into broader funding opportunities.

REFERENCES

Biswas, A., Rangachari, R., and Tortajada, C. eds. (2009). *Water Resources of the Indian Subcontinent*. New York: Oxford University Press.

Braga, B. and Lotufo, J. (2008). Integrated River Basin Plan in Practice: The São Francisco River Basin. *Integrated Journal of Water Resources Development*, 24(1), pp. 37–60.

Cohen-Shacham, E., Walters, G., Janzen, C., and Maginnis, S. (2016). *Nature-Based Solutions to Address Global Societal Challenges*. Gland, Switzerland: IUCN.

EcoAgriculture Partners (2012). *Landscapes for People, Food and Nature: The Vision, the Evidence, and the Next Steps*. Washington, DC.

Falkenmark, M. and Rockström, R. (2004). *Balancing Water and Humans in Nature: The New Approach in Ecohydrology*. Sterling, VA: Earthscan.

FAO (1996). *Report of The World Food Summit 13–17 November*. Rome.

(2002). *Land and Water Linkages in Rural Watersheds. Land and Water Bulletin 9. Proceedings of the Electronic Workshop*. Rome.

(2006). *Livestock's Long Shadow*. Rome.

(2011a). *The State of the World's Land and Water Resources for Food and Agriculture (SOLAW): Managing Systems at Risk*. London: Rome and Earthscan.

(2011b). Yahara Pride Farms Conservation Board. In *2nd International Symposium on Agroecology Scaling-Up Agroecology to Achieve the Sustainable Development Goals*. Rome.

(2015). *Revised World Soil Charter*. Rome.

(2016). *AQUASTAT*. Rome.

(2017a). *Conservation Agriculture*. Rome.

(2017b). *Sustainable Land Management (SLM) in Practice in the Kagera Basin. Lessons Learned for Scaling Up at Landscape Level: Results of the Kagera Transboundary Agro-ecosystem Management Project (Kagera TAMP*. Rome.

(2017c). *The Future of Food and Agriculture: Trends and Challenges*. Rome.

(2017d). *Voluntary Guidelines for Sustainable Soil Management*. Rome.

(2018). *The State of World Fisheries and Aquaculture*. Rome.

GWP (2000). *Integrated Water Resources Management: TAC Background Paper No4*. Stockholm: Global Water Partnership.

(2015). *Linking Land and Water Governance*. Stockholm: Global Water Partnership.

Hodgson, S. (2004). *Land and Water – The Rights Interface FAO Legislative Study*, 84. Rome: FAO.

Hoogeveen, J., Faurès, J. M., Peiser, L., Burke, J., and van de Giesen, N. (2015). GlobWat – A Global Water Balance Model to Assess Water Use in Irrigated Agriculture. *Hydrology and Earth System Sciences*, 19(9), pp. 3829–3844.

IWMI (2007). *Water for Food, Water for Life: A Comprehensive Assessment of Water Management in Agriculture*. London: Earthscan, and Colombo: International Water Management Institute.

Kibaroglu, A (2002). *Building a Regime for the Waters of the Euphrates–Tigris River Basin*. The Hague: Kluwer Law International.

Lenton, R. and Muller, M., eds. (2009). *Integrated Water Resources Management in Practice*. Stockholm: Global Water Partnership.

Lowder, S. K., Skoet, J., and Raney, T. (2016). The Number, Size, and Distribution of Farms, Smallholder Farms, and Family Farms Worldwide. *World Development*, 87, pp. 16–29.

Lundqvist, J. and Unver. O (2018). Alternative Pathways to Food Security and Nutrition: Water Predicaments and Human Behavior. *Water Policy*, 20(5), pp. 871–884.

Miller, A. and Reidinger, R. (1998). *Comprehensive River Basin Development: The Tennessee Valley Authority. WTP410*. Washington, DC: World Bank.

OECD-FAO (2018). *OECD-FAO Agricultural Outlook 2018–2027*. FAO, Rome/OECD, Paris.

Ramalho, C. (2019). *Namibia Launches NDC Partnership Plan Climate Action*. Available at ndcpartnership.org/news/namibia-launches-ndc-partnership-plan-climate-action

SERIDAS (2021). *Sustainability of Engineered Rivers in Arid Lands*. Available at www.harcresearch.org/work/SERIDAS

Shah, M. (2013). Water: Towards a Paradigm Shift in the Twelfth Plan. *Economic and Political Weekly*, 48(3), pp. 40–52.

Shah, T. (2014). *Groundwater Governance and Irrigated Agriculture*. Stockholm: Global Water Partnership.

(2016). *Increasing Water Security: The Key to Implementing the Sustainable Development Goals*, 22. Stockholm: Global Water Partnership.

SIWI (2018). Speech by Amina Mohammed World Water Week Daily 28 August, Stockholm.

United Nations (1992). *UN Earth Summit*. Rio Agenda 21 Programme.

(2018). *Sustainable Development Goal 6: Synthesis Report 2018 on Water and Sanitation*. New York.

Ünver, O. (2007). Water-Based Sustainable Integrated Regional Development. In *Water Resources Sustainability*. New York: McGraw-Hill Education, pp. 235–266.

Wada, Y., van Beek, L. P. H., and Bierkens, M. F. P. (2012). Non-sustainable Groundwater Sustaining Irrigation: A Global Assessment. *Water Resources Research*, 48 (1).

WEF (2020). *Global Risks Report*, 15th ed. World Economic Forum. Geneva.

Weitz, N., Huber-Lee, A., Nilsson, M., and Hoff, F. (2014). *Cross-Sectoral Integration in the Sustainable Development Goals: A Nexus Approach*. SEI Discussion Brief, Stockholm.

Woodhouse, P. (2012). Foreign Agricultural Land Acquisition and the Visibility of Water Resource Impacts in Sub-Saharan Africa. *Water Alternatives*, 5(2), pp. 208–222.

WWAP (2012). The United Nations World Water Development Report 4: Managing Water under Uncertainty and Risk. Knowledge Base Vol. II. World Water Assessment Programme. UNESCO Paris.

(2016). The United Nations World Water Development Report 2016: Water and Jobs. World Water Assessment Programme. UNESCO Paris.

Zimmer, D. and Renault, D. (2003). Virtual Water in Food Production and Global Trade: Review of Methodological Issues and Preliminary Results. *Proceedings of the International Expert Meeting on Virtual Water Trade, Value of Water-Research Rapport Series*, No. 12, 93–109. Delft: IHE Delft.

Notes

1 Green' water comes from rainfall stored in the soil and transpired by crops. 'Blue' water is sourced from surface or groundwater and used for agriculture, industry and domestic purposes.

2 Slaking is when large and dry soil aggregates break down into smaller sized aggregates as they are immersed in water. More information can be found at http://vro.agriculture.vic.gov.au/dpi/vro/vrosite.nsf/pages/soil health_soil_structure_slaking

6 Declining Environmental Flows

Stephanie Glenn and R. James Lester

6.1 ENVIRONMENTAL FLOWS

Environmental flows are defined as "flows that maintain the biophysical and ecological processes of river corridors" (Arthington, 2012). In this chapter we will focus on the impact of engineered features on environmental flows, specifically impacts on the quantity, timing, and quality of the water required to meet the demands of native plants and animals living in and along rivers and estuaries. If too much water is withdrawn from a river in the process of meeting human demands, then the aquatic ecosystem in that river is damaged. If upstream cities do not return much of what they withdraw from rivers, then the freshwater inflow to the coastal estuaries could be insufficient to support historical productive ecosystems. Commercial and recreational fisheries are dependent on appropriate volumes remaining in rivers to support food webs. One challenge for the regulation of environmental flows is the difficulty in proving the relationships among the pattern of freshwater flow, the water quality (i.e. nutrients and contaminants in the water), and the reproduction and survival of the species in the rivers and estuaries. Estimating how much water should be left in the river for environmental purposes is difficult and requires complex data analysis and stakeholder consensus. Each river system has a different environmental flow requirement; therefore, this chapter will primarily describe general scientific consensus on the issue and deal with specifics for one river only, the Rio Grande/Rio Bravo.

Alterations to a flow regime can be characterized in several quantitative ways. Flows can be described according to multiple time periods – the most common is an annual pattern. In this annual pattern are different types of flows that differ in their magnitude (discharge rate), frequency of occurrence, duration (of discharge event), timing (seasonal predictability), and rate of change (rapidity of rise and fall in magnitude) (Arthington, 2012). Many studies of the ecological impacts of alteration in flow regimes have used magnitude as the causal parameter. The other parameters are associated with pulse flows that can be highly variable and difficult to compare.

6.1.1 Arid Lands and Environmental Flow Seasonal Patterns

River flow patterns change seasonally and annually depending on precipitation, snowpack melt, and groundwater supply. Animals and plants adapted to a river ecosystem can cope with historical fluctuations but can be extirpated from a river when the flow pattern deviates too far from the historical. Rivers in arid lands are commonly altered by high levels of water extraction, which has drastic effects on the ecology of the river. The impact on biodiversity of dewatering a river is often obvious; however, as climate change produces a stronger effect on the historical pattern of precipitation and flow, it will become more difficult to assign cause and effect to changes in riverine biodiversity. A 2013 study by Grafton et al. projected climate change impacts on flows for four river basins, three of which are Sustainability of Engineered Rivers in Arid Lands (SERIDAS) basins: the Colorado, the Yellow, and the Murray Darling (Grafton et al., 2013). The study noted that the impacts from current water use strategies for these basins will be exacerbated with climate change impacts in the basins. Climate change is projected to result in higher temperatures and possibly increased evaporation, resulting in reduced flows. The different scenarios the authors modeled showed projected reductions in flows ranging from 4 to 18 percent for the Colorado (projected out to 2050); 9 percent projected out to 2020, 21.9 percent projected out to 2050, and 29 percent projected out to 2080 for the Yellow; and up to 69 percent for the Murray-Darling projected out to 2030. Climate change and increasing demands on water combine to impact environmental flows.

Figure 6.1 shows information related to Environmental Flows for the current condition of a river system. According to the International Water Management Institute, the current condition of rivers as displayed in Figure 6.1 is based on the river health indicators developed by Vorosmarty et al. (2010). This is just flow required to maintain the current condition. As the figure shows, some river basins are currently in peril of low-flow conditions; future management scenarios need to plan for strategies to increase these inflows.

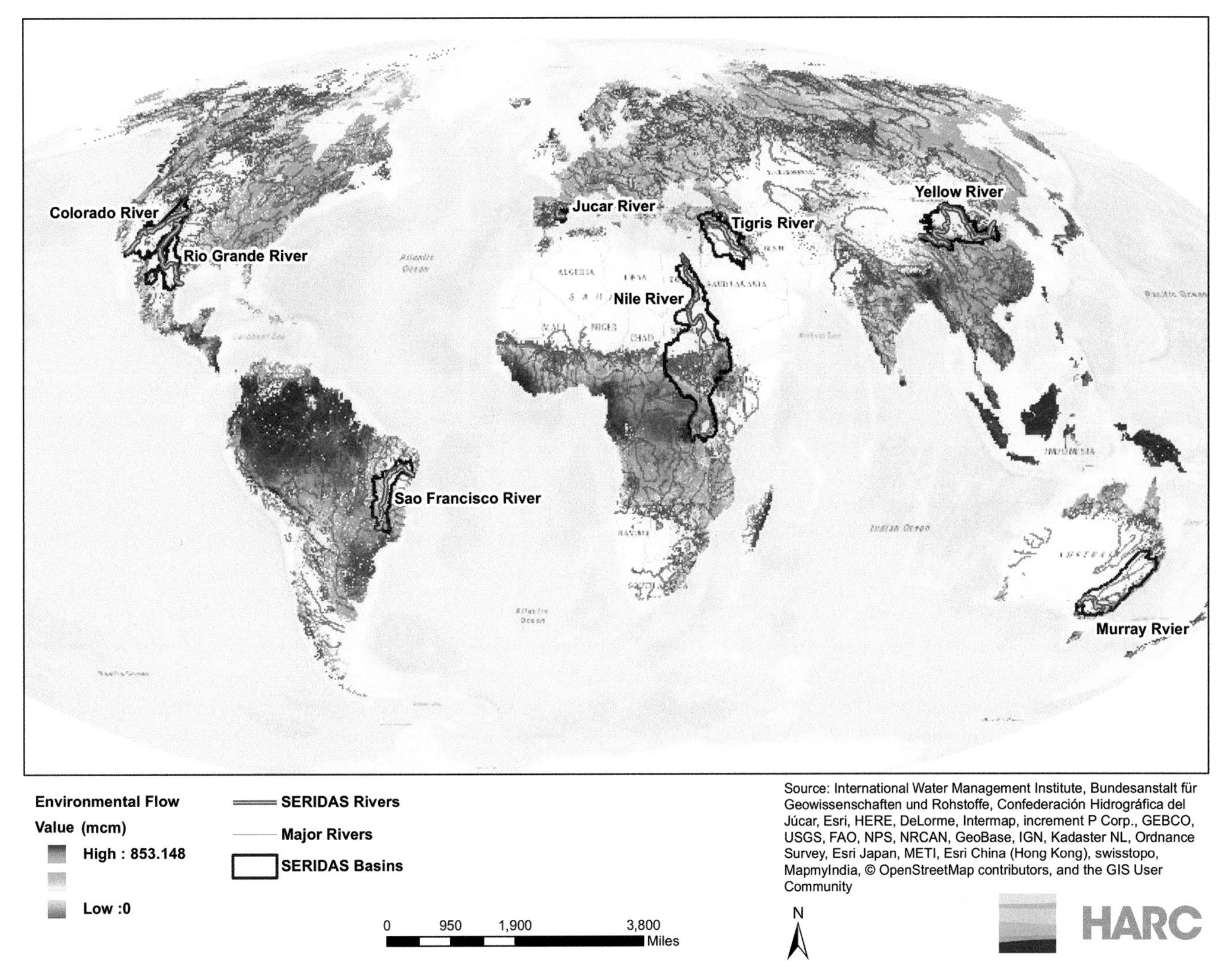

Figure 6.1 SERIDAS river basins and current probable environmental flow in million cubic meters (mcm).
Map developed by HARC using model and data from International Water Management Institute (IWMI) as cited in the figure

Researchers with the Transboundary Waters Assessment Programme (TWAP) are tasked with creating the first baseline assessment of global transboundary water resources. As part of a larger study, TWAP is studying projected Environmental Stress for the 2030s and 2050s induced by Flow Regime Alteration. Figure 6.2 shows the results for the four SERIDAS river basins that were included in this study. The environmental water stress indicator, as shown in the figure, addresses environmental stress in the 2030s and 2050s induced by flow regime alterations due to anthropogenic impacts such as dam operation, water use, and climate change, with a score of 1 being low stress and 5 being high stress (UNEP and UNEP-DHI, 2015).

6.1.2 Engineered Alterations

Humans have developed an ethic regarding water and land that shows a poor understanding of the complexities and interconnectedness of natural systems. Across the world, populations have developed in areas that are water-stressed; as a result, people have attempted to use their engineering capabilities to alter the natural water budget. Land modification has occurred on a massive scale to convert forested land and semi-arid grassland into cropland; crop management then requires irrigation and reduces the potential for the soil to hold moisture. Both agricultural and municipal uses are threatened by growing demand, shrinking supply, and climate change. In addition, the energy of flowing water is often harnessed by hydroelectric dams to produce electricity for the population and industries that may exist far from the river providing this resource. Humans' need for water and power leads to engineering of rivers using dams and diversions.

There is strong motivation to create reservoirs and diversions along rivers in arid watersheds. There are three reasons to construct a dam: water supply, flood control, and generation of

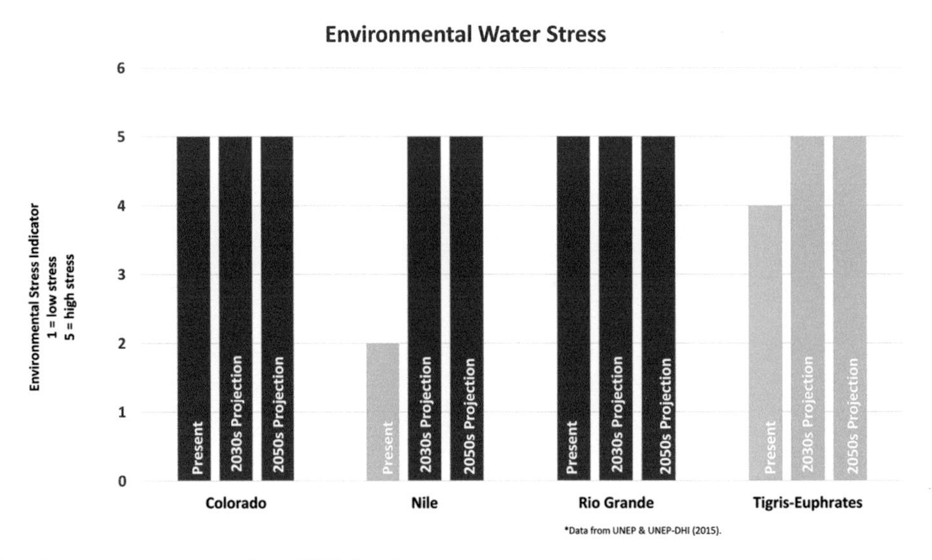

Figure 6.2 Projected environmental water stress from TWAP project. Figure by HARC with TWAP data (UNEP & UNEP-DHI, 2015)

electricity, and each purpose is associated with different management strategies. Water supply reservoirs are maintained with as much volume as possible; flood control reservoirs are managed with little water and high retention capacity; and hydroelectric dams have inconsistent releases depending on electricity demand, often providing extreme daily fluctuations. Each of these reservoir management strategies leads to alteration of flow regime and changes in the downstream ecology.

Under some conditions of water excess, i.e., floods, humans are killed or injured, and their property is destroyed or damaged. Fear of future harm from floods is another cause of engineering of rivers and is expressed in levees, channelization, and flood control reservoirs (more dams). Engineering for flood control can lead to a problematic cycle. To the extent that the engineering is successful in mitigating flooding it permits development in at-risk locations that increases impermeable surfaces and exacerbates flooding in high precipitation events. Thus, subsequent floods can be justification for more flood control engineering, which alters environmental flows.

When rivers, including those in arid regions, are engineered to reduce flooding, the usual strategy is to limit overbank flows using levees and channelization. Levees and channelization lead to disconnection of the river flow from adjacent riparian habitat. Typical river flow regimes include overbank flows that provide multiple benefits to the plants and animals living in the adjacent riparian zone. These benefits are derived from the water, nutrients, and sediments provided to the riparian ecology. Engineered rivers often have riparian ecology that is very different from their historical ecosystems.

Diversions also have multiple reasons: municipal supply, irrigation, and industrial supply, especially power plant cooling. Municipal diversions result is some consumption and some return flows; irrigation is usually highly consumptive with little to no return flows; and industrial uses are usually for cooling purposes and have high return flows reduced by evaporation. The timing of flow reductions from diversions for municipal supply and irrigation is quite different. Municipal needs are more consistent than irrigation needs, which are episodic. Diversions for irrigation usually have the largest impact on the magnitude of flow in the river. These diversions often increase as precipitation decreases, which amplifies the impact on river flow and riverine ecology. In extreme cases (e.g., the Rio Grande), diversions can result in dewatering of a river segment and loss of aquatic life in that section. Diversions and return flows associated with municipal and industrial use can have significant effects related to water quality as well. Pollutants in municipal and industrial wastewater or thermal pollution can dramatically alter the biota of a river. The Rio Grande has a history of water pollution stored in the sediments behind dams showing contributions of toxics from agricultural and industrial chemicals (Van Metre et al., 1997).

Withdrawal of water from shallow groundwater aquifers that occur in the river's alluvial sediment has an impact similar to diversion of surface water. Due to the bidirectional connection between the water in the river and the water in the aquifer, withdrawal from either impacts the other. One example of this is found along the Rio Grande in New Mexico.

Once reservoirs and diversions are engineered, the environmental flow regime of the river is changed. First, the magnitude of flow is reduced by consumption and increased evaporation, also the frequency of high flow pulses, particularly those that tend to occur at the beginning of a rainy season, are reduced. Reservoirs tend to support downstream water rights that can reduce the frequency of very low flows because water must be released from the reservoir to satisfy downstream water rights.

Rivers provide water to the coastal region of oceans, usually to an estuary, where the river and ocean waters mix and nutrients are more abundant than in the deep ocean. The riverine supply to estuaries is called freshwater inflow and is the subject of debate about the relative value of water for coastal ecosystems and for human uses. Engineered rivers often exhibit significantly reduced levels of freshwater inflow unless the engineering includes interbasin transfers or is supplemented by groundwater sources unconnected to the river. In some extreme cases, water withdrawals in arid regions result in no flow at the river mouth. Lowering freshwater inflow usually causes reduced nutrient and sediment supply to the coastal area.

6.2 ALTERED FLOW REGIMES AND BIODIVERSITY, GENERAL PRINCIPLES

Flow determines physical habitat in a river and altering the flow regime will change the quantity and quality of habitats. Aquatic species and many riparian species have evolved life history strategies in response to the natural flow regime and alteration will impact reproduction and survival. Many life histories depend on connections between habitats that occur along the length of the river and across the floodplain. Engineering features that alter flow regimes reduce connectivity and will reduce biodiversity. Alteration of flow regimes increases stress on native species and facilitates invasion and colonization by exotic species. (Arthington, 2012).

Chapter 2 of this volume discusses the impacts of climate change and climate variation on engineered rivers in arid lands; one notable impact is the increased intensity of droughts in the arid lands. Droughts can have catastrophic results for environmental flows.

Droughts can completely dewater some habitats, such as riffles and shallow runs, and water velocity is likely to decline but could be critical for species with floating eggs or adaptations for use of surfaces free of algae. High flow pulses are the signals for reproduction in some species and they are absent under drought flows. Riparian communities depend on the water table in the alluvium of a river that will lie at greater depth under drought conditions. Also, overbank flows will not occur, which will deprive the riparian community of water for seed dispersal and germination. This is only a brief selection of the potential impacts of drought on riverine and riparian ecologies.

Drought will also stress the estuarine community at the mouth of the river. Rivers deliver nutrients and sediments to the estuary zone where their discharge mixes with oceanic water. The base of one food web in an estuary is the detritus that river flow brings to the estuary. Under drought conditions, the quantity of detritus is greatly reduced. Some species are adapted to certain salinity regimes during portions of their life cycle. The aquatic plant *Vallisneria* occupies a narrow salinity zone at the mouths of rivers in eastern North America. It can disappear under drought conditions.

6.2.1 Altered Nutrients and Sediments

The fauna and flora of the river and riparian zone are adapted to the flow regime existing before engineering and will likely be changed by alterations to the historical flow regime. The existence of a reservoir along a river changes the distribution of sediments and nutrients that underlie the dynamics of riverine ecologies. A reservoir develops an artificial ecosystem that traps both sediment and nutrients that would have flowed downstream. Phosphorus is usually the limiting nutrient for freshwater systems. Very limited amounts of phosphorus dissolve in water, and most are found in particulate matter. Thus, the retention of sediments behind dams depletes downstream areas of a limiting nutrient. The other essential nutrient, nitrogen, occurs in higher concentrations in its dissolved forms, but is utilized by the reservoir ecosystem. Again, this can deplete the nutrient availability for downstream communities.

The presence of a water supply reservoir always implies altered nutrients. Diversions to municipalities usually mean return flows from sewage, which may or may not have been well treated to reduce nutrient concentrations. In addition to nutrients, there are often toxic contaminants that can affect aquatic life or cause a risk for humans consuming contaminated aquatic organisms. Diversion to agriculture for irrigation implies that any return flows will have elevated levels of nutrients from fertilizer and contain pollutants associated with pest and weed control. Diversions for municipal and agricultural use can increase or decrease nutrient concentrations depending on the quantity and concentration of return flows. In arid regions where return flow is low, this alteration of water quality will be of limited impact.

Dams usually cause sediment retention and reservoirs lose capacity over time. Rivers self-correct their sediment load, if possible, depending on flow volume and rate. Engineered rivers typically have reduced flow volumes that result in reduced bed and bank erosion and correlate with lower sediment loads. This can have dramatic impacts on the river delta where the balance between aggradation and degradation will shift toward degradation. Changes in sediment load will also change the topography of the river related to bank erosion, sediment deposition, and substrate particle size.

Agricultural diversions are associated with irrigation of disturbed land, which means that return flows are likely to have elevated levels of sediment. The combination of high sediment input and reduced flow volume results in more deposition along the riverbed.

River segments that have levees or are channelized will have altered nutrient regimes because overbank flows are inhibited.

Overbank flows are major sources of nutrient spikes associated with detritus picked up from riparian soil.

Dams trap nutrient rich sediments, and reservoirs have ecosystems that assimilate nutrients; channelization reduces acquisition of nutrients from overbank flows; and diversions reduce flow and the capacity of a river to carry nutrients and sediments. As these engineered features act along a river, the result is a more oligotrophic (low nutrient) estuary.

6.2.2 Altered Topography and Habitat

The topography of a river is shaped by the flow regime; therefore, engineering of a river changes the topography. A natural flow regime produces a distribution of riffle, run, pool, backwater, alternate channels, and substrate types (Arthington, 2012). While a natural flow regime is dynamic and changes over time, it will exhibit greater variance in flow and thus greater diversity of habitat than an engineered flow regime. Some engineering changes a river's topography directly, such as levees and channel dredging; some changes it indirectly, such as by pumping shallow groundwater from a connected aquifer. As the altered flow regime continues over time, the shape of the river will change, and the quantity and quality of habitats will change.

In a metanalysis of the impact of altered flow regimes on the ecology of engineered rivers, Poff and Zimmerman (2010) found significant but variable ecological responses to all types of flow alteration. However, there was no general quantitative relationship that would be predictive for future engineering of rivers.

All of the rivers considered in this volume have been engineered to supply water for human activities, which will reduce average flow volume. A reduction of high flows will alter the erosional and depositional patterns along the banks. This will affect plant species that are adapted to the habitat on the riverbank. Also, the change in flow will alter the distribution of substrate types. Different fish species require different particle sizes in their spawning areas. Absence of high flows usually leads to an increase in the portion of stream bottom covered by fine particles. Those species that require large particle size (e.g., gravel) may not spawn if gravel beds have been covered by fine particles. Some fish species have floating eggs that must remain in the water column for a period of development and require a minimum flow rate. If the river flow is too low, these species cannot reproduce.

Low flow and channelization alter multi-channel, ribbon-type reaches and replace them with single channels. Some species are adapted to the small channels and backwaters of such sections.

Reduced flow and loss of high flow events results in more vegetated banks, less bare bank, and fewer undercut banks. There are species associated with bank habitat, especially those that nest in holes in the bank, such as some species of catfish, which will be impacted by such changes.

There are also ecologies that depend on detritus swept into the river by overbank flows. If the engineered river experiences fewer overbank events, as is likely, then the resource for detritivorous species is reduced and their population declines. Reduction of overbank flows decreases the connection between the river and its riparian zone. Riparian zones are typically areas of high biodiversity but depend on the water and nutrients delivered by the river in overbank flows. Many riparian species have life histories that have adapted to the flow regime of the river(s) where they occur. Seed dispersal, seed germination, and seedling growth are likely to depend on the timing and level of overbank flow. Decline of cottonwood trees along western rivers in North America is related to changes in river flow regime (Rood and Mahoney, 1990). Altered flow regimes allow for shifts from native species to the spread of exotic species, often leading to invasive species problems, such as the Russian olive, saltcedar (Tamarisk), water hyacinth, and hydrilla that are problematic in western US watersheds.

Many rivers, including some described in this volume, have well-connected floodplains adapted to relatively predictable inundation. Such systems have fish species that depend on floodplain inundation for habitat used for spawning and early development (Zeug et al., 2005). When floodplains are separated from rivers by levees or flow is reduced and prevents regular overbank flow, then these floodplain-adapted species decline or disappear.

These are examples of species life history characteristics that will be negatively impacted by reservoir construction and flow control. Certainly, there are other species that will benefit from the alterations of the river. There have been multiple studies of the effect of dams and flow regime changes on freshwater mussel species. Following dam construction on rivers in the southeastern United States, species diversity of freshwater mussels declined, in some studies by over 50 percent, but the reservoirs support a small but consistent assemblage of mud-tolerant mollusk species (Neves et al., 1997).

Reservoirs in the southern United States are famous for their largemouth bass fisheries. This native species is adapted to large pools in rivers and has characteristics that pre-adapt it to reservoirs. Many native river species do not possess the characteristics to survive and reproduce in a reservoir; therefore, reservoir mangers create species assemblages to populate these artificial lakes. It is not surprising that such artificial ecologies are prone to invasion by exotic species, which often reach problematic densities. Such invasions differ by geographic location, but the process and results are similar.

Engineered rivers designed for stability of flow usually have altered habitats supporting larger and denser populations of plant species because plants are not removed by scour from high flow events. The roots of these aquatic plants trap sediment and alter habitats, leading to reduction in aquatic invertebrates and plant diversity. Some of these successful plant species are native, but some are exotic (e.g., Hydrilla).

Some fish are unable to complete their life cycle under stable flow conditions because high flow rates are necessary to keep their eggs afloat. As sediments accumulate under stable flow or moderate flow conditions, some species lose their spawning habitat, which is dependent on large particle size substrate. Rivers with regulated flows have lower biodiversity than undisturbed or lightly disturbed rivers. Poff and Zimmerman (2010) found that alteration of river flow resulted in loss of fish species diversity whether the flow increased or decreased.

Invasion of riparian zones in North America and Australia by tamarisk (salt cedar) has been especially problematic. Tamarisk is favored over native cottonwood and willow when low flow results in a lower water table in the riparian soil (Stromberg et al., 2007). The evapotranspiration rate of these shrubs is high enough to affect flow volume and riparian aquifer recharge when they are abundant. The connection between river flow and alluvial aquifers is a significant determinant of riparian plant communities.

Hydroelectric dams are usually associated with highly variable flow regimes, which are stressful to native species adapted to predam conditions. Species differ in their response to rapid changes in flow rate and depth, which results in differences in tendency to be stranded on substrate or isolated in backwaters. This type of environmental situation usually selects against habitat specialists and favors habitat generalists.

The detention of water in reservoirs, augmented evaporation levels, and changing patterns of migration of water from reservoirs into groundwater all change the timing and quantity of flow from the river mouth into the ocean. Coastal waters are the most productive in oceans due to the nutrients contained in freshwater inflows. To the extent that engineering alterations reduce the nutrient content of the inflow, coastal productivity will be reduced.

6.2.3 Altered Connectivity and Migratory Life Histories

Various species of vertebrates and invertebrates evolved life cycles that include required migration between different habitats. The familiar example is salmon, which are anadromous and migrate as young (smolts) from rivers to the ocean and return to their river habitat to spawn. There are also catadromous species that mature in rivers and migrate to the ocean to spawn (e.g., eels). Some species simply migrate from river to the estuary at the river's mouth to spawn (e.g., Macrobrachium shrimp). Texas rivers, including the Rio Grande, historically contained populations of four species of river shrimp. Three of these species are now limited to the lower reaches of rivers below the first dam due to their required migration between habitats. One species actually leaves the water and walks around dams to reach suitable upstream habitat (Horne and Beisser, 1977). The common requirement of all the life histories of migratory species impacted by dams is connectivity between the habitats that are critical to completion of their life cycles. Dams are the primary cause of severance of this connectivity but dewatering of a river segment due to excessive diversion can have the same effect.

6.3 RESERVOIRS AND EXOTIC SPECIES

Reservoirs, although constructed for water supply or flood control, are often managed in the United States as recreational resources, primarily for fishing. Therefore, natural resource management agencies commonly stock species of interest to fishermen in water supply reservoirs. In the past, there was little concern about whether a species was native or exotic. Intentional stocking of exotic species in reservoirs has created a variety of problems. Most recreational fishers prefer large predatory species, which can upset a native food web and change the composition of the fauna. In other cases, exotic fish have been stocked for control of aquatic vegetation in these reservoirs and subsequently colonized the connected rivers. For example, grass carp and other Asian carp species are invasive in many rivers around the world and change the ecology of the systems they invade.

There are several invasive mussel species that threaten even the successful reservoir species. The zebra mussel and quagga mussel are exotic species that invade both reservoir and riverine habitats, develop dense populations, and severely impact native species. These invasive species were introduced and are spread by human transport and water transfers. They have remarkably high reproductive potential and can occur in densities sufficient to consume a high proportion of the plankton in the water of a reservoir; thereby starving another plankton-dependent biota. Reservoirs with large populations of this species have experienced a decline in populations of freshwater mussels, many of which are threatened or endangered due to other aspects of river management. At very high densities, these mussels can clog intake pipes and prevent municipalities and industries from accessing the water supply.

6.4 DECLINING ENVIRONMENTAL FLOWS CASE STUDY: RIO GRANDE

As discussed in Chapter 14, the Rio Grande River flows from the Rocky Mountains to the Gulf of Mexico and forms the boundary between the United States and Mexico for much of its length. We will examine the impact of engineering alterations on the environmental flow regime of this river and the impact of the altered flow on the flora and fauna of the river, the human uses of the watershed, and the receiving waters of the Gulf of Mexico.

6.4.1 Physical Geography of the River

The Rio Grande begins at its headwaters in the San Juan Mountains of southern Colorado, flows through New Mexico, and forms the international boundary between Texas and Mexico before flowing into the Gulf of Mexico. Engineering features and altered flows occur on every branch (Ward and Schmandt, 2013).

Construction of dams began in the second half of the nineteenth century and continued for most of the twentieth century. On the Upper Rio Grande there are numerous small dams that function as water supply and flood control structures. Six large reservoirs have been constructed on the Middle Rio Grande with both water supply and flood control functions. The largest reservoir on this section is Elephant Butte with a storage capacity of 2,495 acre-feet. This dam has all three functions: water storage, flood control, and power generation. On the Lower Rio Grande there are two large reservoirs that combine water supply and flood control functions: Amistad and Falcon (Ward and Schmandt, 2013).

In addition to dams, the river also has many diversions along its length. Diversions for irrigation of over one million acres of cropland cause the largest alteration in flows. The greatest demands for irrigation are in the Colorado headwaters, below Elephant Butte reservoir to El Paso and in the Lower Rio Grande Valley below Amistad. The large cities in the upper basin, Albuquerque, El Paso, and Juarez, meet their municipal needs primarily with groundwater. Above Elephant Butte the demand for irrigation can result in the total dewatering of the river (Booker and Ward, 2002).

The environmental needs of the Rio Grande watershed are often at odds with the needs of its human inhabitants. Human uses of the Rio Grande include agricultural irrigation, municipal and tribal water needs, tourism, hazardous waste disposal, and border protection. Competing uses such as irrigation and municipal water demand, increasing human development, drought conditions, and transboundary issues make the management issues of this binational resource complex. Along with water quantity, other important issues in the Rio Grande Basin include threatened and endangered species, water quality degradation, and exotic species introductions (Booker and Ward, 2002).

6.4.2 Topography and Habitat Changes

A number of diverse habitats exist along the Rio Grande. The watershed of the headwaters in southern Colorado and north-central New Mexico includes the San Juan Mountains, plains, wetlands, and several designated wilderness areas. The middle to upper reaches of the Rio Grande from Santa Fe to the Rio Conchos consist of mountains, grasslands, high plains, conifer forest, the Chihuahuan Desert, and Big Bend National Park. The southern reach of the Rio Grande from Del Rio to Boca Chica is home to riparian habitat as well as sabal palm forest, Chihuahuan thorn forest, clay lomas, coastal prairie, and the mouth of the Rio Grande where it enters the Gulf of Mexico (Dahm et al., 2005).

The hydrology of the upper river is dominated by a few, large perennial streams fed by snowmelt originating in adjacent mountainous areas and many smaller intermittent and ephemeral streams. High flow normally occurs in spring months due to snow melt, and this pattern continues above Cochiti Reservoir near Santa Fe.

The Middle Rio Grande from Cochiti Reservoir to Elephant Butte Reservoir has been extensively modified by engineering and resultant changes in sediment distribution. Padilla and Young (2006) discuss the impacts of diversions, levees, and jetties on the topography of this river segment. The engineered features have greatly degraded habitat along this segment caused by sediment deposition in overbank areas and scour in the main channel. Alterations in flow and morphology have resulted in reduction of mean flows, channel widths, floodplain widths, and sediment transport rates. Channel depth, channel velocities, and number of bars have increased due to the same causes (Dahm et al., 2005).

Flows below Elephant Butte Dam are low in volume and intermittent in places, especially between El Paso–Juarez and the mouth of the Rio Conchos, where the river is often dewatered (Ward and Schmandt, 2013). Below Elephant Butte Dam there have been no significant flood events since its completion and the absence of scouring has caused the substrate to shift to predominately sand and silt from larger particle sizes. Discharge from this dam is lower than from Cochiti.

From El Paso to Presidio, river topography has changed from a wide multi-channeled river to a narrow single, deeper channel with greater tendency to overbank. No flow from the upper and middle Rio Grande currently passes Fort Quitman. All the flow from Presidio to the Gulf of Mexico is from the Rio Conchos, groundwater, and precipitation (Dean and Schmidt, 2011). River flow in the channel of the Rio Grande is currently restored by discharge from the Rio Conchos. However, outflow from the Rio Conchos has been reduced by engineered alterations similar to those along the Rio Grande.

The Lower Rio Grande (Figure 6.3) annual flow has declined from over 6 billion m^3/year early in the last century to less than 1 billion m^3/yr at the start of this century. Large, long duration flood flows occurred in the river along the international border with frequency of once every 5 years pre-1950 but have occurred only five times in the last 60 years. Such floods increase channel width, scour vegetation, and erode small grain sediment. In the absence of these events, the channel has narrowed by 50 percent from 1950 to 2008 (Dean and Schmidt, 2011).

Engineering of the Lower Rio Grande below Laredo and its floodplain began early in the twentieth century to support large agricultural acreage. Falcon Dam was closed in 1954 and changed the hydrology of the lower river to reduce periodic widespread flooding and deltaic processes. As urbanization increased, more structures were introduced to reduce and divert flow from the river channel to other channels (e.g., the Arroyo Colorado) (LRGBBEST, 2012).

Historically, the waters of the Rio Grande flowed into the Gulf of Mexico. However, in recent years, reduced flows in the Rio

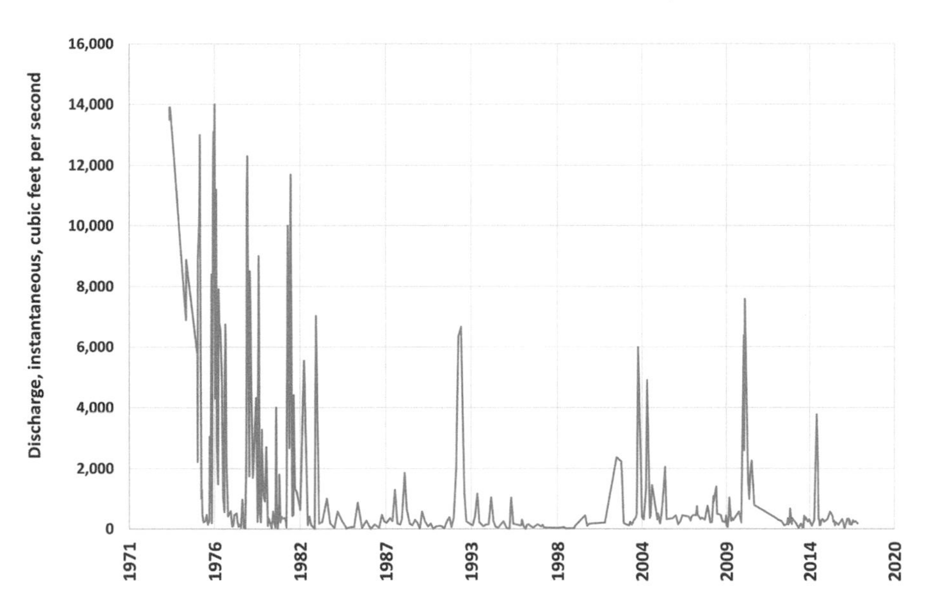

Figure 6.3 Discharge at USGS gage 08475000 on the Rio Grande near Brownsville, Texas, from 1973 to 2017. Figure by the authors using data from USGS

Grande due to water diversions and flow management have resulted in siltation at the mouth. At low flows, the along-shore currents are depositing sand bars that can extend across the mouth of the river and close it off from the Gulf of Mexico, as happened in 2001–2002.

Normal flow below Brownsville is currently so minimal that there is in effect no estuary at the mouth of the Rio Grande (Dahm et al., 2005). There is a saltwater wedge that extends normally beyond river mile 10 along the channel of the river (LRGBBEST, 2012). The effect of engineering on this reduction in flow is clear. In addition to Falcon and Amistad reservoirs, which reduce flows, there are two dams that serve to divert water into canals and other floodplains. The flood control and irrigation diversions return little to no water to the Rio Grande because return flows enter other water bodies, not the river. Since 1934 when flow gaging began at Brownsville, annual flow has declined from 2.5 million ac-ft to 1.1 million ac-ft. Seasonality has been reduced such that the difference between second quarter and the high flow third quarter flows has declined from >100,000 ac-ft to <10,000 ac-ft (LRGBBEST, 2012). Figure 6.3 shows reductions in discharge since 1973. The reasons are various, due to increasing demand (both agricultural and municipal) as well as storage and dams upstream.

6.4.3 Ecological Effects

The diversity of habitats existing historically in the Rio Grande lends itself to a rich collection of ichthyofauna, including the endangered Rio Grande silvery minnow (*Hybognathus amarus*). Over 150 species of fish have been collected from the Rio Grande and its estuary. Thirty-four of these are now considered threatened or endangered. These include shovelnose sturgeon, Rio Grande trout, and a number of suckers, chubs, minnows, and shiners. Several species of fish from the basin are considered extinct, including phantom shiner and Amistad gambusia. Dahm et al. (2005) have found that "[d]ams along the river have greatly restricted the natural range of the American eel and greatly expanded the range of other species."

The reduction in flow and variability of flow has resulted in loss of habitat diversity along the river. Studies of riverine fish habitats usually differentiate a variety of habitats based on water depth, substrate type, and water velocity. Some species require riffles with large particle size substrate for spawning. Other species require low velocity backwaters for larval or juvenile development. In the Rio Grande, there is a diversity of minnow and shiner species that are adapted to different habitats in the river. Darters are adapted to riffles and shallow runs, but not pools. Bass need deep pools and the Mexican tetra requires shallow and deep runs (URGBBEST, 2012). As the topography of the river changes, new habitat will provide advantages to some species and disadvantages to others, but as the diversity of habitats declines, so will the biodiversity.

One of the morphological changes caused by flow alteration and directly related to fish habitat is the change in substrate particle size. Between Cochiti and Elephant Butte there has been a decrease in sand substrate and increase in gravel and cobble substrate (Dahm et al., 2005). Different species are adapted to these substrates for feeding and reproduction.

In other reaches of the river, low flow has increased the area of sand and silt substrate.

In the section of the river between Big Bend National Park and Amistad Reservoir, eight of forty-one native fish species have been extirpated (Hubbs et al., 2008). This loss of species was caused by dams, habitat alteration, and competition from nonnative species (URGBBEST, 2012). Three of five mussel species documented in this reach have only been collected as empty shells in recent years by the National Park Service (URGBBEST, 2012). Freshwater mussels are considered excellent bioindicators of environmental quality in rivers. Some river segments above Amistad Reservoir have impaired water quality due to high salinity, but groundwater contributions mitigate this problem as river flows out of Big Bend National Park (Bennet et al., 2012). High salinities are associated with algal communities adapted to brackish water conditions, which would not normally occur in the Rio Grande (Porter and Longley, 2011).

During the twentieth century, nonnative tamarisk and giant cane have become the dominant riparian vegetation from Big Bend to Amistad (Moring, 2002). Their dominance continues below Amistad as well. In the river itself, nonnative plants have become dominant in many reaches. Water hyacinth, water lettuce, and hydrilla are the most common infestations of invasive aquatic plants. Engineering for prevention of flood flows has benefited these exotic species and supported their continued dominance of sections of the Lower Rio Grande.

In the case of Falcon and Amistad Reservoirs in the Rio Grande, the stocked species include *Oreochromis aureus*, a nonnative tilapia species that can have profound effects on aquatic food webs. Unlike the native species of fish in the Rio Grande, this tilapia species is almost entirely planktivorous and competes with filter feeding mussels and the larvae of other fish for food.

The effects of flow alterations in the lower river and its discharge to the Gulf of Mexico have been to degrade the riparian and wetland ecologies historically associated with the river delta (LRGBBEST, 2012). Dams and diversions have reduced river flow and freshwater inflow so that most freshwater inflow is due to flood events. The former estuary resembles a lagoon with near oceanic salinity. The river below Brownsville is usually infested with dense mats of floating nonnative plants (water hyacinth and water lettuce) that impede flow and cause water loss through evapotranspiration. Native vegetation in the riparian zone has been replaced with invasive *Tamarisk* species (Figure 6.4). The proportion of freshwater fish species has declined, and estuarine and marine species have increased in abundance (Edwards and Contreras-Balderas, 1991). Since the construction of major dams began, two species have been extirpated from the river below Falcon Dam and one has gone extinct from the entire river (LRGBBEST, 2012).

In some recent years there has been no outflow at the mouth of the Rio Grande. If the flow velocity falls below approximately 1 ft/sec at the mouth, then the long shore current will close the mouth with sediments (LRGBBEST, 2012). This would eliminate the estuarine function of the lowest section of river and stop the use of this section as a nursery for marine species, e.g., shrimp, crabs, and snapper.

Figure 6.4 Near-entire saltcedar infestation of the riparian zone of a stream.
Photo by Steve Dewey, Utah State University, Bugwood.org, licensed under Creative Commons Attribution 3.0 License

6.5 CONCLUSIONS

Engineering alterations of rivers will alter natural flow regime and these changes will affect ecosystems in and around rivers. Restoring the environmental flows needed to support the natural ecosystems of engineered rivers may not be possible, particularly for rivers in arid regions with substantial human populations dependent on their water. Anthropogenic alterations of rivers in arid regions have resulted in changes to the topography and hydrology of these rivers and the extinction and extirpation of many aquatic species. Changes in flow regimes have provided opportunities for the establishment of exotic species that replace or compete with the native species. We must conclude that much of the damage done to native ecosystems in engineered rivers is irreparable and the best one can hope for is river management that mitigates the damage while maintaining reasonable benefits to the human populations dependent on these water supplies.

REFERENCES

Arthington, A. H. (2012). *Environmental Flows: Saving Rivers in the Third Millennium*. Berkeley: University of California Press, p. 406.

Bennett, J., Brauch, B., and Urbanczyk, K. M. (2012). *Estimating Ground Water Contribution from the Edwards Trinity Plateau Aquifer to the Big Bend Reach of the Rio Grande, Texas*. South-Central Section – 46th Annual Meeting (8–9 March 2012).

Booker, J. F. and Ward, F. A. (2002). Restoring Instream Flows Economically: Perspectives from an International River Basin. In L. Fernandez and R. T. Carson, eds., *Both Sides of the Border: Transboundary Environmental Management Issues Facing Mexico and the United States*. Dordrecht: Kluwer Academic Publishers, pp. 235–252.

Dahm, C. N., Edwards, R. J., and Gelwick, F. P. (2005). Gulf Coast Rivers of the Southwestern United States. In A. C. Benke and C. E. Cushing, eds., *Rivers of North America*. Burlington, MA: Elsevier Academic Press.

Dean, D. J. and Schmidt, J. C. (2011). The Role of Feedback Mechanisms in Historical Changes of the Lower Rio Grande in the Big Bend Region: Geomorphology. *Journal of Geomorphology*, 126, pp. 333–349.

Edwards, R. J. and Contreras-Balderas, S. (1991). Historical Changes in the Ichthyofauna of the Lower Rio Grande (Río Bravo del Norte), Texas and Mexico. *Southwestern Naturalist*, 36(2), pp. 201–212.

Grafton, R. Q., Pittock, J., Davis, R. et al. (2013). Global Insights into Water Resources, Climate Change and Governance. *Nature Climate Change – Perspective*. https://doi.org/10.1038/NClimate1746

Horne, F. and Beisser, S. (1977). Distribution of River Shrimp in the Guadeloupe and San Marcos Rivers of Central Texas, U. S. A. (Decapoda, Caridea). *Crustaceana*, 33, pp. 56–60.

Hubbs, C., Edwards, R. J., and Garrett, G. P. (2008). An Annotated Checklist of the Freshwater Fishes of Texas, with Keys to Identification of Species. *Supplement in 2nd Edition of Texas Journal of Science*, 43(4), pp. 1–87.

Lower Laguna Madre Basin and Bay Expert Science Team (2012). *Rio Grande, Rio Grande Estuary, Environmental Flows Recommendation Report*. Austin, TX: TCEQ.

Lower Rio Grande Basin and Bay Area Expert Science Team (LRGBBEST) (2012). *Environmental Flows Recommendation Report*. Final Submission to the Environmental Flows Advisory Group. Austin, TX: TCEQ. Available at www.tceq.texas.gov/assets/public/permitting/watersupply/water_rights/eflows/lowerrgbbest_finalreport.pdf

Moring, B. J. (2002). *Baseline Assessment of Instream and Riparian-Zone Biological Resources on the Rio Grande in and Near Big Bend National Park, Texas*. Water-Resources Investigations Report 02-4106. Austin, TX: U.S. Department of Interior, Geological Survey, p. 33.

Neves, R. J, Bogan, A. E., Williams, J. D., Ahlstedt, S. A., and Hartfield, P. W. (1997). Status of Aquatic Molluscs in the Southeastern United States: A Downward Spiral of Diversity. In G. W. Benz and D. F. Collins, eds., *Aquatic Fauna in Peril: The Southeastern Perspective*. Decatur, GA: Southeastern Aquatic Research Inst. Lenz Design and Communications, pp. 43–85.

Padilla, R. and Young, C. (2006). *Monitoring Aggradational and Degradational Trends on the Middle Rio Grande, NM*. In Proceedings of the Eighth Federal Interagency Sedimentation Conference, Reno, NV.

Poff, N. L. and Zimmerman, J. K. H. (2010). Ecological Responses to Altered Flow Regimes: A Literature Review to Inform the Science and Management of Environmental Flows. *Freshwater Biology*, 55, pp. 194–205.

Porter, S. D. and Longley, G. (2011). *Influence of Ground and Surface Water Relations on Algal Communities in the Rio Grande Wild and Scenic River*. Texas State: Edwards Aquifer Research and Data, p. 54.

Rood, S. B. and Mahoney, M. (1990). Collapse of Riparian Poplar Forests Downstream from Dams in Western Prairies: Probable Causes and Prospects for Mitigation. *Environmental Management*, 14, pp. 451–464.

Smakhtin, V. and Eriyagama, N. (2008). Developing a Software Package for Global Desktop Assessment of Environmental Flows. *Environmental Modelling and Software*, 23, pp. 1396–1406.

Sood, A., Smakhtin, V., Eriyagama, N. et al. (2017) *Global Environmental Flow Information for the Sustainable*

Development Goals. Colombo: International Water Management Institute, p. 37. https://doi.org/10.5337/2017.201

Stromberg, J. C., Lite, S. J., Marler, R. et al. (2007). Altered Stream Flow Regimes and Invasive Plant Species: The Tamarix Case. *Global Ecology and Biogeography*, 16, pp. 381–393.

UNEP and UNEP-DHI (2015). *Transboundary River Basins: Status and Future Trends*. Nairobi: UNEP.

Upper Rio Grande Basin and Bay Expert Science Team (URGBBEST) (2012). *Environmental Flows Recommendation Report*. Austin, TX: TCEQ. Available at www.tceq.texas.gov/assets/public/permitting/watersupply/water_rights/eflows/urgbbest_finalreport.pdf

Van Metre, P. C., Mahler, B. J., and Callender, E. (1997). *Water Quality Trends in the Rio Grande/Rio Bravo Basin Using Sediment Cores from Reservoirs*. USGS Fact Sheet FS-221-96.

Vörösmarty, C. J., McIntyre, P. B., Gessner, M. O. et al. (2010). Global Threats to Human Water Security and River Biodiversity. *Nature*, 467(7315), pp. 555–561.

Wada, Y., Wisser, D., and Bierkens, M. F. P. (2014). Global Modelling of Withdrawal, Allocation and Consumptive Use of Surface Water and Groundwater Resources. *Earth System Dynamics*, 5, pp. 15–40.

Ward, G. and Schmandt, J. (2013). *The Rio Grande/ Rio Bravo of the U.S. and Mexico*. Available at www.harcresearch.org/publication/984

Wohl, E. E. (2004). *Disconnected Rivers: Linking Rivers to Landscapes*. New Haven, CT: Yale University Press, p. 301.

Zeug, S. C., Winemiller, K. O., and Tarim, S. (2005). Response of Brazos River Oxbow Fish Assemblages to Patterns of Hydrologic Connectivity and Environmental Variability. *Transactions of the American Fisheries Society*, 134, pp. 1389–1399.

Part III Engineered Rivers

Past, Present, and Future

7 The Nile River Basin

Muhammad Khalifa, Sephra Thomas, and Lars Ribbe

7.1 INTRODUCTION

The River Nile is well-known not only because of the long history of great civilizations that have flourished along its banks but also because of the geopolitical complexity and the recent transboundary conflict between its riparian countries. It was likely the change in climate during the middle Holocene period that led to the creation of the first civilizations along the Nile. Increasing aridity during the fifth millennium BC led tribes in north-eastern Africa to migrate toward the Nile, creating more densely populated and complex societies along the river (Brooks, 2006). After reaching hyper-arid conditions in the 2nd Millennium BC, climate variability is believed to have played a major role in the rise and fall of civilizations along the river (Yletyinen, 2009). Historically, the River Nile was always of great interest. From the days of ancient mythology, Herodotus is said to have found the waterfalls near Aswan, while as history continues, Grecian researchers found the lake sources of the two main tributaries but did not explore the surrounding lands. Emperor Nero is the first recorded explorer of the Nile's headwaters. In 66 AD, to expand and capitalize on the wealth of his empire, Nero's men found themselves blocked by the waterfalls along the White Nile, and later resisted by the people of As-Sudd. Historical records do not speak of major explorations after Nero until the rise of European empires. In 1618, Pedro Paez successfully led an expedition on behalf of the Spanish Empire and located the source of the Blue Nile: a small spring above Lake Tanganyika in north-western Ethiopia. The British Empire also joined in quests to find the source waters of the Nile and by the late 1600s, both empires were established in territory within the basin. The discovery of the Nile's source is accredited to John Hanning Speke, alongside Sir Richard Francis Burton, during an expedition in the nineteenth century funded by the Royal Geographical Society of London (Mark, 2014).

In the twenty-first century, the River Nile has entered a new era, as new challenges and opportunities have emerged mainly due to drivers such as population growth, increased water demand, climate change, and development in the upstream countries (Abtew and Dessu, 2019). Many of the riparian countries of the Nile Basin started to rely heavily on the river's water to foster their development. With all the benefits that the Nile and its tributaries provide to local people, the environment, and economic development, there are also problems that arise. Due to demand for socioeconomic development and lack of effective cooperation between the riparian countries of this transboundary river basin, the River Nile has been a particular source of tension between the downstream countries (Egypt and Sudan) from one side and the other upstream countries from the other side. Based on an analysis conducted by Yoffe (2002), who studied 122 transboundary river basins and classified them according to the conflict risk level, the River Nile is considered one of the hot spots of conflicted river basins, which represent only 3 percent of the total number of the studied basins. This category includes the Jordan, Tigris-Euphrates, and Aral Sea basins. Currently, several of the Nile countries are planning unilateral water schemes along the river to cover their increasing demand. Additionally, the Nile Basin is highly vulnerable to climate change impacts and it is witnessing several environmental problems and development challenges.

As the effects of climate change intensify, population growth continues, and development needs increase, there are major disputes concerning the River Nile. Water, energy, and food supply now stand as security issues, as they are all dependent on the limited water resources in the basin. Demand for these three services increases as many riparian countries are experiencing their unique industrial revolutions, resulting in more engineering feats and political tension along the river. In the basin, environmental advocacy and conservation have become a low priority as each state is concerned with the survival of its own people, changing the ecological landscape of the basin. With large land degradation and deforestation (EPA, 2010), climate variability with frequent extremes (Coffel et al., 2019), water pollution (Kim and Kaluarachchi, 2009), and endemic poverty (Oestigaard, 2012), along with transboundary conflict (Li, 2005), managing water resources in this basin is a challenging task. The complexity of current issues makes the Nile Basin an interesting case to learn from, while the solutions to growing challenges in the future will be decisive for hundreds of millions of inhabitants.

7.2 THE RIVER UNTIL TODAY

7.2.1 Physical Setting

The Nile River Basin is one of the largest transboundary basins in the world. It occupies an area that equals approximately 10 percent of the total area of the African continent (Barnes, 2017). With a total length of 6,695 km, the River Nile is the longest in the world. The Nile Basin is shared between eleven countries, namely, Burundi, DR Congo, Egypt, Eritrea, Ethiopia, Kenya, Rwanda, South Sudan, Sudan, Tanzania, and Uganda (Figure 7.1). The two main tributaries of the Nile River, the Blue Nile and the White Nile, originate, respectively, from the Ethiopian Highlands and the Equatorial Lakes. These two major tributaries meet at Khartoum, the capital of Sudan, to form the main Nile River, which flows northward through northern Sudan and then to Egypt to drain into the Mediterranean Sea.

According to estimates of the Population Division of the United Nations Department of Economic and Social Affairs (UN DESA), the current (2019) total population of the eleven riparian countries of the Nile is around 0.54 billion people. Among all the riparian countries, Ethiopia, Egypt, and DR Congo have the largest populations with estimates of 115, 102, and 95 million people, respectively. Within the Nile Basin countries, there are six cities with a population of 3 million inhabitants or more. Only Cairo (Egypt) is classified as a megacity – with a population greater than 10 million people. Other large cities such as Kinshasa (DR Congo), Khartoum (Sudan), and Alexandria (Egypt) exhibit the typical characteristics of megacities. These include large peri-urban populations at their fringes, with dominant illegal settlements and inadequate housing, sanitation, and other essential services. Other features of megacities include traffic congestion, water and air pollution, smog, unemployment, lack of open spaces, and high population densities. The speed of growth of the megacities, driven by rapid rural–urban and urban–urban migration, is faster than the rate of expansion in economic infrastructure. This problem stretches city resources and threatens to overwhelm city administrators. Although urbanization may alleviate pressure on the rural environment by offering alternative income and livelihoods, it brings with it a different set of challenges. Rapidly growing and unplanned urban centers often lack the infrastructure and institutions needed to protect human and environmental health, supply adequate water and sanitation, or provide affordable housing and transportation. Most of the large urban areas in the basin have conventional municipal wastewater treatment systems. However, the systems were built many decades ago, and their capacity has not grown in tandem with population growth. Therefore, many large urban centers in the basin do not possess proper treatment systems for domestic wastewater, which ends up eventually contaminating the environment with organic matter, plant nutrients, suspended solids, and pathogenic organisms.

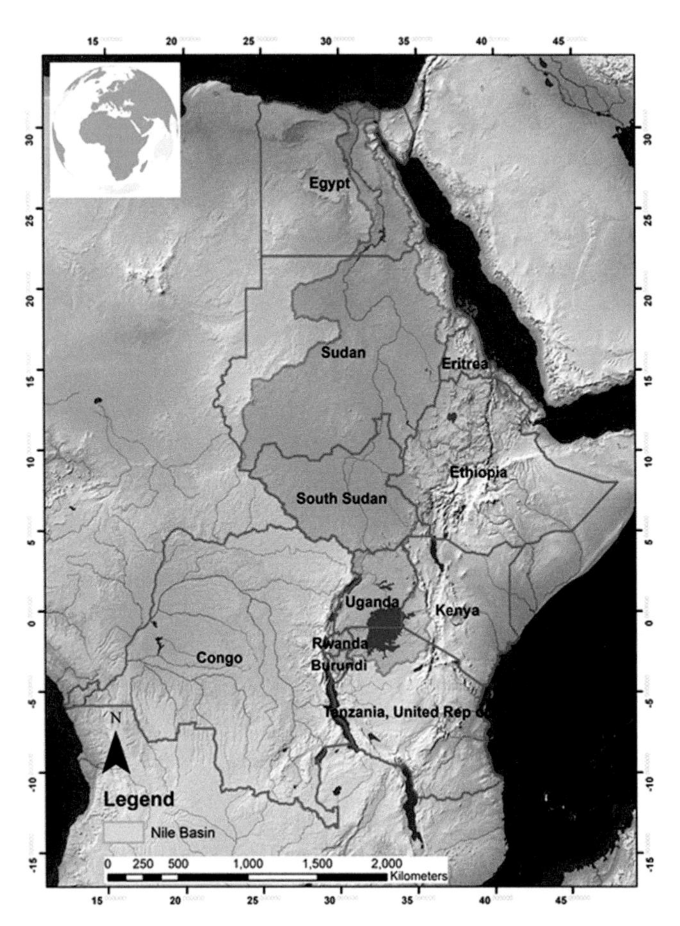

Figure 7.1 Location map of the Nile Basin with the 11 riparian countries.
Map created by the authors using data from GADM and Natural Earth Data. River basin boundaries were delineated using SRTM Digital Elevation Model

The Nile Basin has diverse land cover types (Figure 7.2). Most of the area in the basin is bare lands, which represent around 28 percent of the total area of the basin. Bare lands extend in the northern arid and semi-arid regions, mainly in Egypt and northern Sudan (Figure 7.3). Forests, shrublands, grasslands, and croplands cover vast areas in the basin. While large irrigated croplands are in Egypt (Delta) and central Sudan, rainfed agriculture characterizes the central and southern countries of the basin, where rainfall is higher. Savannas, grasslands, shrublands, and forests are in the central part of the Nile Basin. Wetlands represent 1.1 percent of the total basin area, and include montane bogs, lowland herbaceous swamps, seasonally flooded grasslands, swamp forests, riverine wetlands, and lake-fringe wetlands. The basin wetlands are concentrated in two areas, namely, the Equatorial Lakes region and the Sudd area in South Sudan. The Nile Delta in north Egypt, once an area of lush natural wetlands, has now been almost entirely converted into agricultural lands.

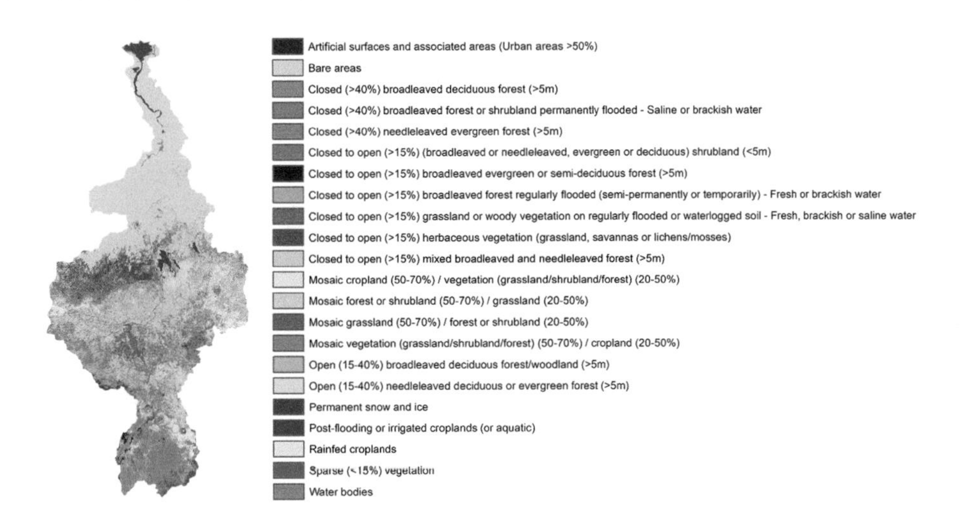

Figure 7.2 Land cover map of the Nile Basin.
Figure created by the authors using data from GlobCover (2009)

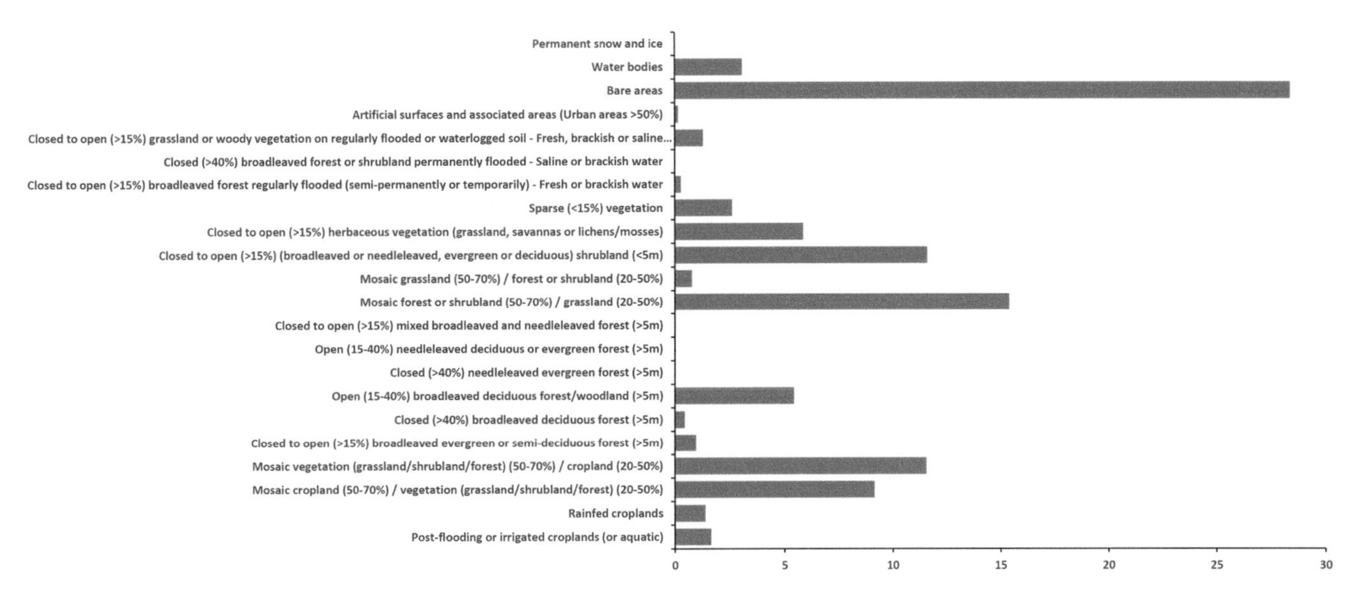

Figure 7.3 Percentage area of each land cover type as regards to the total area of the Nile Basin.
Figure created by the authors using GlobCover (2009) data

The Nile Basin is characterized by large spatial and temporal variation in climatic conditions such as rainfall and temperature. Spatially, the northern regions of the basin receive negligible rainfall levels (Figure 7.4) and are characterized by high temperature (Figure 7.5). High rainfall occurs generally in the eastern and southern regions of the basin, especially the Ethiopian highlands. The average monthly rainfall ranges between 0 and 450 mm. The basin receives little rain from November to April, except in some locations in the region of the Equatorial Lakes, and the main rains fall during the period from June to October (Figure 7.4). The average monthly temperature ranges between 1°C and 35.5°C.

7.2.2 Water Resources and Water Use Patterns

As mentioned earlier, the Nile River originates in two main regions: the Ethiopian highlands (Lake Tana) and the equatorial lakes (Lakes Victoria, Koyoga, and Albert), which are the sources of the Blue Nile and White Nile rivers, respectively. The main tributaries of the Blue Nile river are the Rahad and Dinder, which originate in the Ethiopian highlands and join the Blue Nile river upstream from Khartoum, Sudan. Several major tributaries contribute to the White Nile, namely, the Bahr el Jabal, Bahr el Gazal, Baro, Pibor, and Sobat rivers. Compared to the high seasonality of the water flow in the Blue Nile river,

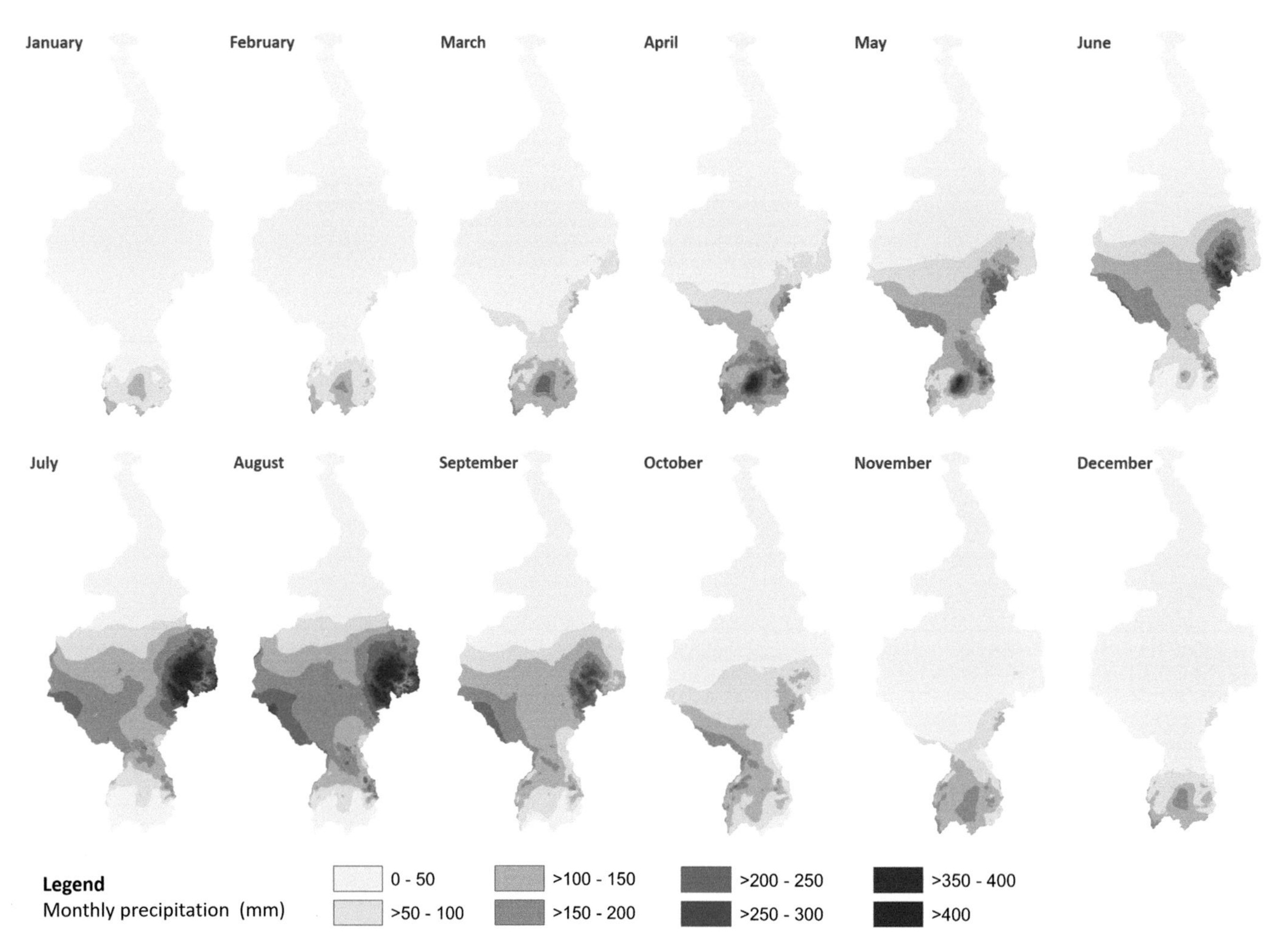

Figure 7.4 Long-term monthly average rainfall (1970–2000) over the Nile Basin.
Figure created by the authors using data from WorldClim (Fick and Hijmans, 2017)

the White Nile is regulated by the large Sudd swamp in South Sudan. The only major tributary of the Main Nile (downstream of the confluence of the Blue Nile and the White Nile) is the Atbara and Tekeze River, which contributes around 12 km^3 per year. The historical flow of the Nile is 84 km^3 as measured at the High Aswan Dam in Egypt (Sutcliffe and Parks, 1999), of which approximately 62 percent is contributed by the Blue Nile river (Amdihun et al., 2014).

As an adaptation measure against insufficient water availability temporally and spatially and to provide water for food and energy production, several dams were built on the Nile tributaries during the last decades. The existing major water control structures are listed in Table 7.1. Sediment resulting from soil erosion taking place in the upstream parts of the Nile Basin presents a serious problem for water management of the reservoirs in the basin. For instance, some of the dams in Sudan, i.e., the Sennar, Roseries, and Khashm El Girba, have lost more than 60 percent of their original storage capacities (Shahin, 1993; Ahmed, 2004; Williams, 2009). The total sediment yield at the outlet of the Blue Nile basin is estimated to be around 131×10^6 tonnes per year (Betrie et al., 2011). From 1980 to 2009, researchers found a 5 percent increase in the sediment yield in this region (Gebremicael et al., 2013).

Natural lakes and artificial reservoirs show large evaporative losses. For instance, approximately 94.5 km^3, 30.9 km^3, and 10.5 km^3 of water are evaporated annually from Lake Victoria, the Sudd swamps, and the Aswan Dam, respectively. An estimated 3 percent (95,926 km^2) of the Nile's open water is in the form of lakes. Notable large lakes include Victoria, Kyoga, Albert, George, Edward, and Tana. The lakes are primarily located in the Equatorial Lakes Plateau region. The only major lake in the desert biome is Lake Nasser/Nubia, which resulted from the damming of the Nile at Aswan. The lakes in the basin have various functions, including acting as a habitat for aquatic plant and animal species, buffering the discharge of outflowing rivers

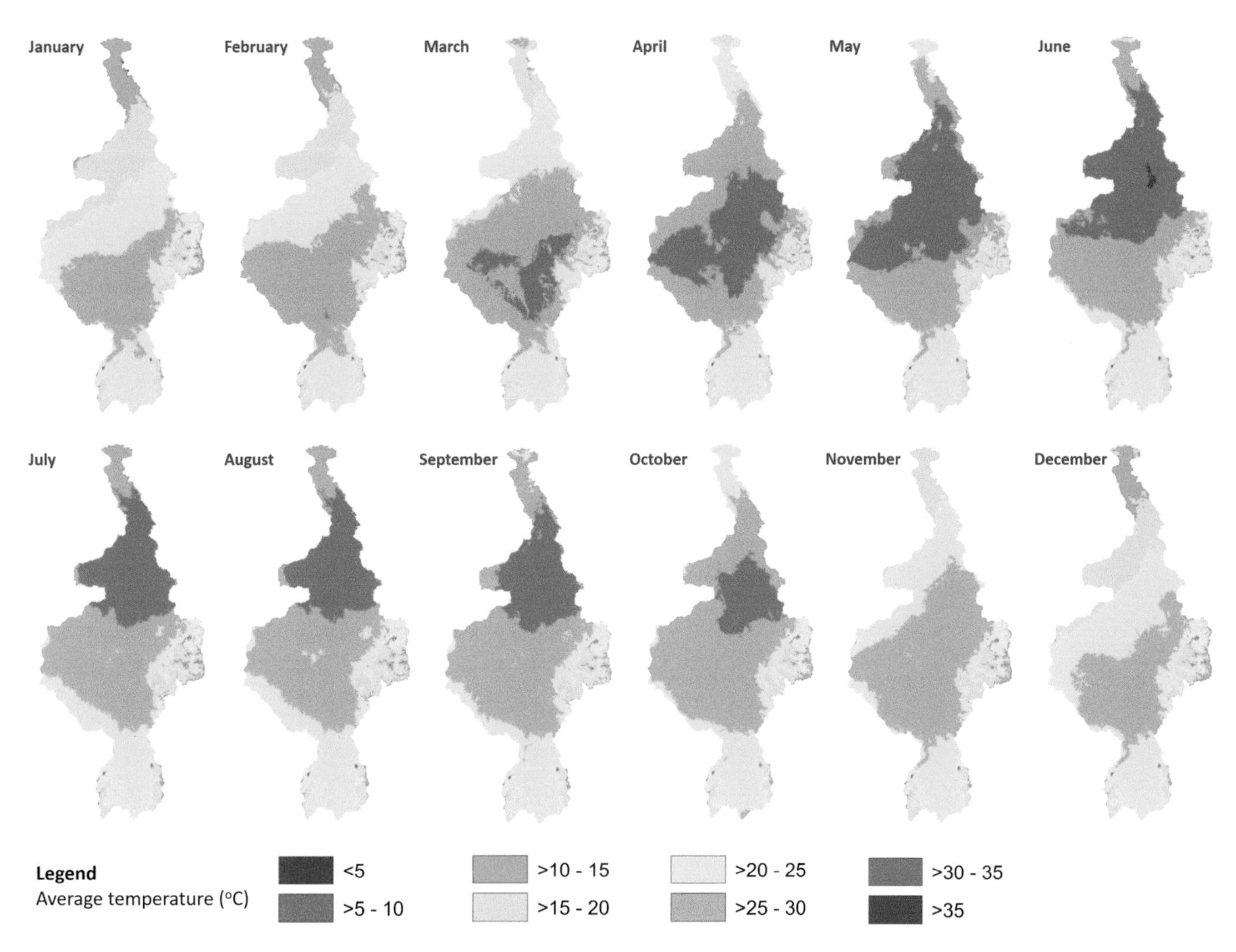

Figure 7.5 Long-term monthly average temperature (1970–2000) over the Nile Basin.
Figure created by the authors using data from WorldClim (Fick and Hijmans, 2017)

against seasonal extremes, and acting as a trap for sediments from the headwater areas.

As well as surface waters, the Nile Basin countries have considerable groundwater resources occurring in localized and regional basins. Groundwater is an important resource, supporting the social and economic development of the Nile riparian countries and making an important contribution to water and food security in the region. The degree to which groundwater is relied on varies from country to country, but commonly it is the most important source for drinking water, especially for rural communities in the basin. Groundwater in the Nile Basin mainly occurs in four rock systems or hydrogeological environments, namely, Precambrian crystalline/metamorphic basement rocks, volcanic rocks, unconsolidated sediments, and consolidated sedimentary rocks. Water in these four hydrogeological systems occurs in confined and unconfined conditions. The main groundwater aquifers in the Nile Basin are (1) Victoria artesian aquifer, (2) DR Congo hydrogeological artesian aquifer, (3) the Upper Nile artesian aquifer, (4) volcanic rock aquifers, (5) the Nubian sandstone aquifer system, and (6) the Nile Valley aquifer.

Using remote sensing data validated with ground measurement of different water balance components, Bastiaanssen et al., (2014) estimated the water balance in fifteen sub-basins in the Nile Basin between 2005 and 2010. According to this study, while total rainfall over the basin is 2013 km^3/yr, the total evapotranspiration is around 1987 km^3/yr. (Table 7.2), which implies that most of the available water in the Nile Basin is lost by the evapotranspiration process. The contribution of the upstream sub-basins to the flow in downstream basins differs widely. Among the upstream sub-basins, the Blue Nile sub-basin is contributing most of the water that flows to Egypt. In addition, inter-basin transfer of surface water and groundwater between the sub-basins is a small fraction compared to other water balance components.

Table 7.1 *List of major dams in the Nile Basin with their storage capacity.*

No.	Country	Dam	Storage capacity (km^3)
1	Egypt	Aswan Low Dam	5
2		Aswan High Dam	162
3	Sudan	Merowe	12.5
4		Jebel Awlia	3.5
5		Sennar	0.9
6		Roseries	3
7		Atbara	1.3
8		Upper Atbara and Setit Dam Complex	2.7
9	Ethiopia	Grand Renaissance Dam (GERD)	74 (under construction – 2019)
10		Mendaia	15.9
11		Tekeze	9
12		Megech	1.8
13		Rib	0.2
14	Uganda	Owen falls	80

Agriculture is the largest water consumer in the Nile Basin countries compared to the industrial and municipal uses (FAO, 2005) (Figure 7.6). For instance, agriculture accounts for nearly 97 percent of the total water use in Sudan. The annual irrigation requirement rate varies widely in the Nile basin countries, ranging from nearly 8,000 m^3/ha in Kenya and Uganda to 13,700 m^3/ha in the arid and semi-arid regions of Sudan (Awulachew et al., 2012). Egypt and Sudan alone account for around 82 percent (87.26 km^3) of the total agricultural water withdrawal in the basin. Current agricultural, industrial, and municipal water uses in countries such as Burundi, Rwanda, South Sudan, and Eritrea are negligible (Figure 7.6).

7.2.3 Agriculture and Energy Production

Agriculture is the main source of livelihood for most of the population and a main contributor to GDP (approximately 20 percent); it accounts for approximately 40 percent of all employment in the Nile Basin (Appelgren et al., 2000). The Nile Basin has large arable lands, of which around 8 million hectares have potential for irrigation (FAO, 2005). Sudan alone has approximately 105 million ha of arable land, and only 17 percent of this

Table 7.2 *Annual water balance (2005–2010) as estimated by Bastiaanssen et al. (2014) using remote sensing over different sub-basins of the Nile basin.*

Sub-basin	Inflow (km^3/yr)	P (km^3/ year)	ET+I (km^3/yr)	Net GW interbasin (km^3/yr)	Net SW interbasin (km^3/yr)	ΔS (km^3/ yr)	Outflow (km^3/yr)
Main Nile 1	36	2	19	4	1	−0.09	14
Main Nile 2	55	3	16	4	1	−0.22	36
Main Nile 3	79	51	71	4	1	0.10	55
Tekeze- Atbara	0	121	105	1	2	1.19	12
Main Nile 4	87	4	6	4	2	−0.07	79
Blue Nile	0	299	237	5	6	1.54	50
Lower White Nile	25	122	14	−11	−7	−0.25	25
Bahr el Ghazal	0	435	446	−3	−10	1.00	1
Sudd	35	162	201	−9	−6	−1.25	12
Baro-Akobo-Sobat	0	242	232	−1	0	−1.17	13
Albert - Bahr al Jabal	33	90	91	−2	−1	−0.08	35
Victoria Nile	28	100	93	3	5	−0.56	28
Semliki - Lake Albert	0	78	72	0	0	0.43	5
Lake Victoria	5	246	208	6	8	2.00	28
Kagera	0	57	52	0	0	0.78	5
Nile		2013	1987	5	2	3	14

Abbreviations: P = precipitation, ET = evapotranspiration, I = interception, GW = groundwater, SW = surface water, ΔS = change in storage.

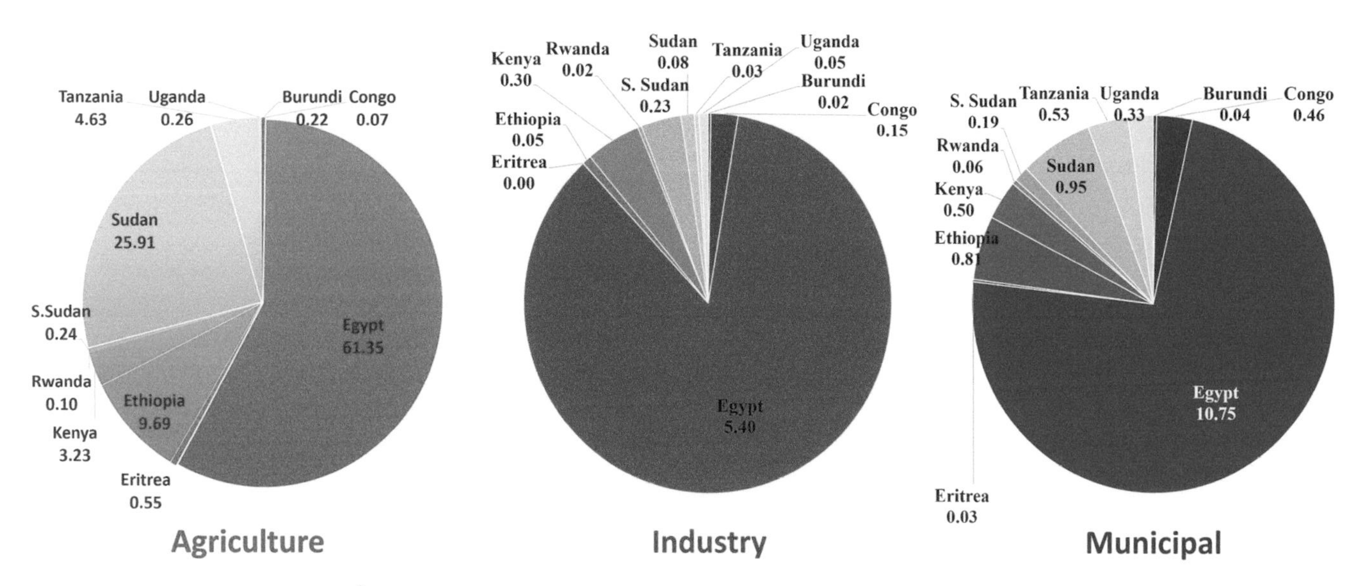

Figure 7.6 Water use quantities (km^3/year) in the Nile Basin countries. The water use is subdivided into agricultural, industrial, and municipal uses. Figure created by the authors using data from AQUASTAT database (FAO 2019a)

large area is currently utilized. Rainfed and irrigated agriculture are both practiced in the basin, though rainfed production is far more common. While only 3.8 percent (on average) of the total arable land in the Nile's sub-Saharan countries is irrigated, this percentage is higher in Sudan and reaches 100 percent in Egypt (Oestigaard, 2012), where rainfall is negligible.

Cereal crops dominate cultivated land in the basin, and among all the Nile Basin countries, Sudan and Ethiopia have the largest area of cereal harvest (Figure 7.7a), with nearly 9.5 and 9.2 million ha, respectively (NBI, 2012). The yield of cereal crops shows high variability among the Nile riparian countries. Egypt stands out, exhibiting higher cereal yield compared to the other Nile's countries (Figure 7.7b). This high yield results in higher production of cereal crops compared to other countries, such as Sudan, which also have large harvested areas (Figure 7.7c).

Land grabbing by foreign countries and large companies is a major issue in some Nile Basin countries. According to a review conducted by Rullia et al. (2013), land grabbed in four of the Nile Basin countries, namely DR Congo, Sudan, Ethiopia, and Sudan, accounts for more than 33 percent of total global grabbed lands. Land grabbing is associated with large water withdrawal. The average withdrawal of Blue Nile water is estimated to be around 3.18 km^3 in these four countries, which represents 27.8 percent of the total average of globally grabbed blue water (blue water can also be understood as freshwater).

Access to electricity is a great challenge in the Nile Basin. Apart from Egypt, all the riparian countries are still far behind in providing sufficient access to electricity for their people (Table 7.3). Egypt exhibits the highest per capita electric power consumption in the region, using more than 1,600 kWh per capita, energy in the region is produced from multiple sources (e.g., hydropower, thermal, and geothermal plants). Hydropower is an important source of energy in the basin. With values of 530,000 MW and 162,000 MW, respectively, DR Congo and Ethiopia own the largest hydropower potential in the basin. This huge potential for hydropower generation in the basin is still largely untapped (Table 7.3).

The main hydro-power dams that are currently in operation are the High Aswan and Merowe dams (on the main Nile River), Rosaries and Sennar (on the Blue Nile River), Khashm El Girba and Atbara and Setit Complex (on the Atbara River), Tekeze (on the Tekeze River), and Gebel Aulia (on the White Nile). Currently, Ethiopia is building the Grand Ethiopian Renaissance Dam (GERD) only 20 kilometers from the border with Sudan. The dam, once completed, will produce a peak of 6,000 megawatts of power, and store 74 km^3 of water in its reservoir (Dessu, 2019), making it the largest hydropower dam in Africa. The GERD consists of a 1.8 km high gravity dam and a 5 km rockfill saddle dam (Abtew and Dessu, 2019).

7.2.4 The Nile's Transboundary Conflict

Use of the Nile waters has for decades been monopolized by the downstream countries, i.e., Egypt and Sudan, which claim "historic right" over the waters. Hegemony over the Nile waters has been under these countries, thus building tensions among the riparian countries. The upstream countries that source the head waters have been impeded from using the Nile for years. Past

Table 7.3 *Demand for electricity and potential and installed hydropower capacities in the riparian countries of the Nile Basin (African Development Bank (2000); NBI (2012); IEA Energy Atlas (2015); World Bank (2019)).*

Country	Access to electricity (% population)	Electric power consumption (kWh per capita)	Peak demand for energy (MW)	Hydropower potential (WM)	Hydropower installed (WM)
Burundi	0.3	–	52	1,366	36
DR Congo	19.1	109	1,050	530,000	2829
Egypt	100	1,683	23,470	3,210	2825
Eritrea	48.4	89	–	–	–
Ethiopia	44.3	69	914	162,000	378
Kenya	63.8	164	1,194	30,000	611
Rwanda	34.1	–	78	3,000	59
South Sudan	25.4	44	–	–	–
Sudan	56.5	190.22	1,360	1,900	225
Tanzania	32.8	104	829	20,000	339
Uganda	22	–	446	10,200	155

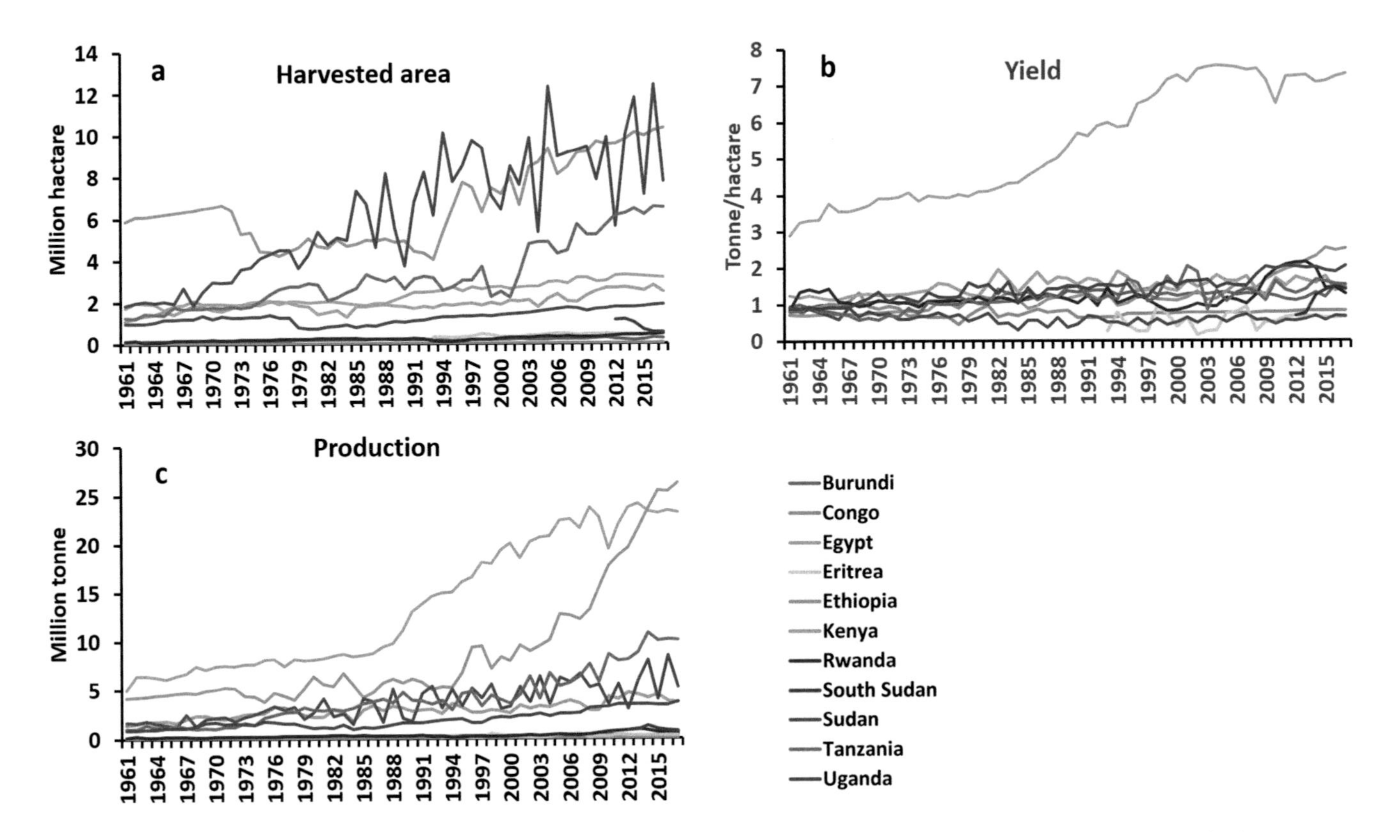

Figure 7.7 Time series (1961–2017) of cereal crops statistics in the Nile Basin countries: (a) harvested area, (b) yield, and (c) production. Figure created by the authors using data obtained from FAOSTAT database

treaties and agreements on the use of the Nile were signed during the colonial period, favoring water rights for Egypt over the other riparian countries, resulting in a history of inequitable rights and use within the basin. Figure 7.8 shows the history of Nile Basin agreements and treaties. These agreements and treaties did not consider a comprehensive view of the impact of water development on the basin's social and biophysical environments and failed to distribute the shares of water properly between the

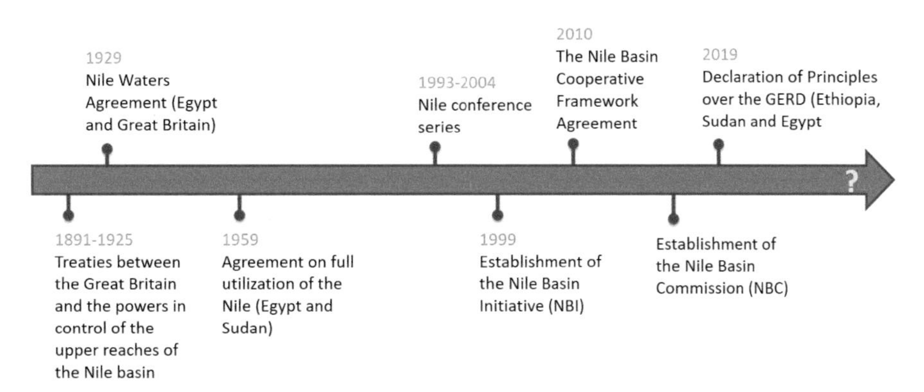

Figure 7.8 History of the Nile Basin agreements and treaties.
Figure modified after the Nile Research Group, ITT, Technical University of Cologne (TH Köln)

riparian countries. Due to the need to support their own socioeconomic development, upstream countries in recent decades have begun to demand the demonopolization of Nile water use by Egypt and Sudan (Ashton, 2002).

Despite the long history of conflict and tension over the Nile waters, currently there is no basin-wide agreement ratified by all riparian states. The 1929 Anglo-Egyptian Treaty gave Egypt the right to veto any project in the Nile Basin, as agreed by Britain and Egypt, to protect British cotton interests in the Nile Valley. The 1959 Bilateral Agreement increased Sudan's and Egypt's water allocation to 55.5 and 18.5 km^3, respectively, and stipulated that any increase in yield should be divided equally among them. Although the 1959 Agreement was not binding or inclusive of other riparian countries, notably Ethiopia, it stated that Egypt and Sudan had veto power on any upstream use or development projects along the Nile. (Salman, 2013).

The Technical Cooperation Committee for the Promotion of Development and Environmental Protection of the Basin (TECCONILE) in 1992 led to the eventual creation of the Nile Basin Cooperative Framework, which served as the initial foundation of the Nile Basin Initiative (NBI). Water ministers of Burundi, DR Congo, Ethiopia, Egypt, Kenya, Rwanda, Sudan, Tanzania, and Uganda have participated since 1999, while Eritrea has observer status. The NBI serves as a transitional institutional framework for managing transboundary trade-offs and opportunities such as sharing hydropower benefits, stronger integration in agriculture markets, and exploiting opportunities for regional trade.

The Cooperative Framework Agreement (CFA) was introduced in 2010 by the NBI to allow for a more equitable distribution of ownership on the Nile between the riparian countries. The CFA contains forty-four articles clearly defining the intention, utilization, sustainability, optimization, benefit sharing, and cost sharing principles of the Nile riparian states. Burundi, Ethiopia, Kenya, Rwanda, Tanzania, and Uganda signed the agreement, but Egypt and Sudan insisted that the wording protect current use and ownership rights held by their respective countries.

The need for socioeconomic development in the Nile countries under the conditions of inadequate cooperation and lack of a comprehensive and agreed framework for water use in this transboundary basin are the driving forces that lead some of the basin's countries to take unilateral actions. The case of GERD in Ethiopia is the most recent example. As of 2016, only a quarter of the population of Ethiopia has access to electricity and, according to the Growth and Transformation Plan II, 22,000 km of distribution lines will be installed by 2020 (Kumagai, 2016). On April 2, 2011, the Ethiopian government began construction on the GERD (Kimenyi and Mbaku, 2015). The almost 5 billion US$ project will be completely domestically funded. The US Bureau of Reclamation originally surveyed and suggested the dam construction in 1956 and 1964, but Ethiopia only submitted a final design in November 2010.

Current construction has been contracted with China Gezhouba Group and Voith Hydro Shanghai for a combined $153 million. The GERD will store 74 km^3 of water and produce 6,000 MW of electricity (Kimenyi and Mbaku, 2015). This large dam has the potential to significantly affect the hydrology of the Blue Nile River and consequently the Main Nile that reaches Egypt. Egypt is afraid that the GERD might decrease the Nile share of water that reaches its territories, especially during the reservoir filling period. Ethiopia has ignored Egyptian claims to complete ownership of the Nile river through the 1959 Bilateral Agreement between Egypt and Sudan. Egypt has also used the 1929 Anglo-Egyptian Treaty as a basis for vetoing any projects occurring along the Nile river. On March 2015, Egypt, Ethiopia and Sudan have signed the Declaration of Principles on the GERD, and since then leaders of the three countries have met multiple times in hope of an agreement on Nile rights considering GERD construction. The United States has acted as neutral mediator in treaty negotiations and has successfully reached an agreement that Egypt and Sudan have ratified. However,

Ethiopia refuses to sign and has delayed agreement talks. Most recently, the African Union has been involved in the negotiations. However, no agreement has been reached yet (at the time of preparing this manuscript).

7.3 THE RIVER TOMORROW

7.3.1 Future Dynamics and Expected Changes

Based on forecasts, the total population in the Nile Basin's countries is expected to increase notably (Figure 7.9). While the low variant scenario forecasts a total population of 0.69 billion by 2030 and 1.06 billion by 2060, the high variant scenario expects a population of 0.72 billion by 2030 and 1.36 billion people by 2060. The medium variant scenario values are 0.73 and 1.20 billion people for the two horizon years 2030 and 2060, respectively. This expected large increase in population of the riparian countries would, consequently, put great pressure on the limited water resources of the Nile.

Environmental problems in the Nile Basin are expected to intensify in the future. For instance, Onencan et al. (2016) stated that there is a high confidence that the basin will suffer from severe shifts in biome distribution, compounded water stress, degradation of marine life, and reduced crop productivity. Climate change impacts could impact the Nile Basin seriously. While studies that have investigated temperature change in the basin are consistent in predicting a warming trend during the twenty-first century, studies focused on rainfall change exhibit inconsistency in their predictions. A review conducted by Barnes (2017) showed that most of the studies on temperature change expect an increasing trend over the Nile Basin of values ranging between 0.3 and 0.6 °C per decade based on the A2 and B1 emission scenarios. Due to high diversity in climatic conditions in the basin, the impact of climate change may have different consequences for different parts of the basin spatially and temporally (Degefu and He, 2015).

Studies on some sub-basin scales (e.g., Upper Blue Nile) reported that hot and dry years are more frequent, and this trend seems to continue in the future and consequently may lead to chronic water scarcity in the Nile Basin (Coffel et al., 2019). The expected increase in temperature is predicted to induce an increase in potential evapotranspiration of 7.8 percent in some headwater sub-basins such as the Upper Blue Nile (Worqlul et al., 2018). Climate changes in the Upper Blue Nile sub-basin, which represents the headwater area for the main Nile, could affect water availability in downstream countries Sudan and Egypt, especially due to their sensitivity to the variability of runoff from this headwater area (Kaluarachchi et al., 2008). Moreover, forecast climate changes in some of the sub-basins in the Blue Nile region would increase the mean annual sediment yield by 2050 by around 16.3 percent for scenario A2 and 14.3 percent for scenario B2 (Adem et al., 2016).

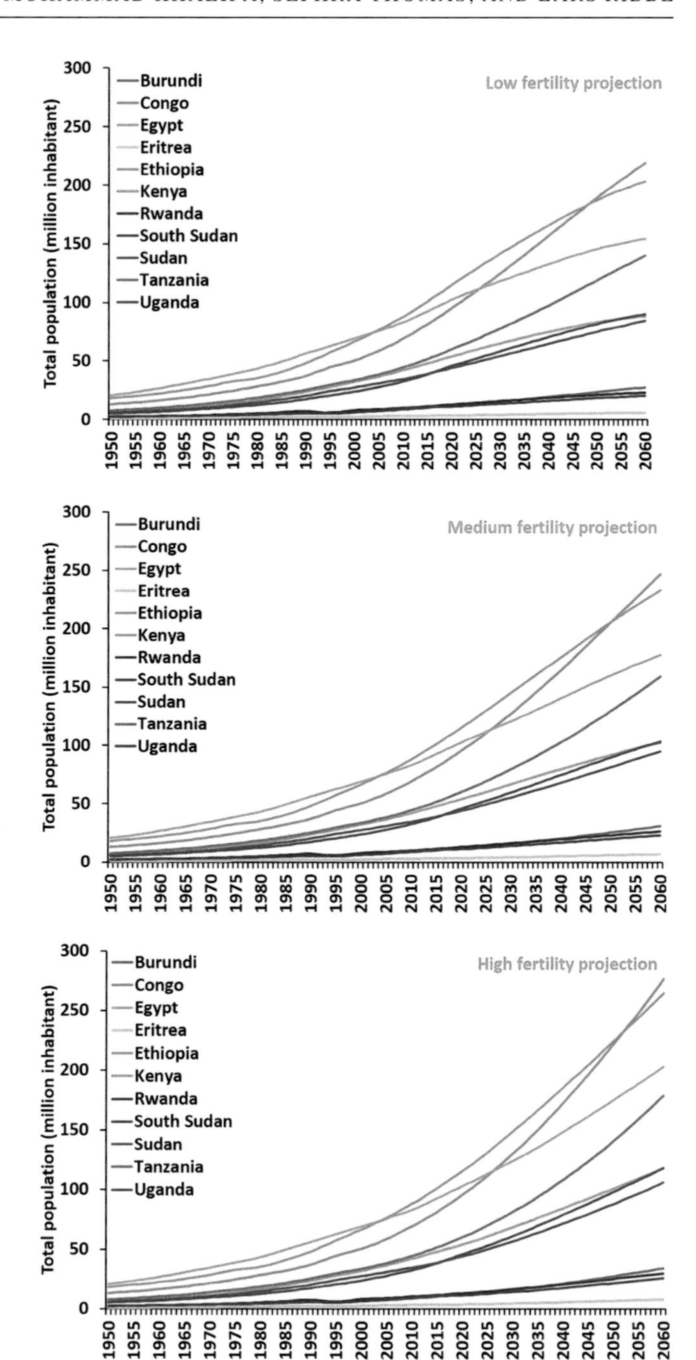

Figure 7.9 Time series (1950–2019) of population estimation in the Nile riparian countries along with future forecasts till 2060. The future forecasts are based on three fertility projections (low, medium, and high).
Figure created by the authors using data from the Population Division of the UN DESA (2019) database

As estimated by FAO (2011), total agricultural water withdrawal in the Nile Basin is 99.19 km^3 (2005). This figure is expected to increase remarkably to reach 107.02 km^3 and 114.77 km^3 by the years 2030 and 2050, respectively. Irrigated

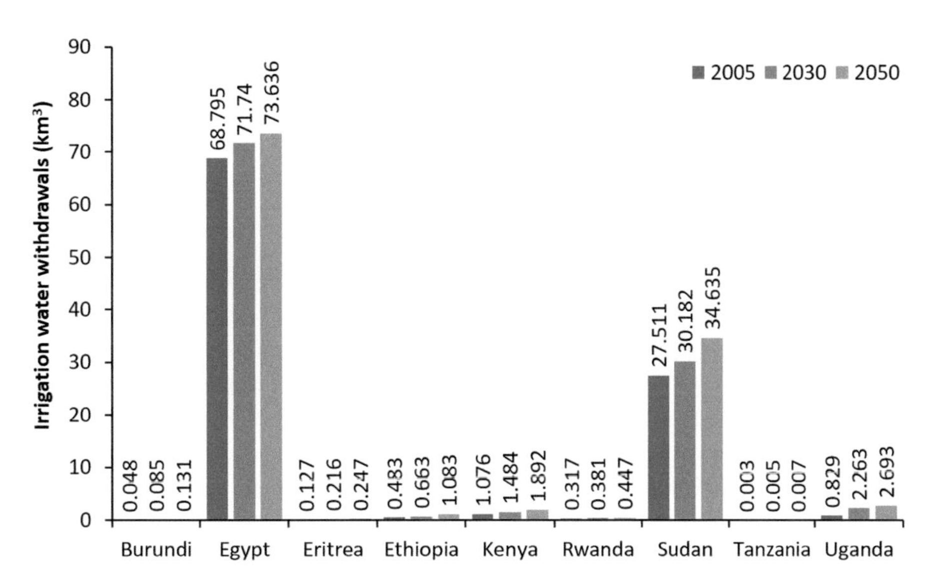

Figure 7.10 Current (2005) and projected (2030 and 2050) agricultural water use in the Nile basin countries.
Figure created by the authors using tabulated data obtained from FAO (2011)

schemes in Sudan and Egypt are assumed to be the main source for this increase in water withdrawal in the Nile Basin (Figure 7.10). Although increasing irrigation efficiency is believed to save substantial amounts of water, some researchers argue that this saved water is insufficient to meet future water demand in the basin (Multsch et al., 2017).

McCarteny et al. (2012) investigated the impact of future development of irrigation and hydropower in Ethiopia and Sudan on the system of the Blue Nile basin – the major source of Nile water – using the Water Evaluation and Allocation and Planning (WEAP) model. Their results indicate that total water storage in Ethiopia will increase to 167 km^3. Irrigation is expected to increase to 13.8 km^3 and 3.8 km^3 in Sudan and Ethiopia, respectively. The annual hydropower generation in Ethiopia is expected to increase to 31,297 GWh.

7.3.2 Implications of Future Changes

As shown earlier, a large increase in population in the future and the need to foster socioeconomic development in the riparian countries will be the main drivers for increasing demand for the main key services, i.e., water, food, and energy. This increase in demand will put natural resources under great pressure. On the other hand, drivers such as climate change and degradation of natural resources are expected to diminish the available natural resources in the basin.

Climate change is predicted to affect the way the water resources will be managed in the Nile basin. Among the conflicting results of the studies conducted to forecast the future of rainfall in the basin, there is a wide acceptance that climate extremes such as drought and flooding will be more frequent and more intense in the future. This calls for immediate measures that enable better adaptation of local people and requires proper policies for better preparedness against such extreme events. A striking example of the expected impacts of climate change on the Nile Basin is coming from the delta region in Egypt. Despite its relatively small area (~2.5 percent of Egypt's area), this region is highly populated, and represents the most important area for food production in the country. The area is highly vulnerable to sea level changes due to climate change. It is estimated that a 1-meter rise in sea level would cause a loss of 4,500 km^2 of land and displace 6.1 million people in the Nile Delta (NBI, 2012).

Negative changes in water supply, development inequity, and population growth are likely to intensify the effects of engineering projects along the Nile and further the transboundary conflict. For example, Ethiopia's construction of the GERD modernizes the energy sector and provides a clean and sustainable source of energy to Ethiopia, but it may put pressure on agricultural output and water security for Egypt and Sudan, especially during the dam's filling period. Ethiopia's construction is projected to reduce flow into Aswan High Dam by 25 percent (Connif, 2017), causing 2 billion US$ in economic losses and approximately 1 million farmers and workers to become unemployed (Lazarus, 2018). The Aswan High Dam currently irrigates around 840,000 ha along the Nile Valley (Heinz, 1983), and produces almost half of the country's electricity (Abu-Zeid and El-Shibini, 1997). Taking into consideration the high demand for water in Egypt and the limited sources of water other than the Nile in the country, combined with the lack of cooperation between the riparian countries, some researchers argue that transboundary conflict over the Nile's water will escalate during the twenty-first century (Rahman, 2012; Keith et al., 2013).

To minimize the GERD impacts on Sudan and Egypt during the filling period, a synchronized dam operation between the three countries is needed (Abtew and Dessu, 2019). Hence, cooperation and communication between the Nile riparian countries is a must. The benefits gained from the Nile Basin could be maximized if a cooperative approach is followed between the riparian countries (Wheeler et al., 2016; Basheer et al., 2018). Potential solutions for Egypt to overcome the problem of water shortage include developing unconventional water resources such as seawater desalination, reuse of treated wastewater, water saving technologies, artificial groundwater recharge, inter-basin water transfer, and virtual water trade (Yihdego et al., 2017; Ashour et al., 2019).

The large increase in water storage in Ethiopia and irrigation water in Sudan and Ethiopia are almost certain to affect the current river regime (McCarteny et al., 2012). These developments are expected to decrease the average river flow at the Ethiopia–Sudan border from 45.1 to 42.7 km^3/year and from 40.4 to 31.8 km^3/year at Khartoum. Despite the significant reduction in river flow at Khartoum, regulating the flow throughout the year – a result of the GERD – would provide Sudan with continuous flow all over the year compared to the current seasonal flow that peaks during the rainy season, enabling Sudan to have more cropping seasons that could help to secure food in the country. Additionally, it might decrease the sediment yield, which accumulates in downstream dams (Ali et al., 2017), and thus slow down capacity loss of these dams and decrease operation and sediment removal costs in the major irrigation schemes.

7.4 CONCLUSION

The findings of this chapter may be summarized into three primary suggestions for enhancing sustainability: (1) political and regional stability, (2) access to modern technology and current information, and (3) a comprehensive and agreed framework to manage the transboundary water.

1. *Political and regional stability*

As climate change may affect access to water and threatens food and energy generation, it is important that governments and regional partnerships remain intact and positive. Political instability is often a result of distrust between the people and ruling governments or a result of low economic opportunity and dismay among citizens. Without first putting these issues to rest, it is difficult to introduce wide-reaching campaigns to conserve and protect water resources. The priority is always stability and protection of humans. If people feel undervalued or ignored, not only will an initiative meant to provide good for both humans and the environment be neglected, but it may cause compounded damage to water resources.

2. *Access to technology and current information*

Numerous studies show that those facing more extreme effects of climate change are often those with less capital and fewer resources. This may be especially true in terms of technology and information. Technological solutions that could help enhance the performance of the agricultural sector in the Nile riparian countries would improve crop production and minimize the need for agriculture expansion, which requires more water and land to be used.

Compared to other transboundary basins, the Nile Basin is considered a data-scarce basin. In many countries on the river, water cycle components are not observed and monitored regularly. The few climate and hydrological stations that exist suffer from serious gaps in their time series. Moreover, gaining access to these data is a challenge. Open source data, including remote sensing, provides a good alternative for this region. Cooperation between the riparian countries and sharing of information and data are critical issues for gaining better understanding and improving management of the basin system. Availability of a basin-wide and mutually recognized database would standardize policy mechanisms within the country and provide quantifiable evidence for transnational advocacy networks along the basin. The role of regional bodies such as NBI is imperative in this regard. Such partnerships build institutional strength across academia but also ensure proper information and sustainable solutions on the ground.

3. *Comprehensive framework for cooperation*

Fostering cooperation between the riparian countries of the Nile is important for the sustainability of water resources in the region. As stated earlier, economic gain in the basin can be maximized by more cooperation between these countries. An agreed framework for cooperation is highly needed. Cooperation between the Nile countries would change the position of Nile water from being one of the major sources of conflict in the region to become one of the most important pillars for a more sustainable future. Shifting from unilateral actions and development toward regional goals and benefit sharing between countries should be the goal of such a framework.

To reach such a comprehensive framework for cooperation it is important to have an open and transparent dialogue. A first step toward such a framework would be building trust between the riparian countries. Within this region, not only are the political agendas of the riparian countries in conflict but now international actors are bringing their own power dynamics to play through land-grabbing and private investment. The best option is to bring in an objective party, such as the World Bank, African Development Bank, or UN, that also has a financial power so that countries do not have to rely on international loans or investments to implement projects. Such institutions must have the power to not only bring expert negotiation skills with full knowledge of regional, hydrological, and ethnic dynamics but also set up financing schemes through partnerships and/or green bonds. This not only escalates water conflict and scarcity to a global agenda but ensures the timeliness and sustainability of the solution across the different dimensions and sectors.

To conclude, in order to ensure sustainability in the Nile Basin, there is a need to create an enabling environment for transboundary cooperation between the riparian countries. This can be achieved through a comprehensive framework that addresses not only water shares but also encourages transboundary cooperation beyond political boundaries, promotes benefit-sharing, and fosters effective communication and data sharing. Additionally, optimizing the use of the Nile's waters is very critical. Integrated management of the total water cycle, developing unconventional water sources, enhancing water use efficiency, and unlocking the untapped potentials for food and hydropower production would play an important role to achieve this goal. Looking at the interlinkages between different sectors (e.g., water-energy-food nexus) would be useful to identify synergies and trade-offs between these sectors, a knowledge that is useful to potential profits and minimize negative impacts. Until these ideas are realized, climate change, as well as demography and land use changes, will continue to intensify, making this transformation a more challenging and costly process.

REFERENCES

Abtew, W. and Dessu, S. B. (2019). *The Grand Ethiopian Renaissance Dam on the Blue Nile*. Berlin: Springer International Publishing AG, p. 173.

Abu-Zeid, M. A., and El-Shibini, F. Z. (1997). Egypt's High Aswan Dam. *Water Resources Development*, 13(2), pp. 209–217.

Adem, A., Tilahun, S. A., Ayana, E. K. et al. (2016). Climate Change Impact on Sediment Yield in the Upper Gilgel Abay Catchment, Blue Nile Basin, Ethiopia. In W. Abtew and A. M. Melesse, eds., *Landscape Dynamics, Soils and Hydrological Processes in Varied Climates*. Berlin: Springer International Publishing, pp. 615–642.

African Development Bank (2000). *Policy for Integrated Water Resources Management*.

Ahmed, A. A. (2004). Sediment Transport and Watershed Management Component, Friend/Nile Project, Khartoum.

Amdihun, A., Gebremaria, E., Rebelo, L., and Zeleke, G. (2014). Modelling Soil Erosion Dynamics in the Blue Nile (Abbay) Basin: A Landscape Approach. *Research Journal of Environmental Sciences*, 8(5), pp. 243–258. https://doi.org/10.3923/rjes.2014.243.258

Appelgren, B., Klohn, W. W., and Alam, U. (2000). *Water and Agriculture in the Nile Basin*. Rome: FAO.

Ashour, M. A., Aly, T. E., and Abueleyon, H. M. (2019). Transboundary Water Resources "A Comparative Study": The Lessons Learnt to Help Solve the Nile Basin Water Conflict. *Limnological Review*, 19(1), pp. 3–14. https://doi.org/10.2478/limre-2019-0001

Ashton, P. J. (2002). Avoiding Conflicts over Africa's Resources. *Royal Swedish Academy of Science*, 31(3), pp. 236–242.

Awulachew, S. B., Demissie, S. S., Ragas, F., Erkossa, T., and Peden, D. (2012). *Water Management Intervention Analysis in the Nile Basin*. IWMI.

Barnes, J. (2017). The Future of the Nile: Climate Change, Land Use, Infrastructure Management, and Treaty Negotiations in a Transboundary River Basin. *WIREs Climate Change*, 8(2). https://doi.org/10.1002/wcc.449

Basheer, M., Wheeler, K. G., Ribbe, L. et al. (2018). Quantifying and Evaluating the Impacts of Cooperation in Transboundary River Basins on the Water-Energy-Food Nexus: The Blue Nile Basin. *Science of The Total Environment*, 630(15), pp. 1309–1323.

Bastiaanssen, W., Karimi, P., Rebelo, L. et al. (2014). Earth Observation Based Assessment of the Water Production and Water Consumption of Nile Basin Agro-Ecosystems. *Remote Sens*, 6, pp. 10306–10334. https://doi.org/10.3390/rs61110306

Betrie, G. D., Mohamed, Y. A., van Griensven, A., and Srinivasan, R. (2011). Sediment Management Modelling in the Blue Nile Basin Using SWAT Model. *Hydrology and Earth System Sciences*, 15, pp. 807–818.

Blackmore, D. and Whittington, D. (2008). *Opportunities for Cooperative Water Resources Development on the Eastern Nile: Risks and Rewards. Nile Basin Initiative (NBI)*.

Brooks, N. (2006). Cultural Responses to Aridity in the Middle Holocene and Increased Social Complexity. *Quaternary International*, 151, pp. 29–49.

CCE News (2019). *Ethiopia Brings in Chinese Firms to Accelerate Mega Dam Construction*. Available at https://cceonlinenews.com/2019/02/21/ethiopia-brings-in-chinese-firms-to-accelerate-mega-dam-construction/

Coffel, E. D., Keith, B., Lesck, C. et al. (2019). Future Hot and Dry Years Worsen Nile Basin Water Scarcity Despite

Projected Precipitation Increases. *Earth's Future*, 7(8), pp. 967–977. https://doi.org/10.1029/2019EF001247

Connif, R. (2017). *The Vanishing Nile: A Great River Faces a Multitude of Threats*, Yale Environment 360. Available at www.e360.yale.edu/features/vanishing-nile-a-great-river-faces-a-multitude-of-threats-egypt-dam

Degefu, D. M. and He, W. (2015). Water Bankruptcy in the Mighty Nile River Basin. *Sustainable Water Resources Management*, 2(1), pp. 29–37.

Dessu, S. (2019). The Battle for the Nile with Egypt over Ethiopia's Grand Renaissance Dam Has Just Begun, Quartz Africa. Available at www.qz.com/africa/1559821/ethiopias-grand-renaissance-dam-battles-egypt-sudan-on-the-nile/

Environmental Protection Council (EPA) (2010). *Africa Review Report on Drought and Desertification.*

ESRI (2019). *GlobCover*. Available at www.due.esrin.esa.int/page_globcover.php

FAO (2005). *Irrigation in Africa in Figures.* AQUASTAT Survey 2005, Rome.

(2011). *Agricultural Water Use Projections in the Nile basin 2030: Comparison with the Food for Thought (F4T) Scenarios – Projections Reports.*

(2019a). *AQUASTAT*. Available at www.fao.org/nr/water/aquastat/data/query/index.html?lang=en

(2019b). *FAOSTAT*. Available at www.fao.org/faostat/en/#data/QC

(2020). *WorldClim*. Available at www.worldclim.org/version2

Fick, S. E. and Hijmans, R. J. (2017). Worldclim 2: New 1-km Spatial Resolution Climate Surfaces for Global Land Areas. *International Journal of Climatology*, 37(12), pp. 4302–4315. https://doi.org/10.1002/joc.5086

GADM (2018). GADM Database. Available at www.gadm.org/index.html

Gebremicael, T. G., Mohamed, Y. A., Betrie, G. D. et al. (2013). Trend Analysis of Runoff and Sediment Fluxes in the Upper Blue Nile Basin: A Combined Analysis of Statistical Tests, Physically Based Models, and Land Use Maps. *Journal of Hydrology*, 482, pp. 57–68. https://doi.org/10.1016/j.jhydrol.2012.12.023

Heinz, S. (1983). Sadd el-Ali, der Hochdamm von Assuan (Sadd el-Ali, the High Dam of Aswan). *Geowissenschaften in Unserer Zeit (in German)*, 1(2), pp. 51–85.

IEA (2015). Energy Atlas. Available at www.energyatlas.iea.org/#!/tellmap/-1118783123/0

Kaluarachchi, J. J. and Kim, U. (2009). Climate Change Impacts on Water Resources in the Upper Blue Nile River Basin, Ethiopia. *Journal of the American Water Resources Association*, 45(6), pp. 1361–1378.

Kaluarachchi, J. J., Kim, U., and Smakhtin, V. U. (2008). *Climate Change Impacts on Hydrology and Water Resources of the Upper Blue Nile River Basin, Ethiopia.* Colombo: International Water Management Institute, p. 27.

Keith, B., Enos, J., Cadets, G. B. et al. (2013). Limits to Population Growth and Water Resource Adequacy in the Nile River Basin, 1994–2100. In *Proceedings of the 31st International Conference of the System Dynamics Society*, Cambridge.

Kimenyi, M. S. and Mbaku, J. M. (2015). *The Limits of the New "Nile Agreement."* Available at www.brookings.edu/blog/africa-in-focus/2015/04/28/the-limits-of-the-new-nile-agreement/

Kumagai, J. (2016). The Grand Ethiopian Renaissance Dam Gets Set to Open, IEEE Spectrum. Available at https://spectrum.ieee.org/energy/policy/the-grand-ethiopian-renaissance-dam-gets-set-to-open

Lazarus, S. (2018). Is Ethiopia Taking Control of the River Nile? CNN. Available at www.cnn.com/2018/10/19/africa/ethiopia-new-dam-threatens-egypts-water/index.html

Li, R. (2005). Transboundary Water Conflicts in the Nile Basin. *Water Encyclopedia.* https://doi.org/10.1002/047147844X.wr152

Mark, J. (2014). *Nile*. Ancient History Encyclopedia.

McCartney, M. P., Alemayehu, T., Easton, Z. M. et al. (2012). Simulating Current and Future Water Resources Development in the Blue Nile River Basin. In S. B. Awulachew, V. Smakhtin, D. Molden, and D. Peden, eds., *The Nile River Basin: Water, Agriculture, Governance and Livelihoods*, pp. 269–291. Abingdon: Routledge – Earthscan.

Multsch, S., Elshamy, M. E., Batarseh, S. et al. (2017). Improving Irrigation Efficiency Will Be Insufficient to Meet Future Water Demand in the Nile Basin. *Journal of Hydrology: Regional Studies*, 12, pp. 315–330.

Nile Basin Initiative (NBI) (2012). *State of the River Nile Basin Report.*

Oestigaard, T. (2012). Water Scarcity and Food Security along the Nile: Politics, Population Increase and Climate Change. *Current African Issues*, 49, Uppsala: Nordiska Afrikainstitutet, Uppsala.

Onencan, A., Enserink, B., Van de Walle, B., and Chelang, J. (2016). Coupling Nile Basin 2050 Scenarios with the IPCC 2100 Projections for Climate-Induced Risk Reduction. *Procedia Engineering*, 159, pp. 357–365.

Rahman, M. (2012). Water Security: Ethiopia–Egypt Transboundary Challenges over the Nile River Basin. *Journal of Asian and African Studies*, 48(1), pp. 35–46. https://doi.org/10.1177/0021909612438517

Rullia, M. C., Savioria, A., and D'Odorico, P. (2013). Global Land and Water Grabbing. *Proceedings of the National Academy of Sciences of the United States of America (PNAS)*, 110(3), pp. 892–897. https://doi.org/10.1073/pnas.1213163110

Salman, M. A. S. (2013). The Nile Basin Cooperative Framework Agreement: A Peacefully Unfolding African Spring? *Water International*, 38(1), pp. 17–29. https://doi.org/10.1080/02508060.2013.744273

Setegn, S. G., Rayner, D., Melesse, A. M. et al. (2011). Climate Change Impact on Agricultural Water Resources Variability in the Northern Highlands of Ethiopia. In A. M. Melesse, ed., *Nile River Basin*. Dordrecht: Springer.

Shahin, M. (1993). An Overview of Reservoir Sedimentation in Some African River Basins. In *Proceedings of Sediment Problems: Strategies for Monitoring, Prediction and Control*. Yokohama: LAHS Publications, pp. 93–100.

Sutcliffe, J. V. and Parks, Y. P. (1999). *The Hydrology of the Nile*. Oxfordshire: The International Association of Hydrological Science (IAHS).

United Nations, Department of Economic and Social Affairs, Population Division (DESA) (2019). *World Population Prospects 2019, Online Edition. Rev. 1*. Available at www.population.un.org/wpp/Download/Standard/Population/

United Nations Economic and Social Council. (2007) *Africa Review Report on Drought and Desertification*. Available at www.un.org/esa/sustdev/csd/csd16/rim/eca_bg3.pdf

Wheeler, K. G., Basheer, M., Mekonnen, Z. et al. (2016). Cooperative Filling Approaches for the Grand Ethiopian Renaissance Dam. *Water International*, 41(4), pp. 611–634. https://doi.org/10.1080/02508060.2016.1177698

Williams, M. A. J. (2009). Human Impact on the Nile Basin: Past, Present, Future. In H. J. Dumont, ed., *The Nile*. Dordrecht: Springer.

World Bank (2019). *World Bank Data*. Available at www.data.worldbank.org/indicator

Worqlul, A. W., Dile, Y. T., Ayana, E. K. et al. (2018). Impact of Climate Change on Streamflow Hydrology in Headwater Catchments of the Upper Blue Nile Basin, Ethiopia. *Water*, 10, p. 120. https://doi.org/0.3390/w10020120

Yihdego, Y., Khalil, A., and Salem, H. S. (2017). Nile River's Basin Dispute: Perspectives of the Grand Ethiopian Renaissance Dam (GERD). *Global Journal of Environmental Science and Management*, 17(2), pp. 1–21.

Yletyinen, J. (2009). Holocene Climate Variability and Cultural Changes at River Nile and Its Saharan Surroundings. Stockholm: Stockholm University, Department of Natural Geography. Available at www.diva-portal.org/smash/get/diva2:400169/FULLTEXT01.pdf

Yoffe S. (2002). Basins at Risk: Conflict and Cooperation over International Freshwater Resources. Ph.D. Dissertation, Oregon State University, Corvallis. p. 79.

8 The Euphrates–Tigris River Basin

Aysegül Kibaroglu

8.1 INTRODUCTION

The Euphrates and Tigris, two mighty rivers of the historical region of Mesopotamia, witnessed the rise and decline of pioneering civilizations. For millennia, millions of hectares of land had been irrigated in lower Mesopotamia (Iraq), which provided food for growing societies but also caused extensive salinization, waterlogging, and desertification. Moreover, in the modern era, Iraq, Turkey, Syria, and Iran developed the river system by building a series of large-scale dams, which responded greatly to the energy, food, and drinking water needs of their growing economies but also caused drastic changes in the hydrology of the river system, with significant environmental and social impacts.

The water question emerged on the international agenda in the Euphrates–Tigris (ET) basin when the three riparian nations, namely Iraq, Syria, and Turkey, initiated major water and land resources development projects in the late 1960s. As national water development ventures progressed, mismatches between water supply and demand occurred throughout the river basin. The political linkages established between transboundary water issues and non-riparian security issues also exacerbated disagreements over water sharing and allocation. In 1987 and 1990 two bilateral water sharing protocols on the Euphrates – acknowledged by all riparian states as being interim agreements – were signed following several high-level meetings of top officials in the basin. However, these bilateral accords failed to include basic components of sustainable water resources management, namely water quality management, environmental protection, and stakeholder engagement.

The ad hoc technical negotiations were unable to prepare the ground for a comprehensive treaty on equitable and effective transboundary water management. In the early 1980s, the Euphrates–Tigris basin countries managed to build an institutional framework, namely the Joint Technical Committee (JTC); however, they did not empower it with a clear and jointly agreed mandate. Instead, the riparian countries continued unilateral and uncoordinated water and land development ventures. The JTC meetings did not make an effective contribution to settlement of the transboundary water dispute nor provide a platform for delineating the priorities and needs of riparian nations as a basis for addressing regional water problems.

On top of this, climate change impacts add to the complexity, as the basin is one of the most affected regions. Scientific findings underline significant decreases in the Tigris and Euphrates flows. Surface flow decreases will have important impacts on the basin: water for irrigation, energy production, and domestic and industrial use will decline drastically, which, in turn, will precipitate conflicts at local and national levels.

A strategic look at sustainability underscores that adopting a water–food–energy nexus approach in the ET basin is important because there are various pressures on the river system due to population growth, agricultural needs, hydropower development, and ecosystem mismanagement. Hence, it is argued that transboundary institutions should apply the nexus approach, which helps to identify key development drivers as well as to unpack and clarify the development challenges and necessary tradeoffs in the basin.

Sustainability of water resources requires stability, cooperation, and peace. Sub-state level conflicts and illegal control of water resources and water infrastructure in the basin deprive people of access to sufficient clean water, energy, and food resources in Syria and Iraq. The prerequisites for establishing or restoring sustainability in a river basin include stability as well as establishing participatory, transparent, inclusive, and accountable governance structures (Figure 8.1).

8.2 THE RIVER BASIN TODAY

8.2.1 Geography, Climate, and Hydrology

The two greatest rivers of the Eurasian landscape, namely the Euphrates and the Tigris, originate in a climatic and topographic zone (i.e., Turkey, Iraq, and Iran) and end up in quite a different one (i.e., the Persian Gulf). The basin is characterized by high mountains to the north and west and extensive lowlands in the south and the east. They begin, scarcely 30 kilometers (km) apart from each other, in a relatively cool and humid zone with rugged 3,000 meter (m) high mountains, visited by autumn and spring rains and winter snows.

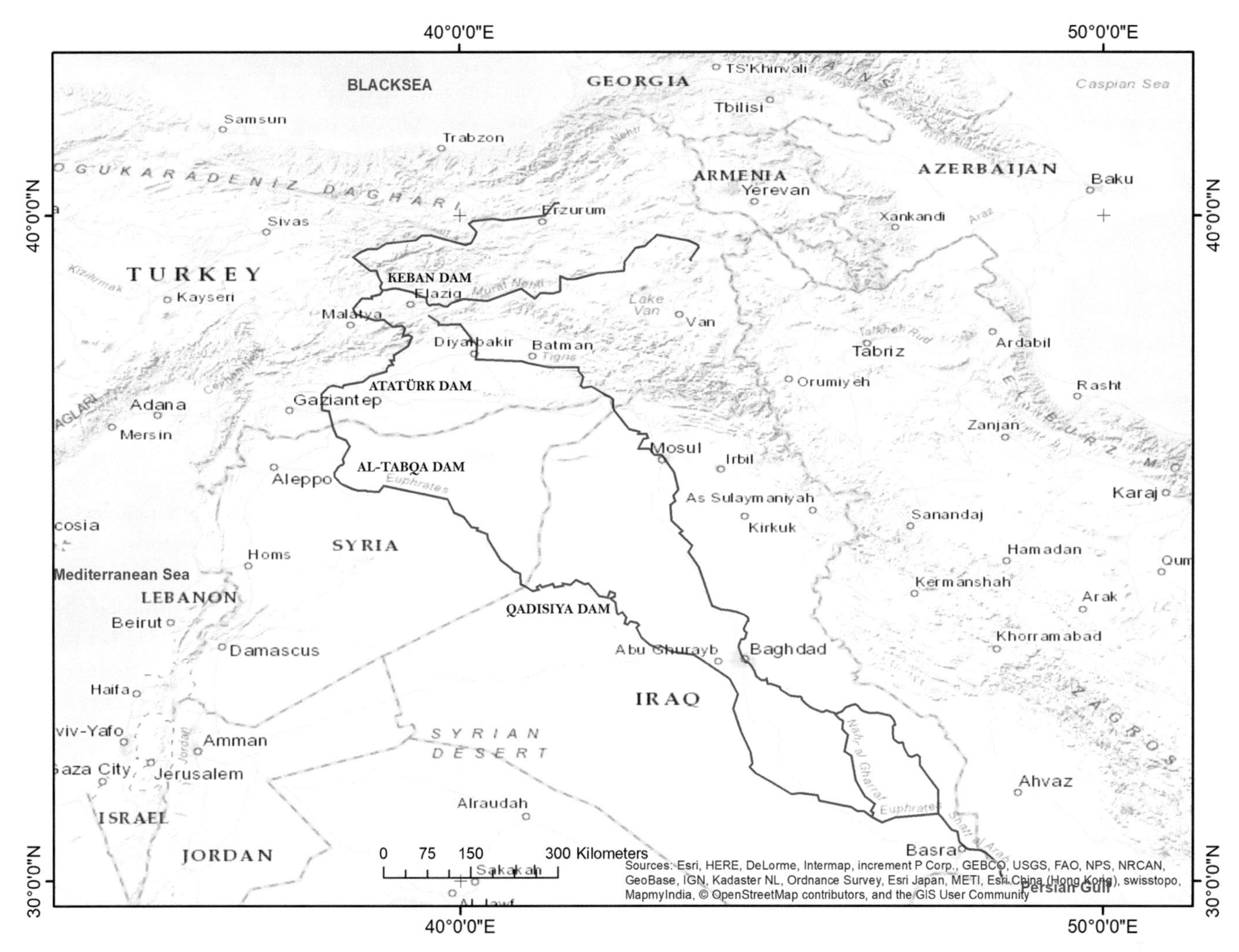

Figure 8.1 The Euphrates–Tigris River Basin.
Map created by Orkan Ozcan (2017) using the data listed within the figure

From there, the two rivers run separately onto a wide, flat, hot, and poorly drained plain. They continue more tranquilly through the plateaus of northern Syria and Iraq, where they cut deep beds in rocks so that their courses have remained stable over the millennia. In their middle courses, they diverge hundreds of kilometers apart, only to meet again near the end of their journey and discharge together into the Persian Gulf. In that section they descend from elevations of about 400 m to little over 50 m above sea level where they enter the alluvial plain. The great alluvium-filled depression containing lakes and swamps, Shatt Al-Arab, has a length of 180 km and constitutes the combined delta of the ET river basin (Kibaroglu, 2002).

In conformity with the expert judgments of geographers, the Euphrates and the Tigris rivers can be considered as forming one single transboundary watercourse system. They are linked not only by their natural course when merging at the Shatt Al-Arab but also as a result of the manmade Thartar Canal, which links the Tigris to the Euphrates through the Thartar Valley in Iraq (Anderson, 1986; Beaumont, 1992; Bilen, 1994; Kliot, 1994).

The Euphrates originates in the eastern highlands of Turkey, between Lake Van and the Black Sea, and is formed by two major tributaries, the Murat and the Karasu. It enters Syrian territory at Karkamis, downstream from the Turkish town of Birecik. It is joined by its major tributaries, the Balik and Khabur, which also originate in Turkey, and flows southeast across the Syrian plateaus before entering Iraqi territory near Qusaybah.

Of the Euphrates Basin, 28 percent lies in Turkey, 17 percent in Syria, 40 percent in Iraq, 15 percent in Saudi Arabia, and just 0.03 percent in Jordan. The Euphrates river is 3,000 km long, divided between Turkey (1,230 km), Syria (710 km), and Iraq (1,060 km) (FAO, 2008).

The Tigris, also originating in eastern Turkey, flows through the country until the border city of Cizre. From there it forms the border between Turkey and Syria over a short distance and then crosses into Iraq at Faysh Khabur. The Tigris river is 1,850 km long, with 400 km in Turkey, 32 km on the border between

Turkey and Syria, and 1,418 km in Iraq. Of the Tigris Basin, 12 percent lies in Turkey, 0.2 percent in Syria, 54 percent in Iraq, and 34 percent in Iran. Within Iraq, several tributaries flow into the river coming from the Zagros Mountains in the east, thus all on its left bank (FAO, 2008).

The Shatt Al-Arab is formed by the confluence downstream of the Euphrates and the Tigris and it flows into the Persian Gulf after a course of 180 km. The Karun River, originating in Iranian territory, has a mean annual flow of 24.7 billion cubic meters (BCM) and flows into the Shatt Al-Arab just before it reaches the sea, bringing a large amount of freshwater.

The mean annual flow of the Euphrates is 32 BCM/year, of which about 90 percent is drained from Turkey, whereas the remaining 10 percent originates in Syria. As for the Tigris, the average total discharge is determined as 52 BCM/year, of which approximately 40 percent comes from Turkey, whereas Iraq and Iran contribute 51 percent and 9 percent, respectively. Estimates for the total flow of the Tigris–Euphrates and their tributaries vary between 68 BCM and 84.5 BCM (Kolars and Mitchell, 1991; Kolars, 1994; Belül, 1996; Altinbilek, 2004).

The upper parts of the ET basin have features of a cold continental climate, whereas the lower parts are classified as hot desert or hot semi-arid (Bozkurt and Sen, 2013). The rivers overflow in spring when the snow melts, augmented by seasonal rainfall, which is at its heaviest between March and May. The summer season is hot and dry, resulting in extensive evaporation and low humidity during the day. Evaporation increases water salinization and water loss in major reservoirs in the three riparian countries (Naff and Matson, 1984; Kliot, 1994).

From their headwaters to the confluence, the discharge patterns of both the Euphrates and Tigris exhibit remarkable dynamics that reflect contribution of runoff from hydrologically different regions, diversion of irrigation water, natural distributaries in the lower parts of the basin, and, more recently, the attenuation of hydrological regimes due to embankment dams that have been built in the main streams and some of their tributaries (Cullmann, 2013). Historically, runoff in the rivers had been characterized by pronounced spring floods. These floods had both positive and negative implications. They fertilized agricultural land with sediments from the upper catchment. They also threatened agriculture and human livelihood whenever flood seasons produced extreme events. After the first large dams were finished, hydrologic dynamics changed drastically. For an integral assessment of the potential benefit of water management it would be necessary to conduct a multi-objective optimization exercise that integrated all existing infrastructure in the basin (Cullmann, 2013).

Kavvas et al. (2011) consider the ET basin as a single hydrologic unit and performed a scientific assessment of its water resources. An inventory of land use/land cover, vegetation, soils, and existing hydraulic structures in the river basin was performed; a regional hydroclimate model of the river basin was developed; and a hydrologic model was developed to route streamflows within the river network of the basin. An algorithm for operating the reservoirs in the basin was developed and utilized to perform dynamic water balance studies under various water supply/demand scenarios to establish efficient utilization of water resources to meet the water demands of the riparian countries in the basin.

8.2.2 Population and Socioeconomy

The waters of the Euphrates and Tigris stand to be significant and strategic for the major riparians: Iraq derives most of its fresh water from the two rivers. Although the Euphrates basin is one of seven river basins in Syria, it is strategically the most important because of its existing and potential uses for agricultural and hydropower purposes. The ET basin is one of twenty-five basins in Turkey but accounts for nearly one third of the country's surface water resources and one fifth of its irrigable land. Hence, when "population" is considered one should not focus only on the population living in the basin but should also consider population projections and growth rates at the country level simply because benefits from water resources development in the basin, that is to say, hydropower generation and agricultural development, turn out to be benefits created for most of the nation.

From the 1940s to the present day the population in all the riparian countries has increased by three- to five-fold, and by 2025, in the cases of Iraq and Syria, they are likely to double again. Even in Turkey almost 2 percent annual growth rate is postulated, especially in southeast (ET basin) Turkey, where some provinces (e.g., Diyarbakir and Sanliurfa) are expanding at the fastest rate.

On the other hand, a survey by the ESCWA-BGR calculates that "the overall ET basin is home to around 54 million people in Iran, Iraq, Syria, and Turkey" (UN-ESCWA and BGR, 2013). The Euphrates Basin has an estimated population of about 23 million, of which 44 percent (10.2 million) live in Iraq, 25 percent (5.69 million) in Syria, and 31 percent (7.15 million) in Turkey.[1] The Tigris Basin comprises a total population of approximately 23.4 million inhabitants, of which more than 18 million live in Iraq, 1.5 million in Iran, and 3.5 million in Turkey. Only 50,000 people reside in the Syrian part of the basin.[2]

Following struggles for independence and liberation, and nation-building efforts in the first half of the twentieth century, the major riparians of the ET basin further consolidated their regimes after the 1960s, and they paid more systematic attention to socioeconomic development based on large scale water and land resources development, which eventually brought them at odds with each other in the regional context.

Turkey had long been dependent on oil imports. Having been hard hit by the oil crises of the 1970s, the government embarked on a program of indigenous resource development, with emphasis on hydropower, with the aim of minimizing the national economy's dependency on imported oil. In this context, Turkey implemented the Lower Euphrates Project, initially a series of dams designed to increase hydropower generation and expand irrigated agriculture. Subsequently, in the late 1970s, the Lower Euphrates Project evolved into a larger, multi-sectoral development project, taking in the Tigris waters as well and known as the Southeastern Anatolia Project (GAP, its Turkish acronym), which included twenty-one large dams, nineteen hydropower plants, and irrigation schemes extending to 1.7 million hectares of land.

The Syrian economy has traditionally been dominated by agriculture. Exploration for oil did not begin until the early 1980s. Even though oil made a significant contribution to export earnings in the following decades as world oil prices fluctuated, Syria focused on agricultural development with the aim of achieving food self-sufficiency. These considerations were reinforced by political goals, which, under the ruling Ba'ath Party, placed the emphasis on the development of rural areas and the organization of peasants as a political power base. Hence, in this context, Syria launched the Euphrates Valley Project under the Ba'ath Party. The government set a number of objectives to be met by the project: irrigation of an area as large as 640,000 hectares, construction of the large, multi-purpose Tabqa or Al-Thawra Dam to generate the electricity needed for urban use and industrial development, and regulation of the flow of the Euphrates to prevent seasonal flooding (Meliczek, 1987). At that time, Syria focused almost exclusively on the Euphrates, prioritizing the completion of the Euphrates Valley Project. As Syrian technocrats eventually encountered technical and social difficulties in reclaiming land in the Euphrates Valley, their attention turned to north-eastern Syria in the late 1990s, where it was possible to expand the amount of irrigated land with the waters to be pumped from the Tigris.

Since 1958, Iraq has changed from being mainly an agricultural country exporting wheat, rice, and other crops to an oil-producing, semi-industrial nation forced to import most of its own food. Yet, after the Iraqi government nationalized oil companies in 1972 and began to receive more income from oil, the focus also turned to agricultural production. This led to an expansion of irrigated areas, with the aim of achieving food security for the Iraqi people. The Ba'ath Party, which came to power under Saddam Hussein's presidency in 1968, adopted the slogan "food security for the Iraqi people," which was to be accomplished through the development of irrigation. To that end, the "Revolutionary Plan" was developed. The Higher Agriculture Council, attached to the presidency; the Soil and Land Reclamation Organization, attached to the Ministry of Irrigation; and many other new departments were established to carry out studies, create designs, and construct and maintain water projects.

In the contemporary context, Turkey's largely free-market economy has been growing steadily (9 percent with $789.3 billion GDP in 2012) despite a drop in the last couple of years. On the other hand, its dependence on imported oil and gas continues at enormous rates (97 percent). Thus, water resources development for hydropower and other purposes, particularly in the ET basin, still stands as a strategic objective.

Despite severe domestic security and structural problems, the Iraqi economy has grown significantly (8.4 percent with $209.6 billion in 2012) in the last decade. Thus, Iraqi (both central and regional-KRG) governments are eager to continue water resources development in the ET basin. Despite modest economic growth ($64.7 billion in 2011) and reform prior to the outbreak of unrest, Syria's economy continues to suffer the effects of the ongoing conflict that began in 2011. The economy further contracted in 2012 because of international sanctions and reduced domestic consumption and production, and inflation has risen sharply. Hence, it is expected that once peace is restored in Syria urgent attention will need to be paid to the efforts of reconstruction and rehabilitation of existing domestic water supply and sewage systems as well as irrigation and energy infrastructure, possibly delaying further expansion and development in the ET basin.

8.2.3 Groundwater

In riparian countries, groundwater is mainly used for irrigation and is supplied from wells, which far exceed the officially approved number (Kibaroglu et al., 2005). Turkey and Syria share a transboundary groundwater resource system, namely the Ceylanpinar aquifer and Ras El Ain karstic springs, which are found in the Urfa-Harran and the Ceylanpinar plains in south-eastern Turkey and in the Lower Balikh and Lower Khabour basins in Northern Syria. The Ceylanpinar aquifer is a karstic carbonate aquifer. The Khabour river, a tributary of the Euphrates, is fed by the Ras El Ain springs, which receive their main discharge from groundwater resources in the Ceylanpinar–Harran–Sanliurfa plains in Turkey. The Ras El Ain karstic springs have an average discharge of 38.66 m^3/s. In other words, precipitation that falls in Turkey is a major source of the aquifer's recharge. According to a World Bank study, "rapid groundwater extraction both in Turkey and in Syria from the transboundary aquifer system (i.e., Ras El Ain) has reduced the spring flow discharges to the (Khabour) river ... Overuse has caused a decline of the flow rate from a long-term average of 50 m^3/s to a few m^3/s at present and down to zero during drought years, as in 2000" (World Bank, 2001).

8.2.4 Water Development Projects

The initiation of large-scale water development projects triggered transboundary water disputes when, in the late 1960s, Turkey and Syria started systematically building dams and irrigation systems in the ET basin (Beaumont, 1992). Iraq was also keen to expand its irrigation schemes (Kibaroglu et al., 2005). In this context, Turkey's GAP Project, Syria's Euphrates Valley Project, and Iraq's Thartar Canal Project were implemented without proper coordination.

The ET basin includes almost one-third of the surface water supply in Turkey. An integrated water and land resources development program was designed with the particular aim of irrigating the fertile lands in south-eastern Anatolia, about one-fifth of the country's irrigable land (Kolars and Mitchell, 1991; Unver, 1997; Tigrek and Kibaroglu, 2011). For this purpose, Turkey implemented the GAP, which included twenty-two large dams, nineteen hydropower plants, and several irrigation schemes (Nippon and Yuksel, 1989). When fully developed, the GAP will provide irrigation for 1.7 million hectares of land, 571,591 hectares of which are now operational (GAP Administration, 2019). Although those water development structures have brought tangible economic and social gains, the GAP, and the construction of large dams, have come in for sharp criticism. Objections concern resettlement issues, environmental and cultural aspects, and the implications of sharing water with Syria and Iraq. Driven by a top-down policy process that failed to fully appreciate the complexities behind the region's socioeconomic underdevelopment, the GAP mostly achieved its technical objectives in terms of hydropower generation and agricultural productivity, but fell short of achieving its social agenda (Harris, 2010; Sayan and Kibaroglu, 2016).

The waters and land resources of the Euphrates river basin are considered of strategic importance in Syria as they include 65 percent of the surface water supply and 27 percent of total land resources. Before oil exploration in the early 1980s, agriculture was the main driver of the Syrian economy. Though oil contributed significantly to export earnings, fluctuations in world oil prices urged Syria to keep guarding its agricultural development for food self-sufficiency (Richards and Waterbury, 1990). So, Syria implemented the Euphrates Valley Project in the early 1960s. A major dam, the Tabqa Dam, was completed in 1973. In the framework of the Euphrates Valley Project, Syria aimed to irrigate 640,000 hectares of land, produce electricity for growing cities and industry, and control flooding (Meliczek, 1987; Bakour, 1992; Wakil, 1993). But these objectives have been only partly realized in the Euphrates basin in Syria over the course of five decades.

The Euphrates and Tigris Rivers provide vital freshwater supply in Iraq. Construction of dams on the Euphrates and the Tigris began in the 1950s with the basic aim of flood prevention. The Samarra Dam was completed on the Tigris in 1954, and the Euphrates Dam in 1956. In 1958, Iraq became a semi-industrialized economy, with large amounts of oil exports and food imports. In the 1970s, the Iraqi government expanded irrigated areas with the aim of providing food security, which was made possible by growing income from oil after the oil companies were nationalized in 1972 (Allan, 1990; McLachlan, 1991). In the 1980s, Iraq constructed a complex network of canals on the ET system, such as the Tharthar Canal. Taking into consideration the constraints of water salinity in the Tharthar Reservoir and the amount of water that can be saved and transferred from the Tigris to the Tharthar, it may be assumed that about 6 BCM of water could be transferred annually from the Tharthar reservoir to the Euphrates River (Cullmann, 2013). Yet, Tharthar Lake alone is presently responsible for more than 50 percent of evaporative losses from Iraq's reservoirs. To better conserve water resources, Tharthar Lake will most likely be used solely for flood control purposes in the future (Tice, 2016).

8.2.5 Transboundary Water Issues

Due to the competitive and uncoordinated nature of these water development projects in Iraq, Syria, and Turkey, disagreements over transboundary water issues surfaced in the late 1960s. The first political crisis occurred in the region in 1975. Turkey began impounding the Keban reservoir while Syria was completing the construction of the Tabqa Dam – during a period of severe drought. The impounding of the two reservoirs triggered a crisis in the spring. Iraq accused Syria of reducing the river's flow to intolerable levels, while Syria blamed Turkey. The Iraqi government was not satisfied with the Syrian response, and the mounting frustration resulted in mutual threats that brought the parties to the brink of armed hostility. A war over water was averted when, thanks to Saudi Arabia's mediation, Syria released additional quantities of water to Iraq. The main cause of this crisis was the mounting political rivalry and tension between the two Ba'athist regimes. In other words, it was not a water-sharing crisis per se, but rather the beginning of the use of water as a political lever in non-riparian issues (Kibaroglu, 2002).

From the 1980s to the late 1990s, transboundary water issues moved into the realm of high politics when non-water issues became decisive factors that led to greater tensions and disputes. Bilateral relations between Turkey and Syria have long been uneasy. Two principal sources of friction were Syria's extensive logistical support to the separatist organisation, the Kurdistan Workers' Party (PKK), and Syrian irredentist claims to the province of Hatay in Turkey. A major diplomatic crisis in the basin took place during the impounding of the Ataturk dam in Turkey. On 13 January 1990, Turkey temporarily interrupted the flow of the Euphrates River to fill the Ataturk reservoir. The decision to fill the reservoir over a period of one month was taken much earlier. Turkey had notified its downstream neighbours by November 1989 of the pending event. In its note, Turkey explained the technical reasons and provided a detailed program for making up the losses. However, the Syrian and Iraqi governments protested officially to Turkey, and consequently called for an agreement to share the waters of the Euphrates, as well as a reduction in the impounding period (Kibaroglu, 2002).

TRANSBOUNDARY WATER NEGOTIATIONS

Ad hoc technical negotiations were held along with recurrent diplomatic crises (Kibaroglu and Unver, 2000). The main theme of these technical negotiations was the impact of the construction of the Keban Dam in Turkey and the Tabqa Dam in Syria on Iraq's historical water use patterns. While Turkey proposed the establishment of a Joint Technical Committee (JTC) to determine the water and irrigation needs of the riparians, Iraq insisted on a guarantee of specific flows and a water-sharing agreement. Though Turkey released certain flows during the construction and impounding of its Keban Dam, no final allocation agreement was reached even after numerous technical meetings (Gurun, 1994). Throughout this period, transboundary water issues were regarded by each country's political leadership as falling between economic and technical objectives, which could be handled by official technical delegations. Water negotiations were held, therefore, by technocrats from the riparians' central water agencies, accompanied by diplomats who advised on and monitored the negotiations, particularly when international legal and political aspects were under discussion (Kibaroglu and Scheumann, 2013). The early technical negotiation processes, however, were unable to prepare the ground for a comprehensive treaty on equitable and effective transboundary water management. And so, even though a hot conflict over water did not occur in the basin, a true institution-based cooperation framework failed to come to fruition.

Yet, over the course of the negotiations, the three major riparians had diverging opinions on transboundary water rights. Syria and Iraq acted together in claiming their historical rights. They also insisted on sharing the Euphrates River based on an arithmetical formula. Turkey, on the other hand, claimed that the waters of the ET basin could only be allocated on the basis of objective calculations of the riparians' needs, and so Turkey put forward the Three-Stage Plan, encompassing studies and examinations of water and land resources in the ET basin and evaluating the riparians' needs according to these studies as the third stage (Turkish Ministry of Foreign Affairs, 1996).

According to Scheumann (1998), the positions of the upstream riparian (Turkey) and downstream riparians (Syria and Iraq) were largely shaped by the water and land resources development plans. That is to say, while Syria and Iraq tried to protect their existing water uses and resisted any change to the flow of the rivers, Turkey emphasized its increasing needs from the rivers and the necessity of developing new water structures.

TRANSBOUNDARY WATER INSTITUTIONS

By the late 1970s, transboundary water relations in the ET basin had become competitive and complex. The accelerated development of the Euphrates by Turkey and Syria caused significant anxiety on the Iraqi side. With water development projects in the ET basin progressing rapidly, there arose a need to establish regular contacts and technical information exchange. In this context, Iraq proposed establishing a permanent technical institution, the JTC, which was established at the first meeting of the Joint Economic Commission between Turkey and Iraq in 1980. With Syrian participation in 1983, it became a trilateral body, whose mandate was to determine ways and means of producing a formula for reasonable and equitable utilization of the ET basin waters (Kibaroglu, 2002).

But in 1993, after sixteen unproductive meetings, the JTC was suspended. Analysis of the minutes of its meetings shows that the JTC could not make any progress due to the riparians' diverging positions on the scope and aim of the negotiations. While Turkey insisted that the negotiations should comprise the entire ET basin as a single river basin, Iraq and Syria maintained that talks should focus on the Euphrates. A common understanding of the aim of the negotiations was also lacking. While Turkey proposed a trilateral plan (the Three-Stage Plan) for determining the "utilization of transboundary watercourses," Iraq and Syria strongly insisted on reaching a "sharing formula" for the "international river." Iraq and Syria considered the Euphrates an international river and suggested that an instant sharing treaty be concluded, based on the demand declared by each country. Turkey, however, considered the Euphrates and the Tigris as constituting a single transboundary river basin, whose waters should be allocated according to objective needs (Kirschner and Tiroch, 2012).

With its flawed structure and functioning, the JTC failed to create a proper setting for fruitful discussions over the riparians' prime concerns and requirements as a basis for addressing regional water problems. In fact, the riparian states did not share any information or experience on their water use and management practices. No progressive exchanges took place over how legislative and institutional structures were to be harmonized. Contrary to the expectations of the institutionalists, the JTC did not become a medium for transboundary Integrated Water Resources Management. One should also add that JTC meetings were closely related to the overall political relations in the region at the time and that the overarching Cold War framework with its tense political atmosphere had a negative impact on the performance of the JTC (Kibaroglu and Scheumann, 2013).

Even though historical bilateral treaties, which were concluded between the young modern Turkish Republic and its neighbours, included some clauses on water usage and development (Kirschner and Tiroch, 2012), including the comprehensive 1946 Protocol between Turkey and Iraq (Kibaroglu, 2002), the current legal framework for transboundary water governance in the ET basin is basically bound by the 1987 Turkey–Syria and 1990 Syria–Iraq bilateral protocols as legally binding instruments of international law (Kibaroglu et al., 2005).

Turkey and Syria signed the Protocol on Economic Cooperation in 1987. It contains, among other things, provisions related to the allocation of the waters of the Euphrates River. Turkey guaranteed to release 500 m^3 of water per second from the Euphrates, with deficiencies in any month to be compensated the next month (Article 6). It was also agreed that Turkey and Syria would invite Iraq to reach an agreement to allocate the waters of the rivers Euphrates and Tigris in the shortest possible time (Article 7). Article 8 sets forth that the two sides agreed to step up the work of the Joint Technical Committee on Regional Waters. Both states also agreed to build and jointly operate irrigation and hydroelectric power projects (Article 9).

In 1989, Turkey had to interrupt the flow of the Euphrates for some weeks when the Atatürk Dam reservoir was filled. This caused anxieties on the Syrian and Iraqi sides. They agreed to determine their bilateral shares from the Euphrates before such interruptions occurred again as the GAP progressed. At the 13th meeting of the JTC in Baghdad, therefore, a bilateral agreement between Syria and Iraq was signed (on April 16, 1990), according to which 58 percent of the Euphrates waters coming from Turkey would be released to Iraq by Syria (Law No. 14 of 1990). The protocol stipulates: "The contingent of water to Iraq passing through the Syrian-Iraqi border is to be a permanent annual total rate of 58 per cent of the river water passing into Syria at the Syrian-Turkish border. The Syrian contingent of the river waters is to be the rest of the waters, totalling 42 percent of the waters passing through the Syrian-Turkish border."

Both protocols are bilateral and pertain only to sharing the waters of the Euphrates River. They do not provide any conditions for efficient and equitable use and management of transboundary water resources in the ET basin. By focusing narrowly on water quantity issues, both protocols fall short of adopting Integrated Water Resources Management. Furthermore, they lack institutional mechanisms for overseeing the implementation of their provisions. The protocols are inadequate for addressing variability in the flow of the Euphrates River. Droughts and floods, which often happen in the basin, produce substantial changes in the river flow regime, but the protocols do not include clauses providing for adjustments under the impact of climate change (Kibaroglu and Scheumann, 2013).

In 2008 and 2009, the governments of Turkey, Syria, and Iraq embarked on cooperative foreign policy initiatives. Cooperative initiatives related to transboundary waters were agreed by signing a series of bilateral memoranda of understanding (MOUs) on the protection of the environment, water quality management, water efficiency, drought management, and flood protection with a view to addressing the adverse effects of climate change.

In this context, Turkey and Iraq signed a protocol on water in 2009 (Memorandum of Understanding, 2009a), covering issues such as sharing hydrological and meteorological data; efficient use and management of regional waters; appraisal of water resources that are under stress due to increasing water use and climate change; harmonization of existing hydrological measurement facilities; modernization of existing irrigation systems; avoidance of losses in the domestic water sector; building water supply and water treatment infrastructure in Iraq with the involvement of Turkish companies; and joint investigation, planning, and implementation of flood control and drought management. The protocol demonstrates that the authorities concerned emphasized relevant aspects of good transboundary water governance rather than insisting on corresponding water rights.

Turkey and Syria signed four protocols involving the waters of the Euphrates, Tigris, and Orontes Rivers. These protocols encompass issues such as jointly building a dam on the border where the Orontes passes from Syria into Turkey, utilization of water by Syria where the Tigris River is the border between Turkey and Syria, drought management, efficient water management, improved water quality management, and protection of the environment (Memorandum of Understanding, 2009b; Memorandum of Understanding, 2009c). In contrast with the 1987 protocol, which concentrated on sharing of the Euphrates waters, these MOUs emphasized the patterns and levels of water development, use, and management and dealt particularly with drought management and environmental protection.

These bilateral MOUs could not be put into practice due to regional instability and increased political tensions between the riparian states. The MOUs also faced the ever-present challenges of incompatibilities in national, institutional, and legal frameworks; complex national water management systems; and uncoordinated water management practices among the basin countries. The existing water protocols, therefore, can only be properly implemented when the riparians' institutional capacities are upgraded and harmonized in more conducive political circumstances (Kibaroglu, 2014).

8.3 THE RIVER BASIN TOMORROW

8.3.1 Climate Change

Future climate change projections indicate substantial reductions in the runoff of the Tigris and Euphrates rivers. According to a high emissions scenario (SRES A2) simulation, surface runoff in these rivers will decrease by 23.5 percent and 28.5 percent for Euphrates and Tigris, respectively, by the end of the present century (these figures are calculated for the Turkish portions of these basins). The same simulation reveals that there will be little snow cover in the headwaters of these rivers in the late twenty-first century, as the increase in regional temperatures will cause precipitation to fall as rain mostly (not as snow). The decreases in surface runoff are primarily related to decreases in precipitation; however, higher evapotranspiration rates in response to increased temperatures also play a role, as they increase water loss into the atmosphere (Bozkurt and Sen, 2013).

In addition to reductions in runoff, which has not been observed in the historical observations, peak flows in future hydrographs will be observed earlier (similar shifts have already been detected in historical observations). The high emissions scenario simulation indicates that the temporal shift will be about 4–5 weeks earlier. These are statistically significant shifts (Sen et al., 2011).

Both changes, i.e., runoff reduction and temporal shifts to earlier, may have important implications for the future of the basin. There will be less water available for irrigation, energy production, and domestic and industrial use. Less water in the rivers will also increase stress on ecosystems along the rivers. The 2008 severe drought in the basin conveyed important messages about what could happen in this area in the future. Such events, which could be more frequent and intense in the future, could threaten water availability and food security, and may cause conflicts in the region (Yucel et al., 2014).

8.3.2 Water Budget

While many innovations may affect the water supply and usage within the coming decades, the full development scenario in 2040 indicates a water deficiency in the Euphrates river (Table 8.1). Projections by various authors indicate a deficiency of 2–12 BCM/y in the Euphrates at full development. It is generally agreed that there will be a surplus of 8–9.7 BCM/y for the Tigris. This picture signals a water shortage that will emerge some time after 2020. In two decades, the requirements of the Euphrates branch will not be met with virgin flow of that tributary alone. Although the transfer of water from the Tigris to the Euphrates is often proposed, this may not entirely solve the water shortage in the Euphrates basin.

8.3.3 Water Quality

Deterioration of water quality and heavy pollution from many sources are becoming serious threats to the ET basin. One problem is the lack of any effective water-monitoring network, so that it is difficult to take measures to address water quality and pollution as it is impossible to identify the causes. The quantity and quality of water entering the Gulf is also an issue to be addressed since fisheries are an important food source for the region. Other environmental issues to be taken into account are the impact of water management and changed flow regimes on migrating fish and terrestrial species and on the viability of riverine and floodplain ecosystems throughout the Tigris and Euphrates basins (FAO, 2008).

The Centre for Environmental Studies and Resource Management (CESAR) study was conducted with publicly available data from Turkey and authorized national data from Syria and Iraq; this led to a comprehensive analysis of the water management systems of the two rivers (Trondalen, 2008). Based on this technical study, but also putting it into perspective, Trondalen claimed that unless the three countries found ways of cooperating, the water quality of the rivers might shortly find itself in a grave condition, particularly that of the Euphrates River in Syria, and subsequently in the southern part of Iraq. Of equal importance was the fact that if water resources were not used effectively, the shortfall between need and availability would grow even larger. Moreover, a serious challenge currently facing irrigation is a high concentration of salt in the topsoil. Highly intensive irrigation as a basis for food production, as well as the area's socioeconomic growth, has characterized all the advanced hydraulic civilizations in Mesopotamia right up to the present day, causing the salinization process to continue, not only in Iraq but also in Syria and even in Turkey. This process is expected to go on unless mitigating measures are taken (Trondalen, 2008).

Salinity issues in the two rivers, especially in the Euphrates, have also been studied by international consultants for the Iraq government. The Iraq Salinity Assessment, for example, is the result of a three-year research project on soil salinity in central and southern Iraq by the government of Iraq and an international research team led by the International Center for Agricultural Research in the Dry Areas (Christen and Saliem, 2012). After conducting historical, regional, and field-scale soil salinity surveys as well as systematic studies on surface water salinity, including longitudinal profiles of and time series trends in river salinity, the research team concluded that action was needed in four areas: Iraq's irrigation and drainage systems need to be upgraded; strategies are needed for farm-level water management, improved salinity control, and irrigation management; management of drainage water and other saline inflows to the river systems is needed; and water use policies and institutions need to be strengthened. The findings of this important study also examined, inter alia, the physical conditions of the river basin and provided a sound basis for discussions of transboundary water governance practices in the basin.

8.3.4 Reservoir Sedimentation

Rivers contain sediment, which partly accumulates in reservoirs. Depending on sediment loads and a reservoir's storage capacity, the accumulation reduces, over time, its safe yield, i.e., the annual runoff of a river, which can be utilized for electricity generation, irrigation, and flood control. If an upstream reservoir

Table 8.1 *Summary of water budgets at full development scenario (BCM/y)*

	Summary of water budgets at full development scenario (BCM/y)				
	Altinbilek (2004)	Kolars[1] (1994)	Kliot[2] (1994)	US Army Corps of Engineers[3] (1991)	Belul[4] (1996)
Euphrates					
Natural flow at Turkish–Syrian border	31.43	30.67	28.20	28.20	31.4
Net withdrawal by Turkey	−14.50	−21.6	−21.50	−21.5	−12.3
Entering Syria	16.93	9.07	6.7	6.7	19.1
Inflows in Syria	2.05	9.484	10.7	4.5	3.1
Net withdrawals by Syria	−5.5	−11.995	−13.4	−4.3	−10.5
Entering Iraq	13.48	6.559	4.0	6.9	11.7
Net withdrawal by Iraq	−15.5	−13.0	−16.0	−17.6	−19.0
Flow into Shatt- al-Arab	−2.02	− 6.441	−12.0	−10.7	− 7.3
Tigris					
Runoff in Turkey	18.87	18.5	18.5	18.500	19.3
Net withdrawal by Turkey and Syria	−8.0	−6.7	−7.2	−6.7	10.2
Entering Iraq	10.87	11.8	11.3	11.8	11.5
Inflows in Iraq by tributaries	30.7	30.7	31.7	30.7	31.0
Net withdrawal by Iraq	−31.9	−33.4	−40.0	−32.8	−33.5
Flow into Shatt- al-Arab	9.67	9.1	8.0	9.7	9.0

Source: D. Altinbilek, "Development and Management of the Euphrates-Tigris Basin," *Water Resources Development*, Vol. 20, No. 1 2004.
[1] J. Kolars (1994) "Managing the Impact of Development: The Euphrates and Tigris Rivers and the Ecology of the Arabian Gulf- A Link in Forging Tri-Riparian Cooperation," *Water as an Element of Cooperation and Development in the Middle East*, All Ihsan Bagis, editor. (Ankara: Ayna Publications and the Friederich Naumann Foundation in Turkey: Ankara.) pp. 129−154.
[2] N. Kliot (1994) *Water Resources and Conflict in the Middle East* Routledge, London, p. 100.
[3] U.S. Army Corps of Engineers (1991) 'Profile: Tigris-Euphrates River' in U.S. Army, *Water in the Sand*, p. 2.
[4] M. L. Belül (1996) *Hydropolitics of the Euphrates-Tigris Basin.* M.Sc. Thesis submitted to the Graduate School of Natural and Applied Sciences, Middle East Technical University, June 1996.

serves as a sediment trap for those downstream, it increases the lifetime over which they provide services; it reduces maintenance requirements in irrigation channels because less sediment deposits therein. In addition, water quality improves downstream with less sediment contents. The Turkish Keban Dam has positive flood control effects for the downstream countries, specifically for Iraq. Positive effects on flood control increased with the construction of the Karakaya and Atatürk dams, which also serve as sediment traps (Scheumann, 1998).

However, the deforested Turkish watershed of the Euphrates has negative effects on sedimentation rates, and estimates assume that yearly sedimentation in the three Turkish reservoirs, i.e., Keban, Karakaya, and Atatürk, could reach a volume of 1,050 cubic meters per square kilometer. The General Directorate for State Hydraulic Works (DSI, in Turkish acronym) assumes 350 cubic meters per square kilometer for each reservoir. Based on the study of upstream geology, experts estimate an annual storage volume loss for Keban dam on the Euphrates of 0.147 percent. The 1975–2060 loss will amount to 13 percent. Diverting, dredging, or dewatering of sediment are possible but these are extremely expensive response strategies.

8.4 CONCLUSION: ROAD TO SUSTAINABILITY

Today, communities in the ET basin face a set of complex, interrelated problems. Many of these problems are related to the issues of water, energy, and food security. The water–energy–food (WEF) nexus has emerged as a useful concept to address the complex and interrelated nature of resource problems. The nexus approach acknowledges those links between water, energy, and food in management, analysis, planning, and implementation. In the context of transboundary basins, water provides a useful point of entry to a nexus analysis. A nexus approach is envisioned to provide a policy framework for the riparian countries to coordinate plans and management measures in the water, food, and energy sectors. However, transboundary river basins represent challenges relating to different national interests, power disparities, and limited national capacities.

The ET basin countries, namely Turkey, Syria, and Iraq, adopted a competitive WEF nexus approach at national level. Water management at transboundary level was conducted in an uncoordinated fashion and failed to strike a balance between various uses and protect a scarce resource. Moreover, the institutions (i.e., JTC and water sharing agreements) that were created to address these complexities failed to coordinate water, energy, and food policies, and the interlinkages among these sectors.

From the 1960s to the turn of the millennium, political confrontation, rising tensions, and economic disintegration occurred, when water became both a national and an international security matter among the basin countries. During this period, riparians adopted unilateral economic growth policies to develop the Euphrates and Tigris waters and land resources for food and energy security without coordination or cooperation among themselves. Hence, large-scale water and land resources development projects were carried out unilaterally and mainly with a development focus short of sufficient care for ecosystem protection.

On the other hand, in the period of rapprochement and cooperation in the first decade of the 2000s, the ET basin riparian states followed a new strategy focusing on practical goals (i.e., the establishment of the High Level Cooperation Councils and signing of a series of memoranda of understandings) by considering the role of water in addressing issues pertaining to energy and food security with a view to contributing to regional economic integration and sustainable development in the basin. Yet, the outbreak of civil war in Syria and the spread of instability in Iraq put a strain on the bilateral as well as trilateral relations among the riparian countries and thus paved the way to an imperfect nexus of water, energy, and food policies and practices in the ET basin (Kibaroglu and Gursoy, 2015).

Adopting a WEF nexus approach in the ET basin is important because there are various pressures on the river system due to population growth, agricultural needs, hydropower development, and ecosystem mismanagement. Impacts of climate change add to the complexity of transboundary water management, as the basin is one of the most affected regions. During the period of rapprochement, steps have been taken to advance cross-sectorial dialogue and planning. However, in the history of transboundary water relations sectorial agreements at the transboundary level such as agreements regarding energy production and distribution were not undertaken.

Hence, when peace and stability is restored in the region, transboundary institutions such as the JTC should apply new approaches, such as the nexus approach, which helps to identify key development drivers, as well as unpacking and clarifying the development challenges and necessary tradeoffs in transboundary river basins. A regional perspective on nexus resources management can provide mutual benefits for riparian countries, and national natural resource constraints can be alleviated through transboundary cooperation, infrastructure development, and trade among countries in the region.

Notwithstanding the failures in interstate water cooperation and the shortcomings and loopholes in existing water agreements, the present overarching challenge in the ET basin is to coordinate water resources management and establish good transboundary water governance in the midst of the current state of affairs. The Syrian civil war and overall political instability, which have had deep impacts and spillover effects in the region, demonstrate that, while the genesis of the conflict is a complicated narrative, water is certainly part of it.

With rising violence and instability in the region, and with no regional coordination and poor security schemes along the rivers themselves, violent nonstate actors such as the Islamic State (IS) have been able to use water both as a resource and as a weapon. Not only have they destroyed water-related infrastructure, such as pipes, sanitation plants, bridges, and band cables connected to water installations, but they have also used water as an instrument of violence by deliberately flooding towns, polluting bodies of water, and ruining local economies by disrupting electricity generation and agriculture (Vishwanath, 2015).

In 2014, for instance, when IS shut down Fallujah's Nuaimiyah Dam, the subsequent flooding destroyed Iraqi fields and villages. In June 2015, they closed the Ramadi barrage in Anbar Province, reducing water flows to the famed Iraqi Marshes and forcing the Arabs living there to flee. The capture of the Mosul Dam, while it was in the group's possession for a few weeks in August 2014, gave IS control of nearly 20 percent of Iraq's electricity generation (Von Lossow, 2016). Since the civil war erupted in Syria, furthermore, IS has seized the opportunity to control territory in the conflicted region by joining the fight against the Assad regime (Hashim, 2014).

IS subsequently lost control of all the dams, but not before using them to flood or starve downstream populations and pressure them to surrender. At the same time, governments and militaries have used similar tactics to combat IS, closing the gates of dams or attacking water infrastructure under their control, thus also causing the surrounding population to suffer. The Syrian government has been repeatedly accused of withholding water, reducing flows, or closing dam gates during its battles against IS or other rebel groups, and of using the denial of clean water as a coercive tactic against many Damascus suburbs thought to be sympathetic to the rebels. Water contamination is widespread, disastrously increasing the incidence of deadly water-borne diseases.

The emergence of IS as the nonstate violent actor in the region means that riparian states must be thoroughly prepared for and responsive to possible attacks on the region's water supply and development infrastructure. This should also convince the riparian states of the need to establish regional security arrangements to preserve and protect their resources. With collaborative management underpinning collective protection, water – often a source of competition and conflict – could become a facilitator of peace and cooperation (Waslekar, 2017).

As the Syrian civil war is pushing the riparian states to develop new water governance principles and practices in conflict and post-conflict situations, the riparian states in the ET basin should improve their understanding of the strategic role that water and water supply infrastructure play in armed conflicts and to reflect on possible ways to improve the protection of water under international law during and after armed conflicts. The linkage between international humanitarian law (Additional Protocols of 1977 to the Third and Fourth Geneva Conventions of 1949) and the law on transboundary water resources (Article 29, UN Watercourses Convention, 1997) may ensure better protection of water during armed conflict. The riparian states should also envisage joint ways of dealing with transboundary water resources during reconstruction and rehabilitation efforts in the post-conflict phase.

REFERENCES

Allan, J. A. (1990). *Agricultural Sector in Iraq*. London: University of London.

Altinbilek, D. H. (2004). Development and Management of the Euphrates–Tigris Basin. *International Journal of Water Resources Development*, 20(1), pp. 15–33.

Anderson, E. W. (1986). *Water Geopolitics in the Middle East: Key Countries*. Conference on US Foreign Policy on Water Resources in the Middle East: Instrument for Peace and Development, Washington, DC: CSIS.

Bakour, Y. (1992). Planning and Management of Water Resources in Syria. In G. Le Moigne, S. Barghouti, G. Feder, L. Garbus, and M. Xie, eds., *Country Experiences with Water Resources Management, Economic, Institutional, Technological, and Environmental Issues*. Washington, DC: World Bank, pp. 151–155.

Beaumont, P. (1992). Water: A Resource under Pressure. In G. Nonneman, ed., *The Middle East and Europe: An Integrated Communities Approach*. London: Federal Trust for Education and Research, pp. 183–188.

Belül, M. L. (1996). Hydropolitics of the Euphrates–Tigris Basin. Unpublished master's thesis. Ankara: Middle East Technical University.

Bilen, O. (1994). Prospects for Technical Cooperation in the Euphrates–Tigris Basin. In A. K. Biswas, ed., *International Waters of the Middle East: From Euphrates–Tigris to Nile*. Oxford: Oxford University Press, pp. 95–116.

Bozkurt, D. and Sen, O. L. (2013). Climate Change Impacts in the Euphrates–Tigris Basin Based on Different Model and Scenario Simulations. *Journal of Hydrology*, 480, pp. 149–161. https://doi.org/10.1016/j.jhydrol.2012.12.02

Central Bureau of Statistics in the Syrian Arab Republic. (2005). General Census, Population in the Areas and Suburbs 2004. http://cbssyr.org/General%20census/census%202004/pop-man.pdf

Central Organization for Statistics in Iraq. (2010). Annual Abstract for Statistics 2008–2009. www.iraqcosit.org/english/section_2.php

Christen, E. W. and Saliem, K. A. (eds.) (2012). *Managing Salinity in Iraq's Agriculture: Current State, Causes and Impacts*, Iraq Salinity Assessment: Report 1.

Cullmann, J. (2013). Hydrology. In A. Kibaroglu, A. Kirschner, S. Mehring, and R. Wolfrum, eds., *Water Law and Cooperation in the Euphrates–Tigris Region*. Leiden: Brill, pp. 183–189.

FAO (2008). *Water Reports 34*, Irrigation in the Middle East Region in Figures, Aquastat Survey.

GAP Administration. (2019). Available at www.gap.gov.tr/gap-ta-son-durum-sayfa-32.html

Gurun, K. (1994). *Akintiya Kürek Cekmek: Bir Büyükelcinin Anıları [Akintiya Rowing: Memories of an Ambassador]*. Istanbul: Milliyet Yayinlari.

Harris, L. and Alatout, S. (2010). Negotiating Scales, Forging States: Comparison of the Upper Tigris/Euphrates and Jordan River Basins. *Political Geography*, 29(3), pp. 148–156. https://doi.org/10.1016/j.polgeo.2010.02.01

Hashim, A. S. (2014). The Islamic State: From al-Qaeda Affiliate to Caliphate. *Middle East Policy*, 21(4), pp. 69–83. https://doi.org/10.1111/mepo.12096

Institute in Turkstat (Statistical Turkey) (2010). Population Statistics. www.turkstat.gov.tr/VeriBilgi.do?tb_id=39&ust_id=11

Kavvas, M. L., Chen, Z. Q., Anderson, M. L. et al. (2011). A Study of Waterbalances over the Tigris–Euphrates Watershed. *Physics and Chemistry of the Earth, Parts A/B/C*, 36(5–6), pp. 197–203.

Kibaroglu, A. (2002). *Building a Regime for the Waters of the Euphrates–Tigris River Basin*. London and The Hague: Kluwer Academic Publishers.

(2014). An Analysis of Turkey's Water Diplomacy and Its Evolving Position Vis-à-vis International Water Law. *Water International*, 40(1), pp. 153–167. https://doi.org/10.1080/02508060.2014.978971

Kibaroglu, A. and Gursoy, S. I. (2015). Water–Energy–Food Nexus in a Transboundary Context: The Euphrates–Tigris River Basin as a Case Study. *Water International*, 40 (5–6), pp. 824–838. https://doi.org/10.1080/02508060.2015.1078577

Kibaroglu, A. and Scheumann, W. (2013). Evolution of Transboundary Politics in the Euphrates–Tigris River System: New Perspectives and Political Challenges. *Global Governance*, 19, pp. 279–307. https://doi.org/10.5555/1075- 2846-19.2.279

Kibaroglu, A. and Unver, I. O. (2000). An Institutional Framework for Facilitating Cooperation. *International Negotiation*, 5(2), pp. 311–330. https://doi.org/10.1163/15718060020848785

Kibaroglu, A., Klaphake, A., Kramer, A., Scheumann, W., and Carius, A. (2005). *Cooperation on Turkey's Transboundary Waters*. Research report, German Federal Ministry for Environment, Nature Conservation and Nuclear Safety, Berlin.

Kirschner, A. and Tiroch, K. (2012). The Waters of the Euphrates and Tigris: An International Law Perspective. In A. von Bogdandy and R. Wolfrum, eds., *Max Planck UNYB 16*. Leiden: Martinus Nijhoff Publishers, pp. 329–394.

Kliot, N. (1994). *Water Resources and Conflict in the Middle East*. New York: Taylor & Francis.

Kolars, J. F. (1994). Problems of International River Management: The Case of the Euphrates. In A. K. Biswas, ed., *International Waters of the Middle East: From Euphrates–Tigris to Nile*. Oxford: Oxford University Press, pp. 44–94.

Kolars, J. F. and Mitchell, W. A. (1991). *The Euphrates River and the Southeast Anatolia Development Project*. Carbondale: Southern Illinois University Press.

Law No. 14 of 1990, *Ratifying the Joint Minutes Concerning the Provisional Division of the Waters of the Euphrates River*. Available at www.informea.org/en/legislation/law-no-14-1990-ratifyingjoint-minutes-concerning-provisional-division-waters-euphrates

McLachlan, K. (1991). *The South-East Anatolia Project (GAP) and Its Effect on Water Supply and Management in Iraq*. London: University of London.

Meliczek, H. (1987). *Land Settlement in the Euphrates Basin of Syria: In Land Reform: Land Settlement and Cooperatives*. Rome: Food and Agriculture Organization Publications.

Memorandum of Understanding between the Ministry of the Environment and Forestry of the Republic of Turkey and the Ministry of Water Resources of the Republic of Iraq on Water. (2009a). On file with the author.

Memorandum of Understanding between the Government of the Republic of Turkey and the Government of the Syrian Arab Republic for the Construction of a Joint Dam on the Orontes River under the Name 'Friendship Dam'. (2009b). On file with the author.

Memorandum of Understanding between the Government of the Republic of Turkey and the Government of the Syrian Arab Republic in the Field of Efficient Utilization of Water Resources and Coping with Drought; Memorandum of Understanding between the Government of the Republic of Turkey and the Government of the Syrian Arab Republic in the Field of Remediation of Water Quality. (2009c). On file with the author.

Naff, T. and Matson, R. C. (1984). *Water in the Middle East: Conflict or Cooperation?* Boulder, CO: Westview Press.

Nippon, K. and Yuksel, P. (1989). *Vols. I–IV, Southeastern Anatolia Project Master Plan Study*. Tokyo and Ankara: State Planning Organization.

Richards, A. and Waterbury, J. (1990). *A Political Economy of the Middle East*. Boulder, CO: Westview.

Sayan, R. C. and Kibaroglu, A. (2016). Understanding Water–Society Nexus: Insights from Turkey's Small-Scale Hydropower Policy. *Water Policy*, 18(5), pp. 1286–1301. https://doi.org/10.2166/wp.2016.235

Scheumann, W. (1998). Conflicts on the Euphrates: An Analysis of Water and Non-water Issues. In W. Scheumann and M. Schiffler, eds., *Water in the Middle East*. Berlin: Springer, pp. 31–45.

Sen, O. L., Unal, A., Bozkurt, D., and Kindap, T. (2011). Temporal Changes in Euphrates and Tigris Discharges and Teleconnections. *Environmental Research Letters*, 6 (024012). https://doi.org/10.1088/1748-9326/6/2/024012

Tice, V. L. (2016). *Water Management and Water Challenges in Iraq*. Available at https://water.fanack.com/iraq/water-management-and-water-challenges-in-iraq/

Tigrek, S. and Kibaroglu, A. (2011). Strategic Role of Water Resources for Turkey. In A. Kibaroglu, W. Scheumann, and Annika Kramer, eds., *Turkey's Water Policy*. New York: Springer Verlag, pp. 27–42.

Trondalen, J. M. (2008). *Water and Peace for the People: Proposed Solutions to Water Disputes in the Middle East*. Paris: UNESCO International Hydrological Programme (IHP).

Turkish Ministry of Foreign Affairs (1996). *Water Issues between Turkey, Syria, and Iraq*. Perceptions: Journal of International Affairs. Retrieved from http://sam.gov.tr/wp-content/uploads/2012/01/WATER-ISSUES-BETWEEN-TURKEY-SYRIA-AND-IRAQ.pdf

UN-ESCWA and BGR (2013). *Inventory of Shared Water Resources in Western Asia*. Beirut: ESCWA. Available at: https://waterinventory.org/

Unver, I. H. O. (1997). Southeastern Anatolia Project (GAP). *International Journal of Water Resources Development*, 13 (4), pp. 453–484. https://doi.org/10.1080/07900629749575

Vishwanath, A. (2015). *The Water Wars Waged by the Islamic State*. Available at www.stratfor.com/weekly/water-wars-waged-islamic-state

Von Lossow, T. (2016). *Water as Weapon: IS on the Euphrates and Tigris. German Institute for International and Security Affairs*. Available at www.swp-berlin.org/fileadmin/contents/products/comments/2016C03_lsw.pdf

Wakil, M. (1993). Analysis of Future Water Needs for Different Sectors in Syria. *Water International*, 18(1), pp. 18–22. https://doi.org/10.1080/02508069308686144

Waslekar, S. (2017). *Wars Will Not Be Fought over Water: Our Thirst Could Pave the Way to Peace*. Available at www.theguardian.com/global-development-professionals-network/2017/jan/19/water-wars-infrastructure-isis-peace

World Bank (2001). *Syrian Arab Republic Irrigation Sector Report*. Report No. 22602-SYR. Washington, DC: Rural Development Department, Water, and Environment Group, World Bank.

Yucel, I., Guventurk, A., and Sen, O. L. (2014). Climate Change Impacts of Snowmelt Runoff for Mountainous Transboundary Basins in Eastern Turkey. *International Journal of Climatology*, 35(2), pp. 215–228.

Notes

1 The ESCWA-BGR study compiled that information from the following sources:

(a) The population estimate for the area of the basin situated in Turkey is based on a 2010 census and includes populations living in the Turkish provinces of Adiyaman, Agri, Bingol, Elazig, Erzincan, Erzurum, Gaziantep, Malatya, Mardin, Mus, Sanliurfa, Sivas, and Tunceli (Turkstat, 2010).
(b) The population estimate for the area of the basin located in Syria is based on a 2010 assessment and includes populations living in the Syrian governorates of Aleppo, Deir ez Zor, Hama, Hasakah, Homs, and Raqqah (Central Bureau of Statistics in the Syrian Arab Republic, 2005).
(c) The population estimate for the area of the basin situated in Iraq is based on a 2009 assessment and includes populations living in the Iraqi provinces of Anbar, Babil, Karbala, Najaf, Ninewa, Qadisiyah, and Muthanna (Central Organization for Statistics in Iraq, 2010).

2 The ESCWA-BGR study compiled information from the following sources:

(a) The population estimate for the area of the basin situated in Turkey is based on a 2010 census and includes populations living in the Turkish provinces of Batman, Diyarbakir, Hakkari, and Siirt, as well as parts of the provinces of Bitlis, Mardin, and Van (Turkstat, 2010).
(b) The population figure for the area of the basin situated in Syria is based on a 2010 estimate and only covers parts of Hasakah Governorate (Central Bureau of Statistics in the Syrian Arab Republic, 2005).
(c) The population figure for the area of the basin located in Iraq is based on a 2009 estimate and includes populations living in the following governorates: Arbil, Baghdad, Diyala, Dahuk, Kirkuk, and Sulaymaniyah. Parts of Basrah, Maysan, Ninewa, Salah ad Din, and Wasit Governorates are also included (Central Organization for Statistics in Iraq, 2010).
(d) The basin population estimate for Iran's share of the Tigris Basin is based on a 2006 assessment and includes populations living in the province of Ilam, and parts of Kermanshah and Kurdistan Provinces (Statistical Center of Iran, 2006).

9 The Yellow River Basin

James E. Nickum and Jia Shaofeng

9.1 INTRODUCTION TO THE PLAYERS

"The cradle of Chinese civilization," "Mother River of the Chinese nation," "China's Sorrow," and, in the earliest records, simply "The River" – even though it is not the largest river in China, and its basin includes less than 10 percent of the country's population, the Yellow River (Huang He) has been central to the identity of the country throughout history.[1] The capitals of its major dynasties for over two millennia from the Zhou (1046–256 BCE) through the Northern Song (to 1127) were located along the river and its tributaries. The Yellow has been engineered for at least as long, providing irrigation to generate agricultural surpluses to support armies and posing a flood threat that reaches back into the foundational legends, notably of Yü the Great (ca 2200 BCE), the first great flood control engineer of the Yellow and emperor-founder of the legendary Xia Dynasty (Ball, 2016, pp. 60–67).

Among the SERIDAS cases, the Yellow is most like the Nile in its length, relatively modest discharge, total area under irrigation, and historically important conveyance of silt. Like the São Francisco, it lies entirely within one large country, with a comparably large basin area (752,000 km^2 compared to the São Francisco's 630,000 km^2), although with twice the length (5,464 km, versus 2,700 km).

The river is conventionally divided into three reaches (Table 9.1). The upper reach, currently inhabited by roughly 30 million people (YRCC, 2013), encompasses over 60 percent of the length of the river and over half its catchment area and runoff, lying within five of the nine province-level administrations that share its basin (Qinghai, Sichuan, Gansu, Ningxia, and Inner Mongolia). It stretches from the river source at an altitude of 4,500 meters in the northeast part of the vast, sparsely populated Tibet Plateau, "Asia's Water Tower," at first flowing eastward. It takes a large, distinctive staple-shaped diversion northward transecting the Yinchuan and Hetao plains, and then flows eastward in a Great Bend before it returns south at Hekouzhen in Inner Mongolia.[2]

Viewed as an engineering challenge, the upper Yellow comprises two distinct parts, divided at the city of Lanzhou. The mountainous upper part, in historical times left in its natural state, is now significantly engineered primarily for hydropower; the flat, lower part, in the most arid part of the river's basin, with mean annual precipitation of 257 mm/year, has long drawn from the Yellow to provide water for extensive irrigated areas, notably the Hetao in Inner Mongolia and Qingtongxia in Ningxia. More recently, it has become the site of a significant share of China's hydrocarbon resources and production.

The southward stretch of the Yellow from Hekouzhen, along the border between Shaanxi and Shanxi Provinces, to Taohuayu in Henan Province forms the middle reach, where the river runs through the world's largest deposit of loess, picking up the silt that gives it its name. In this reach, engineering focuses on watershed control through such means as check dams, terracing, and afforestation. Including the two densely populated major tributaries of the Wei and Fen rivers, this stretch contains most of the basin's population, at close to 80 million.

The middle reach has a catchment area nearly as large as that of the upper. It encompasses most of the cropland of the Yellow River basin, about 30 percent of it irrigated, often in large surface systems (irrigation districts). Over half of the basin's groundwater is in the middle Yellow as well, as are significant hydrocarbon extractive industries.[3]

The lower reach begins with the river's entrance onto the North China Plain in Henan Province and stretches, for the time being at least, as an "elevated river" between high dikes (average about 10 m high, maximum 14 m) built 5–20 km apart, stretching northeast through Shandong Province until what is left of the flow empties into the Bohai Gulf north of the Shandong Peninsula. The right bank dike is 624 km long, divided into four sections. The left, 747 km long, is divided into five sections. The river is thus isolated from the lower-lying Hai River basin to the north and Huai River basin to the south, and has a small catchment area, with only three tributaries (Table 9.1). Since the riverbed is suspended an average of over 3–5 meters above the surrounding land, the river ironically forms the watershed boundary for the Hai and Huai (Liu et al., 2013, p. 229).

With a drastic reduction in its gradient over the plain (Table 9.1), about one-quarter of the silt settles in the main channel, raising it, at least until the 1980s, by 7–10 cm/year. Flooding and even major shifts in river course have frequented

Table 9.1 *The three reaches of the Yellow River*

Reach	Mainstream length (km)	Catchment area (km^2)	Runoff contribution (%)	Silt contribution (%)	Drop in elevation (m)	Gradient	Slope (%)	Number of major tributaries
Upper	3,400	386,000	53	9	3,464	1:982	0.10	43
Middle	1,200	344,000	37	89	880	1:1363	0.07	30
Lower	786	22,000	10	2	95	1:8273	0.01	3

Source: Zhongguo shuili baike quanshu 中国水利百科全书 (China Encyclopedia of Water Resources). 1991. Beijing: Shuili Dianli Chubanshe: 856–857.

Chinese history (summed up in the rule of thumb, "Two breaches every three years and a course change each century," based on over 1,500 recorded breaches and 26 major course shifts in the period since 602) (Liu et al., 2013, p. 230). In this critical portion of the river, engineering is focused strongly on flood mitigation, primarily by diking but also with attention to withdrawals for irrigation and urban uses, including by interbasin transfers. Although only 13 million people reside within the hydrologically insignificant lower reach, consisting mainly of the delta and the land between the dikes, its water supports a much greater population in neighboring river basins, given as nearly 55 million people in 1997 (Fu and Chen, 2006) in irrigation districts covering 2.9 million hectares in eighty counties of Henan and Shandong provinces in 2007 (the latest data available) through irrigating 2.2 million hectares of a total 4.0 million hectares of cultivated land (2007 figures) (Zhang et al., 2013). Including diversions to agriculture and cities in other basins, this stretch has the largest economy. It also contains significant oilfields in the delta.

This final stretch of the Yellow challenges the idea of sustainability. Left on its own, the river would spread its immense load of loess in multiple and shifting channels across the vast North China (Huang-Huai-Hai) Plain. Covering over 400,000 km^2 (comparable in size to Germany or the state of California) and stretching as far north as Beijing, that plain is to a great extent the product of the river's natural engineering, which if left alone it would continue to build.

Since the natural engineering preferred by the river does not match the sustainability needs of human settlement on the North China plain, humans have done their own engineering of the lower Yellow for more than two millennia. This period has been characterized by the conflict between natural and human engineering, with victories by both sides but no ultimate resolution.[4] The Yellow can lay claim to be the most engineered of the cases in this volume. At the same time, if being sustainable means staying in place, while it carried the world's heaviest silt load, the Yellow was by its nature not sustainable. As discussed later, this silt load has fallen considerably in recent years, bringing other sustainability dimensions to the fore – for now.

9.1.1 Enter the Professional Engineers

The late nineteenth century and the twentieth century saw the rise of modern hydraulic engineering, and with it schemes for taming the Yellow River and turning its waters into productive use, especially for large-scale irrigation and hydropower. Pietz (2015) provides a comprehensive overview of the actors (German, American, Japanese, and Russian, as well as Chinese in both the Nationalist [1927–1949] and Communist [since 1949] periods), plans, and historical forces involved in the challenge to solve what imperial dynasties failed to do, in particular to tame the world's most silt-laden river and to expand its economic uses.

Interestingly, the foreign river engineers in the early period wrangled over the same issue that had divided Chinese engineers in historical times: whether to constrain the river within narrow dikes and force it to flow straight and quickly (represented by the American John Freeman and German Otto Franzius, echoing the "Confucian" school of earlier Chinese river managers) or to allow the river to meander within wide dikes (represented by the German Hubert Engels, using a physical hydraulic model in the 1930s, echoing the "Daoist" school) (Pietz, 2015, pp. 94–96).

New ideas came shortly later, with the advent of big dams. An influential Japanese Yellow River Investigation Commission report toward the end of the Pacific War called for a form of integrated river basin control and management flood scheme with retention dams at Sanmenxia, Balihutong, and/or Xiaolangdi, cascades of dams on the mainstem for hydropower, widening a portion of the downstream dikes and reforestation in the loess plateau (Pietz, 2015, pp. 114–115).

9.1.2 Administrative "Stakeholders" under the People's Republic of China

Aside from the dual hierarchies of the government and the Chinese Communist Party, the principal stakeholders in the Yellow River Basin are hydropower companies. In the administrative system, of concern here, the key actors are the provinces and central government ministries. The basin authority is under one of those ministries, the Ministry of Water Resources, although other ministries affect policy and practice in the basin, notably the Ministry of Ecology and Environment (formerly Ministry of Environmental Protection), which is responsible for environmental impact assessments of projects and pollution control. Other ministries of note are those dealing with mining, agriculture, urban development, and agriculture.

Provinces and provincial rivalry. The People's Republic of China was established in 1949 and with it the political and administrative structure that has guided and hampered the engineering of the Yellow River over the past seven decades. Nine province level administrations: Qinghai, Sichuan (not significant), Gansu, Ningxia, Inner Mongolia, Shaanxi, Shanxi, Henan, and Shandong impose their boundaries on the Yellow River basin. While China is not a federal system, it is organized in a complex hierarchical system where provinces play an important role and are capable of articulating their own interests in negotiating usages and allocations of the river (Moore, 2018, pp. 136–182). Competition for water is particularly keen, and the centralized system may exacerbate that competition, by suppressing cooperation between provinces. The largest users are Shandong (mostly diversions out of the basin), Inner Mongolia, and Ningxia (the latter two primarily for irrigation).

Yellow River Commission. Cutting across the provinces is a basin authority, the Yellow River Conservancy Commission (YRCC, or Huanghe shuili weiyuanhui),[5] established in 1949. With a staff of nearly 40,000,[6] it is now one of seven river or lake basin authorities that are agencies of the Ministry of Water Resources, which itself is technocratic and focused on expanding water sources through hydraulic engineering, and increasingly so with the transfer of non-engineering functions to other ministries in a 2018 restructuring (Jia et al., 2019b). Unlike most basin authorities elsewhere, in China "they act more as instruments of control than as instruments of cooperative governance" (Moore, 2018, p. 162), and do not provide any institutional role for subnational governments or civil society. At the same time, the YRCC ranks below provincial governments in China's administrative hierarchy, giving it limited capacity to impose its will on the latter (Zhuang et al., 2015, p. 20). According to Shen (2009, pp. 493), "participation is the weakest aspect in river management in China." River basin organizations are technical in origin and project-oriented, evolving to where they "developed projects to benefit their own interest, acting as the building and management agency of projects with central government investment" (Shen, 2009, p. 495).[7]

One of the first tasks assigned to the YRCC was to formulate a basin development plan in 1954–1955 that proposed the construction of forty-six dams on the mainstem and extensive erosion control measures on the loess plateau, including check dams and afforestation (Pietz, 2015, p. 161). Pietz (2015, p. 162) argues that this plan was heavily influenced by the Soviet "technology complex" of the period. The YRCC has also been tasked with adjudicating interprovincial disputes, with limited results because it "has failed to accumulate the political legitimacy necessary to prevent conflict between basin water users" (Moore, 2018, p. 179). Since the mid-1980s, the YRCC has also been tasked with overseeing interprovincial allocation quotas for the annual flow (discussed later), supported by a real-time monitoring system, the "Digital Yellow River." Nonetheless, in practice the engineering operations of the YRCC continue to prevail, in its unbending advocacy of the continued revenue-generating construction of dams on the mainstem aimed to check silt that is no longer a problem.

9.2 ENGINEERING THE RIVER IN THE MODERN ERA

The engineering of the Yellow River since 1950 has focused on both disaster prevention (shuihai水害) and beneficial uses (shuili 水利). Disaster prevention has focused on dike maintenance and strengthening, which has been successful and will not be covered here, and on control of silt load, which was not successful during the twentieth century, but has made a dramatic turn in the twenty-first century. Beneficial uses have included a considerable expansion of irrigation both within the basin and through transbasin diversions; hydropower generation; mining and industry; and supply to rapidly growing urban areas. In recent years, environmental flows have been considered a beneficial use, but primarily for the purpose of silt discharge. Here we touch on transbasin diversions and large-scale mainstem multipurpose dams.

9.2.1 Sharing the Yellow: Transbasin Diversions from the Lower Reaches

The Yellow River basin is on average one of the most water-short river basins of China. Yet it borders an even more water-deficit basin, the Hai. Roughly one-quarter of the Yellow is transferred out of basin on a permanent basis, nearly all of it in downstream provinces (Ministry of Water Resources, 2016). Transbasin diversion engineering, for showcase projects such as the People's Victory Canal completed in 1953, like dike rehabilitation, preceded dam engineering (Pietz, 2015, p. 147). These diversions, for irrigation, might be considered a rare harmonization of human and the river's natural engineering, except that the silt load was removed at the inlet. Nonetheless, the river could spread and evaporate outside its confining banks, but over a high-water table, which led to secondary salinization of the soil. Consequently, diversions were stopped and focus in the 1960s changed to bringing that salinization under control through drainage. The 1970s were characterized by a series of drought years, and the diversions were resumed for irrigation. Water from the Yellow was occasionally diverted on an emergency basis to the northern coastal megalopolis of Tianjin from 1972, and to nearer urban areas (Cangzhou and Hengshui) after 1997 (Wang et al., 2013, p. 222). In recent years, Shandong Province has constructed more permanent diversions from the right bank, to supply cities on the water-short peninsula (Chen et al., 2020, p. 57).

9.2.2 A Dam Mistake: Sanmenxia, Siltation, and Sustainability

The Sanmenxia Dam, constructed in 1957–1960 during the Great Leap Forward as a multipurpose hydropower and flood control project, quickly became the poster dam of how human engineering can go wrong, especially when hypercharged with rhetoric placing the conquest of nature at the core of the construction of a socialist nation (Seeger, 2014, pp. 174–241). The Great Leap Forward of 1958–1961, when mass mobilization and humans were given precedence over science, statistics, and the exercise of caution, was not a time to consider that nature might win.

The dam is the keystone of Japanese and Russian cascade schemes for multipurpose control and use of the water and silt of the Yellow River, with focus on hydropower generation. Sited toward the end of the middle reaches near the confluence of the Yellow with the Wei, one of its major tributaries, it controls 91.5 percent of the basin, 89 percent of the runoff, and 98 percent of the silt load (Liu et al., 2013, p. 239).

The problem of siltation was recognized, but construction proceeded on the assumption that effective watershed control would occur at the same time – through construction of check dams, reducing cultivation on steep slopes, and revegetation of the ground cover – and that it would be fully effective. Yet the policies of the Great Leap Forward, exemplified in the widespread deforestation to fuel small furnaces to produce iron, and of expanding farmland to strive for self-sufficiency in grain production, operated to expose the loess plateau to even greater erosion. The reservoir reached dangerous levels of siltation within four years, affecting the Wei River, a major tributary flowing from the west, and posing a threat of backflow flooding to the ancient capital and major metropolis of Xi'an, located near that river. Removal of the Sanmenxia Dam was considered, but in the end innovative modifications were made to the dam structure in 1964 and again in 1969 with the addition of eight tunnels to allow discharge of more fine silt through newly opened tunnels, at considerable direct cost and reduction of hydroelectric power generation capacity. The dam's primary purpose became silt regulation (Liu et al., 2013, p. 240; Seeger, 2014, pp. 175–242).

In hindsight, the Sanmenxia Dam appears to reflect high modernist hubris, often linked with nation building or state legitimization more than river control per se, with unforeseen – yet often foreseeable – consequences that do not solve the sustainability problem (Seeger, 2014; Pietz, 2015). It may also reflect that integrated approaches that require an orderly phasing are beyond the capacity of intrinsically dis-integrated human governance systems.

Or it could have been just an expensive learning experience. Its failure as a hydropower dam did not signal the end of the basin development plan. There are now more than thirty hydropower dams on the mainstem of the Yellow River, with a total of fifty ultimately planned (Xie et al., 2018, pp. 4174). Of these, four are significant for silt regulation: the very large Longyangxia (over 20 km^3 capacity, completed 1986) and Liujiaxia in the clearwater upper reaches (over 5 km^3 capacity, completed 1968) through the effect of their releases on flushing downstream silt, and Sanmenxia and Xiaolangdi in the lower (Liu et al., 2013, pp. 244–245). The upstream dams are primarily intended to generate electricity and are not themselves affected by siltation.

9.2.3 Basin Closure: People Get Grain and the River Cedes the Coast to the Sea

The large-scale engineering of the relatively wet later 1950s and 1960s was oriented toward withdrawals for productive uses such as agriculture. The unintended effects of this approach were felt in the subsequent decade in the lower reaches. From 1972, average precipitation and runoff declined, drought years became more frequent with strong ENSO effects (Wang et al., 2006), and the Yellow River did not reach the sea during the dry season in several years. In the worst case, the drought year of 1997, none of the river's water reached the sea for 330 days. The flow trickled out before reaching the Jiahetan Station, 662 km from the outlet, in 1981 and 1995, and in 1997 ran dry at a station near the ancient capital of Kaifeng, 703 km from the sea (Chen et al., 2003, pp. 1–7). The primary cause of closure was the expansion of withdrawals for irrigation, especially in the arid irrigated regions of Ningxia (Qingtongxia) and Inner Mongolia (Hetao) along the lower portion of the upper reaches. Consumptive use in the Hetao irrigated area was quite high, about three-quarters of abstractions, due to evapotranspiration and seepage (Chen et al., 2003, pp. 1–8). The revival of irrigation in the lower reaches was also a significant factor in basin closure (Wang et al., 2013). Wang et al. (2006) also point to regulation of river flow by newly installed dams as a contributing factor, with climate factors, primarily El Niño/Southern Oscillation effects, not global warming, having nearly as strong an effect as anthropogenic change. Others, such as Wang et al. (2013), place much greater onus on human activities, estimating that they led to 83 percent of the reduction in runoff in the river from 1970 to 2008.[8]

Several policy and engineering measures, largely implemented by the YRCC, succeeded in restoring a year-round flow to the sea, although at historically much reduced levels. Foremost among them were water allocation between provinces, a smart monitoring system, incentive systems, and, ultimately, the construction of another large dam downstream of the Sanmenxia.

The Yellow River Water Allocation Scheme (黄河水量分配方案), enacted in 1987 after five years of negotiation between the central government and eleven province-level administrations along the Yellow River and receiving interbasin transfers, set nonbinding quotas allocating 37 km^3 of the historically average flow of 58 km^3 among those administrations, with the remaining 21 km^3 set aside for "ecological" purposes, primarily sediment discharge (Nickum, 2004, p. 131; Wang et al., 2018). The setting of these quotas was based on reported uses in 1980 at the provincial level, with a 40 percent upward adjustment for projected increases in demand.[9]

This allocation scheme failed to fend off a significant deterioration in the frequency and severity of flow cessations in the 1990s. One reason for this was the setting of fixed quotas based on a historical streamflow record from relatively wet years. Natural runoff in the 1990s fell by more than one-quarter, to about 42 km^3 (Yang and Jia, 2008). In addition, since the purpose of the allocation scheme was to guarantee water for use and environmental flows in the lower reaches, it in effect imposed limits on upstream and midstream users, who saw it as a form of expropriation with a significant impact on the revenues of major irrigation districts (Wang, 2003, p. 95; Yellow Basin Project Team, 2010, p. 44). Against this decline was a continuing increase in consumptive use, from 30.7 km^3 basin-wide for 1988–1992 to 37.2 km^3 for 1988–2000, a rise of 21 percent. Just over half of this increase was due to increases in irrigation, especially in the middle reaches, but together industrial (including mining) and domestic consumption more than doubled, especially in the upper and middle reaches (Yang and Jia, 2008, p. 269). Another reason the allocation scheme had limited effect was the lack of effective enforcement of quotas. Downstream Shandong and upstream Inner Mongolia, with the largest quotas, regularly exceeded their allotments by a significant amount (Wang et al., 2018).

Steps were taken to address these problems beginning with the issuance in 1998 of a new Framework Plan giving the YRCC more enforcement power and allowing for annual adjustments in quotas to reflect expected water availability (Yang and Jia, 2008; Wang et al., 2018). The overriding objective for the engineers of the YRCC, however, was to ensure a minimum flow of 30–50 m^3/sec at the estuary.

The existence of quotas was not the end of interprovincial politics, of special pleading by wealthy downstream provinces. In the drought year of 2002, downstream Shandong's irrigation water was threatened with being cut off in order to ensure the minimum flow, so the province requested the central government to provide it with supplemental water from the large reservoirs 4,000 km upstream. The request was granted, but it required restricting irrigation diversions in the upper basin administrations of Ningxia and Inner Mongolia, to the detriment of crop production there (Wang, 2003, p. 95). In addition to this "political negotiation and bargaining process between provinces," there is a high level of uncertainty in estimating water availability in a forthcoming year (Yang and Jia, 2008, p. 270).

Water has flowed continuously from the Yellow into the sea since the beginning of the millennium. Nonetheless, total annual discharge declined in the early 2000s, and remained well below the amount set, somewhat arbitrarily, for ecological purposes. There was no mechanism for compensating the large state-owned reservoirs for the costs of releasing the water or for those irrigation districts that were required to shut their intakes (Yang and Jia, 2008, p. 270).

In 2006, work began on a "Digital Yellow River" to provide the YRCC with real-time digital monitoring and control of water quality and use at local levels (Moore, 2018, p. 177). It was projected on completion to provide control over nearly all facilities in the basin drawing on the River (Watts, 2011). In 2008, provision was made for setting diversion quotas at the sub-provincial prefectural level (Wang et al., 2018). The nationwide institution of the "strictest water resources management strategy" and its implementing criteria – the "Three Red Lines" – provided further enforcement capacity by setting fulfilment of water quantity, efficiency, and quality targets among the performance criteria of administrators ("cadres") (Nickum et al., 2017). The first of the Three Red Lines sets declining caps on total water use at the provincial and sub-provincial level for 2015, 2020, and 2030.

The Xiaolangdi Reservoir, with its dam located 131 km downstream from the Sanmenxia, began filling in 1999. With a total storage capacity of 12.6 km^3, it is the second largest on the mainstem after the first in the cascade, the Longyangxia Reservoir in the upper reaches (24.7 km^3) (Liu et al., 2013, p. 245). Although its primary purposes are silt and flood regulation, it does provide a source of emergency water in the dry season that can be released to ensure perennial flows to the sea that is far more efficient than sources such as the Longyangxia Reservoir, which are several thousand kilometers upstream. The Xiaolangdi Dam, in the "Water and Sediment Regulation Discharge Project," has engaged in the simultaneous release of clear and silty water most years since 2002 shortly before the flood season, simultaneously scouring the channel between the dikes and releasing some of the sediment stored in the reservoir. The scouring, which finally engineers away the problem of bed accretion, has been stronger than the release of sediment, which has accumulated in the reservoir while not providing an adequate sediment load to maintain the delta (Gippel et al., 2012, pp. 67–68; Klein, 2017; Yang et al., 2018).

9.2.4 Water Quality and Ecosystem Degradation

The next major sustainability issue has been improvement of water quality. Moore (2018, p. 177) suggests that addressing water quality problems, which requires interjurisdictional cooperation, is more intractable than ensuring the river flows to the sea, which can be done through changing the operating rules of large state-operated reservoirs. He cites the industrial and urban wastewater releases in the stretch of the upper reaches that includes the industrial city of Lanzhou. Indeed, there was a marked increase in wastewater discharge throughout the basin in the 1980s and 1990s, and a marked decline in water quality in the Yellow. A comprehensive survey of the river system by the YRCC as late as 2007 found that one-third of stretches ("water function zones"[10]) surveyed had water that was classified as unusable for any purpose (worse than Grade V) (Branigan, 2008). This was (and still is) primarily a problem of certain key tributaries, such as the Fen He flowing through the coal country of Shanxi, and downstream tributaries.[11] The fall in actual runoff has often served to increase the concentration of pollutants.

Beginning in 2003, water quality has improved in the lower reaches. Based on point measurements at four stations (Huayuankou, Gaocun, Luokou, and Lijin) from 1983 to 2010,

> [t]he target water quality for the lower Yellow River (Grade III) is currently met most of the time, with a marked progressive improvement apparent after 2000. The improvement is likely partly due to pollution control programs, partly due to trapping of fine sediment (with contaminants attached) in Xiaolangdi Dam, and partly due to dilution and assimilation of contaminants through the release of baseflow water from Xiaolangdi Dam
>
> (Gippel et al., 2012, p. 61).

Nonetheless, compliance with the stricter Grade II, which "offers a lower risk to the health of the aquatic environment," especially during fish spawning and rearing seasons, is quite low (generally below 20 percent) and has not shown significant improvement (Gippel et al., 2012, pp. 56, 61).

By 2016, 65 percent of the stretches along over 22,000 km of the Yellow River and its tributaries were assessed as being of acceptable quality (Categories I, II, and III in the six-tiered Chinese water scale). In 1991, less than 27 percent of 12,300 km surveyed fell into those categories (Zhonghuarenmingongheguo shuilibu shuiwensi, 1997, pp. 31).[12] An even higher share of the water of the mainstem would appear to fall in this range, as do the extensive upstream headwaters (Ministry of Water Resources, 2016, p. 4). In 2018, only 12.3 percent of the river stretches were assessed at the unusable below Category V level, none of them in the mainstem, where water was all at Category IV or better (Ministry of Water Resources, 2018).

One factor in improving water quality for the river as a whole over the past decade may be an outcome of the third of the Three Red Lines, which judges cadres based on their progress in improving performance measured by water quality standards in segments of the River.[13] This accountability was further strengthened with the inauguration of the "River Chiefs" system nationwide in 2017 (Xu, 2017), following a successful pilot in Wuxi City, Jiangsu (Dai, 2015). Related developments are the strengthened position of the environmental administration through the enactment of a water pollution control law (the "Water Ten Law") in 2018 (Xu, 2017), and the creation in the same year of a Ministry of Ecology and Environment that gained oversight of all pollution sources entering a water body (Xu and Chan, 2018).

An engineering approach to water quality is to build water treatment plants. These have become increasingly widespread, although they are not always fully operated due to costs of operation (Tan et al., 2014).

Ecological effects, even when narrowly focused on the impacts of silt discharge or its absence, are hard to sum up in performance objectives. The measurement and implementation of policy related to environmental flows is even more intractable (Jiang et al., 2010; Chen et al., 2019), although fish diversity has declined dramatically (Xie et al., 2018). The reduced water and silt discharge of over two decades of intermittent flow cessations significantly impacted the coastal ecosystem and led to degradation of the delta geomorphology as the natural processes of coastal erosion and subsidence were allowed to reassert themselves (Wang et al., 2006, p. 223). Nonetheless, apart from 2000 to 2003, the delta has continued to grow in area despite the reduction in sediment load (Kong et al., 2014).

The impact on the littoral is not only, or even primarily, due to fluctuations in flow or quality from upstream. The area converted to aquaculture ponds in the coastal area of the Yellow River delta has expanded significantly in recent decades (from 40 km^2 in 1983 to 1407 km^2 in 2015), at first at the cost of natural wetlands, and then, in the 2010s, displacing farmland. This expansion has had serious impacts on the extent and integrity of wetland habitats, water quality, sedimentation processes, offshore fisheries, and land subsidence due to increased pumping of the coastal aquifers (Ottinger et al., 2013; Ren et al., 2018).

Further efforts are in store to "green" the Yellow River, tackling both ecological degradation and water quality. Specific initiatives in the works are in the realm of engineering new policies rather than physical structures. These include ecological development plans specific to the characteristics of each reach in the basin, improvements in water and sewage pricing, prioritizing ecological flows, establishing a Yellow River Protection Law, enforcing water quotas under the Three Red Lines system and expanding water rights trading mechanisms, including compensation payments to upstream areas for ecological services (Dong, 2020).

9.2.5 A Sea Change in the River

The past two decades have witnessed a dramatic change in the river, with both runoff and sediment load declining significantly, with mixed implications for sustainability. As indicated in Table 9.2, over the past century, estimated natural runoff as recorded by key hydrological stations has fallen on average, reflecting a long-run decline in precipitation, at least in the middle and lower reaches.[14] Observed runoff, reduced by human engineering such as diversions for irrigation and storage on a large scale, has declined more precipitously than natural runoff, especially in the lower reaches. As noted previously, Wang et al. (2013) estimate that human activities contributed to 83 percent of runoff reduction in the basin. Figure 9.1 shows trends at Lijin, in the estuary, where the growing impact of upstream withdrawals after the mid-1970s is particularly observable, and a partial rebound is also evident after the turn of the century due to interventions aimed at restoring discharge into the ocean.

Table 9.2 *Natural runoff and silt load over time*

Years	Mean natural runoff (km^3/y) (Huayuankou)	Mean silt load (10^9 t/y) (Tongguan)
1919–1975	55.9	1.63
2000–2015	45.2	0.274

Natural runoff = observed runoff + consumptive uses of surface water (地表耗水还原量) + change in reservoir storage
Source: Liu Xiaoyan 刘晓燕 et al. 黄河近年水沙锐减成因 (Why water and silt drastically declined in the Yellow River in recent years). Beijing: Kexue Chubanshe 科学出版社, 2016, p. 1.
Note: The natural runoff, as a calculated as-if figure, is not the same as the actual runoff that actually carries the silt. The difference reflects anthropogenic effects of consumptive use and changes in storage.

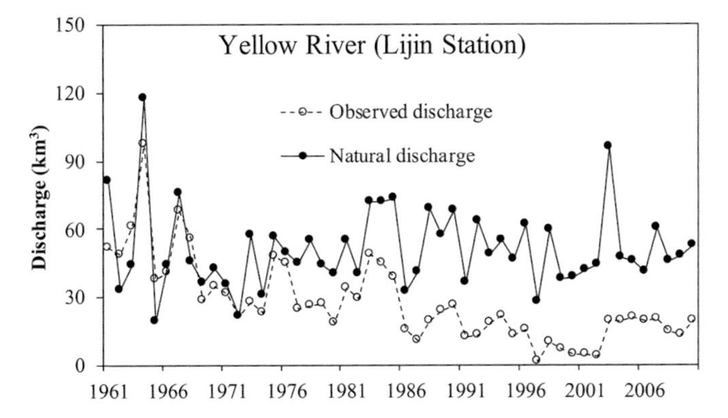

Figure 9.1 Observed compared to natural discharge at the Yellow River estuary, 1961–2010.
Figure provided to the authors by Yang Yonghui, adapted from research published in Zhou et al., 2019, p. 7

More dramatic is the drastic reduction in silt load during the twenty-first century (Table 9.2). The Yellow River, at least for the time being, is not even close to being the most silt-laden river in the world.

At the heart of Yellow River sustainability engineering since the sixteenth century CE has been the containment, flushing, and, usually with limited success, the reduction of the load of loess that gives the river its name (Dodgen, 2001, p. 18).[15] Extensive efforts to address the problem at source, through check dams, terracing, and extending groundcover in the form of trees or grass had limited results before the turn of the twenty-first century in the face of a growing population on the plateau. In the twentieth century, the annual mean silt load of the Yellow River was a world-leading 1.6 billion tonnes, reaching 60 kg/m^3 during the flood season, carried by a mean annual runoff of 55.9 km^3. In the first 15 years of the twenty-first century, this fell by 83 percent, to 274 million tonnes, while runoff declined to 45.2 km^3/year (Table 9.2). The decline in runoff and silt load was not due to a decrease in precipitation. In 2016, average precipitation across the basin was 7.9 percent higher than that of the 1956–2000 period, yet runoff at all stations was lower and, at the estuary, it fell by 76.1 percent (Ministry of Water Resources, 2016, p. 2).

The reasons for this dramatic decline are not entirely clear. A major research study carried out from 2012 to 2015 reported in Liu et al. (2016) sought to identify and quantify the drivers. The cumulative effect of engineering works seems to have been only part of the story. For example, in recent years, silt retention in the numerous small check dams has not been the principal element in reducing the silt load downstream. By far most of the 52,444 check dams have checked out; only 6,371 still have some capacity. Data from 2007 to 2014 indicate that they only contributed about 1 percent of total silt load reduction. The 853 reservoirs[16] provided over 300 million t/year of the total 320 million t/year reduction from impoundments, even though their aggregate storage capacity had already been reduced by 43 percent due to siltation.

A much larger reduction, of 1,254–1,411 million t/year, in that period was attributable to terraced fields (40.3 percent) and vegetation (trees and grass) (56.2 percent). It is possible that increased precipitation and air temperatures stimulated vegetation, which retained both precipitation and silt. This relationship cannot be confirmed or disproven (Liu et al., 2016).

Terraced fields and vegetation have contributed to reducing silt load, in part by markedly reducing runoff. An annual runoff reduction of 4.07 km^3 in the upper and middle reaches (from Xunhua to Tongguan) was allocable primarily to vegetative cover (2.41), followed by terraced fields (1.54), and insignificantly by check dams (0.138).

Further reductions in runoff were attributable to excess groundwater abstraction and evaporation from reservoirs and recreational surfaces. Assuming precipitation of 1966–2014 for the base of 2007–2014, 3.45 km^3 of the fall in natural runoff at

Huayuankou of 10.7 km^3 can be attributed to groundwater abstraction; mining abstractions and reservoir filling, by reversing groundwater–surface flows, reduced flow by nearly 1 km^3; and factors such as terracing and revegetation removed 5.7 km^3.

Liu et al. (2016) do not see much potential for further expansion of vegetative cover beyond another 20–30 years. The installation of two more major reservoirs could reduce silt at Tongguan to 100–460 million t/year by 2050–2060, but even massive reservoirs in the middle and lower reaches are at risk of silting up. Once they reach their silt retention periods, the silt load could rebound to 350–550 million. No other reservoirs are planned. In addition, in their estimation, an extreme event such as the 1919 floods could overwhelm current engineering efforts and lead to a restoration of the original silt load.

Jia (1995) had a more optimistic assessment, showing that in the absence of anthropogenic pressures, vegetative coverage could reach 80 percent of the Loess Plateau and erosion could be reduced by 90 percent. These pressures have eased considerably thanks to both population trends and government policy. With the mass migration of farmers from marginal areas, including the Loess Plateau into China's cities, most of the steep slopes previously cultivated have been converted to forest or grassland, while the availability of fossil fuels such as coal and natural gas to those remaining has reduced the share of rural energy from firewood. The closure of mountainous areas to tree felling or grazing has led to a dramatic increase in vegetative cover (Wang et al., 2016). The increase in ground cover has resulted in the dramatic fall in silt foreseen by Jia, implying that the annual sediment load of the Yellow River can be maintained below 300 million t/year.

9.3 CURRENT USES

In 2016, water abstractions from the Yellow River totalled 39.3 km^3, of which 32.2 km^3 (82 percent) constituted consumptive use, i.e., it was not returned to the river. Groundwater added 12.2 km^3, with 9.1 km^3 (75 percent) consumed. Although its share has declined significantly,[17] irrigation continues to dominate both abstractions and consumptive use, with over three-quarters of surface and over one-half of groundwater use (Table 9.3). Domestic and municipal uses, reflecting urban growth, constitute a relatively small share of surface water use, but rely relatively heavily on groundwater for nearly half their supply.

Groundwater reserves in the basin average 35.6 km^3 per year, with considerable fluctuation due to strong dependence on precipitation and surface flows. Only 11.2 km^3 do not duplicate surface water sources (YRCC, 2013). In the dry irrigated areas in the upper reaches and along the flat lower reaches of the river and its interbasin transfer routes, the water table tends to be high, posing a threat of secondary salinization of the soil. Elsewhere, especially under large cities such as Xi'an and Taiyuan, the water table has fallen, forming cones of depression of varying extents and depths in the shallow aquifer, and creating problems of land subsidence (Jia et al., 2019a, p. 2).

9.4 STRESSORS AND RELIEVERS IN COMING DECADES

The Yellow River basin has the second highest water use-to-availability ratio of the ten major river basins in China and is highly vulnerable to climate change (Xia et al., 2012, pp. 528–529). By some measures, it is already oversubscribed. Factors commonly cited as contributing to this situation are the expansion of irrigation, increases in population, urbanization, industry and mining, climate change, and environmental requirements for flushing sediment. Nonetheless, we cautiously don light rose-colored glasses in looking at the potential effects of these stressors on the future sustainability of the Yellow River.

9.4.1 Irrigation

The breadbasket of China has shifted toward the dry northwest in recent times, with development of urban sprawl in the south,

Table 9.3 *Abstractions and consumption of Yellow River water by sector and source, 2016*

2016	TOTAL (km^3)	Irrigation	Industry	Municipal	Domestic	Environment
Surface						
Abstracted	39.3	30.7 (78%)	4.3 (11%)	0.8 (2%)	1.9 (5%)	1.6 (4%)
Consumed	32.2	24.9 (77%)	3.5 (11%)	0.7 (2%)	1.5 (4%)	1.6 (4%)
Groundwater						
Abstracted	12.2	6.99 (57%)	2.3 (19%)	0.6 (4%)	2.1 (17%)	0.3 (2%)
Consumed	9.1	6.63 (54%)	1.5 (12%)	0.3 (3%)	1.4 (11%)	0.1 (1%)

Source: Yellow River Commission, 2016 黄河水资源公报 (Bulletin of Yellow River Water Resources): 23, 26.

Notes: Irrigation here is primarily for agricultural crops but includes other productive rural uses (forestry, animal husbandry, fisheries). Municipal is public uses. Domestic includes both urban and rural households.

east, and northeast. From the 1970s, much of the increase in irrigation allowing the growth of dryland crops, notably wheat, was provided by tube wells in the plains on either side of the lower Yellow River, augmenting surface diversions (most notably, the People's Victory Canal) built or rehabilitated in the 1950s. Increasingly, the focus of grain production is shifting to surface irrigation areas along the upper reaches of the Yellow River itself. Yet agriculture and other primary industries are not major contributors to the basin's economy, providing 9 percent of value added compared to 47 percent for secondary and 43 percent for tertiary (Wang, 2018). Its share is likely to fall even lower as the basin's per capita income continues to increase, with concomitant changes toward a developed economy structure.

9.4.2 Population

Population in the basin increased significantly in the second half of the twentieth century, from 41 million in 1953 to 84 million in 1982, to 107 million in 1997 (Fu and Chen, 2006, p. 4), to 117 million in 2007 (YRCC, 2013), and to 130 million in 2015 (Wang, 2018). Since the 1990s, with migration out of rural areas, population increase has been entirely in cities.

Population per se is not a major driver of water demand compared to irrigated agriculture and other sectors (notably mining and industry, see later). With the projected peaking of population by 2030, and decline thereafter (Table 9.4), population is unlikely to be a major factor in the foreseeable future.

The population of the Yellow River basin is currently about 120 million, not counting those in nearby areas receiving inter-basin transfers from the Yellow. Official statistics are not available for recent years for the basin, but in 2000 the basin population was 8.7 percent of that of China. Figures published for 2008 (Wang et al., 2013, p. 11) yield a share of 8.5 percent, indicating that the share attributable to the Yellow is stable or falling compared to that of the entire country. China's 2017 population is estimated at about 1.4 billion. United Nations median projections for 2040 and 2060 are 1.4 billion (1.35–1.45 at P = 0.95) and 1.28 billion (1.14–1.4 at P = 0.95), respectively (available at: https://esa.un.org/unpd/wpp/Graphs/Probabilistic/POP/TOT). Taking 8.7 percent as an upper limit for the share of Yellow River basin population in that of the entire country, projections for 2040 and 2060 are 117–126 million (median 122 million) and 99–117 million (median 111 million), respectively. In other words, population pressure per se is likely to ease over the planning horizon.

9.4.3 Urbanization

Urbanization may have an impact on water demand as well as quality and, possibly, siltation. The basin population is slightly more rural than the average for all China, 45.8 percent compared to 50 percent in 2010. This gap may be expected to close over time as China continues to become more urban, due to more rapid economic growth than average in the basin than the country as a whole and programs devoted specifically to the development of new urban areas adjacent to current population centers such as Xi'an, Zhengzhou, and Jinan (Zhang et al., 2019). Both the Yellow River basin and China are projected to be about 80 percent urban by 2050. Urban areas are expected to increase their share of total water use as their populations and geographical extents grow. Even if urban uses double, however, agriculture would still be the largest user, if anyone were left on the farm to do the irrigating.[18] If migration is from poorer rainfed areas upstream, or from midstream loess areas due to stabilization of soil, urban domestic water demand may be offset to some degree by a decline in rural demand. Manufacturing has been a significant user of urban water in the past, but this is unlikely to grow significantly because of improved water efficiency and changes in the economic structure toward a service-based economy.

9.4.4 Mining and Fossil Fuels

Extractive industries and power generation, which are not usually located within cities, may be another matter altogether. The middle and upper reaches of the Yellow River contain a significant portion of China's energy reserves and related chemical industry, which are being developed intensively in the present economic strategies. The demands of this sector on water quantity and quality in the most arid part of the basin have raised concerns over growing conflicts with other priority water uses,

Table 9.4 *Population of China and the Yellow River Basin (YRB) in the first half of the twenty-first century (million persons)*

	2000	2005	2010	2020	2030	2040	2050
China	1,267	1,308	1,334	1,425	1,441	1,417	1,364
YRB	110	114	116	124	125	123	119
YRB/China(%)	8.7	8.7	8.7	8.7	8.7	8.7	8.7

Sources: for China: for 2000–2010, *China Statistics Yearbook*, various years; for 2020–2050, United Nations, 2017, DVD edition. For the Yellow River Basin: for 2000, Kusuda et al., 2009; for 2005 and 2010, unpublished basin planning data; for subsequent years, assuming share of national population is maintained at 8.7%.

such as irrigated agriculture and the environment (e.g., Peng et al., 2011; Tan, 2013). The competition could become particularly acute in mining regions and energy bases in the heavily irrigated arid regions of Ningxia and Inner Mongolia (Xiang et al., 2017). In addition, coal mines have in some areas affected groundwater flows that are linked to runoff (Liu et al., 2016, pp. 284–297).

Nonetheless, there are reasons to be cautiously optimistic that the impact of mining and the thermal power industry may not be as great as feared. Increases in demand from these sectors may be relatively modest. There is considerable scope for reducing water demand in agriculture without reducing production, even if irrigation is maintained at the current level (Xiang et al., 2017). Some of China's pilot projects on water trading in Ningxia and Inner Mongolia are aimed at providing a mechanism for transfer of water from their extensive but high consumption irrigated areas to the energy sector (Zhang, 2012, pp. 46–48; Svensson et al., 2019).

9.4.5 Climate Change

Table 9.5 shows climate change projections for the 2040s and the 2070s using the results of the multi-model ensemble of IPCC RCP4.5 scenario and a base period of 1996–2005. Under this scenario, annual average temperature will increase by 1.4 °C and 2.0 °C in the 2040s and 2070s, respectively, while annual average precipitation will increase by 8.1 percent and 9.7 percent. These changes are somewhat lower than earlier CGIAR projections based on the HadCM3 global circulation model using the SRES B2 scenario (Ringler et al., 2010, p. 685). In both cases, precipitation is expected to increase but not enough to compensate for higher evapotranspiration, resulting in declines in runoff and growing basin water deficits unless new sources are found, or demand reduced.

9.4.6 Aqua Ex Machina: South–North Water Diversion: Solution or Stressor?

Water from the Changjiang (Yangtze River) is the latest, perhaps most massive, engineering approach to providing water to north China that may incidentally relieve the Yellow of some of its transbasin water delivery burdens. Of the set of interbasin transfers called the South–North Water Diversion (Nanshui Beidiao), the Middle Route has crossed under the Yellow River and is supplying water from the Yangtze River basin all the way to Beijing and the coastal megalopolis of Tianjin; the more problematic Eastern Route has also emerged from under the bed of the Yellow River with a terminus in Tianjin. Neither of these supply water directly to the Yellow River, but the Eastern Route is intended to provide an alternative source of water to out-of-basin diversions from the Yellow, especially in Shandong Province and for urban and industrial uses. It failed to do so, at least in the initial two years of operation, due to a number of factors, including the cost of auxiliary conveyance facilities and unplanned improvements in end user water use efficiency (Chen et al., 2020). The technically difficult Western Routes are the only ones intended to directly supplement the water of the Yellow, and if realized, would provide adequate water for the needs of both mining and irrigation in the arid upper and middle reaches, especially in combination with desalination (Jia and Liang, 2020). These even more scaled-up engineering diversions have been put on hold since 2008 due to their expense, however.

9.5 THIS CHAPTER'S ESTUARY

In addressing the issue of sustainability of the Yellow River, it is necessary to consider what is to be sustained and the continuous interplay between human engineering, both structural and non-structural, and "unintended" but sometimes foreseeable consequences as the river follows its own will. The keys to sustainability, not only of water resources but the ecology in the basin, are to avoid the enormous losses that would be caused by silt deposition and excessive diversion of flows.

In the extensive arid[19] region along the lower portion of the river's upper reaches, siltation is not an issue but there is the unresolved question of whether there is enough water for

Table 9.5 *Projected change in temperature and precipitation in the Yellow River Basin*

	2040s temperature change (° C)	2070s temperature change (° C)	2040s precipitation change (%)	2070s precipitation change (%)
Yellow River Basin	1.4	2.0	8.0	9.7
upper reaches	1.4	2.1	9.1	11.0
middle reaches	1.4	2.0	6.8	8.1
down reaches	1.3	1.9	8.1	12.1

Data source: calculated from the simulation results of the IPCC RCP4.5 multi-modal ensemble (https://sedac.ciesin.columbia.edu/ddc/ar5_scenario_process/RCPs.html, accessed March 29, 2020).

extensive irrigation and a rapidly growing hydrocarbon-related mining and industry. Also, some of the flow must go on after that, to provide for the population and economic centers downstream and the delta ecology. On the way there, the river passes through the loessal middle reaches, where the sustainability issues are sedimentation and the abysmal quality of water in many of the major tributaries. In the lower reaches, the river's engineering mission is to build a plain with its heavy silt load. The human engineering mission, possibly made lighter due to recently reduced silt loads, is to prevent it from doing so, thereby sustaining life and property in a vast potential flood plain under the dikes. Due to the tremendous decrease of sediment from the Loess Plateau, this mission appears to be accomplished, with siltation and floods under control. At the same time, an adequate flow of both silt and water is necessary to sustain one of China's economic engines and the ecology and existence of the delta. Dikes and dams have shown their strengths and limitations, as has the top–down governance system. Both need to be at the very least supplemented by nonstructural and institutional innovations, all within a context of unprecedented changes in the economy, population, and climate.

REFERENCES

Ball, P. (2016). *The Water Kingdom: A Secret History of China.* Chicago: The University of Chicago Press, pp. 60–67.

Branigan, T. (2008). One-third of China's Yellow river "Unfit for Drinking or Agriculture," *The Guardian.* Available at www.theguardian.com/environment/2008/nov/25/

Chen, A., Wu, M., Wu, S. et al. (2019). Bridging Gaps between Environmental Flows Theory and Practices in China. *Water Science and Engineering*, 12(4), pp. 284–292.

Chen, D., Luo, Z., Webber, M. et al. (2020). Between Project and Region: The Challenges of Managing Water in Shandong Province after the South–North Water Transfer Project. *Water Alternatives*, 13, pp. 49–69.

Chen, J., He, D., and Cui, S. (2003) The Response of River Water Quality and Quantity to the Development of Irrigation Agriculture in the Last 4 Decades in the Yellow River Basin, China. *Water Resources Research*, 39, p. 1047. https://doi.org/10.1029/2001WR001234

Chen, Z. (1989). Flood Forecasting and Flood Warning System in the Lower Yellow River. In L. M. Brush, M. G. Wolman, and H. Bing-Wei, eds., *Taming the Yellow River: Silt and Floods.* Dordrecht: Springer, pp. 425–449.

China Water Risk (2012). *2011 State of Environment Report Review.* Available at www.chinawaterrisk.org/resources/analysis-reviews/2011-state-of-environment-report-review

(2017) *2016 State of Environment Report Review.* Available at www.chinawaterrisk.org/resources/analysis-reviews/2016-state-of-environment-report-review

Dai, L. (2015). A New Perspective on Water Governance in China: Captain of the River. *Water International*, 49, pp. 87–99.

Dodgen, R. A. (2001). *Controlling the Dragon: Confucian Engineers and the Yellow River in Late Imperial China.* Honolulu: University of Hawaii Press.

Dong, Z. (2020) *Greening the Yellow River for a Beautiful China.* China Water Risk. Available at www.chinawaterrisk.org/opinions/greening-the-yellow-river-for-a-beautiful-china

Fu, G. and Chen, S. (2006). Water Crisis in the Yellow River: Facts, Reasons, Impacts and Countermeasures. *Water Practice & Technology*, 1. https://doi.org/wpt2006028

Gippel, C. J., Jiang, X., Fu, X. et al. (2012). *Assessment of River Health in the Lower Yellow River.* Brisbane: International Water Centre.

Greer, C. (1979). *Water Management in the Yellow River Basin of China.* Austin and London: University of Texas Press.

Han, J., Onishi, A., Shirakawa, H., and Imura, H. (2006). An Analysis of Population Migration and Its Environmental Implications in China: Application to Domestic Water Use. *Environmental Systems Research*, 34, pp. 515–523.

Jia, S. (1995). Calculating Natural Erosion and Accelerated Erosion on the Loess Plain Based on Vegetative Cover: The Case of Anzhai County. *Water and Soil Conservation Bulletin*, 4, pp. 25–32.

Jia, S. and Liang, Y. (2020). Suggestions for Strategic Allocation of Yellow River Water Resources under the New Situation. *Resources Science*, 42, pp. 29–36.

et al. (2019a). *New Technologies, Strategies, Policies, and Institutions.* International Specialty Conference on Water Security. Beijing.

(2019b). A Portfolio of China's Urban Water Governance Sector: Administrative System, Coordination Problems and Policy Evolution. *International Journal of Water Resources Development.* https://doi.org/10.1080/07900627.20191668754

Jiang, X., Arthington, A., and Liu, C. (2010). Environmental Flow Requirements of Fish in the Lower Reach of the Yellow River. *Water International*, 35, pp. 381–396.

Klein, J. (2017). *A New Formula to Help Tame China's Yellow River. The New York Times.* Available at www.nytimes.com/2017/06/02/science/china-yellow-river-xiaolangdi-dam.html

Kong, D., Miao, C., Borthwick, A. G. L. et al. (2014). Evolution of the Yellow River Delta and Its Relationship with Runoff and Sediment Load from 1983 to 2011. *Journal of Hydrology*, 520, pp. 157–167.

Leonard, J. K. (1996). *Controlling from Afar: The Daoguang Emperor's Management of the Grand Canal Crisis, 1824–1826*. Ann Arbor: University of Michigan Center for Chinese Studies. Michigan Monographs in Chinese Studies, Vol. 69.

Liu, J., Cao, L., Liu, J., Wang, J., and Liu, Y. (2013). Sedimentation and Water Use in the Yellow River. In J. Wang, J. Zhao, H. Li, Y. Zhao, S. Peng, and X. Sang, eds., *South–North Water Transfer, and the Comprehensive Arrangement of Water Resources*. Beijing: Kexue Chubanshe, pp. 229–251.

Liu, X. et al. (2016). *Why Water and Silt Drastically Declined in the Yellow River in Recent Years*. Beijing: Kexue Chubanshe.

Ministry of Water Resources (2016). *2016 Statistic Bulletin on China Water Activities*. Yellow River Commission Bulletins, China. Available at www.mwr.gov.cn/english/publs/201806/P020180601369706877305.pdf

(2018). *2018 Statistic Bulletin on China Water Activities*. Yellow River Commission Bulletins, China. Available at www.mwr.gov.cn/english/publs/202001/P020200102601837201385.pdf

Moore, S. M. (2018) *Subnational Hydropolitics*. New York: Oxford University Press.

Nickum, J. E. (2004). Water and Regional Development in the Yellow River Basin. In C. Tortajada, O. Unver, and A. K. Biswas, eds., *Water and Regional Development*. Oxford: Oxford University Press, pp. 114–136.

Nickum, J. E. and Greenstadt, D. (1998). Transacting a Commons: The Lake Biwa Comprehensive Development Plan, Shiga Prefecture, Japan. In J. M. Donahue and B. R. Johnston, eds., *Water, Culture, & Power*. Washington, DC and Covelo, CA: Island Press, pp. 141–161.

Nickum, J. E., Jia, S., and Moore, S. (2017). Red Lines and China's Water Resources Policy in the Twenty-first Century. In E. Sternfeld, ed., *Routledge Handbook on China's Environmental Policy*. London: Routledge Earthscan, pp. 71–82.

Ottinger, M., Kuenzer, C., Liu, G., Wang, S., and Dech, S. (2013). Monitoring Land Cover Dynamics in the Yellow River Delta from 1995 to 2010 Based on Landsat 5 TM. *Applied Geography*, 44, pp. 53–68.

Peng, S., Wang, H., and Zhang, X. (2011). Development of Energy Resources and Heavy Chemical Industry Bases in the Upper and Middle Reaches of the Yellow River and a Strategy for Water Resources Regulation. *Zhongguo Shuili [China Water Resources]*, 21, pp. 28–31.

Pietz, D. A. (2015). *The Yellow River: The Problem of Water in Modern China*. Cambridge, MA, and London: Harvard University Press.

Ren, C., Wang, Z., Zhang, B. et al. (2018). Remote Monitoring of Expansion of Aquaculture Ponds along Coastal Region of the Yellow River Delta from 1983 to 2015. *China Geographical Science*, 28, pp. 430–442.

Ringler, C., Cai, X., Wang, J. et al. (2010) Yellow River Basin: Living with Scarcity. *Water International*, 35, pp. 681–701.

Rogers, P. (1996). *America's Water*. Cambridge, MA, and London: The MIT Press.

Seeger, M. (2014). *Zähmung der Flüsse: Staudämme und das Streben nach produktiven Landschaften in China [Taming of the Rivers: Dams and the Striving for Productive Landscapes in China]*. Berlin: LIT Verlag.

Shen, D. (2009) River Basin Water Resources Management in China: A Legal and Institutional Assessment. *Water International*, 34, pp. 484–496.

Svensson, J., Garrick, D. E., and Jia, S. (2019). Water Markets as Coupled Infrastructure Systems: Comparing the Development of Water Rights and Water Markets in Heihe, Shiyang and Yellow Rivers. *Water International*, 44, pp. 834–853.

Tan, D. (2013). *Water for Coal: Thirsty Miners?* China Water Risk. Available at www.chinawaterrisk.org/resources/analysis-reviews/water-for-coal-thirsty-miners-feel-the-pain/

Tan, D., Hu, F., and Lazareva, I. (2014). 8 Facts on China's Wastewater. China Water Risk. Available at www.chinawaterrisk.org/resources/analysis-reviews/8-facts-on-china-wastewater/

United Nations, Department of Economic and Social Affairs, Population Division (2017). *World Population Prospects: The 2017 Revision*, DVD Edition. New York: United Nations.

UNESCO Representative Office in China (2011). *Climate Change Impacts and Adaptation Strategies in the Yellow River Basin*. Beijing: Popular Science Press.

Wang, D. (2018). Water Pollution Control and Policy in the Yellow River Basin. *Democracy and Science*, 175(6), pp. 26–27.

Wang, H., Yang, Z., Saito, Y., Liu, J. P., and Sun, X. (2006) Interannual and Seasonal Variation of the Huanghe (Yellow River) Water Discharge over the Past 50 years: Connection to Impacts from ENSO Events and Dams. *Global and Planetary Change*, 50, pp. 212–225.

Wang, J., Zhao, J., Li, H. et al. (2013). *Studies on Integrated Water Resources Allocation of the South–North Water Transfer*. Beijing: Science Press.

Wang, S., Fu, B., Piao, S. et al. (2016). Reduced Sediment Transport in the Yellow River Due to Anthropogenic Changes. *Nature Geosciences*, 9, pp. 38–42.

Wang, Y. (2003) Water Dispute in the Yellow River Basin: Challenges to a Centralized System. *Woodrow Wilson Center China Environment Series*, 6, pp. 94–98.

Wang, Y., Ding, Y., Ye, B. et al. (2013). Contributions of Climate and Human Activities to Changes in Runoff of the Yellow and Yangtze Rivers from 1950 to 2008. *Science China Earth Sciences*, 56(8), pp. 1398–1412.

Wang, Y., Peng, S., Jiang, G., and Fang, H. (2018). *Thirty Years of the Yellow River Water Allocation Scheme and Future Prospects*. MATEC Web of Conferences 2018. https://doi.org/10.1051/matecconf/201824601083

Wang, Y., Zhao, W., Wang, S., Feng, X., and Liu, Y. (2019). Yellow River Water Rebalanced by Human Regulation. *Nature Scientific Reports*, 9, p. 9707. https://doi.org/10.10388/s41598-019-46063-5

Watts, J. (2011). Provincial Tug-of-War Waters down China's Yellow River Success Story, *The Guardian*. Available at www.theguardian.com/environment/2011/jun/28/water-yellow-river-china

Xia, J., Bing, Q., and Li, Y. (2012). Water Resources Vulnerability and Adaptive Management in the Huang, Huai, and Hai River Basins of China. *Water International*, 37(5), pp. 523–536. https://doi.org/10.1080/02508060.2012.724649

Xiang, X., Svensson, J., and Jia, S. (2017). Will the Energy Industry Drain the Water Used for Agricultural Irrigation in the Yellow River Basin? *International Journal of Water Resources Development*, 33, pp. 69–80.

Xie, J. X., Tang, W. J., and Yang, Y. H. (2018). Fish Assemblage Changes over Half a Century in the Yellow River, China. *Ecology and Evolution*, 8, pp. 4173–4182. https://doi.org/10.1002/ece3.3890

Xu, Y. (2017) *China's River Chiefs: Who Are They?* China Water Risk. Available at www.chinawaterrisk.org/resources/analysis-reviews/chinas-river-chiefs-who -are-they

Xu, Y. and Chan, W. (2018). *Ministry Reform: 9 Dragons to 2*, China Water Risk. Available at www.chinawaterrisk.org/resources/analysis-reviews/ministry -reform-9-dragons-to-2

Yang, H. and Jia, S. (2008). Meeting the Basin Closure of the Yellow River in China. *International Journal of Water Resources Development*, 24, pp. 265–274.

Yellow Basin Project Team (2010). *Yellow River Basin: Living with Scarcity. Synthesis Report submitted to the Challenge Program on Water and Food (CPWF)*. Washington, DC: International Food Policy Research Institute & Partner Organizations.

Yellow River Conservancy Commission (YRCC) (2013). *The Yellow River Basin General Plan (2012–2030)*. Zhengzhou: Yellow River Water Press.

Yu, L. (2006). The Huanghe (Yellow) River: Recent Changes and Its Countermeasures. *Continental Shelf Research*, 26, pp. 2281–2298.

Zhang, H., Yang, L., and Zhang, X. (2013). An Analysis of Indicators of Socioeconomic Development in the Yellow River Basin. *People's Yellow River*, 10, pp. 15–17.

Zhang, L. (2012). The Practice of Water Rights Reform in Northern China. In S. Jia, L. Zhang, Y. Cao, H. Yan, J. Li, and J. Nickum, eds., *When Water Rights Are Implemented in China: The Case of Gollmud*. Beijing: Shuili shuidian chubanshe, pp. 43–62.

(2016). *The River, the Plain, and the State*. Cambridge: Cambridge University Press.

Zhang, P., Bo, P., Li, Y. et al. (2019). Analyzing Spatial Disparities of Economic Development of Yellow River Basin, China. *GeoJournal*, 84, pp. 303–320. https://doi.org/10.1007/210708-018-9860-9

Zhonghuarenmingongheguo shuilibu shuiwensi [Hydrology Section of the Ministry of Water Resources, People's Republic of China] (1997). *Zhongguo shui ziyuan zhiliang pingjia tuji [Maps of China's Water Resources Quality Assessment]*. Zhengzhou: Yellow River Water Press.

Zhou, X., Yang, Y., Sheng, Z., and Zhang, Y. (2019). Reconstructed Natural Runoff Helps to Quantify the Relationship between Upstream Water Use and Downstream Water Scarcity in China's River Basins. *Hydrology and Earth System Sciences*, 23, pp. 1–15. https://doi.org/10.5194/hess-23-1-2019

Zhuang, C., Xu, J., and Chen, G. (2015). Sustainable Management of Water Resources in the Yellow River Basin: The Main Issues and Legal Approaches. *WIT Transactions on Ecology and the Environment*, 197, pp. 15–23. https://doi.org/10.2495/RM150021

Notes

1 At least in the currently accepted narrative. There are spoilsports, notably Pietz (2015, pp. 15–16), who observes that the "origin story" linking North China, the Yellow River and the origin of Chinese civilization "received powerful sanction from nationalist actors, including the revolutionary CCP [Chinese Communist Party]," yet archeological evidence points to a more widespread polynodal origin. Nonetheless, the dynasties that first linked together those nodes to create a unified empire were headquartered along the Yellow and its tributaries.

2 Some (e.g., Wang et al., 2019) separate the source regions from the upper reaches. This lightly populated area (with 650,000 people) is about 124,000 km^2 in extent and generates by far the bulk of the runoff of the entire river.

3 In 2012, the upper and middle reaches of the Yellow produced 50 percent of China's coal and 35 percent each of its oil and natural gas. (Jia et al., 2019b)

4 One of the most dramatic, and devastating, breaches occurred in 1048, when the river broke from its confinement and returned to its own engineering project of silting up the Huabei plain, laying waste to both economy and ecology for 80 years and probably fatally weakening the Northern Song Dynasty (Zhang, 2016). Eight hundred years later the river defeated control efforts by the emperor (Leonard, 1996) and water bureaucracy (Dodgen, 2001) and shifted its course in the last 200 km from south of the Shandong Peninsula to its present northeastward course. The human engineering has reflected both a "Confucian school," stressing the control of nature through restrictive dikes, and a "Daoist school," which would "allow the river greater freedom to deposit its silt and to find its own course" (Greer, 1979, pp. 33–34; Pietz, 2015, pp. 44–45).

5 A short-lived authority with the same name was established in 1933 under the Republic of China (Pietz, 2015, pp. 89–90)
6 Baidu, entry on the Yellow River Conservancy Commission of the Ministry of Water Resources (水利部黄河水利委员会), accessed 28 January 2020.
7 Elsewhere, this has been identified because of the dominance of technological and political imperatives over scientific or economic imperatives (Rogers, 1996) or an entrenched project-oriented bureaucratic culture (Nickum and Greenstadt, 1998).
8 Some scholars, such as Fu and Chen (2006), also point to population growth in the Yellow River basin, growing from 41 million in 1953 to 107 million in 1997, as a major factor, but as shown in Table 9.2, even in 2016, domestic consumption was only about 7 percent of the total. Population projections are dealt with in a later section.
9 There was an incentive for provinces to overstate their actual use and projected requirements. In the initial round in 1983, the provinces projected their water requirements for the year 2000 to total 74 km^3, well over available runoff and twice the amount eventually set as the limit. (State Council Document No. 16 [1987])
10 The Yellow River Basin is divided into 333 "first level water function zones" and 364 "secondary level water function zones," established beginning in 2000 by the YRCC and province-level administrations based on pollutant carrying capacity (www.docin.com/p-605809797.html). The first level zones are divided into protected zones, conservation zones, buffer zones, environmental restoration zones, and zones for (economic) development. There are about 5,000 of them in China. The second level is divided by uses, such as potable water, industry, agriculture, or recreation, which would be reflected in the five-grade water quality system. To make matters even more complicated, the former Ministry of Environment established its own set of water environmental function zones with different coverage (Jia et al., 2017b, p. 77).
11 In 1991, 5.9 percent of the 3,910 km of the mainstem surveyed were found to have water quality worse than Grade V, while 16 percent of the entire system including tributaries fell into this category. Only 5.8 percent of the surveyed basin had water quality at Grade II (potable without treatment). Only a small stretch of a headwater tributary above Lanzhou was considered to have water of pristine (Grade I) quality (Zhonghuarenmingongheguo shuilibu shuiwensi, 1997, pp. 31–32).
12 Even if all the unsurveyed areas, primarily upstream tributaries, were of high quality, the 2016 data indicate considerable improvement. Cognizance should be made that the water and environmental silos have separate calculations for water quality attainment, with the latter relying on a growing number of monitoring points. The environmental figures do not show dramatic improvement in recent years. In 2016, 59.1 percent of points recorded water with Grade I–III quality, with 13.9 percent below Grade V (China Water Risk, 2017). The share with acceptable quality fell after 2011, then stabilized, while the unfit for any purpose below Grade V category has fallen (Tan, 2013).
13 Standards for the Three Red Lines are restricted to two measures, COD and ammonium nitrate, however, and are linked to the human usage of thousands of "water function zones" (Jia et al., 2017b, p. 77). While these targets, if attained, may in the aggregate lead to improvements in water quality, they only cover a small portion of the twenty-four parameters used in setting the five-grade system, and allow for Grades IV (industrial) and V (mainly irrigation) if the zone is so designated.
14 Wang et al. (2013, pp. 1403–1404) found no decrease over a period from the 1950s to the 2000s at Lanzhou in the upstream, but average decadal declines in annual precipitation of 8.8 mm and 9.2 mm in the middle reaches, and 9.8 mm at Lijin.
15 Before River Commissioner Pan Jixun established this paradigm after 1565, focus was primarily on diverting the flood waters of the Yellow into multiple smaller channels, said to be following the practice of Yu the Great (Dodgen, 2001, p. 14).
16 In Chinese statistics, a reservoir (*shuiku* 水库) is an impoundment with a total storage capacity of at least 100,000 m^3.
17 Between 1988 and 1992, irrigation withdrawals accounted for a much higher 92 percent of surface and somewhat greater 61 percent of groundwater totals than in recent times, as shown in Table 9.3 (Chen, He, and Cui, 2003).
18 Economic benefits of economic growth have tended to be concentrated, at least relatively speaking, in the eastern and already urbanized areas of the basin (Zhang et al., 2019). China's massive migration flows in recent decades have tended to be from west and central areas to the east, including urban to urban, and from rural areas to cities (Han et al, 2006).
19 While this portion has an average precipitation of only 150–200 mm/year, so is marginally arid, average basin-wide precipitation for the Yellow River is 450–500 mm/year, well above 250 mm/year, the upper limit for arid regions (regional precipitation read from Chinese Academy of Sciences, Institute of Geography, *Map of Annual Precipitation in China* (中国年降水量图), Beijing: Ditu Chubanshe, June 1982; basin-wide precipitation from Chen, He and Cui, 2003, 1-1 and *Water Conservancy Encyclopaedia China* (中国水利百科全书) Vol. 2. Beijing: Shuilidianli Chubanshe, 1991, p. 856).

10 The Murray–Darling River Basin

Daniel Connell

10.1 INTRODUCTION

Sustainable river management in Australia's Murray–Darling Basin (MDB) has been the goal of governments since at least the 1980s when growing salinity problems and decreasing security of supply began to cause political conflict between and within the four states that share the catchment. Governments responded with various strategies, ranging from better coordination of the storages in the upper catchment to promoting behavior change by individual farmers and householders. Other strategies were designed to manage demand. As conflicts between stakeholders intensified, particularly during droughts, institutional changes were made to take better account of a wider range of stakeholders. The most recent attempt to balance these tensions was the national government's Murray–Darling Basin Plan ("Basin Plan"), which was approved in 2012. The Basin Plan is based on comprehensive sub-catchment water resource plans that are meant to coordinate the management of surface and groundwater, take account of development and climate change pressures, manage tensions between a diversity of stakeholders and satisfy national and international environmental obligations such as the protection of Ramsar wetland sites. To induce the states to cooperate with the preparation and implementation of the Basin Plan, the national government is investing A$13 billion in infrastructure to improve water security for towns and irrigators and acquire water entitlements for environmental rehabilitation (Figure 10.1).

10.2 PHYSICAL SETTING: THE MURRAY–DARLING BASIN

The Murray–Darling Basin is home to just over 2.6 million people and generates approximately 40 percent of the gross value of Australia's agriculture and pastoral production. Recent estimates calculated that water-related tourism in the basin is worth about A$8 billion per annum and irrigation-based agriculture about A$12-13 billion per annum (MDBA, 2015). In addition, the MDB also supports a substantial population and development outside the catchment in South Australia. In the year ending June 2014 the gross state economic product for South Australia was approximately $95 billon (ABS, 2015). Much of this productivity was only made possible by supplies from the Murray. The catchment is just over 1M square kilometers in size and has a diversity of landscapes, ecosystems, land uses, and climates ranging from the subtropical north to the temperate south with its long dry summers and wet winters. It contains the watersheds of two major rivers, the Darling and the Murray, and many tributaries. Adelaide, a city of nearly 1.5 million people and the capital of South Australia, receives on average about 55 percent of its water per annum from the Murray, a dependence that went up to 90 percent in the drought of 1982/3. The nearly four million people who are fully or partly dependent on the MDB for their water, and the extra economic productivity in South Australia made possible by supply from the same source, use less than 5 percent of all water extracted. The other 95 percent is used by irrigated agriculture. In the water year 2012/13, extractions for urban and industrial use inside and outside the MDB were 365 gigalitres (GL) (1,000 gigalitres equals about 800,000-acre feet), while irrigation used 11,357 GL (NWA, 2013).

10.2.1 Major Hydrological/Infrastructure Features

The main river in the MDB is the River Murray. After the river leaves the Great Dividing Range it drops less than 200 metres as it flows west over 2,500 kilometers to its estuary in South Australia. Most of the twenty-three major rivers of the MDB have their headwaters in the Great Dividing Range, which runs parallel inland from Australia's east and south east coasts. Along the western slopes of that range there is a string of major dams managed by the Murray–Darling Basin Authority (MDBA) and the four state governments, which re-regulate flows during the long dry summers and frequent droughts. For interstate water sharing the key storage is Hume Dam, completed in 1936. It is backed up by the even larger Dartmouth Dam further upstream on the Mitta Mitta River, Lake Victoria on the mid-Murray, and the Menindee Lakes on the lower-Darling. Storages managed by

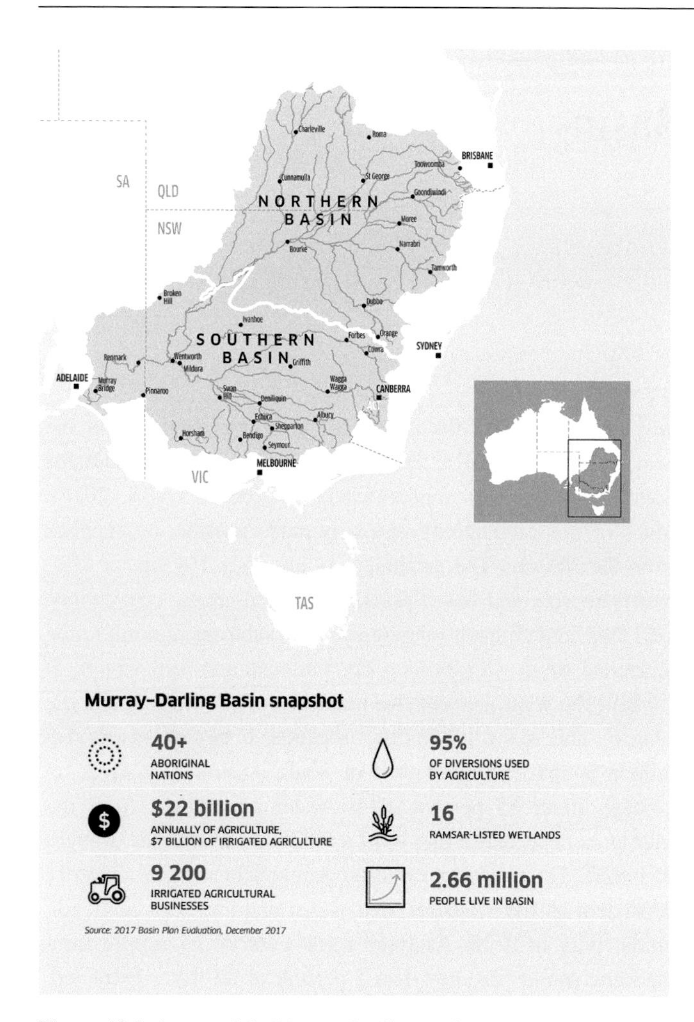

Figure 10.1 A map of the Murray–Darling Basin with a snapshot of key facts, which appeared in the MDB Authority's annual report for 2017/18. It should be noted that irrigated agriculture uses 95% of water extracted during a time of intense public debate about the need to move to sustainable management. The other 5% supplies about 2.6 million people inside the MDB. More than 90% of towns in South Australia outside of the catchment are also partially dependent on water from the River Murray.

the MDBA and the state governments have a combined capacity of just under 24,000 GL. After seepage, evaporation, and other losses, this makes possible surface water extractions averaging just over 11,000 GL per annum (MDBA, 2019b). In many years, the MDB is a closed basin with flows to the sea only maintained by dredging.

In the future under climate change conditions, the MDB is predicted to become drier and more variable. Across the catchment, annual average temperatures have been rising since 1910 and accelerating since 1960. There are two climatic zones. The climate in the north is subtropical with dry winters and wet summers; in the south, winters are wet and summers dry. From the perspective of population, economic activity, and river management, the most important climatic zone is that of the Murray in the south. According to a major study conducted by the Commonwealth Scientific and Industrial Research Organization (CSIRO) and published by the national government in 2015, global warming is pushing the subtropical climate band southwards, resulting in increased uncertainty along the Murray (CSIRO, 2015). The rapidity of change will vary according to the rate of global warming but for all scenarios considered it was projected that there will be,

- continued substantial warming of mean, maximum, and minimum temperatures
- more hot days
- fewer days with frost
- less rainfall in the cool season
- an increase in the number of heavy rainfall events, many of them more intense
- greater frequency of droughts
- increases in solar radiation and decreases in humidity in winter and spring
- increases in evaporation rates and reductions in soil moisture
- a harsher fire-weather climate[1]

To combine these projections with predicted rates of temperature change and different development pathways (such as growth in the number of on-farm storages and increased plantation forestry, both of which are already significantly reducing runoff) to produce numerical predictions of the volumes of water that will be available at particular dates in the future would require some very brave assumptions. However, this list of just some of the variables that will impact on water availability does indicate the importance of building governance capacity and human capital to deal with increased uncertainty and the likelihood of more frequent conflict between stakeholders in the future.

Early in the twentieth century, even before extractions became a significant factor, the River Murray was not a reliable source of potable water due to great variability in both quantity and quality. A central factor was the volume of water flowing down the main channel. High flows cause water to seep from the main channel into surrounding groundwater systems. Low flows result in groundwater seeping back. In the middle and lower reaches of the catchment, however, this groundwater is increasingly saline. In 1914, before major re-regulating dams such as Hume were constructed, salinity at Morgan on the lower Murray in South Australia spiked as high as 10,000 electrical conductivity units (ECs). (Sea water is about 40–50,000 ECs and humans can drink up to 2,500 ECs for short periods.) Since Hume Dam was completed in 1936, however, there has always been a reliable supply of drinkable-quality water all the way along the river (albeit subject to periodic rationing).

Linked but managed separately from the storages just described is the Snowy Mountains Scheme (SMS) situated in the highest section of the Great Dividing Range. The largest construction project in Australian history, the SMS was built between 1949 and 1974 to provide electrical power for the eastern seaboard and supplementary water for irrigation in the MDB. The SMS includes sixteen major dams, seven power stations, and more than 225 kilometres of tunnels (141 miles). Its primary feature is the redirection from east to west of the Snowy River, which drains most of the Australian snowfields. Previously, the Snowy flowed east to the Tasman Sea, but under the scheme it was re-routed west, where it now adds over 2,000 GL a year of reliable water to the Murray and Murrumbidgee Rivers. Modeled on the Tennessee Valley Authority and developed with assistance from the United States Bureau of Reclamation, the Snowy Mountains Scheme was designed to meet power generation and water supply objectives. It was successful from the perspective of the development-focused policy priorities of the post-war era, but the scheme subsequently became politically controversial in Victoria because of its environmental impacts on the Snowy River. The SMS is now being re-engineered to provide pumped-hydro reserve support to the electricity grid suppling major cities such as Brisbane, Sydney, and Melbourne in eastern and southern Australia.

One result of the historical legacy of the reengineering of the Murray in the early twentieth century is the segmentation of flows resulting from the building of a series of weirs and pound locks along the river and some of its tributaries. These were installed to make the system navigable for the paddle steamers that their proponents hoped would fuel economic development in the same way that river transport on the Rhine and the Mississippi rivers had done for their continents. In the MDB, building the network of weirs reflected thinking already obsolete because of the growth of railway networks. An additional major modification to the river system made in the 1930s and 1940s was the construction of five barrages between the islands in the estuary. This was to exclude the sea and maintain Lakes Alexandria and Albert as freshwater systems, partly in response to declining flows resulting from agricultural development upstream. Construction of the dams in the headwaters, weirs along the middle and lower reaches, and barrages at the estuary resulted in the River Murray and its tributaries becoming a long string of weir pools most of the time. Modern engineered rivers provide quite different environmental conditions compared with the highly variable flood- and drought-prone rivers that Europeans first encountered when they came to the region in the early nineteenth century.

10.2.2 Managing the Consequences of Modification

Water management in the Murray–Darling Basin is inseparable from salinity management. Millions of years ago much of the southwestern section of the catchment was inundated by sea water that left vast volumes of salt deposited in the catchment's subsoils. Since then additional salt has been slowly added by rainfall. Before settlement by Europeans starting in the early nineteenth century, most of this salt remained stored within the landscape. The clearing of native vegetation for dryland farming and the development of irrigation along waterways, however, increased the rate of rainfall accession to groundwater systems that, in many regions, mobilized large volumes of salt upward to land surfaces and laterally into streams. Because the rivers of the Murray–Darling Basin constitute a low energy hydrological system, most of this mobilized salt does not flow out to the sea but remains within the catchment, moving to high-value, low-lying agricultural land, and wetlands.

Water salinity is managed in several ways. The importance of maintaining regular flows in the main channel of the River Murray through releases from Hume Dam has already been discussed. In addition, in the central and lower reaches of the Murray, there are eighteen salinity interception schemes managed by the MDBA on behalf of the state governments. Each of these schemes has hundreds of groundwater tube wells intercepting saline groundwater before it seeps into the main river channel. After collection by the tube wells, the dissolved salt is pumped into evaporation basins. In total, these schemes capture over a half a million tonnes of salt each year that would otherwise go into the river. This elaborate network of physical infrastructure is backed by administrative arrangements that require the developers of new irrigation projects that will have salinity impacts to fund additional interception activities so that on balance salinity in-stream, as measured at Morgan in South Australia, is reduced. These arrangements are subject to detailed monitoring and the operation by the MDBA of a registry of salinity credits and debits for the three southern states. In addition, when preparing the water resource management plans required for all extraction programs, state governments must take account of all salinity impacts. Other salinity-related schemes focus on irrigation efficiency (to reduce salinity seepage back into streams), salinity zoning to restrict agricultural activity in districts where it results in high salinity impacts, the promotion of conservation farming to reduce the volume of rainfall going to groundwater, and the use of deep-rooted plants to draw down water tables.

Another problem in the MDB exacerbated by development pressures is that of blue green algae blooms resulting from combinations of nutrient runoff from agriculture, low flows, heatwaves, and other factors. In the summer of 1991/1992, the

Darling River experienced the longest recorded instance worldwide when over 1,000 kilometres were affected. The causes are not fully understood but the episode provided an example of the potential for dramatic unexpected events to occur in severely modified river systems. Another water quality issue that emerged during the drought of the early 2000s resulted from the increase in the extent of acid sulphate soils due to the exposure of lake beds in the estuary.

The impacts of river modification on fish life in the MDB have been drastic (Lintermans, 2009). In the early decades of European settlement, native fish such as Murray cod and golden perch were abundant, but their numbers are now much reduced. The segmenting of the River Murray into a string of stable, low-flow ponds through the building of a sequence of weirs to support navigation, combined with the large number of dams in the upper catchment and reductions in flow volumes due to irrigation extractions, made the riverine environment much more hostile for native fish. Now river depths are stable most of the time rather than highly variable. In addition, re-regulation of the rivers through dams holding back winter rains for release for irrigation during the dry summers has inverted the seasonal pattern of flow. Now the rivers run low in winter instead of flooding and are cold and deep in the summer instead of shallow and warm. These conditions are, however, very suitable for the unpopular carp that now constitute up to 90 percent of the biomass in the rivers of the southern basin.

Why is the condition of riverine environments important? The environment matters for many reasons. For some there is a concern for other species and the continued existence of wild places for their own sake, independent of human utility. For others there is the imperative to maintain the biophysical dynamics of riverine environments as functioning sustainable ecosystems that deliver a range of ecosystem services and positive living environments for people. There are many economic benefits to be gained from high quality river environments, including water-based recreation and tourism that draw large numbers of people to visit each year as well as benefiting residents who live there permanently. Inland river-based urban centers are often significant retirement centers, but they can only hold their populations and avoid the economic impacts of population decline while they continue to be attractive places to live.

10.3 WATER POLICY AND MANAGEMENT

Water policy and management in the MDB is divided between six jurisdictions: the state governments of New South Wales, Victoria, South Australia, and Queensland, the Australian Capital Territory, and the national Commonwealth Government, which has overall responsibility and provides most of the funding for river-related programs (even when they are delivered by other governments).

10.3.1 The Legacy of the Past

Through the late nineteenth and most of the twentieth centuries, water management and entitlement systems evolved largely independently in each of the four states in the MDB with little or no concern for what was happening across borders. Each state controlled the distribution of water through powerful government departments, but this did not result in uniformity between irrigation districts. Managers were encouraged to adapt to local circumstances. Water was heavily subsidized and management practices poorly documented. For each region, the water management regime was the accumulation over time of management responses to specific issues and events working within the very general framework created by state government legislation and departmental principles. The role of water managers was to expand irrigation in their regions to get the best returns on the investment by governments in dams and other infrastructure. The focus was on the distribution of water to irrigators with only marginal concern for environmental values.

A similar set of policy/management regimes developed in parallel for groundwater. Even in those regions where there was high connectivity between surface and groundwater systems, the two were defined and managed separately (with high risk of double counting the resource). Incremental development through the twentieth century put these ad hoc arrangements under increasing pressure. Over time, the rising level of extractions combined with climate variability to significantly reduce the frequency of years in which the full allocation could be supplied. The result of this plethora of water management regimes and growing supply instability was that by the late twentieth century state governments were increasingly unable to manage effectively without greater coordination across borders and more investment by the Commonwealth (the government that collects most of Australia's growth taxes and is best positioned to fund new initiatives).

The water reforms of the 1990s and 2000s were designed to remedy the more obvious defects resulting from over a century of near autarkic water management in the four states. Between the northern and southern parts of the MDB there is great variation in hydrological conditions, types of irrigation, and water governance. This reflects several differences. Irrigation in the southern MDB developed in the late nineteenth and early twentieth centuries with governments leading the way. Governments built the dams and other infrastructure. Irrigation districts were established as government-run operations with water managers exercising detailed supervision of settlers who could be evicted for unsatisfactory performance. Many war veterans were settled along the rivers through these schemes. The relationship between governments and these communities eventually matured and officials pulled back to some degree, but still today the effect of this history is visible in the heavy reliance on government assistance in those older irrigation districts compared with the areas that developed more recently in the north.

Even in the south there were significant differences between the states. The governments of Victoria and South Australia have always been much more involved in the details of irrigation management and invested more heavily in both infrastructure and supervision than New South Wales. In part, this reflected the types of irrigation in each state. Victoria and South Australia concentrated more on viticulture, horticulture, and dairy. They allocated lower proportions of flow so they could provide higher security of supply. New South Wales has traditionally had a greater emphasis on annual crops such as rice, where intermittent supply is less of a threat. If there is a year every now and again when rice cannot be grown, that is not really a problem. A new crop can be planted in the following year. If an orchard dies from lack of water, however, it will be several years before re-plantings will once again provide a harvest. Consequently, New South Wales has traditionally allocated a higher proportion of flow because its risk profile is different (this contrast also reflects differences in soils). Its supervision of individual irrigators has also been more minimal than that of the other two southern states.

Irrigation in the northern part of the MDB developed in the final decades of the twentieth century – much later than in the south. The north was less suited to the building of large dams, and governments were no longer interested in that type of investment. By contrast with the southern MDB, rainfall in the north comes in the summer and rivers have a different hydrological character. Most of the storage is private and off-river and much of the water is captured on the floodplain before it enters designated streams where it comes within the official management framework. The largest of the northern irrigation farms is Cubbie Station in southern Queensland, which has the capacity to capture more than 400 GL from overland flows (Davies et al., 2018) (the annual usage of Sydney, a city of over 4M people is about 600GLs). Cubbie pays very little for its water – in part reflecting the fact that it has no dependence on government infrastructure – and it has been successful in the courts and through the political system in deflecting a number of government attempts to rein it in or take it over. It is arguable that under the legal systems and within the political cultures of Victoria and South Australia the results of such contests would have been different and an agricultural business such as Cubbie Station would never have been allowed to develop in the first place.

In the southern MDB during the late nineteenth century, determination to develop the Murray River corridor through the semi-arid desert of the interior was a prime factor behind the push for national federation, which was achieved in 1901. Sharing the waters of the Murray was a challenge because the river was the border between the up-river states of New South Wales and Victoria, and in turn they had to supply South Australia, now empowered through its membership of the new national government – the primary long-term source of development funds. Water sharing arrangements also had to take account of the great variation in annual flows. Eventually, an interstate water-sharing framework based essentially on proportional shares of available flow was negotiated. This formula has proved robust through more than a century of intermittent droughts, some very severe. During wet and average years, the framework gives the two upper river states equal shares of the water in Hume Dam with a requirement to supply South Australia with a set volume. Whenever drought conditions are declared, however, the water remaining in Hume Dam is shared equally between the three southern basin states (in addition, each state retains full rights to water in their tributaries flowing into the Murray). Within states, although there was great variation in the details from district to district, water entitlements for individual license holders had different levels of security of supply reflecting the type of activity being supported. Urban centers have the highest priority. Irrigated orchards and vineyards, which take a long time to replace if devastated, had high security entitlements, while annual crops such as rice and cotton had much lower security entitlements. In recent years, water entitlements have been given stronger legal definition and there is now a well-developed water market in the MDB, particularly within states. The water needs of different activities are now mediated through the dynamics of the market rather than the decision of government officials. In times of crisis, however, it is still possible for governments to intervene and give priority to urban centers (as a result of the recent reforms, water for "critical human needs" is the most secure of all water allocations (MDBA, 2019a)).

10.3.2 Government Responses to Crisis

The 2000s drought was the most severe since records began in the 1890s. At the height of the crisis the six governments in the MDB were able to maintain economic returns for irrigated agriculture at near pre-drought levels by imposing drastic restrictions and promoting large-scale water trading (Kirby et al., 2014). Institutional change was widely accepted as necessary to strengthen basin-wide management. There were several options. The one chosen by the national government was that of a takeover of high-level water planning through the enactment of the Water Act 2007 and the preparation of the Murray–Darling Basin Plan. Under this new legislation, the states were relegated to implementation of the national plan each within their own borders. This was a significant break from the past. Since 1914/15 policy and implementation of the interstate water sharing program had been the joint responsibility of all governments operating through a confederate arrangement underpinned by identical legislation passed in all parliaments (Connell, 2007; Guest, 2017).

To understand the underlying principles of the Water Act 2007 and the Plan it is useful to examine the Blueprint for a National Water Plan, an influential statement published in 2003 by a group of senior biophysical and social scientists known collectively as the Wentworth Group and coordinated by World Wildlife Fund Australia. Their blueprint brought together a large body of research in a form that lent itself to translation into government policy. The Wentworth Group proposed a reformed water entitlements system within a strong regulatory framework meant to encourage water trading but also to ensure environmentally sustainable management. Publication was followed rapidly by the National Water Initiative (NWI), which was approved in 2004 by the Council of Australia Governments (COAG), a high-level consultative body that brings together the national government's Prime Minister and the premiers of the state governments (COAG, 2004).

According to the NWI, the tensions between the many demands that are placed on hydrological systems should be managed through the development of comprehensive regional water plans. It is through their preparation that sustainable management, managing climate change, the interests of competing stakeholders, and the demands for economic growth are to be reconciled. Water plans are to include secure water access entitlements; statutory-based planning; statutory provision for environmental and public benefit outcomes; plans for the restoration of over-allocated and stressed systems to "environmentally sustainable levels of extraction"; the removal of barriers to trade; clear assignment of risk for future changes in available water (governments would be responsible for the financial impacts of policy change but drought and climate change would be the responsibility of water entitlement holders); comprehensive public water accounting; policies focused on achieving water efficiency; capacity to address emerging issues; and much more. The water plans are also to provide for "adaptive management of surface and groundwater systems," with their connectivity recognized where it is significant. In addition, water plans should take account of indigenous issues by planning for indigenous representation in water planning "wherever possible" and provision for indigenous social, spiritual, and customary objectives "wherever they can be developed." They should also include allowance for "the possible existence of native title rights to water in the catchment or aquifer area" (COAG, 2004).

With some modifications the NWI was incorporated into the national government's Water Act 2007 and its MD Basin Plan approved in 2012. States retain control over their water resources but must ensure that their water management activities are consistent with the overall Basin Plan. Compliance is rewarded by access to substantial national government-funded programs. Working within the parameters established by the Basin Plan, each of the four basin states is required to develop its own subplan (underpinned by regional water resource plans for each of thirty-six designated sub-catchments in the MDB), subject to accreditation by the relevant Federal Minister. This allows for different approaches in each of the four states (and the Australian Capital Territory). Central to the Basin Plan is a "sustainable diversion limit" set for the MDB as a whole. At the basin level there are plans for the environment, water quality, and salinity. The Basin Plan identifies key environmental assets (such as Ramsar wetlands) and core ecological functions that must be maintained. It is designed to suit a wide range of circumstances – as is appropriate for a highly variable climate – and is meant to allow for potential risks such as climate change, bush fires, and new agricultural activities that could change run-off patterns.

10.4 PROGRESS REGARDING SUSTAINABLE MANAGEMENT

10.4.1 Factors and Context

The current water management framework now in place in the MDB is the product of more than a century of institutional development. To summarize a complex situation, the basin context includes,

1. An ongoing administrative framework with a central coordinating organization for the whole MDB catchment. Since 1917 a catchment-wide organization responsible for coordinating the management of key storages, major infrastructure, and cross border flows has been in place (previously the River Murray Commission, then the MDB Commission, now the MDBA). This body, in its various forms, has a record of successfully taking on new functions such as environmental management and the research and communications projects that have preceded major expansions in its responsibilities over the last half century (i.e., the 3-year investigation of water salinity dynamics published in 1972, the Water Audit in 1995, and various projects examining threats to future inflows in the 2000s).
2. A coordinated network of engineering structures that allow for efficient water sharing across the catchment and across borders. However, despite expansion in the 1980s to include the River Darling and its tributaries, arrangements in the southern section based on the River Murray are far more developed.
3. Integrated surface and groundwater management (or at least serious plans to introduce it). Before the Water Act 2007 and the Basin Plan, groundwater was administered by

the states separately from surface water. Groundwater was the exclusive responsibility of the states while surface water was coordinated to a significant degree by the cross-border MDB arrangements. Now, under the Basin Plan, the states are preparing integrated regional plans combining both types of water for submission for approval by the national government's Murray–Darling Basin Authority.

4. A robust proportions-of-available flow approach to water sharing, ranging from interstate sharing arrangements to individual extractors who have water entitlements of varying levels of security that take into account the characteristics of differing types of consumption (now somewhat modified by the operation of a cross-basin water market for irrigation water). This allows for a reasonably well-ordered scaling back in times of drought with a distribution of costs and benefits during times of scarcity that is socially well accepted.). Even during the record drought year of 2006/2007, flows to towns throughout the catchment (and beyond it in South Australia) were still maintained. This history of reliable delivery also reflects arrangements that allow governments to intervene decisively in times of crisis to distribute water in accordance with what they define as public policy priorities (i.e., to supply "critical human needs").
5. A well-developed system of water markets for irrigated agriculture, including across state borders (even if the original vision of seamless trading across borders and between uses envisaged by the Council of Australian Governments in 1994 is still a long way from being realized). An indication of the size of the water market is provided by the water accounting figures for the water year 2012/2013 (NWA, 2013). In that year extractions for irrigation were 11,357 GL and the volume of water traded was 6,058 GL (i.e., over 50 percent). As already noted, during the peak of the severe 2000s drought in 2006/2007, the water market across the southern MDB allowed water to be moved from low to high value uses and largely preserved economic returns from irrigation at near average annual levels (Kirby et al., 2014). The water market has also provided a vehicle through which governments have been able to purchase and move around substantial volumes of water for environmental rehabilitation managed by the national government's Commonwealth Environmental Water Holder.
6. Institutionally embedded processes for achieving environmental and sustainability goals (even if the volumes currently being restored are widely regarded by researchers such as the Wentworth Group as insufficient). The national government's Commonwealth Environmental Water Holder provides the operational capacity to manage such waters according to basin-wide priorities. It also consults with regional communities about how that should be done.
7. A 10-year review process in the Basin Plan to allow adjustment to the level of extractions after reviewing the needs of the environment considering changing policy priorities and climatic circumstances. (When the Basin Plan was prepared, climate change predictions were excluded, a decision made after intense debate. Proponents of leaving them out argued that they would eventually be incorporated through the 10-year review process. Their critics argued that this built in a long delay for responses to predicted crises.)
8. An experienced water management workforce in the states and nationally with strong traditions of public service. Water management in the MDB has been a major function of state governments for over a century, with many employees spending a lifetime working in the sector.
9. Well-developed and tested relationships between governments, communities, and commercial irrigation organizations. Many irrigation-based communities in the southern basin were established by state government water management agencies in the late nineteenth and early twentieth centuries and have long had large numbers of locally based staff and ex-staff living within them. In most cases, the regional water distribution organizations now owned by private irrigation corporations were previously state government agencies.
10. Independent monitoring and auditing of water use in relation to water markets and other issues. Independent auditing was a significant feature of water management in MDB in the 1990s and early 2000s. In recent years budget reductions and irrigation industry pressure in favor of self-regulation have resulted in the loss of many of the most important programs. These losses include the elimination of the annual audit of the Cap on extractions introduced in 1996; the cancellation in 2012 of the Sustainable Rivers Audit, which was to track river conditions over time; and the abolishment in 2014 of the National Water Commission's obligation to report on implementation of the National Water Initiative. However, much of the enabling legislation is still in place and could be re-activated if governments decide once again that it would be appropriate.
11. A well-established practice of providing public information to support debates about policy. There has been a long tradition of commissioning major investigations to provide the foundations of future policy development. Examples of subjects investigated include salinity, the level of extractions, and development/climate change pressures. These knowledge projects have not invariably led to policy change – the decision by governments not to follow through

on the large investment in climate change research for the MDB is just one example – but the pattern over the last century is well established, and the MDBA website has reports on a wide range of issues.

12. A long tradition of public consultation. Even if frequently contested in practice, public consultation is an established feature of the governance landscape. The expansion in the 1990s of the MDB Commission's ambit to include more comprehensive catchment management was effectively supported by a raft of communications and education activities that made the environmental condition of the MDB a national issue. In recent years there have been several substantial communications campaigns carried out to support important policy initiatives (not always with success). One example of this chequered record was the effort to explain the various versions of the Basin Plan in the lead up to its final approval by the national parliament in 2012.

Central to the water reform programs of recent decades has been the need to define the requirements necessary for environmentally sustainable management. This gives the ongoing debate about the meaning of "environmental sustainability" a new urgency. There would seem to be two minimal criteria that need to be met for a modified environmental system to be defined as environmentally sustainable: Its environmental condition should be stable from a system-wide perspective and also politically acceptable to society in general (to allow long term continuity despite changing governments). Such a socially acceptable foundation has not been easy to create, however, essentially because of long-term opposition from irrigation-based communities. Since the institutional reforms in the 1980s, there have been at least eight major attempts to increase the proportion of flows that would be left in-stream after extractions (Connell, 2007). The first was the Salinity and Drainage Strategy, approved by the new MDB Ministerial Council in 1989. It was thought that by requiring a reduction in the salinity impacts of irrigation on the River Murray the water management practices and the volumes of water that mobilized salt into streams would also be restrained and reduced. While successful in controlling salinity impacts on water quality in the Murray, the strategy ultimately failed to achieve that wider goal. Subsequently a few other strategies were developed with the same aim. The second was the Natural Resources Management Strategy in 1990. That was followed by the Council of Australian Governments' rural water reform framework in 1994, the cap on further extractions in the mid-1990s, the Integrated Catchment Management Policy Statement in 2000, the Living Murray First Step project in 2004, the National Water Initiative in 2004, and most recently, the eighth attempt, the MD Basin Plan approved in 2012. Despite great effort over decades, however, it has proved difficult to make much progress in restoring the rivers of the MDB to anything that matches the oft-repeated government commitments to establish healthy, attractive riverine ecosystems. Why? One explanation is that water policy has been captured by the irrigation sector that consumes about 95 percent of the water extracted.

Some of the possible consequences of a narrow focus on providing water for irrigation were outlined in the second companion paper to the 5-year review on the MDB cap on extractions undertaken in 2000 (Marsden Jacob Associates, 2000). Discussing the implications should decision makers fail to implement an effective cap and reduce extractions to sustainable levels, the authors predicted that resource sustainability in the MDB would become a major issue. Under those circumstances they thought that continued irrigation growth would undermine the security of established producers and provide a disincentive to new entrants to irrigation-based farming. Degradation of the riverine environment and water quality would proceed at an accelerating pace and exacerbate tensions between irrigation groups and surrounding regions as water supply declined. Water trading would become more aggressive, and irrigated enterprises and communities across the basin would be increasingly sensitive to seasonal and climatic variation. Ultimately, as end-of-valley flows continued to fall and damage to riverine environments became starker, irrigation communities would become alienated from the wider Australian society, leading to protracted social conflict and their political isolation. An alternative to this bleak vision would see extractions for irrigation substantially reduced and economic productivity maintained through water trading and greater efficiency (as occurred during the 2000s drought). The water saved could then be used to restore riverine environments and promote the many social and economic interests that would benefit from sustainably managed, ecologically healthy rivers. Under those circumstances irrigation-based industries would have a more secure future.

10.4.2 The Human Factor

Some of the difficulties involved in managing large modified river systems were outlined early in the national debate about the future of the MDB by policy analyst Stephen Dovers (Dovers and Lindenmayer, 1997). He argued that the environmental sustainability challenges involved in river management are fundamentally different from other policy issues. They occur over much longer time scales and often cut across established administrative boundaries. Poorly defined but finite limits are common, but it is difficult to take them into account. Environmental systems are frequently subject to thresholds that result in significant loss when they occur, but which are hard to predict and difficult to reverse. There is also great uncertainty about the likely effects of policy choices when the costs and benefits often become clear only in the long term. Many impacts are cumulative and interact with each other, and long-established patterns of management can suddenly produce vastly different results compared with the past. The sheer novelty of sustainability problems makes them difficult to handle within traditional modes of management and governance.

For governments, contemporary water management creates divisions that ignore traditional political divisions. Robust criteria for determining whether specific water policy issues should be treated as public or private are hard to develop. The predictive capacity of the various water-related sciences continues to be limited. Trends and general bio-physical processes can be documented at the larger scale but it is often difficult to identify links between particular actions and specific consequences with enough certainty to give managers, the courts, and the people who will be affected by their decisions confidence in the process or the results. Political systems also tend to have a low capacity to penalize those who are responsible for negative environmental impacts or to provide rewards to those who incur extra costs by adopting more sustainable management practices. Despite the growing interest in the so-called triple bottom-line approach, accounting systems are still quite ineffective in capturing the full environmental and social costs and benefits involved. Even more problematic, as the level of anthropogenic pressure grows, it is difficult to take account of ethical and moral considerations created by conflicting stakeholder demands.

In many if not most regions in the MDB, water management is more complex now than it was only a few decades ago. Experience with these systems when they were less modified is not a reliable guide for the present or the future. In addition to longstanding issues related to the level of extractions and salinization, the list of water management issues in the MDB in the early twenty-first century now extends to indigenous issues, acid soils, nutrient pollution, carbon depletion, changing patterns of rainfall, run-off and recharge, loss of native vegetation, threatened biodiversity, declining connectivity between floodplains and stream channels, changes to the seasonal pattern of flows, thermal pollution downstream of dams, degraded amenities, the social impacts of economic and environmental change, climate change, and more. Management of these issues is made even more complex by the fact that many involve different levels of government, occur on private land, or are influenced by the activities of commercial companies.

Some of the skills required to deal with such challenges were listed in 2002 by David Dole, then the General Manager of River Murray Water, the operational arm responsible for the management of storages for the MDB Commission (the precursor to the MDB Authority). Dole explained that the water manager of the future should have high level understanding of

- the hydrology and the hydraulics of whole river systems including their floodplains
- whole catchment land and water processes
- skills relevant to constructing, operating, and managing physical works
- processes that convey water from storage to root zone
- managing and treating drainage waters and how to achieve effective surface or sub-surface drainage
- the biophysical relationships between water, land, and environment, including skills in assessing the impacts of changing flow regimes on river ecosystems
- the water needs of natural systems as well as those of consumptive users
- the need for commitment to create sustainable natural resource systems while also achieving reasonable economic outcomes
- working with communities to jointly build a sound knowledge base that will underpin the negotiation of future actions
- the confidence to recognize the limits of current knowledge of the impacts of society on natural systems
- and the integrity to recognize and promote the need for change.

Where will water managers with this range of skills come from? This personnel gap is emerging at the same time as similar shortages are becoming evident in other spheres of Australian life. Whether the subject is social welfare, transportation, medicine, engineering, business, or sports administration, the level of skills required has increased dramatically in recent years. The shortage of skilled personnel to manage Australia's highly modified hydrological systems, which is already making itself felt, could well prove the greatest source of risk in the medium and longer term. It is likely that a similar assessment could be made for many other major river systems elsewhere.

10.5 SIGNIFICANCE OF THE MURRAY–DARLING BASIN: LESSONS FOR THE INTERNATIONAL COMMUNITY

What is the relevance of policy development in the Murray–Darling Basin for researchers and policy makers outside of Australia? This has been a sustained attempt to respond to a challenge experienced in many other countries, albeit one with quite different manifestations place to place. Since the 1980s there has been a serious attempt by the national and four state governments, with periodic changes in the political complexion of the parties in power, to implement sustainable river management in the MDB. The results are far short of that goal, but the saga reveals some of the difficulties that any society undertaking such a task is likely to face. Despite early confidence, it has proved difficult to define what "sustainable management" might mean in practice. Even more challenging is to link that concept to the predictions and early evidence of climate change. There is also the major issue of policy capture, in this case by the irrigated agriculture sector. It has been noted that preceding the current struggle to implement the Murray–Darling Basin Plan there were a few failed attempts to restore water to riverine environments (and, so far, there have been minimal attempts to learn from previous failures before trying again). Improved institutional design to empower a wider range of stakeholders who are able and willing to contest the development of policy appears crucial. The importance of comprehensive monitoring and auditing programs to support policy development also stands out as an issue. As the MDB experience indicates, without good information as a shared reference point debates can all too easily degenerate into ideological wrangles poorly linked to riverine realities. It is also important to recognize that discussions about sustainable management must involve more than just scientific environmental and ecological knowledge. If the needs, ambitions, and fears of communities are ignored, discussions between professionals regarding different policy options will rapidly be overwhelmed by raw politics.

REFERENCES

Australian Bureau of Statistics (ABS) (2015). *Gross State Product*. Statement for Year Ending June 30, 2014. 52220.0 Australian National Accounts: State Accounts, 2014/15.

Bruntland, G. (ed.) (1987). *Our Common Future: The World Commission on Environment and Development*. Oxford: Oxford University Press.

Commonwealth Environmental Water Office (2019). Available at www.environment.gov.au/water/cewo

Commonwealth of Australia (2009). *Water Act 2007*. Canberra: Commonwealth Parliament.

Connell, D. (2007). *Water Politics in the Murray–Darling Basin*. Sydney: Federation Press.

Council of Australian Governments (COAG) (2003). *Intergovernmental Agreement on Addressing Water Overallocation and Achieving Environmental Objectives in the Murray–Darling Basin*.

(2004). *Intergovernmental Agreement on a National Water Initiative. Attachment A: A Water Resource Policy* (1994).

CSIRO (2015). *Murray Basin Cluster Report: Projections for Australia's NRM Regions*. Canberra: Australian Government.

Davies, A. et al. (2018). Murray–Darling: When the River Runs Dry. *The Guardian*. Available at www.theguardian.com/environment/ng-interactive/2018/apr/05/murray-darling-when-the-river-runs-dry

Dole, D. (2002). Managers for All Seasons. In D. Connell, ed., *Uncharted Waters*. Canberra: Murray–Darling Basin Commission.

Dovers, S. R. and Lindenmayer, D. B. (1997). Managing the Environment: Rhetoric, Policy, and Reality. *Australian Journal of Public Administration*, 56(2), pp. 65–80. https://doi.org/10.1111/j.1467-8500.1997.tb01547.x

Guest, C. (2017). *Sharing the Water: 100 years of River Murray Politics*. Canberra: Murray–Darling Basin Authority.

Jones, G., Arthington, A., Gawne, B. et al. (2003). *Ecological Assessment of Environmental Flow Reference Points for the River Murray System: Interim Report Prepared by the Scientific Reference Panel for the Murray–Darling Basin Commission*. Canberra: Living Murray Initiative, Cooperative Research Centre for Freshwater Ecology.

Kirby, M. et al. (2014). Sustainable Irrigation: How Did Irrigated Agriculture in Australia's Murray–Darling Basin Adapt in the Millennium Drought? *Agricultural Water Management*, 145, pp. 154–162.

Lintermans, M. (2009). *Fishes of the Murray–Darling Basin: An Introductory Guide*. Canberra: Murray–Darling Basin Authority.

Marsden Jacob Associates (2000). Companion Paper 2 – Economic and Social Impacts. Review of the Operation of the Cap: Overview Report of the Murray–Darling Basin Commission. Canberra: Murray–Darling Basin Ministerial Council.

Murray–Darling Basin Authority (MDBA) (2015). *Water Markets in the Murray–Darling Basin*. Available at www.mdba.gov.au

(2019a). *Critical Human Water Needs*. Available at www.mdba.gov.au/river-information/water-sharing/critical-human-water-needs

(2019b). *Water Storage*. Available at www.mdba.gov.au/managing water/water-storage

National Water Accounts (NWA) (2013). *Murray–Darling Basin Water Access and Use*. Available at www.bom.gov.au/nwa/

Painter, M. (1998). *Collaborative Federalism: Economic Reform in Australia in the 1990s*. Cambridge: Cambridge University Press.

Powell, J. M. (1989). *Watering the Garden State: Water, Land, and Community in Victoria 1834–1988*. Sydney: Allen & Unwin.

Wallis, P. J. (2015). Governing Irrigation Renewal in Rural Australia. *International Journal of Water Governance – Special Issue*, pp. 1–18. https://doi.org/10.7564/14-IJWG41

Young, M. et al. (2008). A Future Proof Basin. In R. Manne, ed., *Dear Mr Rudd*. Melbourne: Black Rock Inc.

Young, W. J. (ed.) (2001). *Rivers as Ecological Systems: The Murray–Darling Basin*. Canberra: Murray–Darling Basin Commission.

Notes

1 Bushfires pollute storages with ash and reduce runoff by promoting forest regrowth in their catchments.

11 The São Francisco River Basin*

Antônio R. Magalhães and Eduardo Sávio P. R. Martins

11.1 INTRODUCTION

The São Francisco is an entirely Brazilian river. Its sources are in the southeast of Brazil, in the state of Minas Gerais. From there, it runs for 2,700 km until it meets the Atlantic Ocean, in the Northeastern region. Starting in Minas Gerais, it crosses the states of Bahia, Pernambuco, Alagoas, and Sergipe. The river basin also includes parts of the state of Goiás and the Federal District, Brasilia. From its source, the river runs northward to the city of Barra, in Bahia, and then north-eastwards, until it reaches the city of Cabrobó, in Pernambuco. Finally, it goes south-eastward until meeting the Atlantic Ocean between the states of Alagoas and Sergipe.

The whole basin covers 630,000 km^2 – this is, for example, twenty times the size of Belgium and about the same size as France. The basin is divided into four subregions, known as physiographic regions, which are the following (Figure 11.1):

(1) High São Francisco, running from the river's sources in Serra da Canastra in Minas Gerais, until the city of Pirapora, still in Minas Gerais. This means an extension of 702 km and an area of 100,076 km^2, representing 16 percent of the total basin. It has a population of 6.2 million inhabitants, most of whom live in the metropolitan area of Belo Horizonte.
(2) Medium São Francisco, from Pirapora, Minas Gerais, to the city of Remanso, Bahia. This part of the river runs for 1,230 km and covers the biggest area of the basin, 402,351 km^2, representing 53 percent of the total basin. Its population totals 3.2 million inhabitants.
(3) Sub-medium São Francisco, from Remanso, Bahia, to Paulo Afonso, Bahia, with 440 km of extension and 110,446 km^2 of area. It represents 17 percent of the total basin and has a population of 1.9 million inhabitants.
(4) Lower São Francisco, going from Paulo Afonso, Bahia, to the Atlantic Ocean, between the states of Alagoas and Sergipe. This part has 214 km, an area of 25,523 km^2, and 1.4 million inhabitants.

* We would like to express our very great appreciation to Marilia Castelo Magalhães for her valuable contribution to information provided in this chapter. We are also in debt to Juliana Lima Oliveira and Bruno Zaranza for their help in creating the figures and maps.

The São Francisco Basin lies in three of Brazil's main ecological regions. These are the Cerrados area (high and part of medium basins), the Semi-arid (part of medium, sub-medium, and part of the lower river basin), and the Mata Atlântica (Atlantic Forest, part of the lower basin). In the high basin there is an area that is, in fact, a transition between the Mata Atlântica and the Cerrados. Most of the river, however, runs through the Semi-arid lands of Minas Gerais, Bahia, Pernambuco, Sergipe, and Alagoas. Most of its

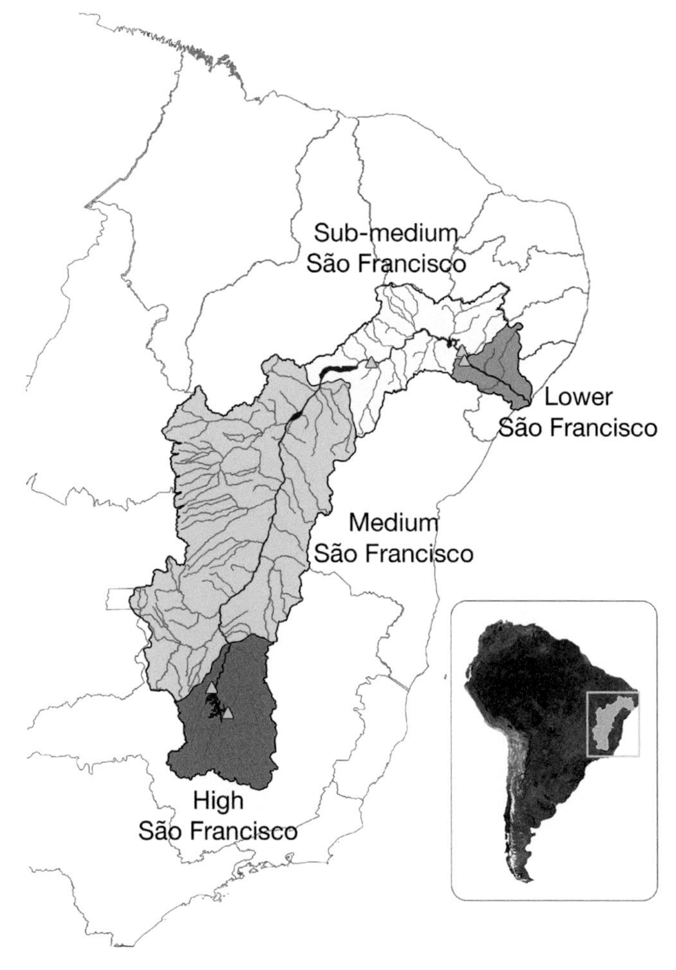

Figure 11.1 The São Francisco River basin and its rivers and hydrographic regions.
Figure by the authors

waters – approximately 85 percent of the river flow – originate in the Cerrados, a vast savannah region, with 72 percent coming from the state of Minas Gerais. The river crosses some of the driest parts of the Brazilian Semi-arid and brings life to ecosystems and to millions of people. The São Francisco River provides 69 percent of all the surface waters of northeast Brazil.

The São Francisco Basin contains 504 municipalities, which represent 9 percent of total Brazilian municipalities. The municipality is the smallest political division of Brazil but, given the dimension of the country, some may be rather big geographically. Most of these municipalities are in the states of Bahia (48.2 percent) and Minas Gerais (36.8 percent). Others are in the states of Pernambuco (10.9 percent), Alagoas (2.2 percent), Sergipe (1.26 percent), Goiás (0.5 percent), and the Federal District (0.2 percent).

The São Francisco River was (and continues to be) particularly important in Brazilian history and economy. It was called the "river of national integration," because at a time when transportation was scarce it allowed communication between the southeast, center, west, and northeast. Climate-wise, the São Francisco River is under different regimes and has different characteristics. Average precipitation, according to the Brazilian Institute of Meteorology (INMET), is 1,003 mm per year. These data are available on the website of the Brazilian National Water Agency (ANA, for Agência Nacional de Águas). Precipitation is much higher in the high São Francisco, lower in the middle, and especially in the submiddle, which coincides with the Semi-arid Northeast, and then higher again in the small area comprising the Atlantic Forest (or the remnants of it), near the mouth of the river.

The São Francisco is a large river. It has 168 tributaries, some of them large rivers as well. Of these, ninety-nine are perennial rivers. There are ninety tributaries coming from the right margin, and seventy-eight coming from the left margin. Among the main tributaries are the Rio das Velhas, with 689 km of extension; Rio Verde, with 458 km; Rio Paracatu, with 448 km; Rio Urucuia, with 381 km; Rio Parnamirim, with 345 km; Rio Pajeú, with 333 km; Rio Preto, with 315 km; and Rio Jacaré, with 297 km.

On average, the São Francisco River discharges 2,846 m^3/s (cubic meters per second) into the Atlantic Ocean. According to the ANA, discharges vary between 1,077 m^3/s and 5,290 m^3/s, depending on the season (ANA, 2017). The main contributors to the flow of the São Francisco River are the following tributaries: Rio das Velhas, Rio Paracatu, Rio Grande, Rio Urucuia, and Rio Corrente.

The São Francisco River is normally subject to floods and droughts. One important feature is that, as most of the waters come from the southeast region, which has a wetter climate, river discharges do not depend necessarily on the climate of the Semi-arid, where droughts used to be more frequent and severe. But if a reduction of rains in the upper part of the river, where it usually rains more, happens to coincide with a drought in the Semi-arid, it may become catastrophic for water availability and water users in the Semi-arid parts of the river. Also, floods used to occur with some frequency, inundating marginal lakes that serve as places for reproduction of fish and other animal species. With regularization brought by dams, this has been adversely affected.

A series of dams has been built in the basin and in the river itself, regulating water flow during different seasons. The basin is highly engineered. There are 466 dams in the high subregion of the basin, 214 dams in the medium subregion, 100 dams in the sub-medium, and 14 dams in the lower subregion (CBHSF, 2015a). Most of them are in tributaries. We are interested here in the seven big dams that have been built in the main river, mostly oriented for hydroelectricity production, and that have interfered in the flow and ecological features of the river. The next section has more information on these dams, which are the main engineering interventions in the river.

11.2 THE SÃO FRANCISCO AS AN ENGINEERED RIVER

Between 1954 and 1994, seven dams were built in the São Francisco river basin (Table 11.1).

The river suffered a significant change in the pattern of its seasonal flow variations with the construction of these dams. The biggest of them, Sobradinho, in the middle São Francisco, involved the creation of a lake that covers an area of 4.2 thousand square kilometres of good quality land. It also covered five cities that were seats of municipalities and directly affected 70,000 people. On the other hand, it allowed for regularization of the river (2,060 m^3/s in normal years), energy production (1,050 MW), irrigation, and water supply. Negative impacts included fishing, navigation, rainfed agriculture, and the environment. The second largest dam, Três Marias, created a lake of 19,000

Table 11.1 *Dams in the São Francisco River*

Name of dam and location	Year of completion	Storage capacity million m^3
Três Marias, high São Francisco	1962	19,000
Sobradinho, medium SF	1980	34,000
Itaparica, medium SF	1988	10,800
Moxotó (Apolônio Sales), sub-medium SF	1977	1,200
Paulo Afonso I-III and IV (two dams), sub-medium SF	1954–1979	0.150
Xingó, lower SF	1994	3,800

Source: CTEC-UFAL-University of Alagoas. Quantificação Preliminar do Aporte de Sedimentos no Baixo São Francisco. www.int-tmcocean.com.br.

million m^3 in the high São Francisco, with a major function of river regularization but also irrigation and water supply. The third largest dam, Itaparica, had a direct impact on 36,000 people that were dislodged from their living places. Irrigation projects were created to employ the dislodged population. In the Hydroelectricity Complex of Paulo Afonso, composed of several smaller dams (Paulo Afonso I, II, III, IV, and Moxotó), probably the socioeconomic and environmental impact was less important, because the production of energy results from natural water fall. The Xingó dam, though a big one, was built in the lower São Francisco, at the end of the São Francisco Canion, and the inundated area is negligible and unpopulated. Socioeconomic impacts are not high, but environmental impacts may still be considerable.

Três Marias is administered by Centrais Elétricas de Minas Gerais (CEMIG, Minas Gerais Power Plants), an energy company in the state of Minas Gerais that is in charge of energy generation and distribution. All the other dams in the São Francisco are controlled by the São Francisco Hydroelectricity Company (CHESF) , a federal energy corporation that belongs to the federal government. Since 1997, with a new water law, and the creation of the ANA in 2000, management of water has changed from sector (energy) to multi-use orientation. The ANA has a decision role in management. In fact, when water accumulation reaches less than 30 percent of total capacity in Três Marias and 20 percent in Sobradinho, decisions are made exclusively by the ANA. Over those levels, responsibility is divided between the ANA and the Operator of the National Electricity System (ONS) (ANA, 2017).

11.3 GROUNDWATER

There is also a variety of aquifers along the whole basin, which constitute a source of water that complements the surface water. There are forty-four aquifer systems in the São Francisco Basin, with a total annual recharge of 1,827.9 m^3/s, of which 20 percent corresponds to exploitable water availability, that is, 365.6 m^3/s, distributed along the whole basin. The rest of the recharge either feeds the rivers or goes to deeper areas of the aquifer. The Urucuia Aquifer is the main one, corresponding to about 41 percent of total underground water availability (148.2 m^3/s). There are interactions between streamflow and groundwater and interactions of karstic nature in some portions of the river. Water from aquifers is extracted through different kinds of wells throughout the São Francisco Basin and is used mostly for irrigation and water supply (Figure 11.3).

11.4 DROUGHTS AND FLOODS

Droughts are a recurrent problem in the Northeast region (De Nys et al., 2016).[1] According to the ANA, between 2003 and 2012, 273 municipalities entered drought emergency. In this 10-year period, 150 municipalities had more than five drought years. From 2010 to 2016, there were six years of drought in the Semi-arid region. Most of the population of the region are poor and family agriculture depends on rain, so drought has important

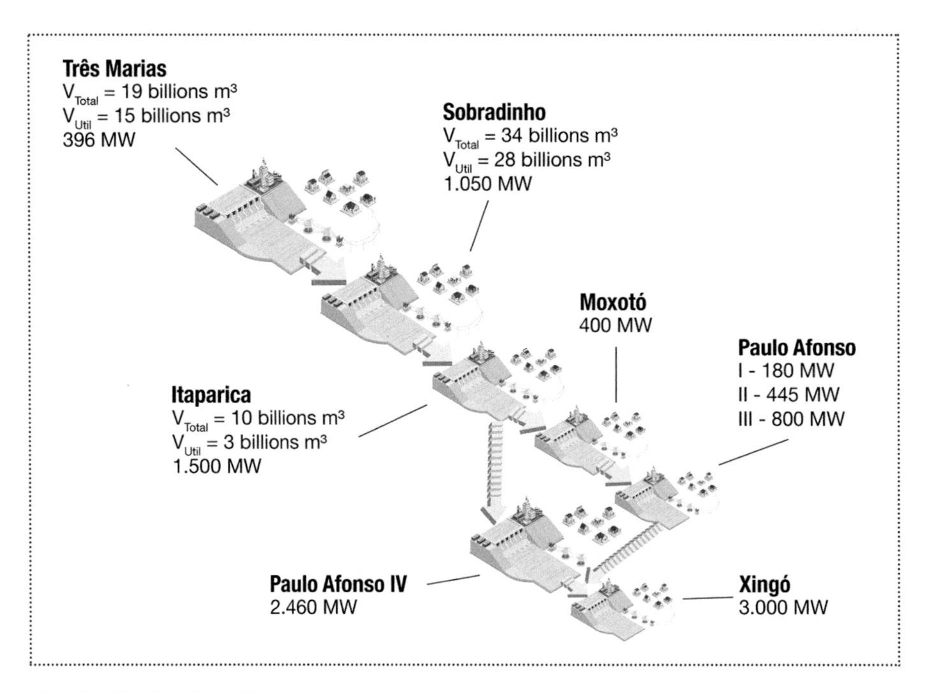

Figure 11.2 Dams alongside the São Francisco River.
Figure by the authors using graphics from Shutter Stock

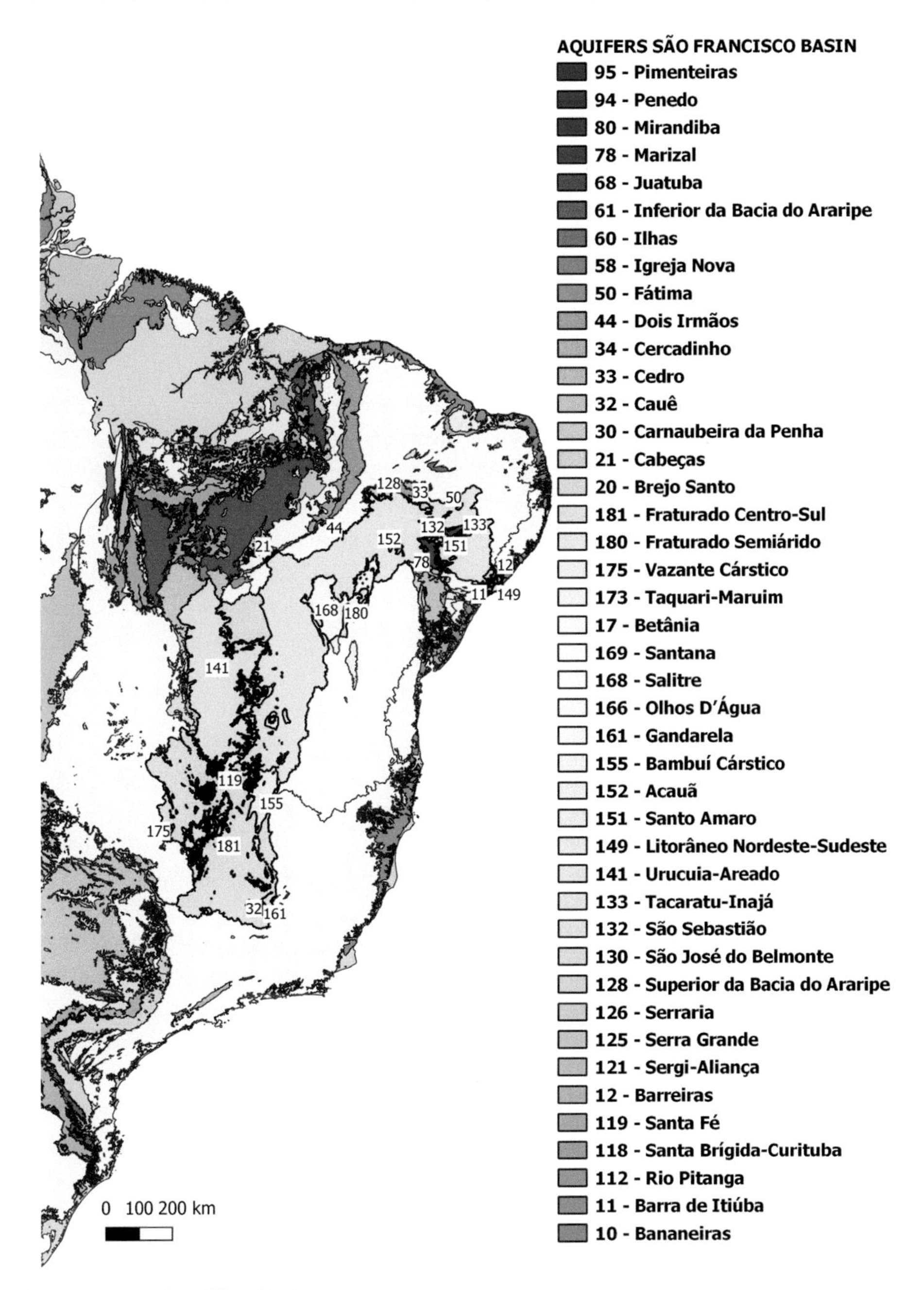

Figure 11.3 Aquifers in the São Francisco River Basin.
Figure by the authors using data from CPRM: Brazilian Geological Survey Service

social and economic impacts. Drought risk is a serious impediment to socioeconomic development of the Semi-arid region.

Droughts are frequent in the part of the São Francisco Basin that is in the Semi-arid region, in the medium, sub-medium, and lower basins. This area covers 377 municipalities, with 343,105 km^2. There is a long history of droughts in this area. As we write this chapter (2020), this area faces the longest drought in history. The drought mostly affects rain-fed agriculture and cattle raising, with severe human and economic impacts. It also affects river flow and reservoirs, with serious consequences for the capacity of reservoirs to regulate water flow in the river. In such situations, the minimum river discharge out of the Sobradinho reservoir, which is 1,300 m^3/s, may be decreased to as low as 550 m^3/s, with serious consequences for the capacity of the river to serve its different uses, including energy generation and irrigation.

Floods are also frequent, though they are more controlled now because of the system of dams in the main river. Presently, floods may be more severe in cities of the lower São Francisco and in the high São Francisco. From 2003 to 2012, 169 municipalities were under flood emergency. Floods occur during the months of

December to March, when river flow may achieve its maximum. Reservoirs that were built with the aim of providing regulation for river flow, especially Queimados (in the upper São Francisco), and Sobradinho, in the medium São Francisco, have shown good capacity to reduce the risk of floods.

The effects of floods may be heavy and are intensified by anthropogenic pressures. The problem of floods is very much linked to bad land use policies and practices, especially in urban areas where construction has been permitted in places that are normally more vulnerable. Some of these unsustainable practices include uncontrolled urbanization on the margins of the river; insufficient or mismanaged drainage systems; reduction of the river bed due to sedimentation; deforestation; and inadequate management of solid waste – all of which contribute to erosion and bring solid materials to the river (CBHSF, 2015b).

11.5 SEDIMENTATION AND SILTING

Pollution, sedimentation, and silting are problems that occur in the São Francisco basin. Deforestation and, particularly, the destruction of riverine forests are causes of erosion and sedimentation. Zellhuber and Siqueira (2007) indicate that 96 percent of riverine forests in the São Francisco basin have already been destroyed. Presently, rangelands reach the margins of the riverbed.

Increasing sedimentation reduces the accumulation capacity of dams. According to Barbosa et al. (2008), Brazilian dams lose on average 0.5 percent of their capacity each year because of silting. Sedimentation causes serious impacts such as reduction of hydroelectrical potential, reduction in navigable capacity, floods, downstream erosion, and aquatic plants. In the longer run, silting and sedimentation may reduce the quantity of water in dams and thus also affect consumptive uses such as irrigation and water supply. Most importantly, sedimentation abbreviates the lifespan of a dam.

The problem of sedimentation, mainly because of its impact on hydroelectricity, has been a concern for the National Agency on Electric Energy (ANEEL). ANEEL has produced a handbook on evaluation of sedimentation (Carvalho, 2000). According to ANEEL, erosion has been increasing in Brazil. Several dams situated in the tributaries of the São Francisco River are already silted.

On the São Francisco River itself, the problem of sedimentation is present. One of the studies was conducted by Silva et al. (2010). In fact, the transportation of eroded materials from land to the ocean is a natural function of rivers. However, this process is altered by human action: On one hand, the increase in eroded material that reaches the river, due to deforestation and other land use changes induced by human activities, and on the other hand, the building of dams in the rivers, which keep eroded materials in reservoirs and prevent them from reaching the ocean. Depending of the size of the reservoir, immense quantities of sedimented materials are deposited in such dams. The denser materials go to the bottom of the dams. This process has important impacts and crucially affects the life systems that depend on the river. Many fish species have disappeared.

Table 11.2 *Sediments transported by the waters of the São Francisco Basin*

Year	Sediments (ton/year)
1975	6,900,000
1993	2,100,000
2003	228,000
2007	262,000

Source: Silva et al. (2010).

Because of the construction of dams, the quantity of sediments transported by the river has been diminishing over time. An estimation made by Silva et al. (2010) shows the following reduction in sediments in the São Francisco River (Table 11.2).

Most sediments are concentrated, however, in the high and medium São Francisco region of the basin. Lima et al. (2001) prepared a study for Embrapa (Brazilian Agricultural Research Corporation) and ANEEL that shows the distribution of sediments in the high, medium, sub-medium, and lower São Francisco (Lima et al., 2001). According to these authors, the São Francisco River presents remarkably high rates of solid sediments in the high and medium parts of the basin, up to the reservoir of Sobradinho. Most of the sediments come from the Rio Paraopebas and Rio das Velhas, both tributaries located in Minas Gerais. In the sub-medium and in the lower São Francisco areas, the rate of sedimentation is incredibly low. According to these authors, there was impact in the Três Marias Reservoir (where the Paraopebas River ends) and a much higher impact in the Sobradinho Dam, which accumulates sediments brought by the das Velhas river and other tributaries of the São Francisco. As a fisherman from the lower São Francisco said: "The muddy waters die in Sobradinho" (Gomes, 2017). Thus, the reservoirs of Três Marias and Sobradinho keep most sediments, especially those that are heavier and tend to go to the bottom of the lakes. It is estimated that, in Sobradinho only, there is a discharge of 40,000 t^2 of sediments every day, in average.

The level of sedimentation in the sub-medium and in the lower river is much smaller. In fact, this is an environmental problem that has been caused by the construction of the dams, with a change in the composition of river waters and impacts on fishing and coastal zones.

Lima et al. (2001) estimated the lifetime of the Três Marias and Sobradinho reservoirs using multiple methodologies.

According to these authors, at present levels of sedimentation, the Três Marias reservoir has a lifetime of 107 years, while the Sobradinho reservoir has a lifetime of 110 years. Depending on the methodology and the assumptions made, these numbers may be different, often higher.

The problem of sedimentation reflects other big challenges presently facing the river, especially unsustainable land use in the basin, with remarkable deforestation, loss of biodiversity, desertification, and poverty. This is presently a big concern both at the governmental and societal levels. A program for the revitalization of the São Francisco Basin, with reforestation and sanitation measures, is a current priority. Revitalization of the river has received great attention and was considered by all stakeholders, including the government, as a condition for inter-basin water transfer from the São Francisco to other rivers in the Semi-arid. The revitalization program mainly involves reforestation, water, and sanitation, aiming at increasing water availability and improving its quality. This program is carried out by the Ministries of National Integration (MI) and the Environment (MMA).

11.6 SOCIOECONOMIC ASPECTS OF THE BASIN (URBANIZATION, MUNICIPALITIES, EMPLOYMENT, POVERTY)

According to the Brazilian Institute of Geography and Statistics (IBGE), the São Francisco Basin had a population of 14.3 million people in 2010, of which 50 percent lived in the high São Francisco (which includes the Metropolitan Region of Belo Horizonte, capital of the southeastern state of Minas Gerais). Seventy-seven percent (77 percent) of the population lived in urban settings, but many small cities in the Semi-arid still depend mostly on agriculture and cattle raising. The basin is scarcely populated: only 22.5 inhabitants per km^2 (CBHSF, 2015c).

It is estimated that in 2035 there will be 20 million people living in the basin. Population will continue growing further into the future and representing an increased pressure from human activities on the region. At the same time, there will be a redistribution of population within the region. Presently, for instance, there is a population increase in Petrolina (PE) and Juazeiro (BA), two cities that are linked by a bridge over the São Francisco River. Public and private irrigation projects in this area have caused an economic surge with the creation of sectors that have attracted both capital and labor. Presently, Petrolina and Juazeiro are important producers of fruit, grapes, and wine that are exported to other parts of Brazil and the world.

Like Petrolina and Juazeiro, there are other areas of the basin that show a special vocation for the development of certain activities and that will probably have their population increased in the future. On the left margin of the São Francisco, in the western part of the state of Bahia, a large area of Cerrados has been undergoing a development process since the 1990s with a considerable contribution to the production of irrigated soybeans, corn, and cotton. This is one of the most promising development areas in Bahia. However, many environmental problems have occurred with widespread irrigation and deforestation, with serious impacts on biodiversity and water resources.

In the high São Francisco, industry and mining are very much developed. Increased urbanization also has significant impact on the basin, both in terms of erosion and transport of sediments and pollution.

Other development regions along the São Francisco where there may be population increase are the north of Minas Gerais (city of Montes Claros), Formoso, and Guanambi, in the high São Francisco; Barreiras, in the medium São Francisco; Moxotó-Pajeú (besides Petrolina-Juazeiro), in the sub-medium; and Bacia Leiteira de Alagoas, in the lower São Francisco. The Table 11.3 shows the biggest cities in the basin and their respective current population (ANA, 2016).

The majority of the 504 municipalities (or 521, according to the ANA) of the basin are small municipalities with low levels of economic development and high poverty indices, especially in the Semi-arid region (the medium and sub-medium basin and part of the lower basin). In these municipalities, the main source of activity is rain-fed agriculture, which is very much subject to frequent droughts that bring loss of production and employment. Most of the population survive on transfers made by the federal government in terms of provision of retirement for the oldest people and a cash transfer called *bolsa família* to families that keep their children at school. Also, local municipalities are major employers, with resources transferred by the federal and state governments through a Municipal Fund created by the Brazilian Government. Together, public employment (in the local government), rural retirement (paid by the federal government), and cash transfers (such as the *bolsa familia* program) are responsible

Table 11.3 *Biggest Cities located in the São Francisco Basin*

City, state	Population (2010)
Belo Horizonte – Minas Gerais (State Capital)	2,375,151
Petrolina, Pernambuco	293,962
Juazeiro, Bahia	197,965
Barreiras, Bahia	137,427
Paracatu, Minas Gerais	84,718
Serra Talhada, Pernambuco	79,232
Pirapora, Minas Gerais	49,970
Penedo, Alagoas	52,385

Source: IBGE – Censo 2010

for most of the income of the families that live in the small cities and rural areas of the Brazilian Semi-arid region.

Table 11.4 shows the number of municipalities, and the rural and urban populations in each of the sub-basins of the São Francisco.

11.7 WATER USES (ENERGY, IRRIGATION, WATER SUPPLY)

Water uses of the São Francisco River include irrigation, hydroelectricity, navigation, water supply and sanitation (urban and rural supply, industry, and infrastructure), aquiculture, tourism, and leisure, besides inter-basins transfers. Some are consumptive uses, such as irrigation, water supply, and inter-basin transfers. Others are nonconsumptive uses, such as energy production, navigation, fishing, aquiculture, tourism, and leisure. All are important for creating economic activity and employment but also have impacts on water availability and quality. Table 11.5 shows the actual abstraction by river segment and purpose, which are greater than the average annual consumption granted by the ANA shown in Table 11.6.

11.7.1 Irrigation

Since the second part of the twentieth century, irrigation has been promoted in the São Francisco Valley. It took some years for public irrigation projects to start working. In cases such as the area around the cities of Petrolina (in the state of Pernambuco) and Juazeiro (in the state of Bahia), irrigation has played an especially important role in regional development, with increases in employment and income. More recently, irrigation projects, known as "perimeters of irrigation," were required to function as irrigation districts, under the command of the irrigators, responsible for the working and maintenance of the irrigation perimeter. Irrigators in public irrigation projects pay a fee to the Company for Development of the São Francisco and the Parnaiba Valleys (CODEVASF), a public company in charge of public irrigation in the São Francisco Valley: one part as amortization of the investments, and other part as payment for maintenance. More recently, a fee was approved to pay for water. This fee is paid by irrigators and other water users to the ANA. The value is fully transferred to the river water agency – the Peixe Vivo

Table 11.4 *Municipalities and population per physiographic region (2010)*

Physiographic region of the SF Basin	Area km^2	Number of cities (chief lieu located at the basin)	Urban population	Rural population	Total population
High	100,085	151	6,706,784	368,803	7,075,587
Medium	402,491	156	2,189,862	1,349,447	3,539,309
Sub-Medium	110,473	73	1,340,371	893,532	2,233,903
Lower	25,417	72	775,351	665,803	1,441,154
Total	638,466	452	11,012,368	3,277,585	14,289,953

Source: IBGE – Censo de 2010. Data organized by the ANA – Agência Nacional de Águas, Conjuntura de Recursos Hídricos no Brasil. Regiões Hidrográficas Brasileiras. Brasilia, 2015.

Table 11.5 *Average annual streamflow (m^3/s) of abstraction by river segment*

	Upstream from Três Marias			Downstream from Três Marias to Sobradinho			Downstream from Sobradinho to Itaparica			Downstream from Itaparica			Total		
Purpose	Fed.	State	Total	Fed.	State	Total	Fed.	State	Total	Fed.	State	Total	Fed.	State	Total
Water supply	0.1	22.0	22.2	3.5	23.3	26.8	3.4	0.24	3.6	5.7	0.1	5.8	12.7	45.7	58.4
Irrigation	2.8	4.0	6.8	88.6	34.3	122.9	63.3	0.00	63.3	16.4	2.2	18.6	171.0	40.6	211.5
Industry	0.3	11.0	11.3	0.8	6.5	7.4	0.1	0.09	0.2	0.0	0.6	0.6	1.3	18.2	19.5
Others	0.1	0.0	0.1	0.2	0.7	0.9	26.5	0.01	26.5	1.8	0.3	2.1	28.5	1.1	29.6
Total	3.3	37.0	40.3	93.1	64.8	158.0	93.3	0.35	93.6	23.9	3.3	27.1	213.6	105.5	319.1

Source: ANA (2016)

Table 11.6 *Average annual consumption granted by river segment and by purpose (m^3/s)*

Purpose	Upstream from Três Marias	Downstream from Três Marias to Sobradinho	Downstream from Sobradinho to Itaparica	Downstream from Itaparica	Total consumption
Water supply	4.4	5.4	0.7	1.2	11.7
Irrigation	5.4	98.3	50.6	14.9	169.1
Industry	2.3	1.5	0.0	0.1	3.9
Others	0.0	0.5	26.5	1.0	28.0
Total	12.1	105.6	77.8	17.2	212.7

Source: ANA (2016)

Agency – which acts on behalf of the Committee of the Hydrographic Basin of the São Francisco (CBHSF). This allows the CBHSF to perform its roles in the whole river, especially in bringing all stakeholders together whenever necessary. In the case of private irrigation, there is wide usage of pivotal irrigation in the left margin of the river. As one of the frontiers of modern agriculture in Brazil, this area of Cerrados has increased sharply. Irrigators get water rights grants (from the ANA or from the state water agency) and pay a fee for the use of water. This is a large area of grain and cotton production in Brazil.

It is estimated by the ANA that 77 percent of all consumptive use of water of the São Francisco River goes to irrigation, both public and private. In 2012, there were 626,000 hectares of irrigated land, and this number continues increasing. The potential for irrigation, considering both land and water restrictions, is about 3,000,000 hectares. So, there is still space for irrigation to grow – and it is growing – but there is already conflict with other uses (especially nonconsumptive uses linked to hydroelectricity production).

According to CODEVASF, in 2016 there were six irrigation projects in process of implementation, thirty-four projects already functioning, and three projects in the study phase. Data regarding these projects are shown in Table 11.7. All these projects together irrigate about 390,000 hectares, of which 135,000 hectares are already under operation. Though several projects are in tributary rivers, the largest projects, which account for most of the water use, take their water directly from the São Francisco River, most of them in the medium São Francisco. Table 11.7 presents public irrigation projects, both operational and planned, and their required water discharge. The total required water discharge shown in this table is not comparable to either the actual abstractions from the São Francisco River in Table 11.5 or the average annual consumption granted by the National Water Agency in Table 11.6.

Data on water used by irrigation projects already under operation are not available. However, considering that, on average, based on projects in preparation or under study, 1,000 hectares use about 1 cubic meter of water (not considering returns of water to the river basin), there is a need for about 436 m^3 of water for public irrigation projects in the São Francisco Basin.

In addition to public irrigation projects, there has been a boom of private irrigation in the São Francisco Basin, especially in the Cerrados located in the west of Bahia. Private irrigation has increased constantly during recent years. According to the CBHSF (2015a), in 2013 there were about 807,000 hectares of irrigated lands in the São Francisco Basin, including private and public irrigation. In 2004, according to the same source, this number was only 342,700 hectares. Recent growth is due practically to private irrigation – most of it on the left margins of the river, in the west of Bahia, for agriculture (soybeans, corn, and cotton) and cattle-raising.

11.7.2 Water Supply and Sewage

There are 126 cities that are seats of municipalities and that have their water supply provided by the São Francisco River. This represents 8.5 m^3/s. However, there is a serious problem of sanitation, especially in the poorer areas of the basin. Presently, there is an effort to increase the availability of water and sanitation in the basin. The number of cities with public sanitation is projected to increase to 215 by the year 2025. This would serve a population of 5.6 million people by that year, still well below the total population of the basin.

Most of the water supply systems do not treat water that is used and returned to the river. The biggest concentration of organic load is in the high São Francisco, because of the Metropolitan Area of Belo Horizonte, which constitutes about 30 percent of all the sewage of the basin.

In most of the river, used water is returned without any form of treatment. Such effluents are diluted by the river, but they end up deteriorating water quality.

11.7.3 Industry and Mining

The most economically developed region of the basin is in the high São Francisco, around the Metropolitan Region of Belo Horizonte. There is a traditional mining industry, including ore mining for export, generating a large portion of residues that are

Table 11.7 *Public irrigation projects in the São Francisco Basin, 2016*

Name of project, state	Sub-basin of the SF Basin	Irrigated area – 1,000 Ha	Required water discharge – m^3
In process of implantation		171.4	>188
Jaiba, MG	High	65.8	75
Baixio de Irecê, BA	Sub-medium	59.3	60
Manituba, AL	Lower	4.2	n.i.
Jacaré-Curituba, SE	Lower	3.1	3.2
Salitre, BA	Sub-medium	31.3	42
Pontal, PE	Sub-medium	7.7	7.8
Under operation		134.956	n.i. (+- 135)
Apolonio Sales, PE	Sub-medium	0.825	n.i.
Barreiras I, PE	Sub-medium	0.316	n.i.
Barreiras II, PE	Sub-medium	0.416	n.i.
Barreiras Norte, Ba	Medium	1.6	n.i.
Bebedouro, PE	Sub-medium	2.4	n.i.
Betume, SE	Lower	2.86	n.i.
Boacica, AL	Lower	2.7	n.i.
Brigida, PE	Sub-medium	1.4	n.i.
Ceraima, BA	Medium	0.48	n.i.
Cotiguiba-Pindoba, SE	Lower	2.2	n.i.
Curaça, BA	Sub-medium	4.2	n.i.
Estreito, BA	Medium	8.0	n.i.
Glória, BA	Sub-medium	0.368	n.i.
Piloto Formoso, BA	Medium	0.408	n.i.
Formoso, BA	Medium	11.7	n.i.
Fulgêncio, PE	Sub-medium	4.7	n.i.
Gorotuba, MG	Medium	4.7	n.i.
Icó-Mandantes, PE	Sub-medium	2.18	n.i.
Itiúba-AL	Lower	0.9	n.i.
Jaiba I, MG	Medium	26.0	n.i.
Lagoa Grande, MG	Medium	1.5	n.i.
Mandacaru, BA	Sub-medium	0.45	n.i.
Maniçoba, BA	Sub-medium	4.1	n.i.
Manga de Baixo, PE	Sub-medium	0.093	n.i.
Mirorós, BA	Medium?	2.16	n.i.
Nupeba/Riacho Grande, BA	Medium	2.8	n.i.
Pedra Branca, BA	Medium	2.37	n.i.
Pirapora, MG	High	1.2	n.i.
Propriá, SE	Lower	1.17	n.i.
Rodelas, BA	Sub-medium	1.2	n.i.
Salitre I, BA	Sub-medium	5.1	n.i.
São Desidério/Barreiras Sul, BA	Medium	1.7	n.i.
Senador Nilo Coelho, PE/BA	Sub-medium	18.56	n.i.
Tourão, BA	Sub-medium	14.2	n.i.
Under Preparation		84.1	112.96
Canal Sertão PE	Sub-medium	33	71.5
Canal Xingó, SE	Lower	16.5	33
Jequitaí, MG	Medium	34.6	8.46
TOTAL		390.456	435.96

Source: CODEVASF – Companhia de Desenvolvimento dos Vales dos Rios São Francisco e Parnaíba. Webpage: www.codevasf.gov

accumulated in dedicated dams. In the same area there are many industrial companies that depend on the waters of the basin.

According to the CBHSF (2015b), water rights granted to mining and industry amounted to 34.7 m^3 and 19.8 m^3, respectively. Most water rights go to the high São Francisco, which contains the populous Belo Horizonte Metropolitan Region. In fact, industrial use of water is concentrated mainly in the high São Francisco sub-region. The distribution of water rights for the several sub-regions of the São Francisco basin is the following (Table 11.8),

Table 11.8 *Water rights for mining and industry in the São Francisco Basin in cubic meters (M^3) 2014*

São Francisco Basin sub-region	Mining	Industry
High	19.0	17.9
Medium	8.6	0.9
Sub-Medium	0.2	0.6
Lower	6.9	0.4
TOTAL	34.7	19.8

Source: CBHSF (2015b)

In terms of industrial sectors, the following are predominant in each sub-region. In the high São Francisco, there are steel industries, cement, automobile, and sugar and ethanol; the medium, which includes the north of Minas Gerais, contains sugar and ethanol, biodiesel, vegetal oils, cement, textile, agro-industries, and the production of spirits (cachaça) made out of sugar cane; the sub-medium contains sugar and ethanol, soybeans (vegetal oil), cement, and agro-industry; and in the lower São Francisco, sugar and ethanol, cement, and milk products are produced.

11.7.4 Water Transfers to other Basins

Four projects are planned, under construction, or functioning that will take waters from the São Francisco Basin to other basins in the Northeast region of Brazil (Figure 11.4). Of these projects, two of them are completed or in the final stages of completion, while the other two are still planned. These are,

1. The "Eixo Oeste" (West Axis) project, that will take 30 m^3/s through an artificial channel of 450 km to the state of Piauí, to the North of the river (planned).

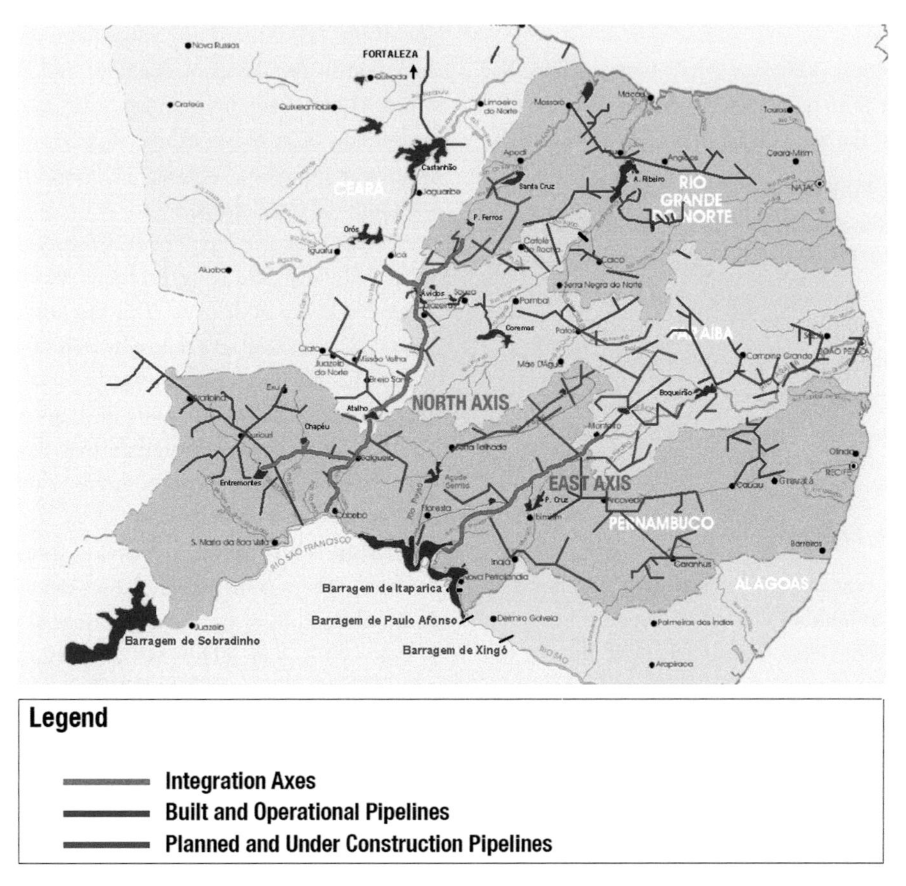

Figure 11.4 Brazil diversion case of study: São Francisco Integration Project (PISF).
Figure by the Brazilian Ministry of National Integration, modified by the authors

2. The "Eixo Sul" (South Axis) project, also called the "canal do sertão baiano" (channel of the Bahia backlands), with 350 km. This channel will take waters from the Sobradinho reservoir to the rivers Itapecuru and Tauípe, in Bahia (planned).
3. The PISF – Projeto de Interligação do São Francisco com outras bacias do Norte do Nordeste, linking the São Francisco Basin with rivers from the states of Ceará, Rio Grande do Norte, Paraíba, and Pernambuco. The PISF has two axes: North Axis, which, through artificial and natural channels, will bring water from the São Francisco to the states of Ceará, Rio Grande do Norte, and Pernambuco. And East Axis, which brings water to the states of Paraíba, Rio Grande do Norte, and Pernambuco. The PISF has a total extension of 477 km, including four tunnels, fourteen aqueducts, nine lifting stations, and twenty-seven dams. Total withdrawal of water will be 26.4 m^3/s, in normal conditions. However, in times of excess waters in the São Francisco Basin, it may withdraw as much as 127 m^3/s. As of 2020, the project was 98 percent completed. As of 2020, the water of the São Francisco Basin is already increasing water security in other basins of the recipient states. By 2021, it will do so in the Metropolitan Region of Fortaleza, with its 3 million inhabitants. Overall, the PISF may serve a population of 12 million people in 390 municipalities (MI, 2018).
4. The DESO transposition. This is a small transposition that takes 2.8 m^3 of water from the lower São Francisco to the city of Aracaju, capital of the state of Sergipe (operational).

Of the four projects, the PISF is now the most important because its East Axis is already functioning, and the North Axis will start functioning early in 2019. Impacts of the project are negligible or small in the donor and correspond to less than 5 percent of the minimum flow release from Itaparica. For the recipient basins, the expectation is that the diverted flow secures water supply to the main cities, increasing the security of other uses. More details are discussed in Chapter 17.

11.7.5 Hydroelectric Energy

Until recently, the São Francisco River has been developed in order to assure its use for hydroelectricity production. The main purpose of the seven dams that have been built in the São Francisco River (Table 11.1 and Figure 11.2) is energy production. In fact, before the 1997 water law, and the creation of the ANA in 2000, regulation of the water sector was done by the energy sector and, of course, had as its objectives the maximization of hydroelectric potential. Present hydroelectricity production capacity in the São Francisco Basin is 10,690 MW. However, total hydroelectricity potential in the basin is more than double that figure. It amounts to 22,596 MW (this includes all the tributaries to the main river). Additional infrastructure is required to reach this potential.

Though hydroelectricity production is a nonconsumptive use, since it makes no withdrawal of water from the river, it competes directly with other uses that are consumptive, such as irrigation or water supply. If water needs to remain in the river to go through turbines, it cannot be withdrawn for consumptive uses beforehand.

Hydroelectricity in the São Francisco is managed by the CHESF.

11.7.6 Other Nonconsumptive Uses

Historically, the São Francisco River has played an important role in navigation. The Sobradinho Dam has a sluice gate that allows the passage of boats and barges. Presently, it is possible to navigate from Pirapora, in Minas Gerais, to Petrolina-Juazeiro, in Pernambuco and Bahia, except for a short section from Pirapora to São Francisco. However, problems of shoals and silting in the river, plus priority for transportation by truck, have been a cause of difficulties for navigation in the São Francisco. Navigation faces several problems in the basin, which is why its importance has been decreasing. Among these problems are the transport of sediments and silting of the riverbed, mostly due to the construction of the Três Marias and Sobradinho dams. This has caused a reduction in the depth of the river and creation of sandbanks, which require many route changes. Unsustainable use of soil and destruction of riverine forests, together with widespread deforestation in the basin, increase the transport of sediments and the instability of river margins.

Another nonconsumptive use is fishing. In the basin, fishing is an activity that has been practiced mostly for subsistence by fishermen living in the area. This has also suffered with environmental degradation and the reduction of fish population in the river. The potential for fishing has been decreasing. Human intervention in the basin, mostly the construction of dams, causes a reduction in organic materials that would allow creation of biomass to feed fish. In fact, this is a main concern of policy makers dealing with the basin.[3]

Tourism is also an activity that is spread along the river and the basin. There is high potential for the development of tourism in all sub-regions of the São Francisco. In this case, the presence of large dams is an additional tourist attraction. The volume of activity is, however, still low. It is an area that can be further developed, with the creation of jobs and implementation of activities that do not harm nature.

11.7.7 Ecological Discharges

There is a conflict between the many consumptive and nonconsumptive uses of the river and the maintenance of conditions for living species that depend on the river. In the first place, for ecological purposes, it is necessary to maintain a reasonable volume of water discharge. Minimum discharge at the mouth of the river is defined as 1,300 m^3/s. Also, water quality and the presence of organic materials are important for life in the river. Both water quality and transport of materials are problems in the basin. Reduced streamflow in the lower São Francisco has caused the sea to invade the river, with serious consequences for fishing and for water supply to coastal cities, because of increased salinity.

11.8 WATER QUALITY

The main negative impacts on water quality in the São Francisco Basin are,

(1) Domestic sewage discharged into the river without any kind of treatment.[4]
(2) Pesticides, insecticides, and fertilizers from agriculture and cattle raising.
(3) Solid waste.
(4) Mining and industry waste.
(5) Sediments due to deforestation, erosion, and solid waste from urban areas.

The principal urban loads come from the Metropolitan Region of Belo Horizonte, in Minas Gerais (high São Francisco), the city of Montes Claros, north of Minas Gerais (high), the city of Petrolina, in Pernambuco (sub-medium), and the city of Arapiraca, in Alagoas (lower São Francisco). On the other hand, the upper São Francisco region is responsible for around 50 percent of all polluted discharges into the river system. Water quality is a major problem in the upper region.

11.9 WATER RIGHTS

Withdrawal of water from the river depends on the granting (*outorga*, in Portuguese) of water rights by the ANA, in the case of federal waters (rivers that cross more than one state, like the São Francisco), or by state water agencies (some tributaries that cross only one state). According to the CBHSF, irrigation is the main water user. Water rights to irrigation amount to 556.6 m^3/s. In second place comes public water supply, with 52.4 m^3/s. In third place comes industry and mining, with 34.7 m^3/s. These are the three main consumptive uses of São Francisco water (CBHSF, 2015a).

According to the ANA, the situation regarding water rights for the many uses in the São Francisco Basin (not only the São Francisco River) is presented in Table 11.9.

Table 11.9 *Water Rights per sector in cubic meters (m^3) 2010 (except for irrigation, which is 2013)*

User sector	Water rights (m^3)
Irrigation	556.6
Public water supply	52.4
Industry and mining	34.7
Fishing and aquaculture	3.5
Dilution of effluents	0.4
Energy production	0.1
Others	73.8

Source: ANA. Cited by CBHSF (2015a, p. 24).

11.10 WATER BALANCE

The water balance exercise considers not only the main bed of the São Francisco River but also its tributaries. The average natural streamflow of the São Francisco is 2,846 m^3/s, varying from 1,077 m^3/s in the dry season to 5,290 m^3/s in the wet season. The regularized streamflow after the Sobradinho dam, in dry periods, is 2,060 m^3/s (ANA, 2016). This streamflow needs to cater for all the different uses of river water: (a) consumptive uses, such as urban and rural water supply, agriculture and cattle raising (including irrigation), and industrial supply; (b) nonconsumptive uses, such as hydroelectricity production, navigation, fishing, tourism, and dilution of polluted waters that are discharged into the river; (c) and ecological water flow, which is the water needed in the streamflow of the river to maintain ecological conditions, including biodiversity.

According to Brazilian water law (Brazil Presidency, 1997), some water uses must prevail over other uses. The usage priority is the following, in this order: (1) human water supply; (2) animal water supply; (3) industry; (4) irrigation; (5) energy. This means that, in situations of water scarcity, priority uses must be met in the first place.

In 2010, the withdrawal of water for consumptive uses was estimated to be 309.4 m^3/s, of which 277.8 m^3/s was from surface water and 31.6 m^3/s was from underground water. Between 2000 and 2010, demand for water resources in the São Francisco Basin, by types of water use, evolved as shown in Table 11.10.

According to the Water Resources Plan, water demands that depend on the main riverbed of the São Francisco are met almost in totality. The major problem regards tributaries within the basin.

The Water Resources Plan for 2016–2025, prepared by the CBHSF, estimates scenarios for future water withdrawals in 2025 and 2035. Starting with 312 m^3/s of withdrawal for consumptive uses in 2013, it estimates a minimum of 458 m^3/s in 2015 and 539 m^3/s in 2035, and a maximum of 786 m^3/s in

Table 11.10 *Evolution of the water resources demand for consumptive uses, by sector, in the São Francisco Basin in m³/s*

Sector	2000	2006	2010
Urban water supply	26.0	27.3	31.3
Rural water supply	3.8	3.7	3.7
Irrigation	114.0	123.3	244.4 (2013)
Animal husbandry	6.7	9.1	10.2
Industrial supply	15.3	17.4	19.8
Total	165.8	180.8	309.4

Source: ANA (2015) and CBHSF (2015b).

2015 and 1,073 m^3/s in 2035 (CBHSF, 2015b). If one also considers the need for water in the riverbed for energy production, it will be necessary to maintain a flow of 600 m^3/s at the Três Marias dam and 2,000 m^3/s at the Xingó dam. The energy requirements for streamflow are also intended to meet ecological needs.

Comparing water availability with 95 percent of permanence of streamflow (Q95), the Water Resources Plan defines scenarios for water balance in the entire basin, considering the legal priorities that define human consumption in the first place. Some conclusions indicate that in each scenario consumption in the main river may be met, but surface water resources will not be sufficient to satisfy the projected demands. The situation regarding underground waters is somewhat better. In all scenarios, however, there will be a deficit regarding lower priority uses such as irrigation, especially in the tributaries.

Main causes of conflict in the main river are between consumptive uses and energy production uses. In some sub-basins, there will also be conflict among consumptive uses, such as irrigation versus human water supply. This clearly points to the need for a more efficient integrated water management policy in the region.

11.11 INSTITUTIONAL ARRANGEMENTS

Brazilian water law 9433 is rather recent, dating from January 1997. The law established the National Water Resources Policy (PNRH) and the National System for Water Resources Management (SINGREH) and defined water as a limited natural resource with an economic value. Historically important was the definition that water resources management will consider its multiple uses, as opposed to dominant sectorial uses, such as the production of energy. According to the law, water management will be decentralized, with the participation of all stakeholders such as water users, civil society organizations, and the government. The law also defines the instruments of water policy: water resources plans, water classification, water rights granting, water fees, and water information systems. Water institutions, such as the National Water Resources Council (CNRH), the ANA, state water agencies, and the river water committees are part of the National Water Resources System (SINGREH).

As previously mentioned, the ANA oversees the implementation of water resources management in Brazil. Water bodies that are completely within a state are managed by the state, which also has its own water resources plan. That is why all the states are supposed to have state water agencies that manage water instruments. If a water body, such as the São Francisco River, crosses two or more states, then it is regulated by the ANA. In the case of the São Francisco Basin, there are tributaries that are completely within a state. In this case, these tributaries are managed by the respective state agencies and by their own river committees. That is why, in the case of the São Francisco Basin, water rights are granted either by the ANA, in federal rivers, or by states located in the basin, such as Bahia, Pernambuco, Minas Gerais, Alagoas, and Sergipe.

Local management of the river basin, including adjudication of user conflicts, is done by a basin committee with the participation of users, civil society, and government. The CBHSF was created in 2001 and is responsible for management of all aspects of the river, including the implementation of water resources policy in the basin, establishment of rules of behavior locally, and management of local conflicts and interests (CBHSF, 2017). The CBHSF has a mandate to act over the whole basin. It also acts with four regional technical bodies, to facilitate its access to each part of the basin. Besides the main CBHSF committee, there are also eighteen committees that belong to tributaries of the main São Francisco river, on both margins.

At the federal level, major institutions are the aforementioned CNRH, MMA, and ANA; at the state level, the State Secretariats of Water Resources, and state water companies or agencies; and at the basin level, the CBHSF and the Basin Water Resources Agency.

11.12 SCENARIOS FOR THE FUTURE

11.12.1 São Francisco Streamflows: A Recent Past Analysis

In this section, an analysis of how streamflow has changed in the recent past is presented. The analysis is performed using a method called wavelets, applied to naturalized streamflow gauge station data of Sobradinho and Paulo Afonso. The goal is to identify variability patterns in different frequencies and whether there is any medium- or long-term ocean–atmosphere oscillation that modulates such variations.

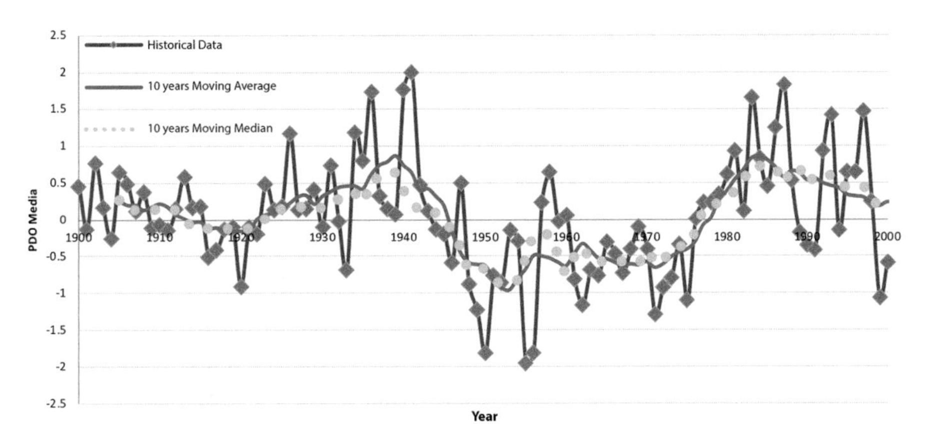

Figure 11.5 Pacific Decadal Oscillation: 1900–2010.
Figure by the authors using data from http://research.jisao.washington.edu/data_sets/pdo/

Figure 11.5 shows the transformed bands (1–8 years, 9–39 years, and low frequency) in wavelets for the Sobradinho and Paulo Afonso stations. The variance in each band is in percentage points. The analysis shows that there is a change in phase for the low frequency band in the period 1970–1975. It is possible that such change is related to recurring ocean–atmosphere patterns such as Pacific Decadal Oscillation (PDO).

PDO can be described as a change in sea surface temperature (SST) of the North Pacific Ocean, in the same fashion as El Niño. There is increasing evidence related to the impacts of PDO over the Southern Hemisphere with important climate anomalies in middle latitudes of the South Pacific Ocean, Australia, and South America. In the twentieth century, PDO fluctuations were most energetic in two general timespans, one from 15 to 25 years and the other from 50 to 70 years. The mechanisms that control such fluctuations are still unknown (Mantua and Hare, 2002).

The systematic records for Sobradinho show variability that is somehow related to the PDO anomaly, which may indicate that such variability may be responding to climate variability. This relationship suggests that streamflow during the PDO Cold Phase (periods with negative PDO anomalies) tends to be smaller. See Figure 11.6.

Wavelets analysis for naturalized streamflow gauge station data for both Sobradinho and Paulo Afonso showed that,

(1). The mean streamflow regime (9–39 years and low frequency bands) and droughts/floods (1–8 years band) vary according to the region where the basin (station) is located and also to climatic phenomena, such as El Niño/La Niña and PDO.
(2). Wavelets analysis showed that there is a high frequency variability associated to just a few years (1–8 years) in the naturalized streamflow series.
(3). The results also show that there is a long-term trend (low frequency trend) of reduction in the naturalized streamflow series.

These characteristics can be explained, at least partially, by changes in the thresholds of PDO.

11.12.2 São Francisco Streamflows: Future Expectations

Before discussing further what to expect in terms of changes in São Francisco streamflows, we examine the RCP4.5 and RCP8.5 climate scenarios in the region of the basin for the period 2041–2070 (SAE/BRAZIL, 2015). The results shown here were obtained by using the regional climate model (ETA) forced by two global climate models MIROC5 and HG2ES for the two climate (RCP4.5 and RCP8.5) scenarios. Figures 11.7 and 11.8 show the anomalies of average temperature (°C) and precipitation for the 2041–2070 period, respectively. The results for the basin show that the average temperature can increase from 1.5 to 2.5°C for the RCP4.5 scenario and from 2 to 4°C for the RCP8.5 scenario. The precipitation changes for the basin show a possible reduction of between −15 and −25 percent for the RCP4.5 scenario and −15 percent and −45 percent for the RCP8.5 scenario. The possible increase in average temperature combined with a possible decrease in precipitation will certainly result in runoff generation reduction, and consequently streamflow, for the São Francisco River basin. The impacts of these changes on temperature (consequently evapotranspiration) and precipitation can result in significant changes in the streamflow of the basin. Figure 11.9 shows these impacts for several streamflow gauge stations in Brazil for the period 2041–2070: you can see the results for Três Marias, Sobradinho, and Xingó. Except for Xingó, with ETA/HG2ES and the RCP4.5 scenario, which indicates a possible significant increase (+35 percent) in streamflow, the future of the basin is concerning. For the RCP8.5 scenario, Xingó may experience significant reductions that range from −40 percent to −70 percent. The reduction for Três Marias ranges from −15 percent to −40 percent in the RC4.5 scenario, and from −20 percent to −55 in the RCP8.5 scenario. For Sobradinho, the results are even more worrying since the

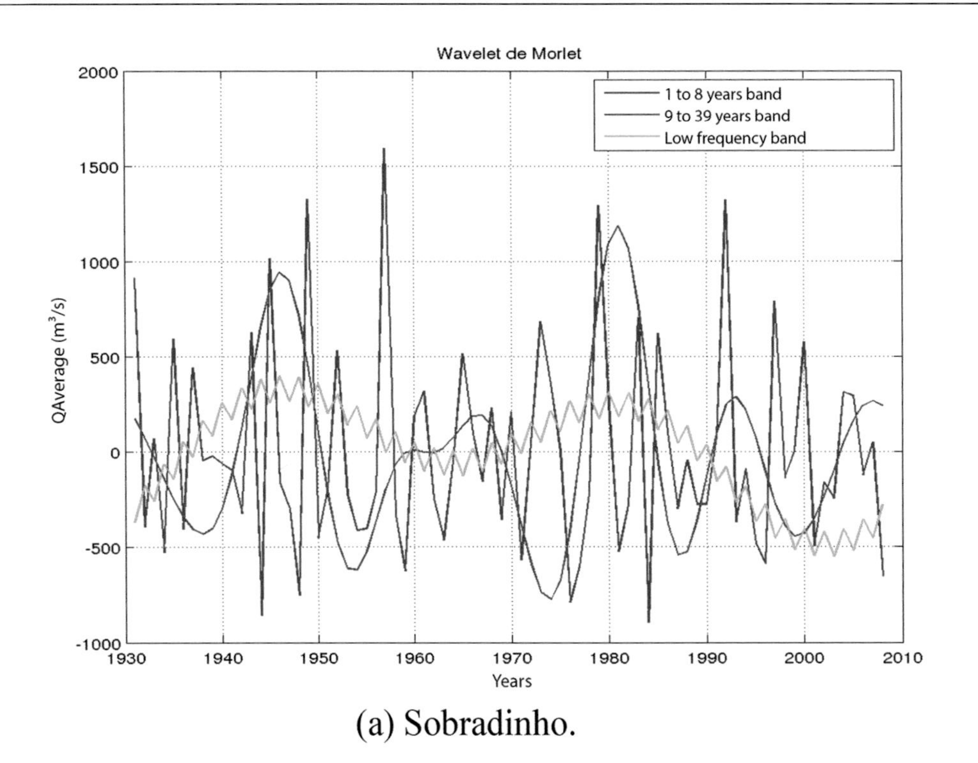

(a) Sobradinho.

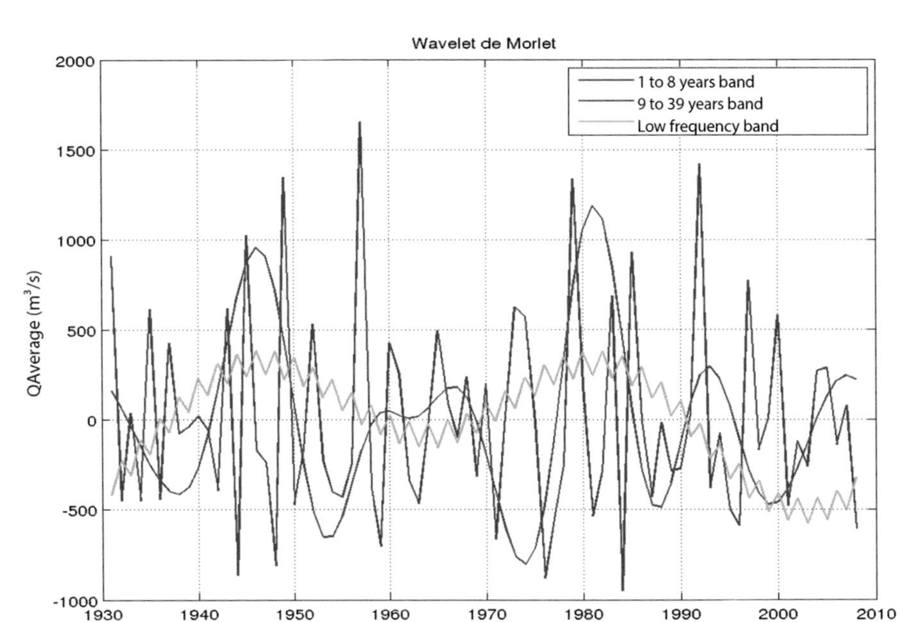

(b) Paulo Afonso.

Figure 11.6 Wavelets bands from 1 to 8, from 9 to 39 years, and low frequency for streamflow data: (a) Sobradinho; (b) Paulo Afonso. Figure by the authors

reductions range from −30 percent to −40 percent in the RCP4.5 scenario and from −35 percent to −70 percent in the RCP8.5 scenario (See Figure 11.10).

We highlight that the aforementioned analysis used downscaling of only two global climate models and, as such, the numerical results should be interpreted qualitatively: indicative of reduction of streamflow. There is a lot of uncertainty in these results for the low portion of the basin (Xingó station). To complement this analysis, we used twenty-three global climate models directly to determine the impact on streamflow in the basin at several

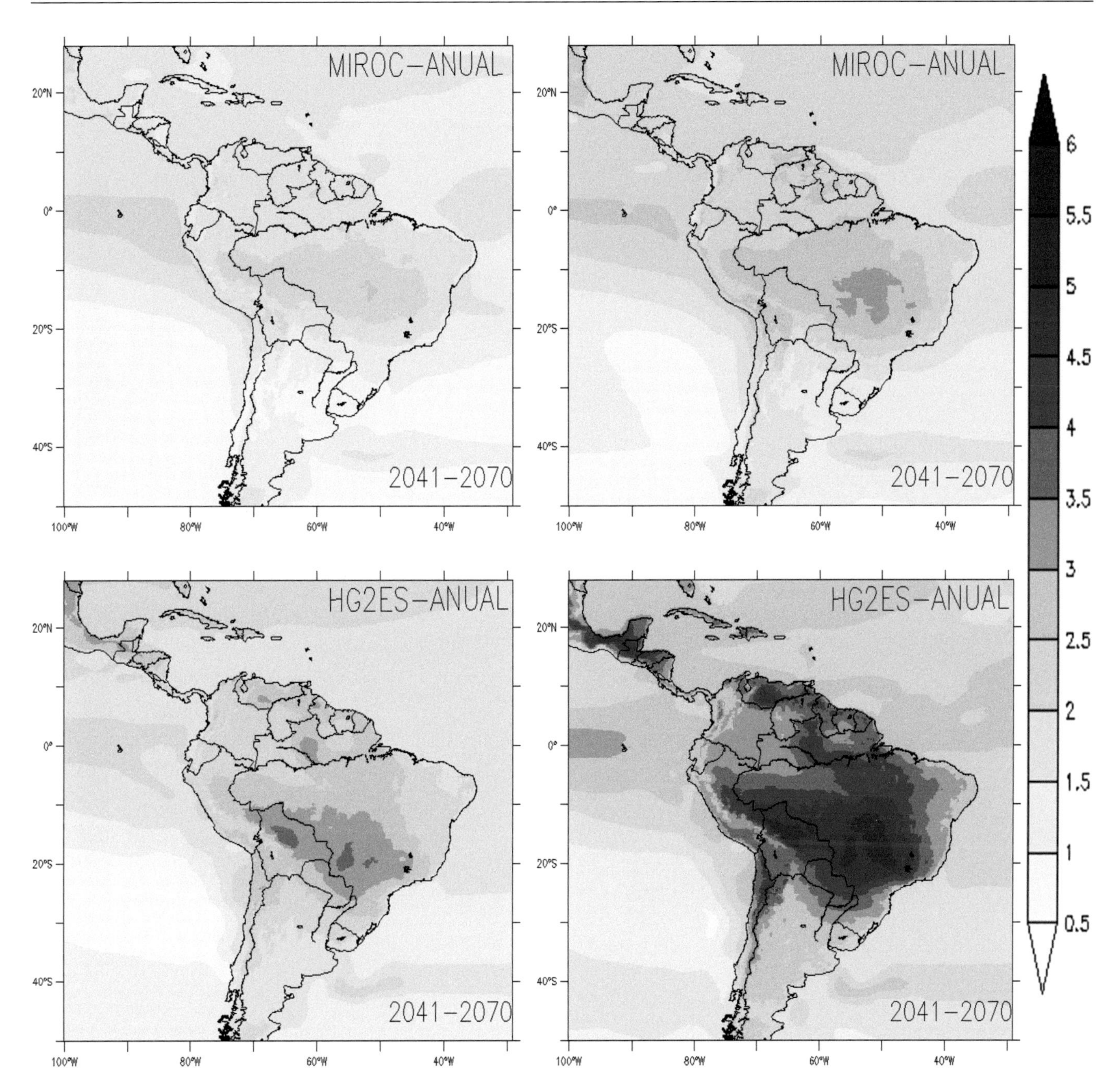

Figure 11.7 Average temperature anomaly (°C) for the period from 2041 to 2070 and models ETA/MIROC5 (top) and ETA/HG2ES (bottom) and scenarios RCP4.5 (left) and RCP8.5 (right).
Source: SAE/BRAZIL (2015)

stations around the country (See Figure 11.11). The indication of reduction persists, but at much lower levels: (i) RCP4.5: most models indicate reduction levels ranging from 0 to 20 percent for Três Marias and 5 to 20 percent for Sobradinho, while for Xingó the results are too uncertain; (ii) RCP8.5: most models indicate reduction levels ranging from 0 to 20 percent for Três Marias and 10 to 25 percent for Sobradinho, while for Xingó the results are again too uncertain but not as much as in the RCP4.5 scenario.

11.13 FUTURE UNCERTAINTY OF WATER MANAGEMENT: CLIMATE CHANGE AND POPULATION GROWTH

The São Francisco River Basin is the most important river in Northeast Brazil. It is the only permanent river that crosses the Brazilian Semi-arid and is responsible for nourishing life – flora and fauna, including humans – in a large part of interior Brazil. Since the second part of the twentieth century, engineering works

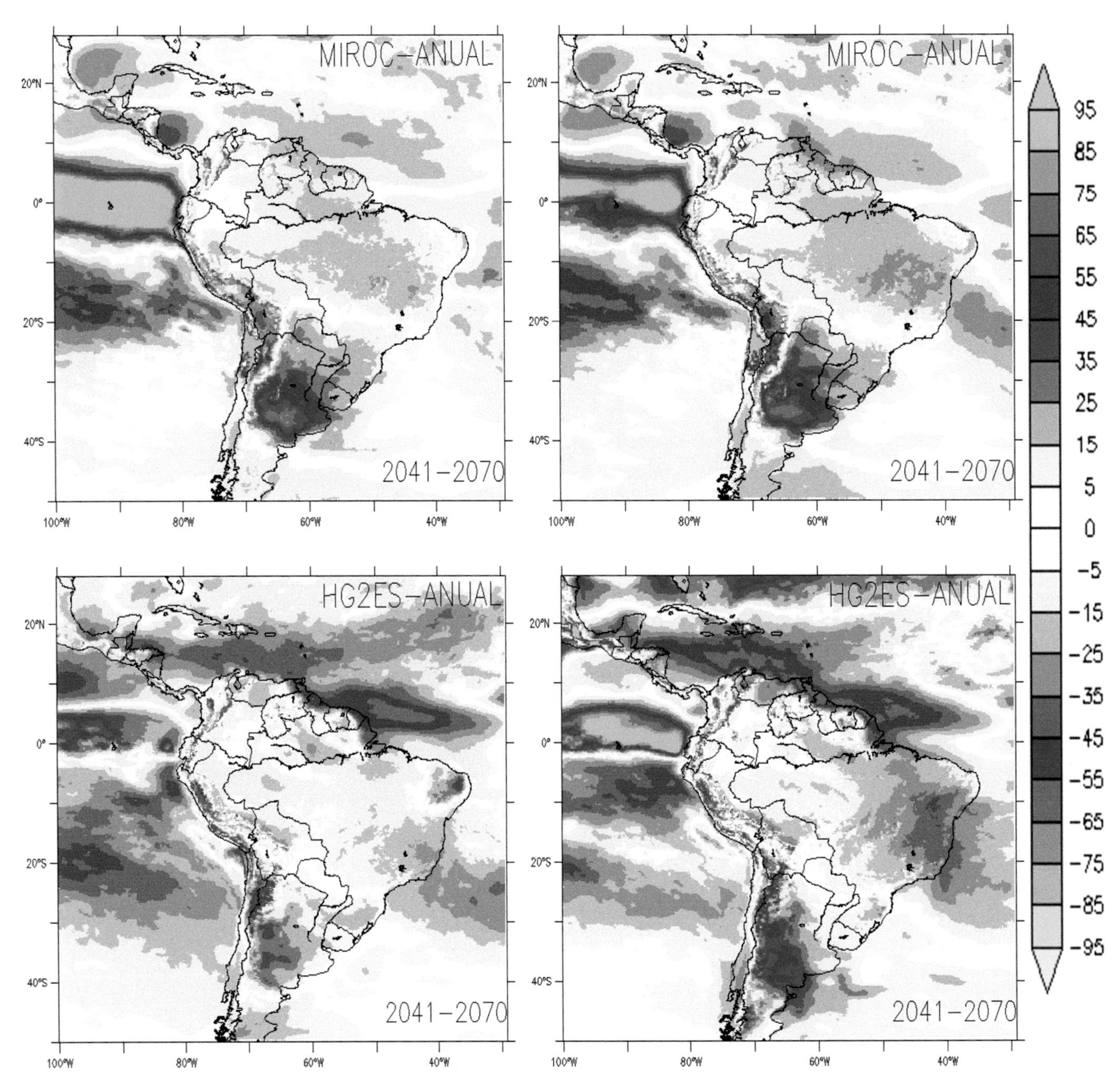

Figure 11.8 Average precipitation anomaly (%) for the period 2041–2070 and models ETA/MIROC5 (top) and ETA/HG2ES (bottom) and scenarios RCP4.5 (left) and RCP8.5 (right).
Source: SAE/BRAZIL (2015)

along the river have transformed it, to produce hydroelectricity. Several dams, irrigation projects, and water and sanitation systems were built, and all this has impacted river conditions, such as streamflows, sedimentation and silting, and water quality. Water conflicts in the river involve mainly consumptive water uses, such as irrigation, and nonconsumptive water uses, such as hydroelectricity.

Regarding water balance, on average there is still some equilibrium in the main river, but problems appear in most tributaries. According to the 1997 Brazilian water law, the highest priority water use is for human and animal supply. Lower priority uses, such as irrigation, must be reduced when there is a water deficit. During drought periods, such as in 2016 and 2017, water authorities reduced water rights for irrigation. Water discharges from the main dams, which regulate energy production and ecosystem uses, were reduced. In December 2017, discharges from the Xingó Dam in the lower São Francisco were reduced to 550 m^3/s from a "normal minimum" of 1,300 m^3/s.

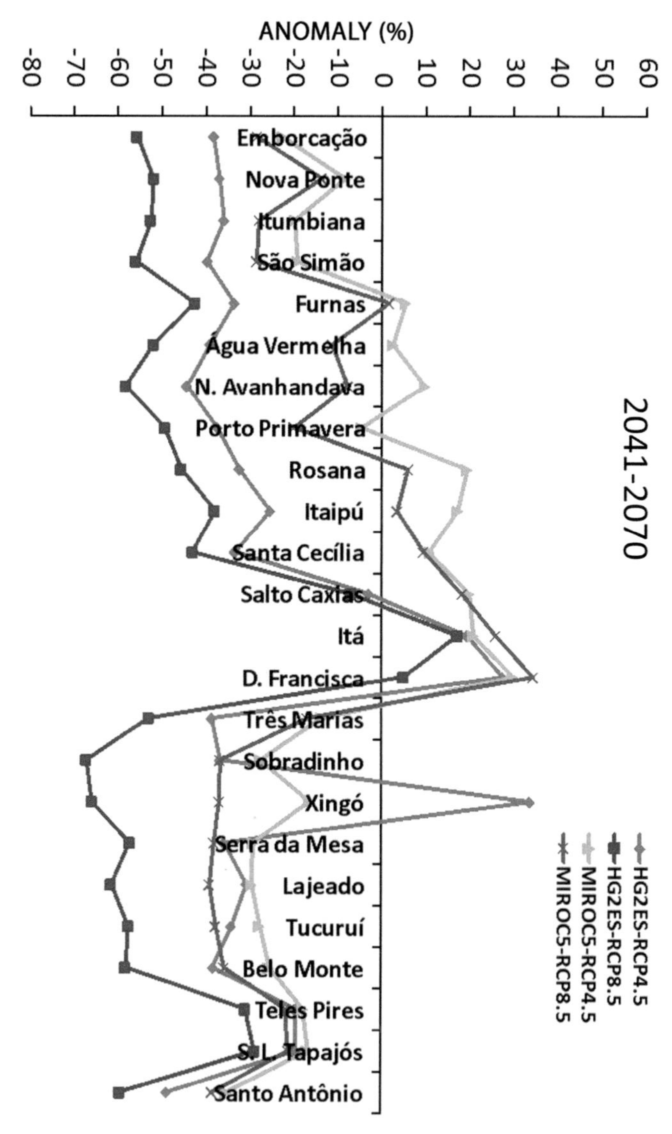

Figure 11.9 Anomaly of average annual streamflow in several streamflow stations of Brazil for the period 2041–2070 and regional climate models ETA/MIROC5 and ETA/HG2ES under scenarios RCP4.5 and RCP8.5. The Sobradinho and Três Marias stations belong to the São Francisco River basin.
Source: SAE/BRAZIL (2015)

Scenarios indicate that, in the future, precipitation and streamflows may be reduced significantly due to climate change. In extreme cases, streamflow reduction may reach 70 percent. At the same time, water demand will increase with population growth and development. River water will become scarcer and will have to be managed more carefully.

11.14 RECOMMENDATIONS

The present and future challenges for the São Francisco River and its basin must be faced with urgency to reduce negative impacts on society and the environment. Most of the recommendations are not new and some are already addressed by water agencies, but in general they need more political support and resources.

(1) Revitalization of the river basin to protect the river, enhance its water quantity and quality, and assure its priority uses. This has several important components, such as forestation or reforestation of river margins (not only of the main river but also all tributaries), and margins of lakes (natural and manmade) and sources of water. This represents a huge challenge, given the size of the area and the barriers to changing land-use, including the legal and bureaucratic systems that govern land property.
(2) Sanitation systems that assure sewage treatment in all cities along the river. Water devolved to the river stream should be treated to avoid pollution.
(3) Recuperation of degraded and desertification-susceptible areas.
(4) Sustainable land management in the São Francisco Basin, to reduce sedimentation and excessive use of pesticides that end up polluting the river.
(5) Increased efficiency in all water uses, such as irrigation and water supply systems, ending enormous losses of treated water that are presently observed.
(6) Enhancing the use of Integrated Water Resources Management (IWRM) to assure the sustainable management of water resources in the whole basin, including water resources plans, water rights, water information and monitoring, and crises management.

In all these areas, there has been considerable progress in recent decades. A São Francisco River revitalization program is in place, coordinated at the national level by CODEVASF, which includes most of the necessary actions and the collaboration of states and municipalities. There are some local experiences from which to extract useful lessons, for instance in Minas Gerais and in Bahia, but the program needs more political support, as well as technical and financial resources. So far, results are limited.

There has also been much progress in IWRM in the basin, which is closely managed by the ANA, the State Water Management bodies, and the River Water Committees, such as the CBHSF, with the participation of relevant stakeholders. There is a great need for knowledge and information, especially of a climatic and hydrologic nature.

In the medium and long run, water uses must be discussed and a conclusion about what should be the best uses of the river waters must be reached. So far, the development of the river, and its engineering solutions, have prioritized the hydroelectric and agricultural sectors. This has started to change with the new water law of 1997 and the creation of the ANA in 2000.

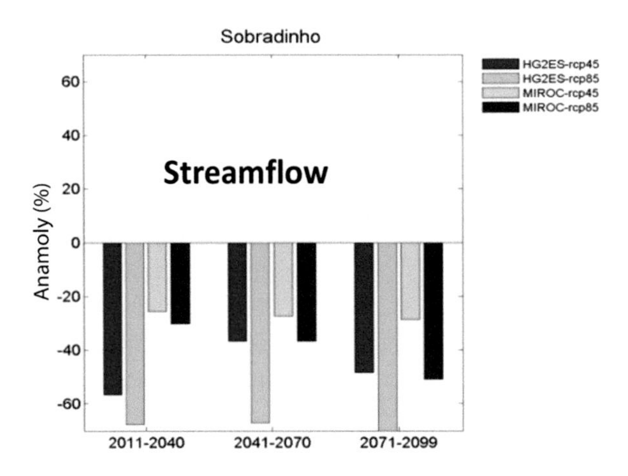

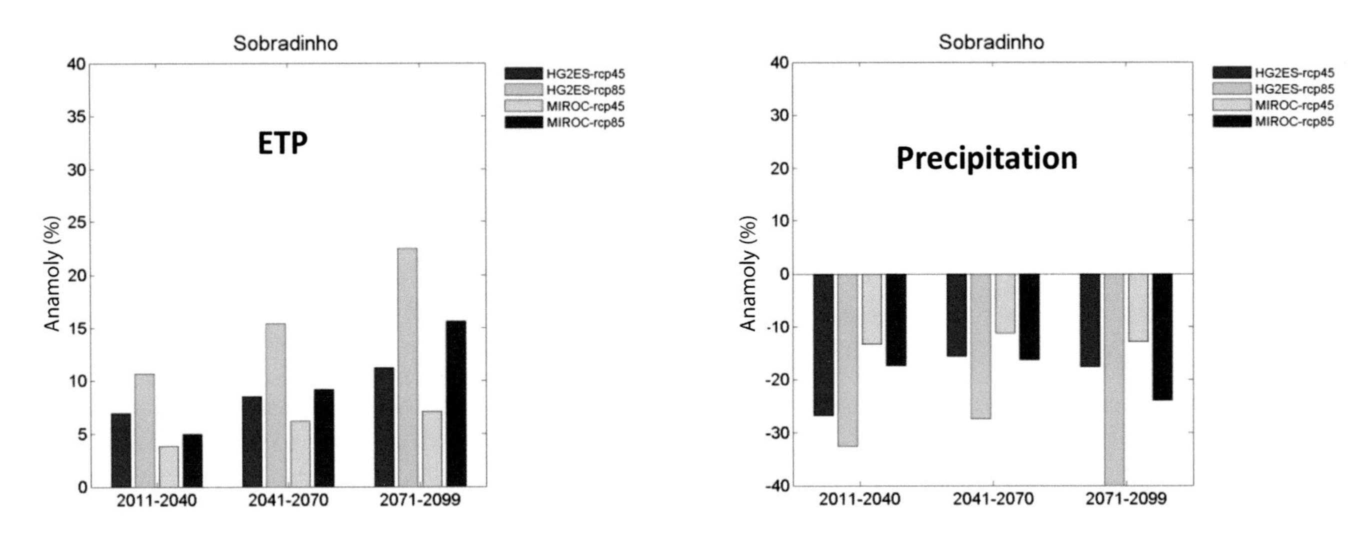

Figure 11.10 Anomaly of average annual streamflow, evapotranspiration, and precipitation for the Sobradinho uncontrolled basin during the periods from 2011–2040, 2041–2070, and 2071–2099.
Source: SAE/BRAZIL (2015)

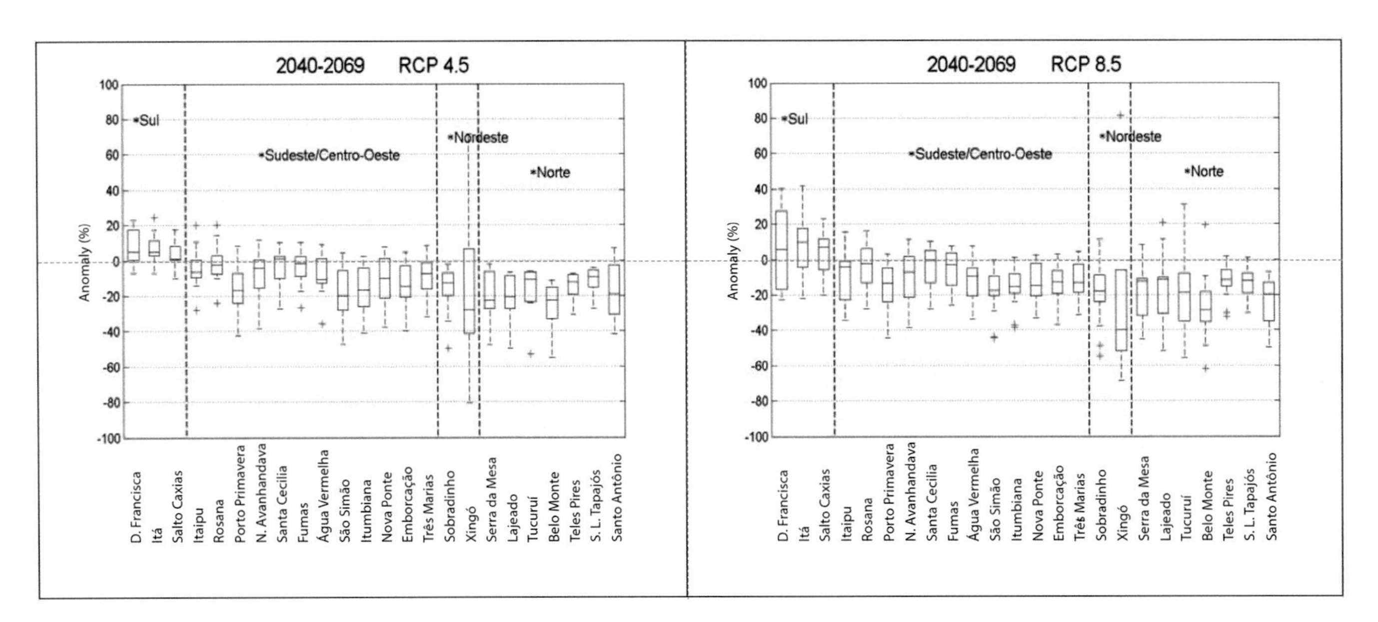

Figure 11.11 Box-plot of the anomaly of average annual streamflow in several streamflow stations of Brazil for the period 2041– 2070 and 23 global climate models under scenarios RCP 4.5 and RCP 8.5.
Source: SAE/BRAZIL (2015)

REFERENCES

Agência Nacional de Águas [National Water Agency] (ANA) (2010). *Atlas Brasil*. Brasilia.

(2016). *Conjuntura de Recursos Hídricos no Brasil. Regiões Hidrográficas Brasileiras [Water Resources in Brazil. Brazilian Hydrographic Regions]*. Brasilia. Available at http://conjuntura.ana.gov.br/docs/regioeshidrograficas.pdf

(2017). *Ficha RH_SF*. Available at www.ana.gov.br

Barbosa, J. M. C., Pinto, M. R., and de Castro, M. A. H. (2008). *Erosão e Assoreamento em Reservatórios [Erosion and Sedimentation in Reservoirs]*, ABRH. Available at www.abrh.org.br

Brazil Presidency (1997). *Law 9433*. Brasilia, 8 January 1997.

Carvalho, N. O. (2008). *Hidrossedimentologia Prática [Practical Hydrosedimentology]*. Rio de Janeiro: CPRM-Eletrobrás.

Carvalho, N. O., dos Santos, P. M. C., and Lima, J. E. F. W. (2000). *Guia de Avaliação de Associamento de Reservatórios no Brasil [Reservoir Sedimentation Assessment Guide]*.

Comitê da Bacia Hidrográfica do Rio São Francisco [Committee of the Hydrographic Basin of the São Francisco] (CBHSF) (2015a). *Plano de Recursos Hídricos da Bacia Hidrográfica do Rio São Francisco – 2015–2016 [Water Resources Plan for the São Francisco River Basin – 2015–2016]*.

(2015b). *Diagnóstico da Dimensão Técnica e Institucional [Diagnostic of the Technical and Institutional Dimensions]*. Vol. VII of *Usos, Balanço Hídricos e Síntese do Diagnóstico [Use, Water Balance and Synthesis of the Diagnostic]*.

(2015c). *Relatório Consolidado da Bacia do Rio São Francisco [São Francisco River Basin Consolidated Report]*.

(2017). Available at www.cbhsaofrancisco.org.br

De Nys, E., Engle, N. L., and Magalhães, A. R. (2016). *Secas no Brasil: Política e Gestão Proativas [Droughts in Brazil: Proactive Management and Policy]*. Brasilia: Centro de Gestão e Estudos Estratégicos and The World Bank.

Gomes, A. (2017). *As águas barrentas morem em Sobradinho [Fisherman in the Lower São Francisco]*. Available at www.cbhsaofrancisco.org.br

Lima, J. E. F. W., dos Santos, P. M. C., Chaves, A. G. M., and Scilewski, L. R. (2001). *Diagnóstico do Fluxo de Sedimentos em Suspensão na Bacia do Rio São Francisco [Diagnostic of the Suspended Sediment Flow in the São Francisco River Basin]*. Embrapa/Aneel.

Mantua, N. J. and Hare, S. R. (2002). The Pacific Decadal Oscillation. *Journal of Oceanography*, 58, pp. 35–44. https://doi.org/10.1023/A:1015820616384

Ministério da Integração Nacional [Ministry of National Integration] (MI) (2018). *Integração do Rio São Francisco [Integration of the São Francisco River]*. Available at www.mi.gov.br/web/projeto-sao-francisco

Secretaria de Assuntos Estrategicos [Secretariat for Strategic Affairs] (SAE) (2015). *Brasil 2040 Resumo Executivo [Brazil 2040 Executive Summary]*. Governo Federal, Presidencia Da Republica [Federal Government, Presidency of the Republic]. Available at www.mma.gov.br/images/arquivo/80182/BRASIL-2040-Resumo-Executivo.pdf

Silva, W. F., Medeiros, P. R. P., and Viana, F. G. B. (2010). *Quantificação Preliminar do Aporte de Sedimentos no Baixo São Francisco e seus principais impactos [Preliminary Assessment of Sediment Inflow in the Lower São Francisco River and Its Impacts]*.

Universidade Federal de Alagoas (2010). *Sedimentos no Baixo São Francisco e seus principais impactos [São Francisco River in the Wrong Path: Degradation and Revitalization]*. Presentation in the X Symposium of Water Resources in the Northeast.

Zellhuber, A. and Siqueira, R. (2007). *Rio São Francisco em Descaminho: Degradação e Revitalização [São Francisco River in Descaminho: Degradation and Revitalization]*. Salvador: Cadernos CEAS. Available at www.cptba.org.br

Notes

1 There is a large literature on droughts in the Northeast region.

2 t is shorthand for metric tonne, which is equal to 1,102 US tons.

3 CODEVASF has implemented a program to support pisciculture in the basin, with the creation of the Integrated Centers for Fishing and Aquiculture Resources.

4 According to the ANA (ANA, 2010), considering municipalities located in the São Francisco Basin, 211 municipalities depend on superficial or underground waters. In general, there is no treatment of water that is returned to the river; 114 municipalities have isolated systems using subterranean waters; 126 municipalities have integrated systems. The capacity of the river to dilute polluted effluents is particularly important. There is a bigger problem in the semi-arid region of the basin, where rivers are intermittent. In this case, there is no dilution.

12 The Limarí River Basin

Alexandra Nauditt, Justyna Sycz, and Lars Ribbe

12.1 INTRODUCTION

Mountainous headwater catchments are of key importance for supplying water to semiarid and arid regions. Mountains, as "water towers," can provide up to 95 percent of total basin discharge because of higher precipitation as rain or snow at higher elevations (Viviroli et al., 2007; Adam et al., 2009; Price and Egan, 2014). They supply more than one sixth of the world's population with drinking water as well as with water resources for irrigation, hydropower, other industries, and ecosystems (Barnett et al., 2005; Cooper et al., 2018).

Over the course of the last century, most of these mountainous catchments have been changed by engineering infrastructure that stores water for agriculture, drinking water supply, and electricity production (Schmandt et al., 2013). This infrastructure has improved water security, allowed economic development, and enabled large-scale agriculture in snowmelt-driven basins of semi-arid lands.

However, in recent years, irrigated agricultural regions located in snow-melt-dominated, semi-arid catchments have turned out to be highly vulnerable to climate variability and change (Barnett et al., 2005), as streamflow relies entirely on snowmelt-driven hydrological regimes (Bates et al., 2008; Pepin et al., 2015). While demographic and socioeconomic growth are leading to increasing water demand, precipitation and water availability are expected to decrease (Sheffield et al., 2012). To adequately respond to these challenges, it is necessary to analyze the current conditions of the basins and assess their capacities to face decreasing water availability.

This chapter analyzes the conditions of the semi-arid Central Chilean region, with the Limarí as an example. We selected this region due to its capacity to face water stress thanks to an extended irrigation infrastructure and an active water market that interact as main water management instruments to reduce climate vulnerabilities, deal with drought cycles, and enable intensive agriculture in a semi-arid climate (Bauer, 2004; Álvarez et al. 2018). Irrigation reservoirs with a total storage capacity of 1,000 million m^3 and the possibility to buy and sell water rights dependent on water demand have provided farmers with the flexibility to manage risks caused by uncertainties in water supply and promoted intensive agriculture with high-value crops. Export-oriented irrigated agriculture is the main economic activity (Álvarez et al., 2006; Oyarzun, 2010; Observatorio Laboral Chile, 2019).

However, north-central Chile since 2008 has faced a long-term drought and is expected to become warmer and drier because of anthropogenically driven climate change (Garreaud et al., 2017; Alvarez-Garreton et al., 2018). Empty reservoirs (Kretschmer et al., 2014), groundwater pumping, illegal drilling of boreholes, and resulting overexploitation for irrigation security constitute increasing problems (Oyarzun et al., 2015).

Although the Limarí Basin is a widely discussed example of water markets and irrigation management in semi-arid environments, a holistic analysis of the challenges related to water management, including monitoring, data management, and governance aspects, has not been provided. Hence, in this chapter, we provide an overview of the Limarí water management system – including its hydrology, engineering infrastructure, and governance mechanisms – to derive recommendations on how to cope with decreasing water availability in the basin. The analysis is based on a literature review evaluating the most recent scientific publications, reports published by Chilean public institutions, and experiences of the authors gained during 20 years of work with water management projects and related education in north-central Chile.

12.2 THE LIMARÍ RIVER BASIN IN SEMI-ARID NORTH-CENTRAL CHILE

12.2.1 Biophysical and Hydro-climatic Characteristics

The Limarí Basin has an area of 11,760 km^2 and forms part of the semi-arid northern-central Coquimbo region, one of sixteen administrative regions in Chile (Figure 12.1). The Limarí River extends from the Andes at 5,550 m of elevation east of the Pacific coast in the west and is a snowmelt-dominated hydrological system. It is formed by tributaries originating in the Andean Cordillera as the Grande and Hurtado rivers, with headwaters at 4,500 m above sea level; thus snowfall makes an important contribution to discharge. Groundwater is available in a gravel-sand dominated aquifer formed by shallow sediments of the flood plains of valleys between Ovalle downstream along the rivers,

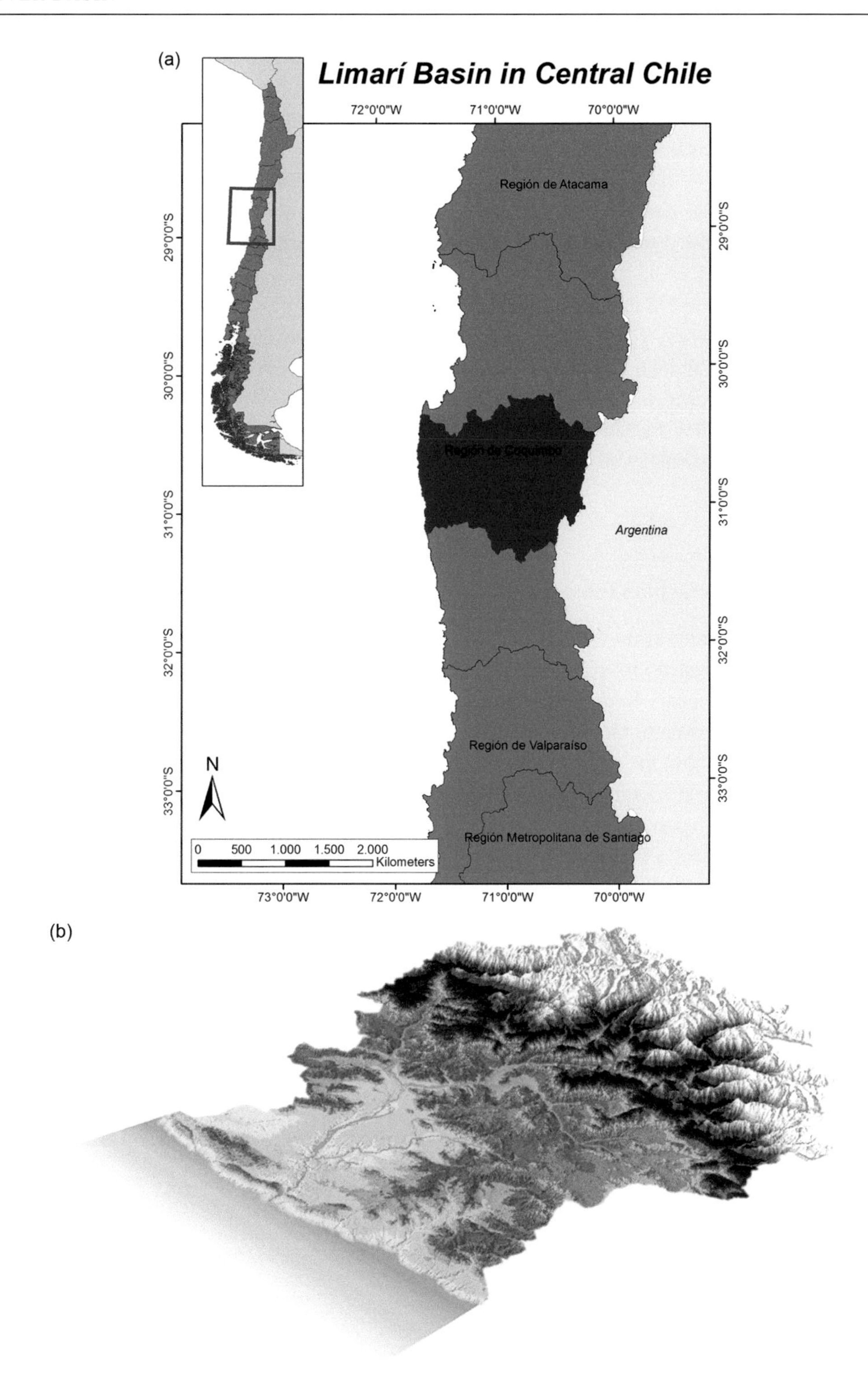

Figure 12.1 (a) Map of Central Chile, Limarí Basin (size 11,760 km^2), and (b) illustration of the topography and river network of the basin. Figures created by the authors

with groundwater levels below 10 m (INE, 2007). The dominant vegetation is "Matorral," dry steppe, mainly consisting of bushes of sparse distribution on bare ground. Agricultural land – approximately 80,000 ha corresponding to about 7 percent of the total area – is found mainly in the valleys and flood plains of the lower areas.

The climate of the Limarí catchment is arid to semi-arid with a marked Mediterranean seasonality (Finlayson et al., 2007). It strongly varies along the 140 km between the Pacific coast and the eastern Andean Mountains due to the south-eastern Pacific anticyclone, the cold Humboldt Current along the Pacific coast, and the steep topography of the mountain range of the Andes (Kalthoff, et al., 2006).

Precipitation ranges from 100 to 300 mm per year from the coastal area to the highest station in the Andes (recorded up to 1,250 m of elevation) and also from north to south, with mean annual values of 70 mm in the north and 275 mm in the south at lower elevations (Oyarzun, 2010; Favier et al., 2009). Interannual precipitation is characterized by high variability; years with high precipitation are typically linked with a High Oceanic Nino Index (ONI) representing so-called El Nino years, while "La Niña" years are usually drier (Garreaud et al., 2009; Meza, 2013).

12.2.2 Economic Activities and Irrigation Infrastructure

Agriculture is the major economic activity in the Coquimbo Region with about 80,000 ha of irrigated crops (INE, 2007). Within the Coquimbo Region, the Limarí Basin is the most important agricultural area with 33 percent of the entire surface of the Region of Coquimbo, 42 percent of the regional agricultural surface, and 70 percent of regional exports. According to the agricultural census of 2007, the irrigated area of the Limarí province amounts to 41,760 ha. Industrial activities are of minor importance and are mainly associated with wine and food processing industries, specifically the Pisco industry (INE, 2007). Mining activity is dispersed, with one major plant "Panulcillo" – but even here production is marginal within the region (Obervatorio Laboral Chile, 2019).

The Limarí Basin is the most engineered basin in Chile. The core of the water infrastructure is the Paloma System, a network of three dams with a total storage capacity of 1,000 million m^3 and a complex channel network extending over more than 700 km allowing for water storage and distribution (Figure 12.2). The reservoirs were built by the national government between 1930 and 1972 with the objective of reducing climate vulnerabilities and enabling the development of intensive, export-oriented agriculture. The three reservoirs in the Limarí are La Paloma, with storage capacity of 740 hm^3; Recoleta, with 100 hm^3; and Cogotí, with 150 hm^3 (Vicuña et al., 2014). The system was designed originally to increase irrigation security for up to four years.

Figure 12.3 shows the Limarí Basin with its hydrological network, topography, and agricultural area and the La Paloma Irrigation System consisting of three reservoirs.

12.3 WATER GOVERNANCE IN CHILE

12.3.1 The 1981 Chilean Water Code (WC): Water Rights and Water Markets

Water rights and water market mechanisms are key characteristics to describe water management and allocation in the Limarí

Figure 12.2 The La Paloma reservoir below the El Palqui irrigation scheme.
Photo by the authors

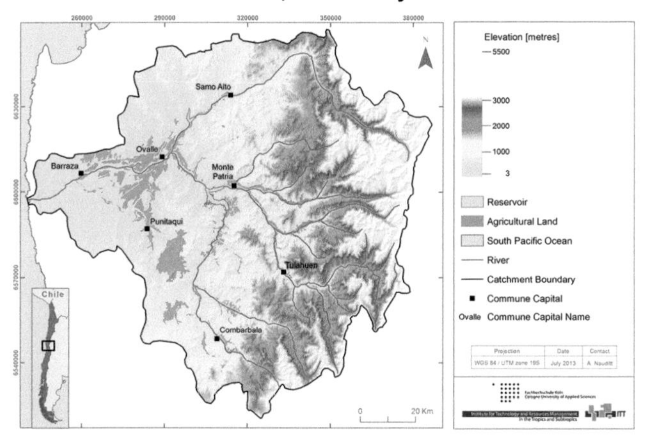

Figure 12.3 The Limarí basin with its hydrological network, topography, the La Paloma reservoir system and irrigated agricultural area.
Figure created by the authors

Basin (Hearne and Easter, 1998). The most important legal basis of water management in Chile is the 1981 Water Code (WC), which sets up the legal rules and preconditions for water market trading. Most water rights have been allocated to water users as farmers, mining industries, and water supply companies since 1951, when the first water code was launched in Chile (Donoso, 2006).

The WC strengthens private water use rights and declares them freely tradable. Water resources access should be treated as a commodity delivered to a free market regime.

This extreme free market approach makes Chile a unique case around the world. In other countries with water markets, such as Australia or the United States, markets are regulated by a wide framework of water use regulations and policies. In Chile, by contrast, water law and policy are dominated by the free market (Bauer, 2004; Endo et al., 2018). Since its establishment in 1981, the WC has been modified several times. A major reform took place in 2005 through law Nº 20.017. The main features of the Water Code 1981 and the regulations of the 2005 reform are presented in Table 12.1.

12.3.2 Key Institutions Involved in Water Management

Introducing the Chilean Water Code 1981, the Chilean State reduced its possibilities to intervene in water management activities and established that water rights owners are responsible for water management. Thus, the structure of water management is characterized by the coexistence of centralized and decentralized institutions (Vergara and Rivera, 2018).

Public (centralized) organizations comprise the administrative bodies of the State. The leading public institution is the General Directorate of Water (DGA, in Spanish initials), which is the Chilean water authority. The DGA is responsible for water resources planning and monitoring; issuing and regulating water rights under the WC; granting permission for major works; implementing policies; supervising the operation of water user organizations; and developing the Public Water Registry (OECD, 2017). Further important institutions are the Directorate of Hydraulic Works (DOH), responsible for the planning and construction of major hydraulic infrastructure, and the National Irrigation Commission (CNR) of the Agricultural Ministry that establishes policies and programs for the irrigation subsector (OECD, 2017; Donoso, 2018).

Private (decentralized) organizations – next to the water supply and sanitation companies – are water user associations (WUAs) whose members are water rights holders. WUAs manage and distribute water at the local level (Donoso, 2012). There are three types of WUAs in Chile: water communities that are not connected to a larger irrigation system (comunidades de aguas); canal user associations (asociaciones de canalistas) that share water from a specific canal coming from the reservoir; and vigilance committees (juntas de vigilancia) that manage larger systems as reservoirs or river sections (Vergara and Rivera, 2018). Although "*confederaciones de agua*," as coordination entities for entire rivers and river systems, are manifested in the Water Code, their implementation in most river basins has not been successful or even attempted.

In the Limarí Basin, water management is carried out by ten canal user associations responsible for canal maintenance and water distribution and nine vigilance committees responsible for natural water sources in sub-basins (DGA, 2004; Alvarez, 2018; Universidad de Chile, 2018). There are no regulations related to the cooperation between the private water user organizations and the DGA, which is responsible for integrated water resources management at catchment level. Consequently, as public institutions due to their limited power and financial capacities have only limited possibilities to intervene in local water management, water is managed in a very fragmented way (Donoso, 2018). Figure 12.4 shows the large number of administrative entities of the Limarí River, each indicated by a different color.

12.4 CHALLENGES RELATED TO SUSTAINABLE WATER RESOURCES MANAGEMENT IN THE LIMARÍ BASIN

12.4.1 Intensification of the Agricultural Activities in the Limarí Basin

Historically, scarce water resources and climatic variability hampered socioeconomic and agricultural development in the Limarí Basin. Hence, in 1972, after a strong drought, the Chilean

Table 12.1 *Main features of the WC 1981 (Own elaboration based on the Water Code text published by the Chilean government).*

Main features of the WC 1981

The WC 1981 defines water rights as private property that

- is separated from land and not tied to the sale of land.
- can be freely traded, mortgaged, and transferred.
- individuals can use in a private and exclusive way for the purpose they wish with complete freedom. There are no taxes for the use or no use of water rights (Vergara and Rivera, 2018).

New water rights are granted

- free of charge.
- by the General Water Directorate (DGA), which is the lead water management institution of the governance,
- without requiring a specification of intended use and,
- with the same procedure for surface and groundwater rights.

The reallocation of the once granted or bought water rights takes place only through **voluntary market transactions**.

The WC 1981 reduced the role of the government in water resources management; minimized public regulation to a minimum; and increased the management powers of water rights holders.

The WC 1981 **abolished the water use hierarchy** present in the previous Water Codes of 1951 and 1967; the market and water rights holders navigate the water allocation system; e.g., there is no prioritization of drinking water supply (Bauer, 2006).

The following additional shortcomings of the Water Code 1981 were identified and responded to:

Shortcomings of the Water Code 1981	Recent regulations responding to shortcomings in WC 1981
Missing protection of the ecology of aquatic systems: new water rights can be granted without considering the water requirements of aquatic ecosystems and third party rights (e.g., downstream).	**Introduction of minimum ecological flow thresholds** (MEFs) to reduce damages to aquatic ecosystems. New water rights are granted as long as there is water availability to satisfy the total water flow demanded, without affecting the rights of third parties (Donoso, 2012).
No fees for the non-use of water has led to water use inefficiency and unfair trading conditions. Water rights owners could privately benefit from selling the rights without contributing to society's development.	**Introduction of non-use tariffs (WC 2005):** The holders of water rights need to pay a fee if they don´t use the rights (Donoso, 2012).
No rules to protect groundwater resources were favoring overexploitation.	Introduction of regulations on groundwater exploration and exploitation. Last modification in 2014, Supreme Decree N° 203 of 2013 (Vergara and Rivera, 2018) with extraction restrictions, monitoring, and areas with extraction prohibitions.

government invested in irrigation infrastructure to construct the La Paloma System. The engineered, large-scale irrigation infrastructure was complemented under the Chilean Water Code in 1981, offering an institutional framework that created excellent conditions for intensive agriculture development in the Limarí Valley. The market for water rights – where the only reallocation potential lies in freely tradable market transactions – has fostered water resources security by carefully protecting the precious resource.

According to several scholars, the Limarí River Basin hosts the most active water market in Chile (Hearne and Easter, 1997; Bauer, 2004; Donoso, 2006; Donoso, 2018; Urquiza and Billi, 2018). The Water Code has had the following beneficial effects,

- It facilitates the reallocation of water use from lower to higher economic value activities (e.g., from traditional agriculture to export-oriented agriculture and other sectors such as water supply and mining).
- Water markets contribute to mitigating the impacts of droughts by allowing temporal transfers from lower value annual crops to higher value perennial crops.
- Water as an economic asset leads to a more efficient and environmentally sustainable use of the resource, as it encourages the owner to use it efficiently.

The large-scale engineered irrigation system, on the other hand, has improved water security, ensuring physical availability and

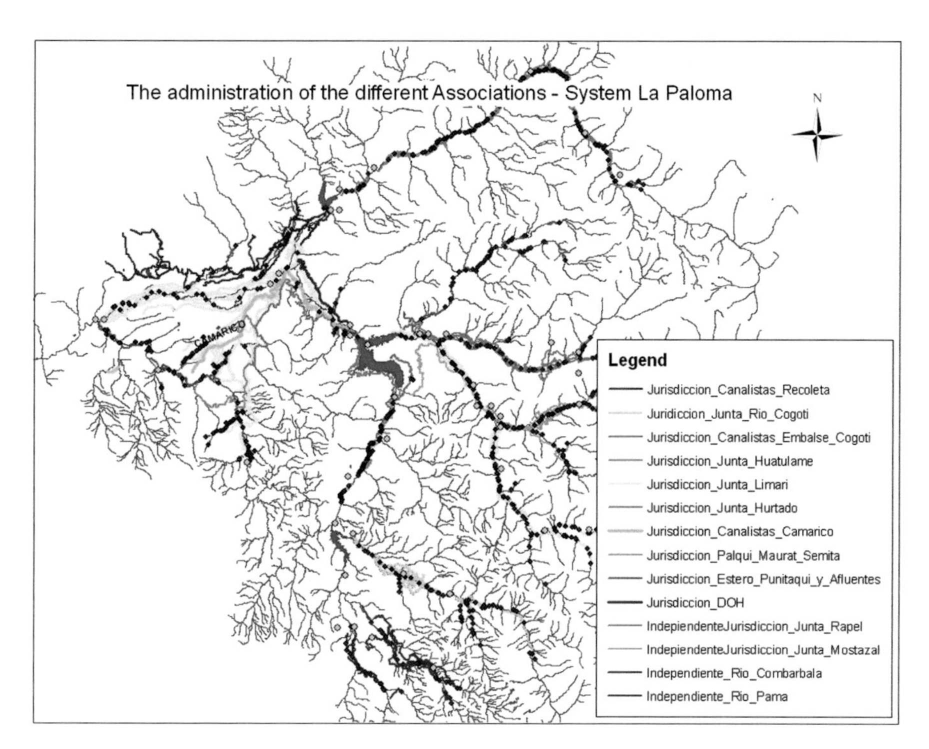

Figure 12.4 The administration of the La Paloma System. River stretches are colored according to abstracting irrigation organizations.
Source: Kretschmer, 2010

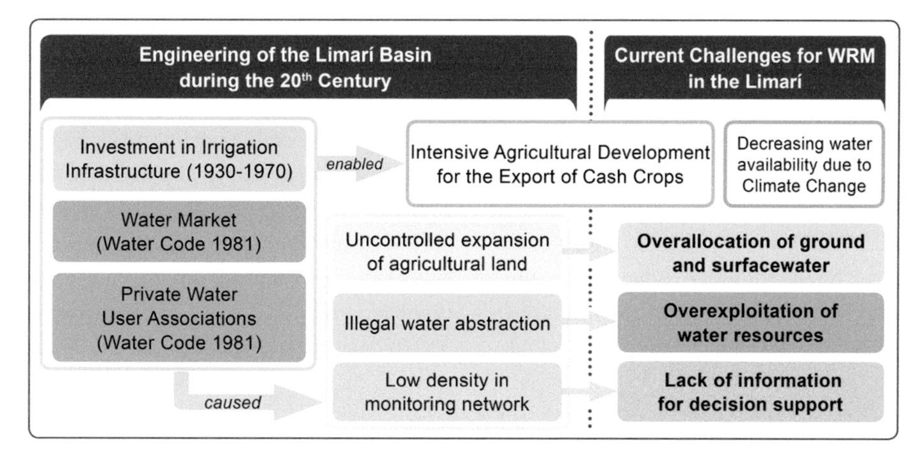

Figure 12.5 Overview of the evolution of agricultural activities in the Limarí Basin and resulting current challenges.
Figure created by the authors

distribution. Local water management carried out by water rights holders has equipped them with institutional and political power to manage the resource according to their demands. Figure 12.5 synthesizes the historical engineering development of the Limarí system and the resulting challenges for water management at present.

These institutional and infrastructural conditions have allowed farmers and water managers to deal with water supply variability and encouraged farmers to cultivate high value crops such as vineyards, orchards, and vegetables. Table 12.2 shows the land area used for various agricultural activities as recorded by the Agricultural Census and the "Fruit" Census (Catastro Fructícola). It shows that since 1962 and the construction of the La Paloma System in 1972, the cultivated area for annual crops had decreased almost by half in 2007, while the area of water-intensive fruit trees in 2011 had increased by a factor of 5.8. To date (2020), no census has been published since 2007; only the Fruit Cataster in 2011 published numbers related to cash crop fruit trees: avocado, citrus,

Table 12.2 *Evolution of agricultural land use in the Limari Province [area in ha]. Data: Agricultural Census, 2007; Catastro Fructicola, CIREN, 2011; SAG, 2018.*

Land use area (in hectares)	1962	1997	2000	2005	2011	2015	2018
Annual crops, horticulture, meadows	39,068	20,211	27,117	25,803			
Orchards (table grapes, avocado, citrus, olives)	3,279	9,661	13,669	15,722	16,256	16,661	16,784
Vineyards (Pisco and wine)	1,414	6,344	7,215	7,475	7,737	8,159	8,082
Total perennial crops	4,693	16,005	20,885	23,197	23,993	24,820	24,866

table grapes, and wine/Pisco grapes (CIREN, 2011; Vicuña et al., 2014). The census conducted in 2011/2012 was withdrawn due to data weaknesses and related disputes (WEIN, 2014).

Continuous technification of agricultural practices, installing sprinkler irrigation in large parts of the irrigation scheme, has increased water use efficiency, leading to a sudden rise in water availability for irrigation and a drastic decrease of groundwater recharge (Álvarez, 2018). This in turn enabled farmers to further expand the cultivation area for perennial crops. According to Vicuña et al. (2014), while in 1997 only 30 percent of irrigated land had drip irrigation, in 2007 already 60 percent of the area had drip irrigation installed.

12.4.2 Decreasing Water Availability: Climatic Trends, Change Projections, and Related Changes in the Hydrological Regime

Central Chile is extremely vulnerable to climate change-related temperature increases and changes in precipitation patterns due to dependency on the Andean cryosphere and melting water in springtime (Barnett et al., 2005; Adam et al., 2009; Vuille et al., 2015). Enhanced snow and glacier melt due to increasing temperatures in the short term will accelerate runoff in springtime and decrease discharge values in summer (Souvignet et al., 2010; Vicuña et al., 2011; Boisier et al., 2018). During the next decade, significant water amounts are expected to fill reservoirs and hence be available for irrigation due to enhanced glacier melt. In the long run, however, reduced precipitation and snow cover along with increasing temperatures are expected to make glaciers and rock glaciers melt away completely, leading to accelerated hydrological processes in the Andes and reduced water availability in downstream areas (Adam et al., 2009; Souvignet et al., 2010; Vicuña et al., 2011). This prediction is supported by recent analyses of the available historical precipitation and streamflow records (50 years), showing a clear trend toward dryer conditions in central and southern Chile (30–48°S) (Boisier et al., 2016, 2018) and an increasing number of warm days, especially for the high elevation Andean regions (Vuille et al., 2015).

12.4.3 Missing Information for Decision Making in Water Management

Decision making in water resources management requires reliable data and robust models to predict spatially and temporally distributed water availability and scarcity.

Figure 12.6 shows the elements needed to develop decision support information and to approach scenario development in the strongly engineered Limarí Basin

To develop scenarios that examine future water scarcity, we need reliable historical data on climate, discharge, land cover evolution, and demand to calibrate hydrological and water allocation models. In the following section, existing and missing elements of this framework are listed.

LACK OF DATA AND MONITORING

The DGA maintains the hydro-meteorological monitoring network in Chile and in the Limarí Basin. Long-term data for climatic variables are available from thirteen reliable climate stations (Souvignet et al., 2012; Kretschmer et al., 2014). However, records for the remote Andean region above 1,000 m are sparse, which leads to extreme data uncertainty, a major problem, given that most precipitation inputs in the semi-arid region occur in the upstream area (Nauditt et al., 2016; Nauditt et al., 2017). Out of forty-eight installed discharge stations, only eighteen stations are continuously measuring, fifteen stations in streams and three in large irrigation canals (Souvignet et al., 2012; Kretschmer et al., 2014). Both climate and discharge stations are vulnerable to climate extremes, and hence often produce gaps in the records. Discharge is not recorded under extreme low or high flows, leading to data gaps when information about hydro-meteorological variability is most needed.

As shown in Table 12.2, land uses, and crop distribution are only recorded by INE to feed into the census, which has not been updated since 2007. Also, real or high-resolution satellite-based images are not available to assess cropping patterns and changes over time. Consequently, land use data and socioeconomic information are sparse and uncertain, hindering modeling and scenario development for decision support (WEIN, 2014).

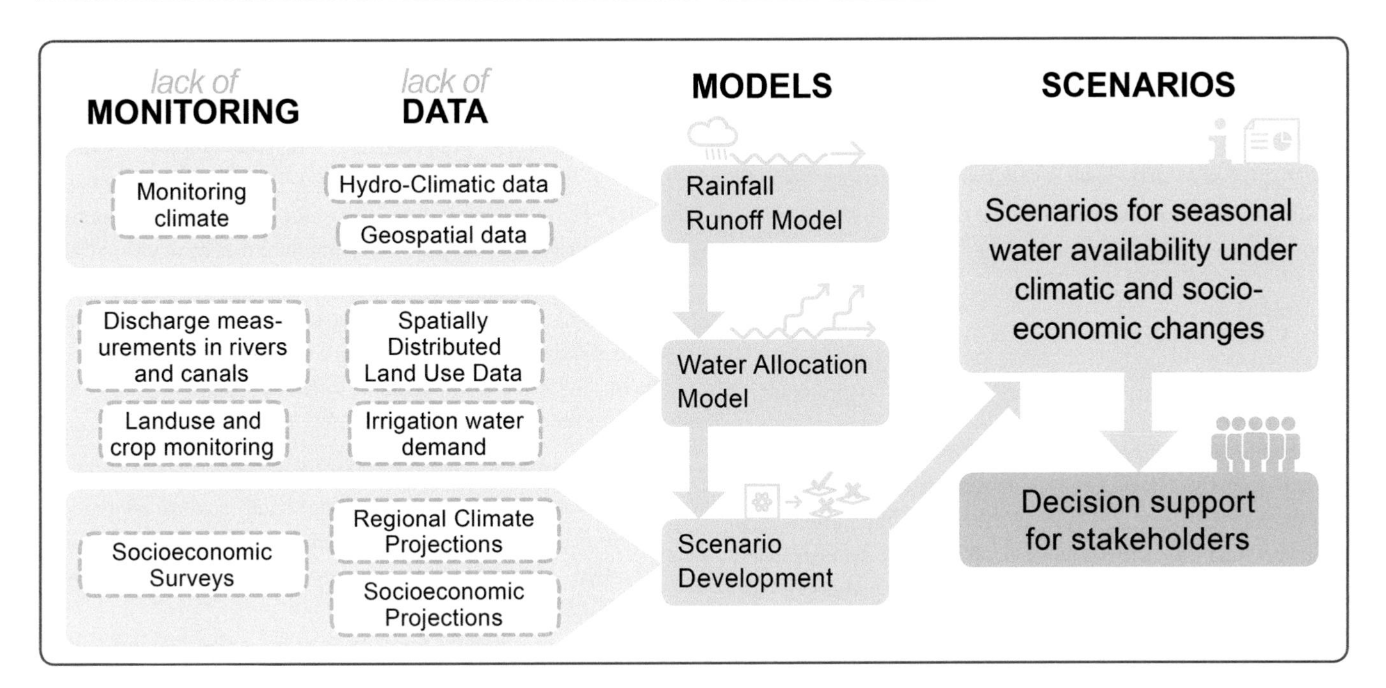

Figure 12.6 Elements needed for scenario development for the Limari irrigation system considering future climate variability.
Figure created by the authors

MODELING AND SCENARIO DEVELOPMENT

Robust rainfall-runoff simulations provide valuable information on long term and seasonal water availability as well as for the management of hydrological extremes such as floods and droughts. In arid and semi-arid environments, conditions for hydrological modeling are unfavorable. This can be mainly attributed to the aforementioned lack of representative rainfall values to force rainfall-runoff modeling, as well as general data scarcity with mostly unknown contributions from near-surface, sub-surface, and groundwater flow paths (Van Loon and Van Lanen, 2012; Nauditt et al., 2016). Hence, few modeling exercises have been carried out in mountainous areas draining to subtropical semi-arid or arid regions like northern-central Chile.

To date and to our knowledge, the following rainfall-runoff and water allocation models have been calibrated to simulate discharge as input to the Limarí irrigation system, each with a different level of complexity and spatial distribution approaches to simulating discharge,

- Vicuña et al. (2011) used the monthly scale Water Evaluation and Planning tool (WEAP) to assess the impacts of a changing climate on variations in streamflow using climate projections based on statistical downscaling. The results suggested that increasing temperatures and changing precipitation patterns are likely to significantly reduce long-term runoff due to accelerated snow and glacier melting and hence lead to lower water availability, especially in the summertime.
- The semi-distributed hydrological model SWAT (Neitsch et al., 2005) was used by Souvignet et al., (2010) to simulate streamflow of the Hurtado, a sub-basin of the Limarí, as inflow to the Recoleta reservoir.
- Penedo-Julien et al. (2019) applied the spatially more distributed model J2000 (Krause and Kraslisch, 2005) to the Hurtado, a Limarí headwater catchment, to help gain understanding about hydrological processes and as a management tool in fragile and vulnerable basins in arid regions facing new challenges such as climate change, overuse, and land use changes.
- The conceptual rainfall-runoff model HBV light (Seibert et al., 2012) was used to test the hydrological processes dominated by snow melt and groundwater movements in Limarí headwaters during both dry and wet periods, to predict seasonal water availability (Nauditt et al., 2016).
- To assess potential effects on water allocation, Vicuña et al. (2012) again applied the WEAP model as a water allocation model to the La Paloma irrigation system, using historical climate data, water infrastructure, irrigation coverage, and crop patterns, and forced it with artificial climate scenarios.

All of these models had to face extreme data scarcity in terms of unknown precipitation inputs, daily discharge in canals (in- and output), no or contradicting information on land uses, lack of historical information on the evolution of crop patterns, and respective water demand.

Recent gridded datasets like satellite-based precipitation estimates (Zambrano-Bigiarini et al., 2017; Baez-Villanueva et al., 2019), reanalysis data, and combined data products (Álvarez-Garretón et al., 2018) have huge potential to serve as input for model calibration and water availability projections (Nauditt et al., 2018; Sheffield et al., 2018).

12.5 CONCLUSION

This chapter summarized the history of overexploitation of surface and groundwater resources in the Limarí Basin, which is the most engineered basin in Chile. It shows that the current legal framework and engineering level of the Limarí system cannot cope with climate extremes like the recent "Megadrought" and further expansion of the agricultural area. In 2013, the La Paloma Reservoir relied on only 3 percent of its storage capacities and the Cogotí reservoir was empty. As a result, illegal groundwater drilling and pumping had to save the large area of valuable fruit trees that need to be irrigated during the whole year. Table 12.3 summarizes the key shortcomings of the system that have led to the overexploitation of surface and groundwater resources and over-allocation of water rights.

This holistic assessment shows that engineering infrastructure, climatic conditions, and institutional capacities in terms of tradable water rights and private water user associations allowed economic development in the Limarí Valley. However, the lack of governmental regulation mechanisms has led to overexploitation of water resources threatening water security, specifically environmental and agricultural sustainability. In the face of climate change and decreasing water availability, the current infrastructural and management system requires reforms.

In line with the conclusions of other scholars, we derive the following recommendations for policy makers,

- Start publicly guided integrated river basin management to plan water resources under drought and increase adaptation capacities (smaller cultivation areas but efficient irrigation system, nature-based storage solutions, groundwater recharge).
- Create incentives for alternative labor opportunities other than agriculture to avoid further expansion of agricultural area.
- Reform the WC81 to prioritize drinking water and ecological flow over productive water uses.
- Increase penalties for not using water rights as decided in the updated water code of WC 2005.

Table 12.3 *Summary of factors leading to overexploitation of surface and groundwater resources and overallocation of water rights in the Limarí Basin*

Governance and legal background	Institutional setup	Information transparency: Monitoring and Data Management	Water management and irrigation efficiency
WC81: only limited public regulation of private water rights, its uses, and water markets; too little penalties for not using water rights (WC 2005) compared to water trade value.	Public institutions related to environment and water resources protection (DGA) are small and financially weak.	Unknown precipitation inputs at high Andean elevations: no climate monitoring from 1,000 to 5,500 m. => uncertain water availability	Illegal abstractions from surface and groundwater that are not monitored ⇨ Unknown demand
WC81: Water rights are disconnected from land rights (incentive to transport water)	Water user associations (irrigation management, NGOs) that regulate the distribution of water are private and led by farmers, little public control of water allocation practices.	Insufficient monitoring in irrigation system and canals ⇨ uncertain and unfair water allocation	Low water use (irrigation) efficiency – losses from canals and to evaporation
WC81: no hierarchical prioritization of water uses (irrigation over other uses); no Minimum Ecological Flow Thresholds are considered (or established)	CNR and DOH (responsible for irrigation infrastructure of the Agricultural Ministry) are only seeking agricultural expansion, not irrigation (water use) efficiency.	Agricultural lands are not spatially monitored: no spatial information on crop patterns and hence on seasonal water demand; no Agricultural Census information since 2007 to monitor expansion of agricultural area.	Lack of capacities in irrigation management and monitoring

- Introduce ecological flow thresholds at different river sections as decided in the updated water code of WC 2005.
- Increase temporal and spatial frequency of monitoring all climate variables, discharge, water fluxes in reservoirs and canals, groundwater levels and quality, land cover and cropping systems.
- Provide funding for monitoring, modeling, scenario development, and integrated river basin planning with most recent data sets and digital facilities.
- Novel data products and tools can facilitate quicker assessments and information for decision making.
- Create and fund educational programs and capacity building materials for Integrated Water Resources Management to educate modern water managers.

REFERENCES

Adam, J. C., Hamlet, A., and Lettenmaier, D. P. (2009). Implications of Global Climate Change for Snowmelt Hydrology in the Twenty-first Century. *Hydrological Processes*, 23, pp. 962–972. https://doi.org/10.1002/hyp.7201

Álvarez, P. (2018). The Water Footprint Challenge for Water Resources Management in Chilean Arid Zones. *Water International*, 43(6), pp. 846–859. https://doi.org/10.1080/02508060.2018.1516092

Álvarez, P., Kretschmer, N., and Oyarzun, R. (2006). *Water Management for Irrigation in Chile: Causes and Consequences, Technology, Resource Management and Development*. Paper presented at the international water fair "Wasser Berlin 2006."

Alvarez-Garreton, C., Mendoza, P. A., and Boisier, J. P. et al. (2018). The CAMELS-CL Dataset: Catchment Attributes and Meteorology for Large Sample Studies – Chile Dataset. *Hydrology and Earth System Sciences*, https://doi.org/10.5194/hess-2018-23

Baez-Villanueva, O. M., Zambrano-Bigiarini, M., and Ribbe, L. et al. (2019). *A Novel Methodology for Merging Different Gridded Precipitation Products and Ground-Based Measurements*. Remote Sensing of Environment. RSE-D-19-00549

Barnett, T. P., Adam, J. C., and Lettenmaier, D. P. (2005). Potential Impacts of a Warming Climate on Water Availability in Snow-Dominated Regions. *Nature*, 438, pp. 303–309. https://doi.org/10.1038/nature04141

Bates, B. C., Kundzewicz, Z. W., Wu, S., and Palutikov, J. P. (2008). *Climate Change and Water, Technical Paper of the Intergovernmental Panel on Climate Change*. Geneva: IPCC Secretariat.

Bauer, C. (2004). Results of Chilean Water Markets: Empirical Research since 1990. *Water Resources Research*, 40, W09S06, https://doi.org/10.1029/2003WR002838

Bauer, C. J. (2015). The Evolving Water Market in Chile's Maipo River Basin: A Case Study for the Political Economy of Water Markets Project.

Boisier, J. P., Alvarez-Garreton, C., and Cordero, R. R. et al. (2018). Anthropogenic Drying in Central-Southern Chile Evidenced by Long-Term Observations and Climate Model Simulations. *Elementa: Science of the Anthropocene*, 6(74). https://doi.org/10.1525/elementa.328

Boisier, J. P., Rondanelli, R., Garreaud, R. D., and Muñoz, F. (2016). Anthropogenic and Natural Contributions to the Southeast Pacific Precipitation Decline and Recent Megadrought in Central Chile. *Geophysical Research Letters*, 43(1), pp. 413–421. https://doi.org/10.1002/2015GL067265

CIREN (2011). *Catastro frutícola [Fruit Crop Records, Land Registry]*, Principales Resultados [Principal Results], IV. Región de Coquimbo: CIREN, ODEPA.

(2015). *Catastro frutícola [Fruit Crop Records, Land Registry]*, Principales Resultados [Principal Results], IV. Región de Coquimbo: CIREN

Cooper, M. G., Schaperow, J. R., Cooley, S. W. et al. (2018). Climate Elasticity of Low Flows in the Maritime Western U.S. Mountains. *Water Resources Research*, 54, pp. 5602–5619. https://doi.org/10.1029/2018WR022816

Dirección General de Aguas [General Directorate of Water] (DGA) (2004) *Diagnostico y Clasificación de los cursos y cuerpos de agua según objetivos de calidad-Cuenca del Rio Limari [Diagnosis and Classification of Water Courses and Bodies According to Quality Objectives – Limari River Basin]*. Sistema Nacional de Información Ambiental [National Environmental Information System]. Available at www.sinia.cl/1292/articles-31018_Limari.pdf

Donoso, G. (2006) Water Markets: Case Study of Chile's 1981 Water Code. *International Journal of Agriculture and Natural Resources*, 33(2), pp. 131–146.

(2012). The Evolution of Water Markets in Chile. In J. Maestu, ed., *Water Trading and Global Water Scarcity: International Experiences*, pp. 111–129.

(2018). *Water Policy in Chile*. Berlin: Springer International Publishing AG. https://doi.org/10.1007/978-3-319-76702-4

Endo, T., Kakinuma, K., Yoshikawa, S., and Kanae, S. (2018). Are Water Markets Globally Applicable? *Environmental Research Letters*, 13(3). https://doi.org/10.1088/1748-9326/aaac08

Favier, V., Falvey, M., Rabatel, A., Praderio, E., and Lopez, D. (2009). Interpreting Discrepancies between Discharge and Precipitation in High-Altitude Area of Chile's Norte Chico

Region (26–32°S). *Water Resources Research*, 45(2). https://doi.org/10.1029/2008WR006802

Finlayson, B. L., McMahon, T. A., and Peel, M. C. (2007). Updated World Map of the Köppen-Geiger Climate Classification. *Hydrology and Earth System Sciences*, 11(5), pp. 1633–1644.

Garreaud, R. D., Alvarez-Garreton, C., and Barichivich, J. et al. (2017). The 2010–2015 Megadrought in Central Chile: Impacts on Regional Hydroclimate and Vegetation. *Hydrology and Earth System Sciences*, 21, pp. 6307–6327.

Garreaud, R. D., Vuille, M., Compagnucci, R., and Marengo, J. (2009). Present-Day South American Climate. *Palaeogeography, Palaeoclimatology, Palaeoecology*, 281 (3–4), pp. 180–195.

Hearne, R. (2018). Water Markets. In G. Donoso, ed., *Water Policy in Chile*. Springer Book Series: Global Issues in Water Policy, pp. 117–127.

Hearne, R. and Easter, K. (1997). An Analysis of Gains-from-Trade in Chile: The Economic and Financial Gains from Water Markets in Chile. *Agricultural Economics Journal*. https://doi.org/10.1016/S0169-5150(96)01205-4

Instituto Nacional de Estadistica [National Institute of Statistics] (INE) (2007). *Censo Nacional Agropecuario y Forestal [National Agricultural and Forestry Census] VII*. Santiago, Chile: Ministry of Agriculture, Government of Chile. Available at www.censoagropecuario.cl/

Kalthoff, N., Fiebig-Wittmaack, M., and Meißner, C. et al. (2006). The Energy Balance, Evapotranspiration, and Nocturnal Dew Deposition of an Arid Valley in the Andes. *Journal of Arid Environments*, 65(3), pp. 420–443.

Krause, P. and Kraslisch, S. (2005). The Hydrological Modelling System J2000-knowledge core for JAMS. In A. Zerger, and R. M. Argent, eds., *MODSIM 2005 International Congress on Modelling and Simulation*, pp. 676–682.

Kretschmer, N.; Nauditt, A.; Ribbe, L., 2013: Basin Inventory, Limarí, Chile. Project Report CNRD – unpublished.

Kretschmer, N., Nauditt, A., Ribbe, L., and Becker, R. (2014). *Limarí River Basin Study Phase I-Current Conditions, History and Plans*. Institute for Technology and Resources Management in the Tropics and Subtropics.

Meza, F. (2013) Recent Trends and ENSO Influence on Droughts in Northern Chile: An Application of the Standardized Precipitation Evapotranspiration Index. *Weather and Climate Extremes*, 1, pp. 51–58.

Ministerio de Justica (1981). *Código de Aguas [Water Code]*. In Diario Oficial de Chile [Official Gazette of Chile]. Available at www.leychile.cl/Navegar?idNorma=5605&idParte=0

Nauditt, A., Gaese, H., and Ribbe, L. (2000). *Water Resources Management in Chile*. Available at www.tt.fh-koeln.de/d/itt/publications/subject_bundles.htm#2002Vol2

Nauditt, A., Soulsby, C., and Birkel, C. et al. (2017). Using Synoptic Tracer Surveys to Assess Runoff Sources in an Andean Headwater Catchment in Central Chile. *Environmental Monitoring and Assessment*. https://doi.org/10.1007/s10661-017-6149-2

Nauditt, A., Thurner, J., and Zambrano-Bigiarini, M. et al. (2018). *Evaluating the Performance of Satellite-Based Rainfall Estimates in Low Flow Modelling in Data Scarce Andean Catchments at Different Latitudes of Chile*, EGU2018–18702.

Nauditt, A., Birkel, C., Soulsby, C., and Ribbe, L. (2016). Conceptual Modelling to Assess the Influence of Hydroclimatic Variability on Runoff Processes in Data Scarce Semi-arid Andean Catchments. *Hydrological Sciences Journal*, https://doi.org/10.1080/02626667.2016.1240870

Neitsch, S. L., Arnold, J. G., Kiniry, J. R., Williams, J. R., and King, K. W. (2005). *Soil and Water Assessment Tool Theoretical Documentation*. Ver. 2005. Temple, TX.: USDA-ARS Grassland Soil and Water Research Laboratory, and Texas A&M University, Blackland Research and Extension Center.

Observatorio Laboral Chile [Chile Labor Observatory] (2019). *Panorama Regional*. Available at www.observatoriocoquimbo.cl/panorama.html

OECD (2017). Gaps and Governance Standards of Public Infrastructure in Chile. Infrastructure Governance Review.

Oyarzun, R. (2010) *Estudio de caso: Cuenca del Limarí, Región de Coquimbo, Chile [Case study: Limarí Basin, Coquimbo Region, Chile]*. Compilación Resumida de Antecedentes, Centro de Estudios Avanzados en Zonas Aridas – Universidad de la Serena (CEAZA-ULS) [Summary Compilation of Background, Center for Advanced Studies in Arid Zones – University of La Serena (CEAZA-ULS)].

Oyarzún, R., Arumí, J. L., Alvarez, P., and Rivera, D. (2008). Water Use in the Chilean Agriculture: Current Situation and Areas for Research Development. In M. L. Sorensen, ed., *Agricultural Water Management Trends*. New York: Nova Publishers, pp. 213–236.

Oyarzun, R., Jofre, E., and Morales, P. et al. (2015). A Hydrogeochemistry and Isotopic Approach for the Assessment of Surface Water–Groundwater Dynamics in an Arid Basin: The Limarí Watershed, North–Central Chile. *Environ Earth Science*, 73, 39–55.
https://doi.org/10.1007/s12665-014-3393-4

Penedo-Julien, S., Nauditt, A., and Künne, A. et al. (2019). Hydrological Modelling to Assess Runoff in a Semi-arid Andean Headwater Catchment for Water Management in Central Chile. In A. Godoy-Faundez and D. Rivera, eds., *Andean Hydrology*. Boca Raton, FL: CRC Press.

Pepin, N., Bradley, R., and Diaz, H. et al. (2015). Elevation-Dependent Warming in Mountain Regions of the World. *Nature Climate Change*, 5, pp. 424–430. https://doi.org/10.1038/nclimate2563

Price, M. F. and Egan, P. A. (2014). *Our Global Water Towers: Ensuring Ecosystem Services from Mountains under Climate Change*. Policy Brief, Paris: UNESCO.

Schmandt, J., North, G. R., and Ward, G. H. (2013). How Sustainable Are Engineered Rivers in Arid Lands? *Journal of Sustainable Development of Energy, Water, and Environment Systems*, 1(2), pp. 78–93. https://doi.org/10.13044/j.sdewes.2013.01.0006

Seibert, J., Vis, M., and Kaser, D. (2012). HBV Light – A User Friendly Catchment-Runoff-Model Software. *Geophysical Research Abstracts, EGU General Assembly*, 2012, p. 14.

Servicio Agricola y Ganadero [Agricultural and Livestock Service] (SAG) (2018) *Catastro Vitivinicola Nacional [National Wine Registry]*. Available at www.sag.cl/ambitos-de-accion/catastro-viticola-nacional/1490/publications

Sheffield, J., Wood, E. F., and Pan, M. et al. (2018). Satellite Remote Sensing for Water Resources Management: Potential for Supporting Sustainable Development in Data-Poor Regions. *Water Resources Research*, 54(12), pp. 9724–9758. https://doi.org/10.1029/2017WR022437

Sheffield, J., Wood, E. F., and Roderick, M. L. (2012). Little Change in Global Drought over the Past 60 Years. *Nature*, 491, pp. 435–438. https://doi.org/10.1038/nature11575

Souvignet, M., Oyarzun, R., and Verbist, K. et al. (2012). Hydro-meteorological Trends in Semi-arid North-Central Chile (29-32º S): Water Resources Implications for a Fragile Andean Region. *Hydrological Sciences Journal*, 57(3), p. 479.

Souvignet, M., Gaese, H., Ribbe, L., Kretschmer, N., and Oyarzun, R. (2010). Statistical Downscaling of Precipitation and Temperature in North-Central Chile: An Assessment of Possible Climate Change Impacts in an Arid Andean Watershed. *Hydrological Sciences Journal*, 55(1), pp. 41–57.

Universidad de Chile, Facultad de Ciencias Agronomicas, Laboratorio de análisis Territorial [University of Chile, Faculty of Agronomic Sciences, Territorial Analysis Laboratory] (2018). *Diagnostico Nacional de Organizaciones de Usuarios [National Diagnosis of User Organizations]*. Resumen Ejecutivo [Executive Summary]. Available at https://snia.mop.gob.cl/sad/ADM5812v1.pdf

Urquiza, A. and Billi, M. (2018). Water Markets and Socio-ecological Resilience to Water Stress in the Context of Climate Change: An Analysis of the Limarí Basin, Chile. In *Environment, Development and Sustainability*, pp. 1–23. https://doi.org/10.1007/sl0668-018-0271-3

Van Loon, A. F. and Van Lanen, H. A. (2012). A Process-Based Typology of Hydrological Drought. *Hydrology and Earth System Sciences*, 16, pp. 1915–1946. https://doi.org/10.5194/hess-16-1915-2012

Verbist, K., Robertson, A. W., Cornelis, W., and Gabriels, D. (2010) Seasonal Predictability of Daily Rainfall Characteristics in Central-Northern Chile for Dry-Land Management. *Journal for Applied Meteorology and Climate*, 49(9), pp. 1938–1955.

Vergara, A. and Rivera, D. (2018) Legal and Institutional Framework of Water Resources. In G. Donoso, ed., *Water Policy in Chile*. Berlin: Springer Book Series, Global Issues in Water Policy, pp. 67–84.

Vicuña, S., McPhee, J., and Garreaud, R. D. (2011). Climate Change Impacts on the Hydrology of a Snowmelt Driven Basin in Semiarid Chile. *Climatic Change*, 105(3–4), pp. 469–488.

(2012). Agriculture Vulnerability to Climate Change in a Snowmelt Driven Basin in Semiarid Chile. *Journal of Water Resources Planning and Management*, 138(5), pp. 431–441. https://doi.org/10.1061/(ASCE)WR.1943-5452.0000202

Vicuña, S., Alvarez, P., Melo, O., Dale, L., and Meza, F. (2014). Irrigation Infrastructure Development in the Limarí Basinte Variability and Climate Change. *Water International*, 39(5), pp. 620–634. https://doi.org/10.1080/02508 060.2014.94506 8

Viviroli, D., Durr, H. H., Messerli, B., Meybeck, M., and Weingartner, R. (2007). Mountains of the World, Water Towers for Humanity: Typology, Mapping, and Global Significance. *Water Resources Research*, 43(7). https://doi.org/10.1029/2006WR005653

Vuille, M., Franquist, E., Garreaud, R., Lavado Casimiro, W., and Caceres, B. (2015). Impact of the Global Warming Hiatus on Andean Temperature. *Journal of Geophysical Research: Atmospheres*, 120(9), pp. 3745–3757. https://doi.org/10.1002/2015JD023126

WEIN (2014). *Increasing Water Use Efficiency in the Limarí Basin*. Available at www.hidro-limari.info

Zambrano-Bigiarini, M., Nauditt, A., Birkel, C., Verbist, K., and Ribbe, L. (2017). Temporal and Patial Evaluation of Satellite-Based Rainfall Estimates across the Complex Topographical and Climatic Gradients of Chile. *Hydrology and Earth System Sciences*. https://doi.org/10.5194/hess-2016-453

13 The Colorado River Basin

Douglas S. Kenney, Michael Cohen, John Berggren, and Regina M. Buono

13.1 INTRODUCTION

The iconic Colorado River and its tributaries may be the most heavily managed and utilized river system in the world. The Colorado provides at least part of the water supply for over 35 million people in one of the driest regions of North America, irrigates millions of hectares of farmland and pasture, and attracts millions of visitors to its scenic beauty, including the Grand Canyon and other landmarks. This intensive use, however, carries a heavy burden. Several aquatic and riparian species have been extirpated throughout most of their former habitat, and humans face increasing challenges as demand continues to grow, potentially surpassing available supplies. It is likely that already strained supplies are beginning to shrink due to the varied impacts of climate change. The Colorado River today is at a tipping point, with water storage at unprecedented low levels, forcing the unwelcome recognition by many water managers and water users that historic patterns of management and use are likely no longer sustainable.

Though decades of management decisions have contributed to the crisis environment that currently surrounds the river, cooperative action has emerged on many levels. Marc Reisner (1986, pp. 125) once characterized the Colorado as the "most legislated, most debated, and most litigated river in the entire world," but the river management system today is marked by a more collaborative approach. Over the past fifteen years, stakeholders and managers have enacted significant changes to the rules and regulations governing the Colorado, and many additional reforms remain close to enactment. Unlike the deal-making that characterized negotiations in the twentieth century, recent efforts focus primarily on strategies to promote coordinated management, reduce consumption, and restore a more holistic watershed perspective to a river that was legally apportioned long ago among a highly complex amalgam of jurisdictions and water users. The Colorado is a river in transition – a newly "closed" basin where demands have exhausted reliable supplies. In many respects, the river management system is a successful example of adaptation to changing social and hydrological conditions. Yet, it remains an open question whether the pace and scale of reform is sufficient to counteract those forces working to undermine the regional water budget.

13.2 PHYSICAL SETTING

The Colorado River Basin extends across 663,000 km^2, predominantly in the western United States, with some 3,100 km^2 in northwest Mexico (Figure 13.1). The mainstem of the river runs 2,300 km from its headwaters in the Rocky Mountains to its mouth in the Upper Gulf of California (also known as the Sea of Cortez). When the headwaters of the tributary Green River are included, the total length of the river extends approximately 2,700 km. Other major tributaries include the San Juan, Little Colorado, Gila, Gunnison, Yampa, White, Duchesne, Virgin, Muddy, Salt, Verde, Dolores, and San Pedro rivers, though several of these are now so heavily developed and diverted that they rarely contribute any flow to the river's mainstem.

Elevation and precipitation both drop sharply as the river rushes downstream.[1] The Colorado is a snowmelt-dominated system, with roughly 72 percent of the river's flow originating from mountainous landscapes above 2,400 m, which comprise less than 10 percent of the total basin area. Downstream, the basin is flat and arid, with many locations receiving less than 10 cm of precipitation annually. These areas contribute negligible flows to the mainstem, but are well suited to irrigated agriculture, resulting in a dramatic spatial dichotomy between the origins of flows and the eventual consumption of water.

Over two dozen dams tightly control and regulate the flow of the river (Figure 13.2). Two primary structures, the Glen Canyon Dam and Hoover Dam, bracket the Grand Canyon, forming the two largest reservoirs in the United States: Lake Powell and Lake Mead. Several additional structures are listed in Table 13.1 in descending order by storage capacity. Overall, the dams on the mainstem of the river (primarily Hoover and Glen Canyon) can store a remarkable four years' average annual flow of the river.[2] Many others, such as Morelos Dam at the United States–Mexico border and nearby Imperial Dam on the border between Arizona and California, lack meaningful storage capacity but exist to facilitate large-scale diversions out of the river channel.

The Colorado River historically carried a tremendous sediment load as it scoured the sandstone canyons of the Upper Basin. The old adage about the Colorado (and several other

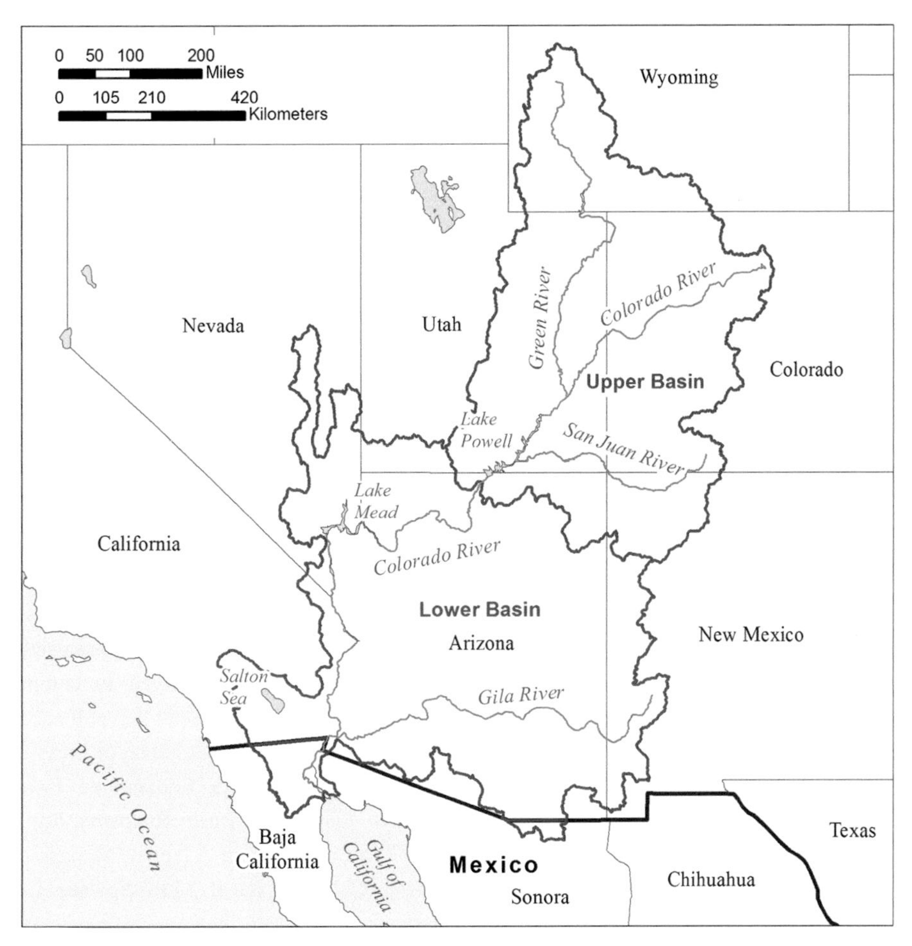

Figure 13.1 The Colorado River Basin.
Figure reprinted with permission from Cohen et al. (2013)

western rivers) was that it was "too thick to drink, too thin to plow." Despite this historical large sediment load, however, the numerous dams and reservoirs described in this section have captured most of this sediment and have sufficient capacity to continue to capture this sediment in the coming decades or longer. Glen Canyon Dam traps most of this sediment, an estimated 60,000 m^3 per year. This is but a small fraction of the reservoir's capacity, though as storage diminishes due to decreasing runoff this sediment will be deposited closer and closer to the dam itself. Some have argued that the Glen Canyon Dam could lose all its live storage within 60 years (Powell, 2010), though the US Bureau of Reclamation (Reclamation) and others contest this projection. Importantly, however, because Glen Canyon Dam is upstream of the Lower Basin dams, many of the downstream reservoirs will not need storage space reserved in their design to capture the expected sediment yield. Indeed, one of the primary sediment issues for the Colorado River system is not losing storage capacity, but the upstream dams trapping this sediment starving the river of its normal load and causing it to scour its channel downstream of the dams. A lack of sediment load downstream – particularly in the Grand Canyon – has decreased fluvial sandbars and beaches, leading to a series of controlled flooding experiments attempting to reverse this trend (Mueller et al., 2018).

In addition to the dams, many canals, aqueducts, and pumping stations also mark the basin, conveying water away from streams to fields and cities that may lie hundreds of kilometers distant. The largest of these, all found in the Lower Basin, take advantage of the flood control and water storage offered by Hoover Dam to divert significant amounts of water to agricultural and urban users in California and Arizona. Collectively, the All-American Canal and Colorado River Aqueduct deliver over one-fourth of the river's annual flow to southern California water users, while the Central Arizona Project moves roughly 10 percent of the annual flow uphill to users as far away as Tucson. Upstream in the headwaters, diversion structures are more numerous but considerably smaller in capacity given their location above major storage facilities. The state of Colorado, for example, boasts some twenty-nine transbasin diversions, conveying water from headwater reaches within the Colorado River Basin across the

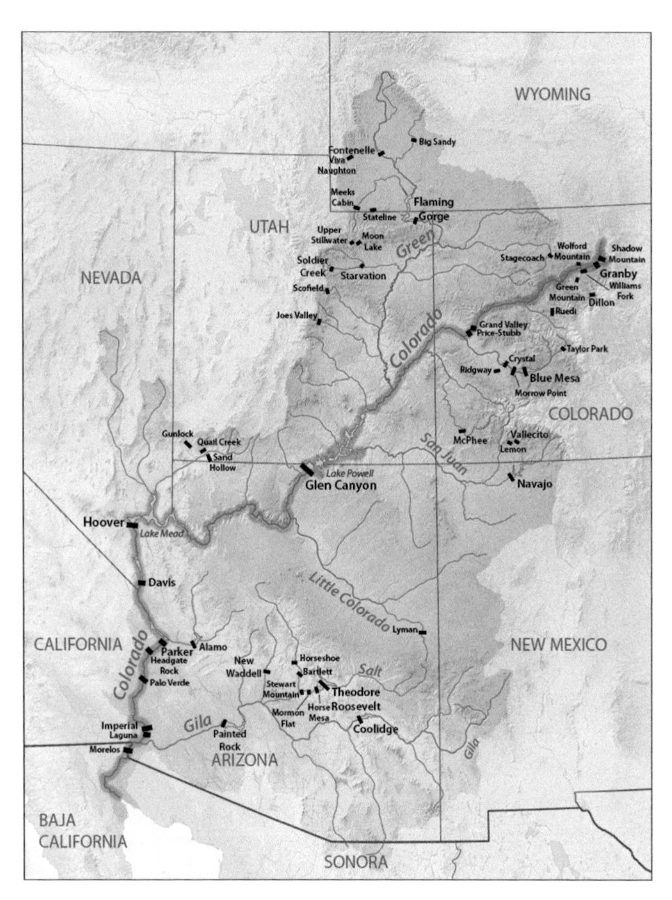

Figure 13.2 Dams in the Colorado River Basin.
Figure by Wikimedia Commons user, Shannon1, and reprinted under the GNU Free Documentation License

continental divide to the eastern plains. One of the oldest of these, the Grand Ditch, conveyed water out of the basin as early as 1890.

The extensive infrastructure in the basin is a response to the highly varied natural hydrologic conditions, characterized not only by dramatic intra-annual fluctuations in flow volumes (as is typical in snow-dominated basins) but significant inter-annual variability as well. The gage record for the mainstem of the Colorado River extends about a century. However, as a practical matter, releases from Hoover Dam dictate mainstem river flows for the last 30 percent of the river's length, and releases from Glen Canyon Dam determine about 95 percent of mainstem river flows, so actual gage measurements do not reflect the undiverted, unregulated flow of the Colorado River, known as the "natural flow." These flows, consequently, must be estimated (reconstructed) by using other data that account for releases, storage, and various types of water consumption.[3] As shown in Figure 13.3, the black line ("Abv Imp Nat'l") represents the reconstructed natural flow of the Colorado River at Imperial Dam, slightly upstream of the confluence with the Gila River. The red line ("Lee's Ferry Nat'l") represents the reconstructed natural flow of the river at Lee's Ferry, roughly the river's mid-point located slightly downstream of Glen Canyon Dam.[4] The blue line shows the actual gage record at Lee's Ferry, reflecting the depletions due to upstream diversions and reservoir evaporation and the influence of infrastructure and governance on river flow. In several years, more water flowed at Lee's Ferry than would have flowed naturally, despite upstream diversions, because reservoir storage in Lake Powell behind Glen Canyon Dam supplemented the natural flow. The large difference in flow between natural and actual flows at Lee's Ferry from 1963 to 1979 reflects the filling of Lake Powell. Figure 13.3 also shows the dramatic inter-annual variability in mainstem flow, with reconstructed natural flow ranging from 7.6 to 31.7 km^3/year (6.2–25.7 maf) above Imperial Dam.[5] Actual recorded flow of the river at the last gauging station on the river is now zero in most years; only rarely does the Colorado reach the ocean.[6]

River development has had a devastating impact on the aquatic environment, due to a combination of flow reductions, shifted hydrographs, increased water temperatures, loss of riverine habitat, and a sharp decline in river-borne sediment. Native species are increasingly disadvantaged in competitions with exotics, some deliberately introduced. Several substantial environmental programs exist to mitigate some of these impacts, including the Upper Colorado River Endangered Fish Recovery Program, the San Juan River Basin Recovery Implementation Program, the Glen Canyon Dam Adaptive Management Program, and the Lower Colorado River Multi-Species Conservation Program. Perhaps best known, however, are recent efforts to address the decimation of the Colorado River Delta, once among the most biologically diverse places on the continent. With the completion of Glen Canyon Dam in 1963, flows to the delta in most years have decreased dramatically or been nonexistent. In 2014, an environmental "pulse flow" was released under a binational agreement to support the ecological health of the delta. While the lasting environmental benefit of that flow is difficult to determine, it has been instrumental in reminding basin leaders – especially those in the Lower Basin – of the environmental and social costs of managing the river merely as a plumbing system. Upstream, many of the Colorado River's dramatic pre-development and post-development flow regimes persist in some Upper Basin tributaries. However, there, as elsewhere, water diversions by agricultural and municipal contractors, as well as hydropower generation schedules,[7] often play a key role in shaping river flows.

13.3 THE INSTITUTIONAL SETTING

Equally stunning and unique as the Colorado's physical setting and development is the river's institutional setting. A dense and complex collection of binational agreements, interstate compacts,

Table 13.1 *Major dams in the Colorado River Basin, by reservoir capacity*

Dam	River	Completed	Reservoir capacity (KM^3)	Power capacity (MW)	Annual generation (MWh)
Hoover	Colorado	1936	35.70	2,079	3,806,000
Glen Canyon	Colorado	1966	32.33	1,296	3,454,000
Flaming Gorge	Green	1964	4.67	153	344,400
Theodore Roosevelt	Salt	1911	3.59	36	123,900
Painted Rock	Gila	1960	3.07	N/A	N/A
Davis	Colorado	1951	2.24	255	1,147,000
Navajo	San Juan	1962	2.11	32	215,000
Soldier Creek	Strawberry	1974	1.39	N/A	N/A
New Waddell	Agua Fria	1994	1.37	45	
Alamo	Bill Williams	1968	1.29	N/A	N/A
Blue Mesa	Gunnison	1966	1.16	86.4	203,000
Coolidge	Gila	1930	1.12	N/A	N/A
Parker	Colorado	1938	0.80	120	457,000
Granby	Colorado	1950	0.67	N/A	N/A
McPhee	Dolores	1984	0.47	1.3	5,300
Fontenelle	Green	1964	0.43	10	67,000
Dillon	Blue	1963	0.32	1.8	
Horse Mesa	Salt	1927	0.30	129	126,900
Bartlett	Verde	1939	0.22	N/A	N/A
Imperial	Colorado	1938	0.20	N/A	N/A
Morrow Point	Gunnison	1968	0.14	173.3	293,000
Morelos	Colorado	1950			
Laguna Diversion	Colorado	1905			
Palo Verde	Colorado	1958			
Headgate Rock	Colorado	1941			
Grand Valley Diversion	Colorado	1916			

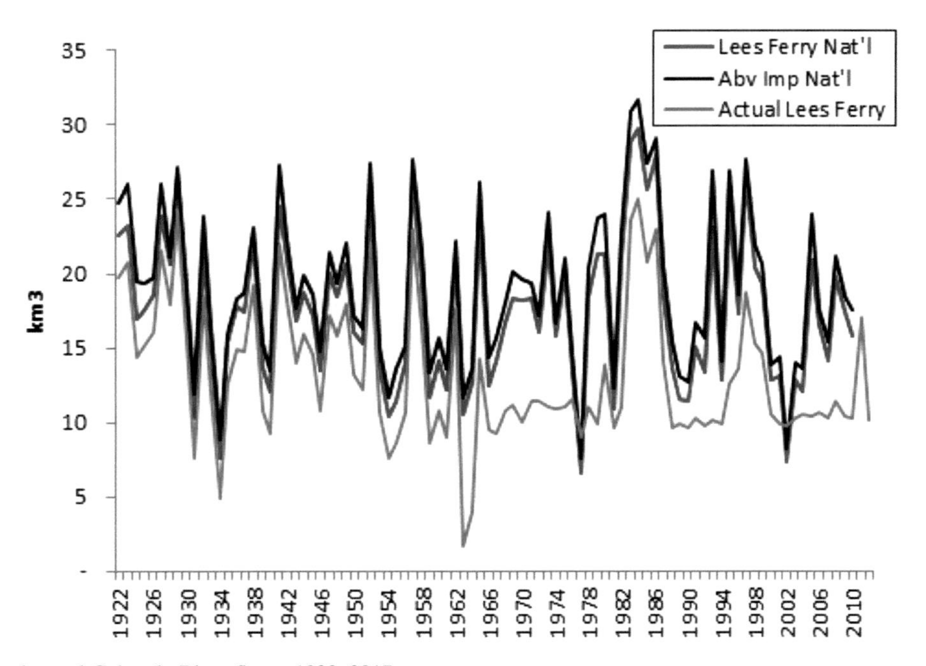

Figure 13.3 Natural and actual Colorado River flows, 1922–2017.

Figure created by the authors using Colorado River Basin natural flow and salt data from the US Bureau of Reclamation (available at www.usbr.gov/lc/region/g4000/NaturalFlow/)

Table 13.2 *Some key elements of the Law of the River*

	Year	What it did (major elements)	Citations
Colorado River Compact	1922	Apportioned river flow between Upper and Lower Basins	45 U.S. Stat. 1057
Boulder Canyon Project Act	1928	Ratified Compact; authorized Boulder Canyon Dam	45 Stat. 1057
Treaty with Mexico	1944	Apportioned river flow between the United States and Mexico	59 Stat. 1219
Upper Colorado River Basin Compact	1948	Apportioned Upper Basin waters among Colorado, New Mexico, Utah, and Wyoming	63 Stat. 31 (1949)
Colorado River Storage Project Act	1956	Authorized several projects, including Glen Canyon Dam	43 U.S.C 620
Arizona v. *California* decision	1963	Settled Lower Basin apportionments among Arizona, California, and Nevada (as first proposed in 1928)	373 U.S. 546 (1963)
Colorado River Basin Project Act	1968	Authorized the Central Arizona Project, and required preparation of Annual Operating Plan (AOP) for the two major dams (Hoover and Glen Canyon)	43 U.S.C. 1501
Minute 242	1973	Established salinity criteria for deliveries to Mexico	Treaty minutes available at www.ibwc.gov/
Colorado River Basin Salinity Control Act	1974	Created basinwide salinity control program	43 U.S.C. 1571
Grand Canyon Protection Act	1992	Established reservoir operating criteria at Glen Canyon to balance hydropower and environmental concerns	106 Stat. 4669
Interim Guidelines (Colorado River Interim Guidelines for Lower Basin Shortages and the Coordinated Operations for Lake Powell and Lake Mead)	2007	Provided for coordinated reservoir operations of Lake Powell and Lake Mead, and defined Lower Basin curtailment schedules. Promoted conservation and storage in Lower Basin, implicitly creating multi-year accounting.	www.usbr.gov/lc/region / programs/ strategies/FEIS/ index.html
Minute 319	2012	Expanded US/Mexico cooperation on river management especially regarding storage, curtailments, and delta recovery	Treaty minutes available from IBWC: (www.ibwc.gov/)
Minute 323	2017	Extended terms of Minute 319, provided for emergency curtailments, funding for conservation projects in Mexico, and further restoration of the delta	Treaty minutes available from IBWC: (www.ibwc.gov/)

and local, state, and federal laws, rules, regulations, ordinances, and court decisions (listed above in Table 13.2), known collectively as "The Law of the River," governs the management and use of the Colorado River and water in the basin.[8] The challenges and opportunities facing the Colorado River can only be understood in the context of the Law of the River and the political environment associated with its application. As noted by Fradkin (1981, pp. xviii): "To me the [Colorado] river, in its present state, is primarily a product of the political process . . . rather than a natural phenomenon." Rosenberg et al. (1991, p. 3) concurred, arguing that the "[p]olitics of water use is a fundamental aspect of life in the Southwest today and is the determinant of the Colorado River's future."

The doctrine of prior appropriation forms the foundation of western water law and was instrumental in shaping the Law of the River. California gold miners first developed the doctrine to resolve disputes over mining claims, using the axiom "first in time, first in right." As miners increasingly turned to hydraulic mining, they applied the doctrine to water as well: The first person to divert and put a quantity of water to "beneficial use" was entitled in perpetuity to that quantity of water before any subsequent diverter could be satisfied, even if that meant that the stream was entirely dewatered and more junior rights-holders received no water at all. Prior appropriation provides clear property rights to water and a recognized order (seniority) for satisfying these rights. It also rests on a concept of forfeiture, commonly known as "use it or lose it," which provides that water rights not put to beneficial use may be forfeited to other users ready to do so.

While the Colorado River basin states had individually accepted prior appropriation (in various forms and degrees) as the basis for state water law, the doctrine was not applied at the interstate scale until 1922, when the United States Supreme Court held that the concept could be applied to the Laramie River shared by two prior appropriation states (*Wyoming* v. *Colorado* (259 U.S. 419)). The prospect that this precedent could be applied by the courts to the Colorado River alarmed the headwaters states, as it would allow the more rapidly developing downstream states, particularly California, to claim extensive senior legal rights to Colorado River water even though they contributed almost no flow to the river. This fear prompted Colorado's Delph Carpenter – a lead attorney on the *Wyoming* v. *Colorado* litigation – to propose a negotiated allocation among the states using the interstate compact provision provided in Article I, Section 10 of the United States Constitution. This was a novel idea – since copied in roughly two dozen other United States basins, mostly in the western part of the country – that succeeded in bringing the seven basin states to the negotiating table, as the downstream states needed resolution of the interstate apportionment question before Congress would authorize construction of the Boulder Canyon Dam (later renamed Hoover Dam[9]). Unprotected by such a dam, primitive diversion structures in the Lower Basin were frequently washed out by the violent, sediment-laden flows associated with the spring snowmelt.

These negotiations led to the landmark Colorado River Compact of 1922 ("Compact"), enacted by Congress in the same legislation that authorized the Hoover Dam in 1928. At the heart of the agreement was an allocation of water between an Upper and Lower Basin (as shown in Figure 13.1). Presumably, the states of the Upper Basin (Colorado, New Mexico, Utah, and Wyoming) and Lower Basin (Arizona, California, and Nevada) were awarded equal shares of the river, although as discussed later, an overestimation of natural flows has ensured this will not occur in practice. Importantly, allocations to the two sub-basins were not tied to the speed of development, thereby reducing Upper Basin state trepidation about the speed of development downstream. Allocations to the other major political sovereigns, namely Mexico and Native American tribes, were deferred until later. With these stipulations, the Colorado River Compact formed the foundation of the allocation among the major jurisdictions, whereas prior appropriation would handle intrastate allocation among individual users. This framework largely sidestepped the need to empower any administrative or planning body (e.g., river basin commission) to further guide basin development, although it was clear that several Compact negotiators thought this step should be considered in the future.

Over time, the Law of the River evolved and built on the Compact, establishing a multi-tiered system for allocating Colorado River water. At the macro scale is the apportionment of water between the United States and Mexico. While such a division was anticipated in the 1922 Compact, this did not occur until the 1944 Treaty with Mexico, which also addressed the shared waters of the Rio Grande.[10] The 1944 Treaty calls for the United States to deliver annually 1.85 km^3 of water (1.5 maf), an amount that may be reduced in cases of "extraordinary drought." As discussed later, only recently has this provision been quantitatively defined, using a treaty implementation agreement known as a "Minute." In 1948, the states of the Upper Basin negotiated an apportionment of the Upper Basin allotment in percentages among the four states: 51.75 percent to Colorado, 23 percent to Utah, 11.25 percent to New Mexico, and 14 percent to Wyoming. Earlier, the Boulder Canyon Project Act of 1928 had proposed an interstate apportionment of the Lower Basin share, an action not confirmed until extensive Supreme Court deliberations concluded in 1963. The Lower Basin allocation is provided as fixed quantities of annual consumptive use: 5.4 km^3 (4.4 maf) for California, 3.5 km^3 (2.8 maf) for Arizona, and 0.4 km^3 (0.3 maf) for Nevada.[11] Thus, by 1963, the river was fully allocated at the international, interbasin, and interstate scales, while a series of settlements and lawsuits (many still ongoing) sought to quantify the obligations to Native American tribes.[12] Legislation in 1956 and 1968 authorized the completion of major infrastructure projects.

While this allocation framework has provided a heretofore stable basis for development and water management in the basin, it is problematic in several ways. Most importantly, the agreements allocate more water than is physically available, a consequence of Compact negotiators severely overestimating the average yield of the river. Also, highly problematic, the agreement reserves no water for instream uses, notably the maintenance of flows necessary to sustain fish and wildlife, as well as the robust recreation economy in the basin. Other problematic omissions include the near total silence of the major allocation agreements on groundwater (which in many cases is hydrologically connected to surface flows), water quality, and climate change. Efforts to address these (and related) management concerns are ongoing, including major actions in 1973 and 1974 addressing salinity concerns, legislation in 1992 to address environmental impacts of hydropower regimes, and more recent action since the turn of the century to address looming water shortages.

Resolving the omissions and shortcomings in the Law of the River are primarily the responsibility of the "principals" (representatives from the seven basin states in the United States) working in conjunction with the Secretary of the Interior ("Secretary") and Reclamation. The role of the Secretary is particularly salient in the Lower Basin, as the 1963 Supreme Court decree established the Secretary as the "Watermaster" responsible for managing and allocating the river and publishing annual accounting reports of Colorado River mainstem use. Reclamation is the Secretary's designee and maintains operational control of the river, making Reclamation a key player in river management. Other key players in the Lower Basin include large municipal water agencies and irrigation districts, including the Metropolitan Water District of Southern California, the Imperial Irrigation District, the Southern Nevada Water Authority, and the Central Arizona Water Conservation District.

Management and river accounting in the Upper Basin differs from the Lower Basin: The federal role is diminished; water accounting is estimated based on irrigated acreage as well as measured deliveries to cities and irrigators; and state water agencies play a much stronger role, supplemented by the Upper Colorado River Commission (comprised of state and federal representation) established in the 1948 Compact. In recent years, Denver Water and the Colorado River District of Colorado have played a stronger role, with diminished roles played by the other Upper Basin states.

Internationally, the United States and Mexican offices of the International Boundary and Water Commission are the key players in binational negotiations. The involvement of tribal officials in basin water issues has historically been negligible – far below what is appropriate given the magnitude of tribal water rights.[13] Finally, nongovernment organizations (NGOs), especially environmental NGOs, have emerged in recent years as major players, often helping to bring new ideas to negotiations and to broker new agreements among more established decision-makers.[14]

13.4 WATER USE

Despite the hydrologic basin being sparsely populated, the waters of the Colorado River are utilized extensively and are critically important to the economic and social fabric of the southwestern United States and part of northwestern Mexico. Surface and groundwater from the Colorado River Basin supply water for more than 35 million people, of which roughly 25 million live outside of the hydrologic basin, primarily in the urban corridor stretching from Los Angeles, California, to Tijuana, Mexico. Other major out-of-basin water users include cities in Colorado's Front Range (including the Denver-metro region); Salt Lake City, Utah; Albuquerque, New Mexico; and parts of northern Mexico. Within the basin, the largest metropolitan areas are Phoenix, Arizona (4.5 million people), Las Vegas, Nevada (2 million), Tucson, Arizona (1 million), and Mexicali, Baja California, Mexico (1 million). None of these urban centers are physically adjacent to the river itself. The largest city along the river is Yuma, Arizona (95,000 people); the largest city in the Upper Basin is Grand Junction, Colorado, with about 62,000 people.[15]

Despite rapid population growth within the basin and service area, the dominant use of Colorado River water is irrigated agriculture, which consumes about 70 percent of the developed water supply. Some 1.3 million hectares (ha) of land are irrigated within the Colorado River Basin, and another 0.14 million hectares of dry-land farming exist within the basin. Water exported from the basin helps irrigate another million hectares outside the basin, primarily in northern Colorado. Alfalfa accounts for about a quarter of all irrigated acreage in the basin.[16] Crop diversity increases markedly in the southern portions of the basin; the Coachella, Imperial, and Yuma valleys account for most of the United States' winter vegetable crop, while Mexico's portion of the basin grows the majority of that nation's wheat crop. The Upper Basin irrigates about 50 percent more land than does the Lower Basin, but the Lower Basin consumes more than three times as much water due to the more favorable climate for agriculture (Cohen et al., 2013). Other important land uses within the basin include resource extractive industries such as coal and hard rock mining, oil and natural gas extraction, and timber operations. The basin also supports a thriving recreation-based economy, including rafting, fishing, and motorized water activities.[17]

Groundwater–surface water interactions in the basin are poorly understood, with estimates of total groundwater contributions to surface flows ranging broadly from roughly 20 to 60 percent.[18] Groundwater withdrawals support irrigation and municipal supply in many areas, especially in Arizona and Mexico, where aquifers are in decline due to over-extraction. Arizona estimates annual state-wide groundwater extraction at about 3.6 km^3. Groundwater extraction within the Mexican portion of the basin approaches 1.0 km^3. In 1980, Arizona adopted an innovative Groundwater Management Act requiring new development in the most populated areas of the state to demonstrate a sustainable 100-year supply of water, though groundwater overdraft continues to be a problem in many areas of the state. Historically, Colorado River high-flow events recharged much of the Mexican aquifer. There have not been any such events in the past 15 years, however, so Mexico is attempting to reduce the amount of irrigated land to minimize groundwater overdraft. California's Coachella Valley, north of the Salton Sea, has suffered from extensive groundwater overdraft and land subsidence for many years.

13.4.1 Future Projections of Water Supply and Demand

The accounting of water supplies and demands in the Colorado River Basin is an imprecise exercise that has taken on a new urgency as water scarcity concerns have grown. The era of abundance on the Colorado came to an abrupt halt with the dawn of the twenty-first century, a fact best illustrated by the rapid depletion of stored water in Lakes Powell and Mead. In just the first four years of the new century, reservoir storage in those facilities dropped roughly by half due to overconsumption and a significant drought event. One response was the "Basin Study," a multi-year research effort of Reclamation as overseen by the seven basin states (Bureau of Reclamation, 2012). The study projects Colorado River Basin supply and demand into the future. Both supply and demand projections are imperfect and controversial, but the Basin Study is nonetheless the best official assessment of the evolving water budget in the basin.

The Basin Study utilized a variety of supply and demand scenarios. The four water supply scenarios were Observed Resampled, based on the historic record (approximately 100 years); Paleo Resampled and Paleo Conditioned, reliant on roughly 1,250 years of flow estimates derived from tree-ring studies; and Downscaled GCM Projected, based on an ensemble of sixteen climate change models and three emissions scenarios. The projections were organized into near-term (2011 to 2040), mid-term (2041 to 2079), and long-term (2066 to 2095) time horizons, with results organized using probability distributions. Supply scenarios based on historic (including paleo reconstructed) data are increasingly recognized as problematic due to the "end of stationarity" – the notion that, due to rising atmospheric greenhouse gas levels, the present does not (and the future will not) look like the past. This is an argument for focusing on projections derived from climate models, which yield wildly diverse – although almost universally negative – water supply futures (Vano et al., 2014).

Six water demand scenarios were utilized, distinguished by the level of expected population growth and the degree of water reserved for environmental and recreational enhancement. These projections also feature high uncertainties, in part because the region has seen a "decoupling" between population growth and water demand (Bureau of Reclamation, 2015). Even though the Colorado River Basin has been among the fastest growing regions in the United States, total water consumption today is roughly the same as thirty years ago. This is largely due to proactive conservation efforts in major cities – including Las Vegas, Los Angeles, Phoenix, Denver, Albuquerque, and many others – that have become increasingly concerned about future water availability.[19] A very similar trend was seen in the agricultural sector over this time, as crop yields have increased roughly 25 percent while consumption has been stable. Whether or not these incremental gains in efficiency (in both sectors) will continue or were simply a one-time savings opportunity that has now gone as best practices have been implemented, is a subject of considerable debate. This question is particularly sensitive in the agricultural sector, where real reductions in consumption are often only possible through the retirement of farmland and the subsequent loss of farming economies and communities. A viable compromise may be the expanded use of ATMs (Alternative Transfer Methods), which are arrangements in which urban users financially compensate farmers for temporarily fallowing lands with the conserved water usually made available for use in cities.

The spread of supply and demand projections is shown graphically in Figure 13.4. A simplified compilation of average values

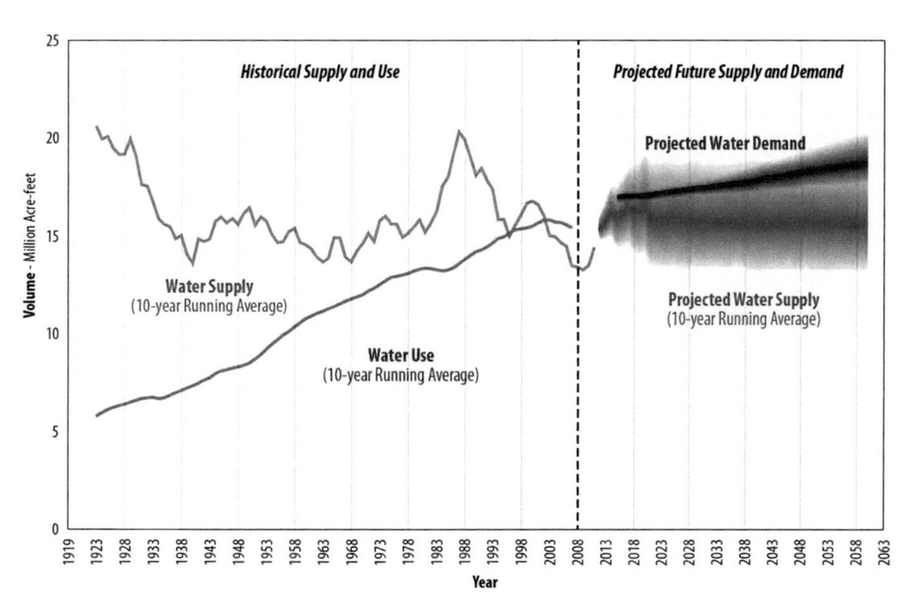

Figure 13.4 Historical supply and use and projected water supplies and demands in the Colorado River Basin.
Figure source: Department of Interior Bureau of Reclamation (2012)

Table 13.3 *Projected demands and supplies in the Colorado River Basin, km³ (maf)*

Future average demands (from the Colorado River mainstem)			
Years used in demand projections	2015	2035	2060
Total demands from the mainstem (*see exclusions**)	15.7 (12.7)	16.6 (13.5)	17.9 (14.5)
Agricultural demands	8.8 (7.2)	8.1 (6.6)	8.3 (6.7)
M&I demands	4.3 (3.5)	5.3 (4.3)	6.3 (5.1)
Other (energy, minerals, environment, tribal uses)	2.6 (2.1)	3.2 (2.6)	3.3 (2.7)
Future average supplies (at Lee's Ferry)			
Time periods used in supply projections	2011–2040 (2025)	2041–2079 (2055)	2066–2095 (2080)
Average Supplies	11.3 (13.9)	10.9 (13.4)	10.6 (13.1)
Decline from 1906 to 2007 mean	−7.5%	−10.9%	−12.4%

* *Values compiled from the Basin Study Technical Reports B (supplies) and C (demands), and may contain some rounding errors. Supplies are estimates of mainstem flows at Lee's Ferry, using the Downscaled GCM Projected scenario (*Bureau of Reclamation, 2012*). Demands are based on the Current Projected scenario, which assumes population increases from 40 million (in 2015) to 62 million (in 2060). Demands do not include Mexico's allocation, reservoir evaporation, and system losses, which in most years collectively account for roughly 5 km³ (4 maf). Lower Basin tributary use is also not included in these values, as this usage is typically considered outside of the Compact apportionment values based on Lee's Ferry flows.*

is featured in Table 13.3, showing values from the Downscaled GCM Projected supply scenario and the Current Projected demand scenario.

As shown in Figure 13.4 and Table 13.3, the results are disconcerting, showing chronic and growing imbalances between supplies and demands. If all obligations and system losses are considered, by 2060, projected demands would chronically exceed average supplies by roughly a quarter – a troubling finding that conceals extreme variability that could make the situation considerably worse or better in any given year. This occurs despite an assumed decline in agricultural water use, additional reductions of water allocated to environmental and recreational purposes, further decreases in per capita water consumption, and reductions in streamflow that are much more conservative that most subsequent climate change investigations (see e.g., Udall and Overpeck, 2017). There are many reasons to believe the chronic shortages projected by the Basin Study may be an optimistic scenario.

13.5 CHALLENGES MOVING FORWARD

Like all major rivers, the Colorado River faces a variety of challenges. Environmental protection and restoration are concerns, especially in the Lower Basin, as the extensive development of the river has turned a wild, warm, and muddy river into a plumbing system of tame, cold, and clear water that rarely reaches the ocean. Salinity is a persistent, but heretofore manageable, problem. Longer term, sedimentation of reservoirs (primarily Lake Powell, the major upstream facility) is of some concern. But without doubt, one problem dominates virtually all public policy discussions: the math problem – i.e., the amount of water demanded of the river exceeds what is likely to ever be available. This is not uncommon for a river in an arid land. But the problem on the Colorado is complicated by the apportionment scheme that assigns legal rights to more water than actually exists, which means that water users not only have to adapt to a limited supply but also have to confront the issue of how those "shortages" will be allocated among users that hold tight to treasured legal agreements promising more water. In this respect, not only is the challenge political and legal in nature – in additional to being technical – but it is also growing.

In short, the Colorado River Compact and the treaty with Mexico annually allocate 20.4 km³ (16.5 maf) of consumptive use from the mainstem; however, the twentieth century average river yield is merely 19.7 km³ (16 maf). System losses further unbalance the water budget, as reservoir evaporation consumes roughly 10 percent of the river's yield, an amount further reduced by phreatophytes and seepage. The origins of this problem stem from the first decades of the twentieth century, which were particularly wet, prompting Compact negotiators to severely overestimate the dependable yield of the river in the 1922 agreement.[20] After apportioning equal annual shares (9.3 km³ (7.5 maf)) of consumptive uses to the Upper and Lower Basins, the negotiators believed they had reserved sufficient excess in the system to account for any future apportionment to Mexico (as occurred in 1944) and to account for any future system losses. In reality, they did not. This problem was masked for many decades until average total basinwide consumption (including system losses) climbed to the level of average river yields. As shown in Figure 13.4, this occurred roughly at the dawn of the twenty-first century – the same time that drought conditions took hold in the basin.

The failure to account for system losses (primarily evaporation) is largely behind the so-called structural deficit, which refers to Lower Basin infrastructure and water delivery agreements that require roughly 10.5 km^3 (8.5 maf) in annual diversions to satisfy the 9.3 km^3 (7.5 maf) of Lower Basin consumptive uses authorized by the Compact. As a practical matter, the Lower Basin states are unlikely in the future to consistently receive more than 9.3 km^3 (7.5 maf) per year from Lake Powell releases, thereby establishing the need to conserve water immediately to preserve water levels in Lake Mead. To the credit of basin water leaders, this work is underway, supported by a variety of conservation programs and the establishment of curtailment schedules should storage levels in Lake Mead decrease further. These management tools were formalized in the 2007 "Interim Guidelines"[21] and were expanded further in the 2019 Drought Contingency Plans ("DCPs"). These efforts dovetail with unprecedented new cooperative arrangements with Mexico, including Minutes 319 and 323, which fold Mexico into the evolving conservation and shortage programs.

The math problem manifests itself differently in the Upper Basin, lying instead in another conflict imbedded in the Compact. While Article III(a) of the Compact reserves 9.3 km^3 (7.5 maf) per year of consumptive uses for the Upper Basin (as it does for the Lower Basin), Article III(d) requires that the Upper Basin allow a ten-year consecutive average of 93 km^3 (75 maf) of water to flow downstream (released from Lake Powell past the Lee's Ferry gauge) to satisfy the Lower Basin's allocation in III (a). If flows are insufficient to satisfy both provisions, then the default is to reduce Upper Basin consumptive uses as necessary to satisfy the III(d) release requirement.[22] Some Upper Basin leaders understood this problem as early as the 1960s, acknowledging that this likely capped Upper Basin water availability at roughly 7.4 km^3 (6 maf), even while the legal apportionment remained at 9.3 km^3 (7.5 maf).[23] Given the low populations and short growing seasons in the Upper Basin, this has not been a real constraint to date. Current annual Upper Basin consumption (including losses) rarely exceeds 5.5 km^3 (4.5 maf). However, the Upper Basin states still have plans for growth, in both population and water consumption, a right they presumably secured in the Compact negotiations. New Upper Basin projects (diversions) come online every year, with many more scheduled. To voluntarily cap development at something below the full legal apportionment is an idea that few leaders are willing to embrace even though a continued increase in Upper Basin consumption makes satisfying the III(d) release requirement more difficult. This conundrum raises the specter of a "Compact call," which would curtail Upper Basin uses to satisfy downstream obligations.

In recent years, the problem has become dramatically more severe and complicated, as river flows have plummeted. Through the first 16 years of the twenty-first century, average flows were almost 20 percent below the twentieth century average (roughly 14.8 km^3 (12 maf)). Given the downstream release requirement of Article III(d), this shortfall falls entirely on the Upper Basin and its III(a) apportionment. If these flows continue for the foreseeable future, then the Upper Basin has already exceeded a sustainable level of river development and use. Given the salience of this new wrinkle, diagnosing the cause of the flow decline is tremendously important.

Most water users and basin leaders attribute the low twenty-first century flows to drought. The current drought began in 2000 and, despite a few wet years, persists to the present day. However, recent research suggests that drought is not the complete answer. In an analysis of conditions from 2000 to 2014, Udall and Overpeck (2017) observe that precipitation declined only 4.5 percent, while flows declined 19 percent. In the last comparable drought, which began in the 1950s, precipitation declined 6 percent and streamflows dropped 18 percent. Thus, despite being a third more severe in terms of the precipitation decline, the impact on flow in the 1950s drought was slightly less. Why is the current drought causing a bigger streamflow decline? As shown for the Upper Colorado Basin in Figure 13.5, it is because the twenty-first century drought is a "hot drought" – an amalgam of drought and climate change. The basin today is 0.9°C (1.6°F) warmer than it was in the twentieth century, and higher temperatures mean increased evaporation and longer growing seasons, resulting in reduced flows. Even if precipitation were to return to the twentieth century average, the heat is almost certain to persist – and likely to increase to roughly 2.8°C (5°F) above the twentieth century average by 2050.[24] This warming is projected to reduce flows by 20 percent even if precipitation returns to twentieth century averages. Note that the Basin Study findings presented in this section assumed a decline in streamflows by 2060 of only roughly 9 percent. That finding was alarming to many water users, but it is increasingly looking like a best-case scenario.

13.6 EVALUATING AND ENHANCING SUSTAINABLE MANAGEMENT IN THE BASIN

The context and content of management of the Colorado River Basin continues to evolve in terms of estimated and actual supplies and demands, available resources, and governance processes. As discussed in Section 13.4, Reclamation's Basin Study projected a potential supply and demand imbalance of 3.95 km^3 (3.2 maf) by 2060. The study also evaluated many proposed

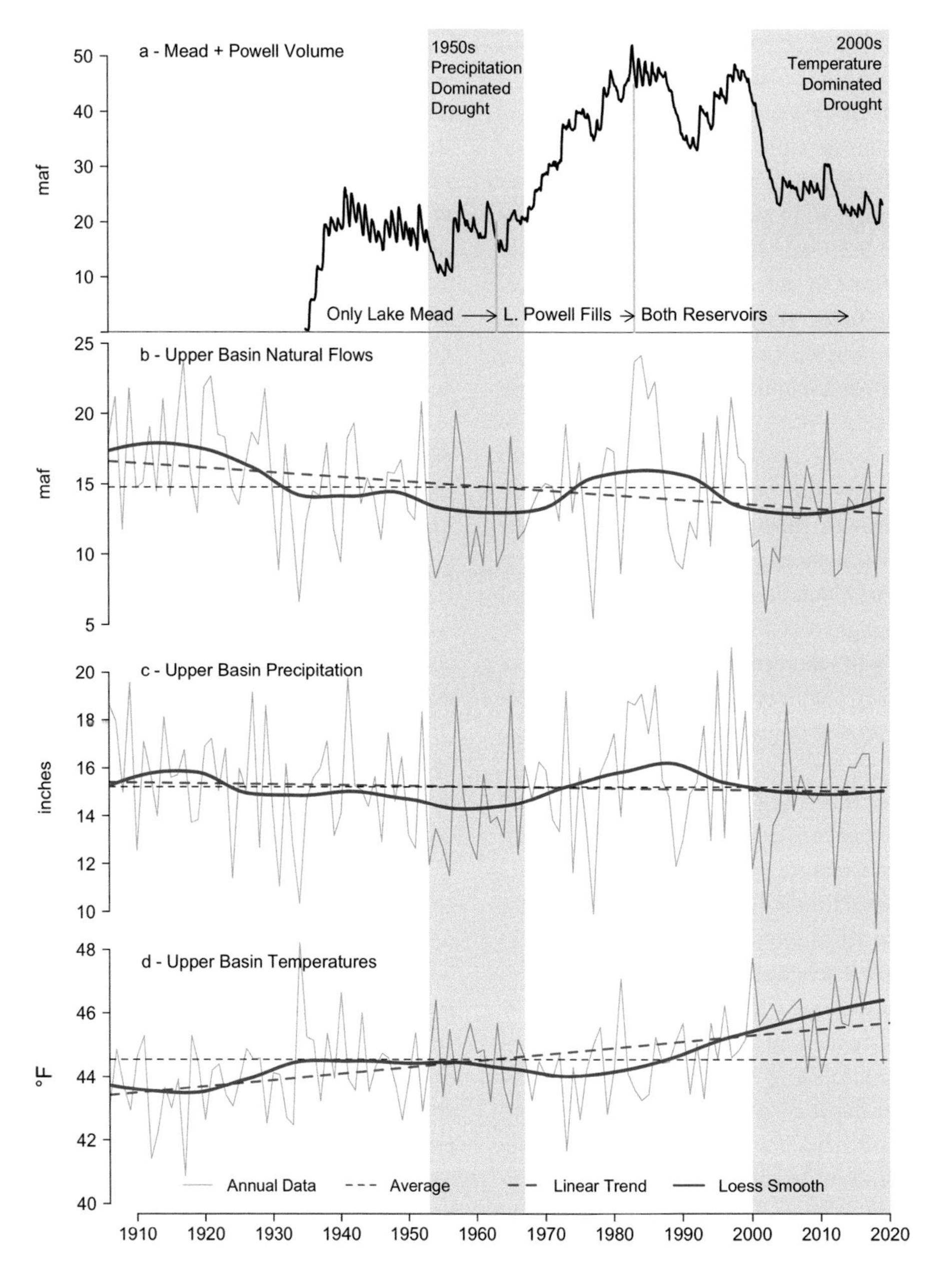

Figure 13.5 Trends in Upper Colorado River Flows, Precipitation, and Temperature. Figure provided by Brad Udall, adapted from Udall and Overpeck, 2017. Precipitation and temperature are from PRISM; natural flow data and reservoir levels are from the Bureau of Reclamation. Data is Water Years 1906 to 2019, except reservoir data, which begins in 1935.

options to bridge this imbalance, grouping these into four broad portfolios. Each of these portfolios includes a set of "common options," projected to generate almost 5 km^3 (4.1 maf) of water annually by 2060. These common options include municipal and industrial water conservation and efficiency projects, increased implementation of water recycling and re-use, and a combination of agricultural water conservation projects and transfer of conserved water from agricultural to municipal users. Many of these options are already emerging in the basin states – the goal now is to accelerate their implementation. While not technically challenging, these options will require significant funding. Given the diversity in scale, sector, and physical location of the options, securing a reliable source or sources of funding will be necessary for successful implementation.

In addition to funding challenges, the problem for Basin water managers and users is whether these options and projects are enough to handle additional drought years or long-term reductions in overall Colorado River water supplies. To their credit, basin state principals and the federal government have embarked on significant efforts to reduce consumptive use throughout the basin. Drought Contingency Plans ("DCPs") – which build upon the 2007 Interim Shortage Guidelines – were finalized in 2019 in both the Lower and Upper Basins. The plans represent a significant step toward resolving the math problem and attempting to bring supplies and demands back into balance.

In addition, the United States and Mexico have recently agreed to a new minute to the 1944 Treaty. Minute 323 is a binational agreement that extends the terms of Minute 319 (signed in 2012) and further defines "shortage sharing" (i.e., emergency curtailments) between the two countries. Additionally, Minute 323 includes funding mechanisms for conservation projects in Mexico and includes a significant ecological component intended to facilitate restoration of riparian habitat in the Colorado River Delta. While Minute 319 was a 5-year agreement used to gain operational and binational cooperation experience, Minute 323 extends these provisions to coincide with the expiration of the 2007 Interim Guidelines in 2026.

It is important to note, however, that the total amount of shortages when combining these current efforts – the 2007 Interim Shortage Guidelines, Drought Contingency Plans, and Minute 323 – are only sufficient to solve the math problem when Lake Mead reaches incredibly dire reservoir elevations. Specifically, the elevation-triggered shortages only exceed a total of 1.5 km^3 (1.2 maf) when Lake Mead drops below 1,025 feet above sea level – an elevation level for which the system would already be facing severe consequences.[25] While shortages in these agreements begin at higher Lake Mead elevations, sufficient reductions to balance the fundamental problem are only implemented once the system is in crisis. Further, even the most significant shortage amounts would not be enough to stabilize the reservoir if a severe, sustained drought impacted the basin (Fulp, 2017). Evidence of such "mega-droughts" has been seen in the paleo-hydrologic record, and the risk for such an event increases as the region continues to warm (Cook et al., 2015; Cook et al., 2016). Thus, while the significance of these historic agreements should not be downplayed – indeed the negotiations themselves were multi-year processes involving enormous effort from a variety of stakeholders (e.g., Grant, 2007) – they still represent primarily a crisis-management approach rather than a true sustainable management approach.

Part of the challenge of achieving "true sustainability" is that the many distinct water users in the basin have very different levels of water security, different risk thresholds, and unique risk management opportunities, and these benchmarks shift in dynamic ways based on the water management practices of fellow water users. Given this, every water user has a slightly different interpretation about what levels of flows are truly dependable, and what patterns of use and management are sustainable. This means that there is no single metric of dependable yield around which to organize action. Current levels of water supply and demand, for example, are likely sustainable for agricultural users in California's Imperial Valley but are not for nearby irrigators in central Arizona holding less secure water rights. Similarly, water security in the city of Las Vegas is among the highest in the basin, but the consequences of shortage are perhaps greater there than any other locale, as alternative water sources are largely nonexistent. In such a heterogeneous setting, it is not surprising that management efforts increasingly are focused around parameters more complex than dependable yield. Two such parameters – neither based directly on river yields – are particularly important. In the Lower Basin, maintaining Lake Mead elevation above 1,075 feet above sea level has become the key target, while in the Upper Basin, the ability to release necessary flows (from Lake Powell) to avoid a Compact call (93 km^3 in every 10-year period) is the critical standard. These are metrics based on reservoir storage and legal obligations – two features that the Colorado River system has in abundance. The result is a complex and often unwieldly management framework, a legacy of an over-allocation "mistake" made a century ago.

One intriguing aspect of these modern management efforts that may herald a more sustainable approach to managing the river is that these reforms were attained using a highly collaborative approach (Mulroy, 2007; King et al., 2014; Pitt and Kendy, 2017). Departing from previous negotiations, Mexico has played a prominent role in both the creation and implementation of these policies. Environmental NGOs were also key collaborators in the binational discussions, another change from previous protocols. Regarding the DCPs, some Native American tribes were more actively involved, most notably in Arizona. While some may argue that important Colorado River stakeholders are still excluded from these decision-making processes, these more recent agreements have demonstrated the benefit of expanding this collaboration beyond the basin states and federal government. We suggest they are a much-needed step in the right direction.

Efforts at improving management in the Colorado River Basin are ongoing, many of them concurrently. To discern which direction is the right one, we offer a roadmap toward sustainability. But first, we consider what "sustainability" means in the context of the basin. Sustainability in the Colorado River Basin has generally been thought about in terms of meeting consumptive use demands.[26] Under this framework, the basin has largely been sustainable: For decades the basin states have been able to utilize their Colorado River allocations and have continue to grow without a declared shortage. The DCPs and binational agreements seem to represent a continuation of this success. But this framework does not include significant components of the Colorado River Basin, namely environmental and certain social aspects. Managing the basin holistically has proven difficult, and often the focus on an individual component can come at the expense of another. In that sense, the basin has not been successful in finding a sustainable approach. This paradox is particularly evident at the Salton Sea in Southern California (Barnum et al., 2017).

The Salton Sea is an approximately 87,000-hectare terminal lake in the most arid region of the basin, sustained entirely from irrigation return flows. As such, any policies or decisions implemented in the region can have a direct impact on the Salton Sea. A water transfer agreement between the Imperial Irrigation District (located just upland from the sea), Metropolitan Water District of Southern California, San Diego County Water Authority, and the State of California was implemented in 2003 to help California reduce its demand for Colorado River water. The Quantification Settlement Agreement (QSA) was necessary because, up until that point, California had been diverting 6.5 km^3 (5.3 maf) annually, well above its 5.4 km^3 (4.4 maf) apportionment, because other states had not yet developed their full apportionments. In addition to bringing California back down to its legal apportionment, the QSA had to navigate a reduction in regional demand with municipal and industry (M&I) users who had junior water rights to the more senior irrigation districts (e.g., Imperial Irrigation District).

Irrigation efficiency upgrades and land fallowing in the Imperial Irrigation District were implemented to conserve water for transfer to M&I users in San Diego County. Because the Salton Sea relies on runoff from Imperial, any conservation measures taken by the agricultural district inevitably reduce the supply into the Sea. As such, the QSA, while laudable in that it reduces California's overall use of Colorado River water and helps balance the regional water budget, largely comes at the expensive of an important environmental resource. This situation took on new urgency at the end of 2017 when, as set out in the original agreement, temporary "mitigation flows" from Imperial to the Salton Sea expired. Consequently, the lake elevation levels have begun to drop precipitously, threatening wildlife, exposing lakebed, and increasing dust emissions in a region already plagued by poor air quality, further impairing public health.

The trade-offs that have characterized previous solutions to challenges in the basin, like those visible at the Salton Sea, underscore the need to think more holistically about defining and measuring what it means to manage the river basin sustainably. Some of the more common criteria for sustainable river basin management include (1) increasing stakeholder participation, (2) incorporating relevant science and information, (3) creating a more fair and equitable process, and (4) better integrating differing stakeholder positions and values (Kenney, 2005; Whiteley et al., 2008; Antunes et al., 2009). While these criteria are broad and diverse, they all reveal the importance of focusing on the process, rather than solely on specific policy or operational outcomes. This has implications for the Colorado River because, as discussed previously, "sustainability" in the basin has largely focused on meeting or reducing consumptive uses – an outcome-based metric rather than a process-related one. While meeting these uses is important, basin managers are excluding large components of the system by limiting sustainability to demand management. The distinction between process and outcomes has important implications for policy and should become a focal point for future negotiations and governance.[27]

The criteria listed offer a clear starting point from which we can begin to evaluate and enhance sustainability in the Colorado River Basin. As noted previously, stakeholder participation in the decision-making process has improved in the last 15+ years. Beyond the basin state principals and the federal government, basin governance negotiations have benefited from increased involvement by environmental NGOs, Mexico, and some Native American tribes. Other stakeholders have proposed specific policy elements, contributed significant resources to the process, and supported greater coordination among stakeholder groups. One such example comes from the Basin Study, wherein the options and solutions evaluated and modelled by Reclamation were solicited from many stakeholders and the public. Moreover, an effort has been made to include these groups earlier in the decision-making process, rather than soliciting feedback and consultation once a draft policy has been created.

Despite these successes in improving stakeholder participation, there remains significant room for additional participants, as well as a need to better integrate differing stakeholder positions and values, and to enhance fairness and equity in stakeholder processes. While several NGOs have been successful in contributing to the process, certain environmental and recreational values are still not included. It is also important to consider not only which stakeholders should be more involved but also how specifically participation could be more formalized. The improvement in participation seen thus far has largely been ad hoc, with specific individuals within the stakeholder groups pushing for greater involvement or someone advocating on their behalf. Other than public comment and consultation requirements in the National Environmental Policy Act (NEPA), there has not been a structured process for inclusivity. NEPA only applies to certain decision-making processes and does not include binational negotiations. Other river basins have established basin commissions, stakeholder advisory groups, or other entities whose purviews include the entire river basin (Blomquist et al., 2005). While such organizations exist in the Colorado River Basin, they are limited in scope to a subset of basin issues and geography.[28]

It is debatable whether establishing a broadly focused commission or another formal process is necessary or appropriate for improving Colorado River management. When such organizations have been proposed for the Colorado (e.g., Getches, 1997), water managers have resisted, fearing a loss of power and autonomy, as well as the burdens of additional bureaucracy. Thus, stakeholder participation, while evolving and improving over the years, is still relatively ad hoc, and truly integrated management remains a distant goal.

Another criterion where the Colorado River Basin has made significant improvements is with the use of science and information. Some of the improvements are due in large part to Reclamation's efforts regarding technical support capabilities. One of the biggest accomplishments from the Basin Study is that it provided a common technical and hydrological modeling platform by which solutions could be evaluated. This common technical platform underlies many recent and ongoing problem-solving efforts, including the recent binational negotiations. Additionally, while there is still significant political controversy regarding climate change, the Basin Study acknowledged and integrated climate change impacts into several supply and demand scenarios – a first for Reclamation in the Colorado River Basin. To many Colorado River managers, the hydrologic impacts of a warming world are now impossible to ignore.

Despite improvements in incorporating science and information into decision-making, there remain shortcomings that need to be addressed to further sustainability goals in the basin. Reclamation's technical capabilities for modeling the mainstem of the Colorado River are impressive, but these capabilities are designed around the delivery of water for consumptive use. The models have difficulty incorporating and projecting ecological needs and the impacts of various operational schemes, especially upstream of the system's major storage sites and diversions (i.e., Lakes Powell and Mead). Moreover, while the Basin Study did help establish a baseline of climate change impacts with which most basin managers can agree, the study was less successful at incorporating some of the related institutional problems and potential solutions. Further, some of the assumptions used in the supply and demand projections were not apolitical (most notably the states' demand projections) and are still contested by some stakeholders in the basin.

In summary, the Colorado River Basin can be characterized by recent collaboration and significant achievements in new interstate operational agreements and binational cooperation. These cooperative efforts and agreements have moved the basin in the direction of greater inclusivity and holistic management; the policy outcomes reflect this shift in governance. Despite these efforts, however, the approach to management of the system remains reactive and crisis-driven, focused on meeting (or reducing) consumptive demands rather than true sustainable management. While this is not surprising given basin manager's goals of meeting customer needs, if sustainability ideals are to be the future of the Colorado River Basin, shifts in how the system is governed are needed. Specifically, a focus on decision-making processes – for example, more formal participatory frameworks, more holistic incorporation of information – would likely create a system with broader stakeholder buy-in and support, increased adaptive capacity, and a greater ability to reflect the diverse set of priorities and values evident throughout the basin.

13.7 CONCLUSION

Colorado River management has changed dramatically in the last 15 years, with stakeholders moving from active litigation and polarized positions to the current period of cooperation across state and international boundaries. Native Americans and NGOs have participated much more actively in policy development, resulting in several water rights settlements and landmark agreements on sharing of surpluses and shortages, and even providing water for environmental purposes. Basin stakeholders have developed several important, forward-looking planning documents to address projected shortages and identify options to offset these impacts, providing a model for basins around the world. Additional negotiations are underway that reinforce and build on this new management model by emphasizing cooperation, conservation, and shared sacrifice.

Tempering this enthusiasm, however, is the fact that the Colorado River's water budget is significantly unbalanced and could be destabilized entirely by climate change-induced shifts in average flows and extreme drought probabilities. While basin leaders are vigilant in the search for solutions that provide benefits for interests across the basin, water allocation is ultimately a zero-sum game that can magnify the widely differing legal, political, and economic resources available to stakeholders. More fundamentally, the allocation regime favors off-stream, consumptive uses over instream values. Any attempt to address these and other legal disparities is shaped by the political reality that California enjoys more representatives in the United States Congress than the other six basin states combined. No fundamental change to the Law of the River will happen politically without California's support, and the current regime is already quite favorable to the state – i.e., it has the largest and most secure allocation. If the parties with the strongest positions currently receive the most benefits, can real change occur? Similarly, within states, the reality is that most water consumption (except in Nevada) is in agricultural uses, whereas most of the population and economic might resides in the cities. Can farms and cities peacefully coexist in an era of extreme scarcity?

These observations all speak to the reality that management of the Colorado is transitioning from a technical engineering challenge to an institutional challenge, emphasizing the critical role of basin decision-makers in charting an innovative path forward that relies on new rules more than new infrastructure. A palpable sense of optimism and pride that such innovation is possible has emerged in recent years, even as reservoir levels remain dangerously low. This sentiment could be boosted further by a few wet years bringing needed relief to a basin on the edge, and more importantly, buying time for difficult negotiations to move forward in a noncrisis environment.

REFERENCES

Antunes, P., Kallis, G., Videira, N., and Santos, R. (2009). Participation and Evaluation for Sustainable River Basin Governance. *Ecological Economics*, 68(4), pp. 931–939. https://doi.org/10.1016/j.ecolecon.2008.12.004

Ault, T. R., Cook, B. I., Mankin, J. S., and Smerdon, J. E. (2016). Relative Impacts of Mitigation, Temperature, and Precipitation on 21st-Century Megadrought Risk in the American Southwest. *Science Advances*, 2(10), e1600873.

Barnum, D. A., Bradley, T., Cohen, M., Wilcox, B., and Yanega, G. (2017). *State of the Salton Sea: A Science and Monitoring Meeting of Scientists for the Salton Sea* (Open-File Report No. 2017–1005) (p. 20). U.S. Geological Survey. https://doi.org/10.3133/ofr20171005

Blomquist, W., Dinar, A., and Kemper, K. (2005). *Comparison of Institutional Arrangements for River Basin Management in Eight Basins.pdf* (World Bank Policy Research Working Paper No. 3636). World Bank. https://doi.org/10.2139/ssrn.757225

Bloom, P. L. (1986). Law of the River: A Critique of an Extraordinary Legal System. In F. Brown and G. Weatherford, eds., *New Courses for the Colorado River: Major Issues for the Next Century*. Albuquerque: University of New Mexico Press, pp. 139–154.

Bureau of Reclamation, US Department of the Interior (2012). *Colorado River Basin Water Supply and Demand Study*. Available at www.usbr.gov/watersmart/bsp/docs/finalreport/ColoradoRiver/CRBS_Executive_Summary_FINAL.pdf

(2015). *Colorado River Basin Stakeholders Moving Forward to Address Challenges Identified in the Colorado River Basin Water Supply and Demand Study*. Available at www.usbr.gov/lc/region/programs/crbstudy/MovingForward/Phase1Report/fullreport.pdf

Cohen, M. J. (2011). *Municipal Deliveries of Colorado River Basin Water*. Pacific Institute. Available at http://pacinst.org/publication/municipal-deliveries-of-colorado-river-basin-water-new-report-examines-100-cities-and-agencies/

Cohen, M. J., Christian-Smith, J., and Berggren, J. (2013). *Water to Supply the Land: Irrigated Agriculture in the Colorado River Basin*. Pacific Institute. Available at http://pacinst.org/publication/water-to-supply-the-land-irrigated-agriculture-in-the-colorado-river-basin/

Colorado River Basin Salinity Control Forum. (2011). *2011 Review: Water Quality Standards for Salinity Colorado River System (October)*. Available at http://coloradoriversalinity.org/docs/2011 percent20REVIEW-October.pdf

Cook, B. I., Ault, T. R., and Smerdon, J. E. (2015). Unprecedented 21st Century Drought Risk in the American Southwest and Central Plains. *Science Advances*, 1(1), e1400082–e1400082. https://doi.org/10.1126/sciadv.1400082

Cook, B. I., Cook, E. R., Smerdon, J. E., Seager, R., Williams, A. P., Coats, S., Stahle, D. W., and Dias, J. V. (2016). North American Megadroughts in the Common Era: Reconstructions and Simulations. *WIREs Climate Change*, 7(3), pp. 411–432. https://doi.org/10.1002/wcc.394

Fleck, J. (2016). *Water Is for Fighting Over: And Other Myths About Water in the West*. Washington, DC: Island Press.

Fradkin, P. L. (1981). *A River No More: The Colorado River and the West*. Tucson: University of Arizona Press.

Fulp, T. (2017). *How Does DCP Address Lake Mead's "Math Problem"? Presented at the Martz Summer Conference*. University of Colorado – Boulder.

Getches, D. H. (1985). Competing Demands for the Colorado River. *University of Colorado Law Review*, 56, pp. 413–479.

(1997). Colorado River Governance: Sharing Federal Authority as an Incentive to Create a New Institution. *University of Colorado Law Review*, 68, p. 573.

Grant, D. L. (2007). Collaborative Solutions to Colorado River Water Shortages: The Basin States' Proposal and beyond. *Nevada Law Journal*, 8, p. 964.

Hundley, N., jr. (1966). *Dividing the Waters: A Century of Controversy between the United States and Mexico*. Los Angeles: University of California Press.

Kendy, E. and Pitt, J. (2017). Shaping the 2014 Colorado River Delta Pulse Flow: Rapid Environmental Flow Design for Ecological Outcomes and Scientific Learning. *Ecological Engineering*, 106, pp. 704–714. https://doi.org/10.1016/j.ecoleng.2016.12.002

Kenney, D. (2005). *In Search of Sustainable Water Management: International Lessons for the American West and beyond*. Massachusetts, MA: Edward Elgar Publishing.

King, J. S., Culp, P. W., and de la Parra, C. (2014). Getting to the Right Side of the River: Lessons for Binational Cooperation on the Road to Minute 319. *University of Denver Water Law Review*, 18, p. 36.

Miller, M. P., Buto, S. G., Susong, D. D., and Rumsey, C. A. (2016). The Importance of Base Flow in Sustaining Surface Water Flow in the Upper Colorado River Basin. *Water Resources Research*. https://doi.org/10.1002/2015WR017963

Mueller, E. R., Grams, P. E., Hazel, J. E., and Schmidt, J. C. (2018). Variability in Eddy Sandbar Dynamics during Two Decades of Controlled Flooding of the Colorado River in the Grand Canyon. *Sedimentary Geology*, 363, pp. 181–199.

Mulroy, P. (2007). Collaboration and the Colorado River Compact. *Nevada Law Journal*, 8, p. 890.

Pitt, J. (2001). Can We Restore the Colorado River Delta? *Journal of Arid Environments*, 49, pp. 211–220.

Pitt, J. and Kendy, E. (2017). Shaping the 2014 Colorado River Delta Pulse Flow: Rapid Environmental Flow Design for Ecological Outcomes and Scientific Learning. *Ecological Engineering*, 106, pp. 704–714. https://doi.org/10.1016/j.ecoleng.2016.12.002

Powell, J. (2010). *Calamity on the Colorado*. Orion. Available at www.orionmagazine.org/index.php/articles/article/5617/

Reisner, M. (1986). *Cadillac Desert: The American West and Its Disappearing Water*. New York: Penguin.

Robison, J. and Kenney, D. (2012). Equity and the Colorado River Compact. *Environmental Law*, 42, p. 1157.

Rosenberg, K. V., Ohmart, R. D., Hunter, W. C., and Anderson, B. W. (1991). *Birds of the Lower Colorado River Valley*. Tucson: University of Arizona Press, p. 416.

Southwick Associates (2012). *Economic Contributions of Outdoor Recreation on the Colorado River & Its Tributaries*. Prepared for Protect the Flows. Available at http://protectflows.com/wp-content/uploads/2013/09/Colorado-River-Recreational-Economic-Impacts-Southwick-Associates-5-3-12_2.pdf

Tipton and Kalbach Inc. (1965). *Water Supplies of the Colorado River*. Report Prepared for the Upper Colorado River Commission. Denver, CO.

Udall, B. and Overpeck, J. (2017). The 21st Century Colorado River Hot Drought and Implications for the Future. *Water Resources Research*. https://doi.org/10.1002/2016WR019638

Vano, J. A., Udall, B., and Cayan, D. R. et al. (2014). Understanding Uncertainties in Future Colorado River Streamflow. *Bulletin of the American Meteorological Society*, 95, pp. 59–78.

Whiteley, J. M., Ingram, H. M., and Perry, R. W. (eds.) (2008). *Water, Place, and Equity*. Cambridge, MA: MIT Press.

Notes

1 Source: The Biota of North America Program, at www.bonap.org/Climate percent20Maps/ClimateMaps.html.

2 By comparison, storage on the Columbia River (United States and Canada) is roughly 0.3 years of flow, a much more typical storage-to-flow ratio even for highly developed rivers.

3 Based on data compiled by Jim Prairie, United States Bureau of Reclamation. The total natural flows of the Gila River, the last major tributary in the basin, are under investigation.

4 Lee's Ferry is the conventional measurement point for the river, marking the legal dividing point between the Upper and Lower basins (discussed later).

5 In scientific settings, river volumes on the Colorado River are generally reported in metric units, whereas in important legal documents and policy settings, the Imperial Unit of million acre-feet (maf) is strongly preferred. Both units are used in this report. One million acre-feet (maf) is equivalent to 1.233 km^3.

6 The only consistent source of water for the Delta is roughly 0.16 km^3 (0.126 maf) of highly saline drainage water from the Wellton-Mohawk Irrigation and Drainage District routed directly to the region by canal to avoid further salt loadings to the Colorado River mainstem (Pitt, 2001).

7 Dams in the basin have an estimated 4,200 megawatts of hydropower capacity.

8 Exactly what is, and is not, a "key element" of the Law of the River is debatable. See Table 13.2 for summaries (and citations). It is well beyond the scope of this brief overview to describe the Law of the River in any detail; for more information, consult Bloom (1986), Getches (1985), and Fleck (2016).

9 The dam was renamed to honor Herbert Hoover, who was the federal chairman of the Compact negotiating team before becoming the 31st President of the United States in 1929.

10 *Treaty between the United States of America and Mexico respecting utilization of waters of the Colorado and Tijuana Rivers and of the Rio Grande*, Feb. 3, 1944, U.S.-Mexico (59 Stat. 1219). Hundley Jr. (1966) provides a detailed history of these negotiations.

11 The very small apportionment to Nevada reflected the lack of any significant population center in Nevada in the region, and the physical difficulty of diverting (raising) the water out of the channel for any potential Nevada water user. However, since 1922, the City of Las Vegas has emerged as a major city almost entirely dependent on Colorado River flows, illustrating the challenge of anticipating future conditions when making resource allocation decisions.

12 There are twenty-two federally recognized tribes in the basin. Tribal rights are an example of "reserved rights"; in this case, water reserved by the federal government for the tribes to allow irrigation on tribal reservations. Tribal water rights, when quantified, are counted against the apportionment of the state where the tribal reservation is located.

13 Tribes in the basin have already established diversion rights of 3.6 km^3 (2.9 maf), with at least thirteen tribes with claims still unresolved (Bureau of Reclamation, 2012).

14 The Grand Canyon Protection Act of 1992 marked an important change in the role of stakeholders and in recognition of the value of water left instream. The Act called on Reclamation to include nontraditional stakeholders in its management and consultation processes, in addition to the usual water contractors. This opened the door for NGOs to participate in subsequent negotiations.

15 Population values are estimates for the metropolitan regions, which in many cases are substantially larger than just the named cities.

16 Alfalfa is primarily a feed crop for beef and dairy cattle and horses.

17 A recent report estimated that the Colorado River generates $26 billion in annual recreation-related revenues (Southwick Associates, 2012).

18 One recent study has suggested that groundwater contributes 56 percent of surface flows in the Upper Basin (Miller et al., 2016).

19 Total municipal and industrial water deliveries (not consumptive use) by agencies relying at least in part on water from the basin increased modestly from about 7.5 km^3 (6.1 maf) in 1990 to about 8.2 km^3 (6.7 maf) in 2008, despite a population increase of about 10 million (40 percent). Municipal deliveries of Colorado River basin water rose from 3.4 km^3 (2.8 maf) to about 4.2 km^3 (3.4 maf) over this period (Cohen, 2011). Per capita water use among agencies relying on basin water varies by more than a factor of four, though in almost all cases such water use remains well above that of other regions of the United States.

20 Based on gage data from 1896, most negotiators thought the river's average annual flow was roughly 21 km^3 (17 maf) or more.

21 United States, Bureau of Reclamation. *Record of Decision: Colorado River Interim Guidelines for Lower Basin Shortages and the Coordinate Operations for Lake Powell and Lake Mead*. Department of the Interior, 2007. Available online www.usbr.gov/lc/region/programs/strategies/RecordofDecision.pdf

22 The delivery obligation to Mexico is not mentioned in this discussion simply because it supersedes the functioning of the interstate allocation – i.e., the delivery to Mexico is the "first priority" on the river. While not explicitly stated, the Compact's III(d) release requirement essentially establishes the Lower Basin as the second priority, leaving the Upper Basin last in line – despite being the origin for almost all the flow. It is an odd situation, as in most transboundary water disputes, upstream is a favored position.

23 The classic studies on this topic are by Tipton and Kalmbach, Inc. (1965) and the papers of the severe sustained drought project (see the special issue of the *Water Resources Bulletin*, 31(5), October 1995).

24 Future warming is a safe bet, as the greenhouse gases (GHGs) that are already in the atmosphere likely ensure a few more decades of warming even if large-scale global reductions in GHGs were to occur. Predicting

future precipitation trends is much more difficult; many models suggest slightly wetter conditions may be coming to the Colorado River Basin. It would require a significant increase in precipitation (roughly 8 percent) to offset the streamflow impact from the projected heating (Udall and Overpeck, 2017).

25 For context, Hoover Dam loses the ability to generate hydropower at 1,050 feet above sea level, which has significant impacts on the electricity grid and the funding generated from the sale of hydropower. Further, one of Southern Nevada Water Authority's (SNWA) Lake Mead intakes becomes inoperable at 1,050 feet, with a second intake going offline around 1,000 feet. Because of this risk, SNWA recently completely construction of an $817 million third intake that is capable of drawing water from Lake Mead at even lower elevation levels (including after Lake Mead reaches "dead pool," where no water can be physically moved through Hoover Dam).

26 Meeting consumptive use demands has been no trivial task and has provided enormous benefits to the economies relying on Colorado River water. The total economic productivity of Colorado River water has not been estimated, though it could be said to support the entire economic productivity of Arizona ($231 billion) and Las Vegas ($96 billion), and contributes to the economies of southern California ($970 billion), metropolitan Denver ($168 billion), and metropolitan Salt Lake City ($72 billion), as well as agricultural productivity inside and outside of the basin.

27 See Robison and Kenney (2012) for a further clarification of this distinction. They note the importance of distinguishing "substantive equity" versus "procedural equity," and the implications for river basin management.

28 For example, the Upper Colorado River Commission formalizes and represents the positions of the Upper Basin States and implements the provisions of the Upper Colorado River Basin Compact of 1948. Another example is the Glen Canyon Dam Adaptive Management Workgroup, which was established as part of the implementation of the Grand Canyon Protection Act of 1992. By mandate, the Workgroup must include certain stakeholder groups, and those groups' positions must be considered when management plans for the Glen Canyon Dam are created or modified.

14 The Rio Grande / Río Bravo Basin

George H. Ward and Jurgen Schmandt

14.1 THE RIVER TODAY

14.1.1 Rio Grande and Tributaries

The Rio Grande, in Mexico the Río Bravo (del Norte), is the fourth largest river in North America by both mainstem length and drainage area (Kammerer, 1990; Benke and Cushing, 2005). It extends from the Rocky Mountains to the Gulf of Mexico, a total length of about 3,100 km, passing through or bordering three states in the United States and four states in Mexico. The drainage basin has an estimated area of 900,000 km^2 (literature values ranging ± 3 percent), lying in five Mexican states and three US states. This is based on tracing the topographic divides between adjacent drainages, including several basins that drain to their interior, so their runoff does not enter the Rio Grande tributary system. These noncontributing drainages make up about 47–49 percent of the total basin. This chapter will be concerned only with the contributing portions of the basin, whose areas total about 460,000 km^2. Figure 14.1 displays the basin showing the main channel and its major tributaries, and other important geographical or physical features. The principal mainstem streamflow gauges are listed in Table 14.1 (where "principal" means gauges strategically located and/or possessing a long record of measurements). The principal tributaries and their downstream-most gauges are listed in Table 14.2 (where by "principal" we mean a substantial contribution to the river flow and/or an important regional supply of surface water). These tables also list representative flows, viz., annual flows averaged over the period 1981–2010 (the latest climatological normal period for the available period-of-record of the majority of these gauges).

The Rio Grande heads in the San Juan Mountains of Colorado and flows within the US interior through New Mexico to El Paso, Texas; see Figure 14.1. Below El Paso it forms the international border between the United States and Mexico, ultimately debouching into the Gulf of Mexico. This 1,600-km reach of the Rio Grande is the longest river border in the world between countries at different levels of development (Herzog, 1990). (By contrast, the other major river of this study in the American Southwest – the Colorado – marks the international border for a short distance only.) Both Rio Grande and Colorado border waters are managed under the same bi-national treaty arrangement between Mexico and the United States (see Section 14.3).

The morphology of the basin alternates between narrow canyons and broad alluvial valleys, until the river emerges onto the coastal plains of the Gulf of Mexico (e.g., Bryan, 1938; Schmidt et al., 2001; Ewing, 2016). In the headwaters' regions, generally above Otowi (Table 14.1), the topography of mountains and gorges is spectacular. Here, the channel falls 500 m from some 2,100 m in elevation, and much of the river is narrow and swift, popular among canoeists and trout fishers. A substantial flow enters the river above Del Norte, even though this is the smallest drainage area of the river. This is essentially orographic, in the form of snow or rainfall, depending on the season. The principal tributaries above Otowi are the Conejos River in Colorado and the Rio Chama in New Mexico. The Chama (a relict channel of the Rio Grande) is the most important tributary above El Paso.

From Otowi southward through New Mexico down to the gauge at Fort Quitman in Texas, the river channel lies in the Rio Grande rift and falls about 400 m. This reach is semi-arid in the north and arid in the south: here the river descends into the great Chihuahuan Desert of North America. Vegetation in this reach is characteristically bosque and desert. The principal tributaries entering this reach of the river are the Jemez River and the Rio Puerco.

Some of the oldest water management practices in North America obtain in the Rio Grande valley of New Mexico, particularly in diversion strategies and the use of ditches for water distribution. Substantial modifications have been made to the river and watershed in Colorado and New Mexico since the early nineteenth century, mainly to supply irrigation needs, especially in the riparian zone of the river and in the San Luis valley. These include installation of more than fifty dams, and the implementation of the San Juan–Chama Project (of the US Bureau of Reclamation) by which a net importation of water is made from the Colorado basin to the Rio Grande via a tunnel beneath the continental divide. At present, from Del Norte in Colorado to Fort Quitman in Texas, the Rio Grande provides much of the water supply for extensive irrigated agriculture as well as for several municipalities.

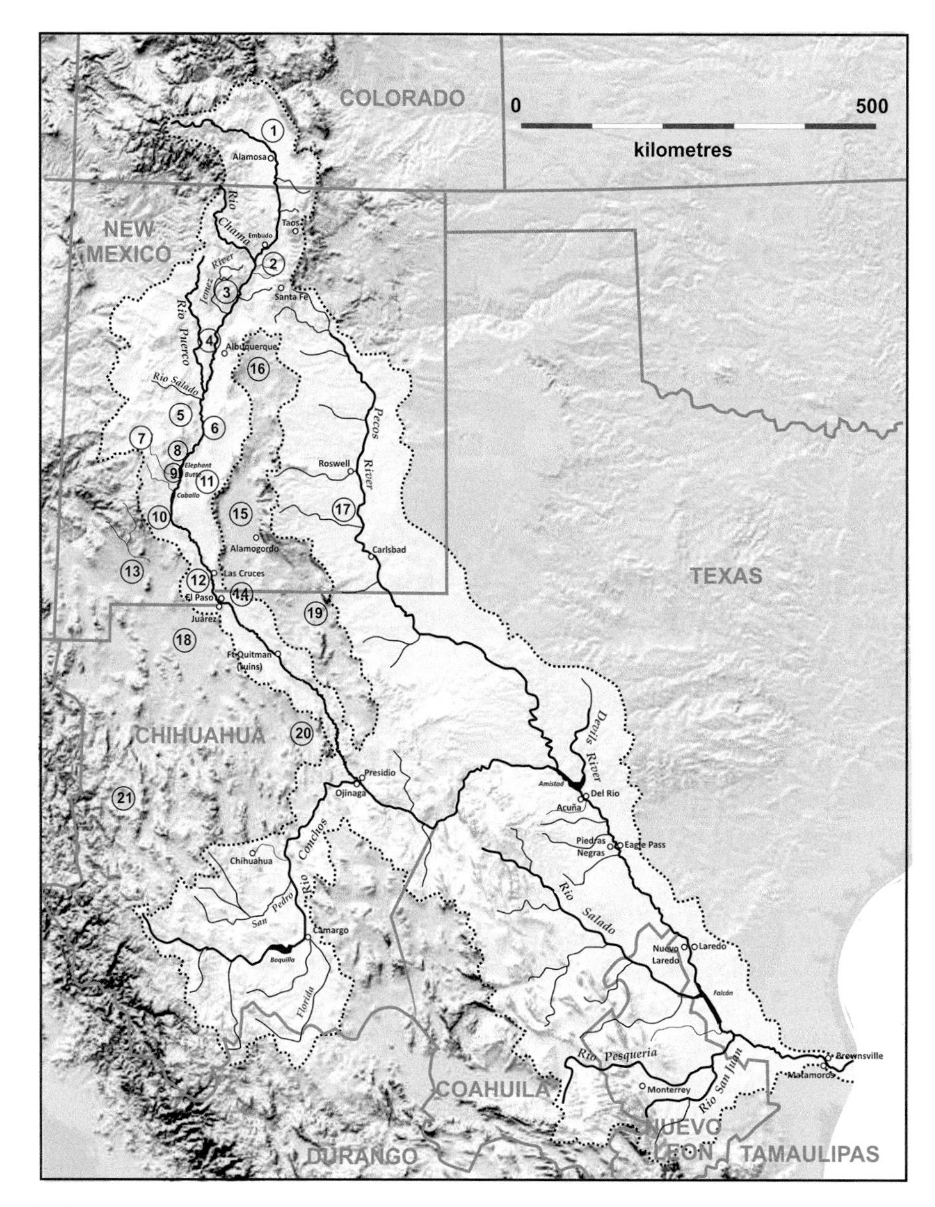

Figure 14.1 Figure 1. Rio Grande basin (broken line) with major tributaries. Bolsóns are numbered as follows: 1) San Luis Basin; 2) Española Basin; 3) Santo Domingo Basin; 4) Albuquerque-Belen Basin; 5) La Jencia Basin; 6) Socorro Basin; 7) San Agustin Basin; 8) San Marcial Basin; 9) Engle Basin; 10) Palomas Basin; 11) Jornado del Muerto Basin; 12) Mesilla Basin; 13) Mimbres Basin; 14) Hueco Bolsón; 15) Tularosa Basin; 16) Estancia Basin; 17) Roswell Basin; 18) Bolsón de los Muertos; 19) Salt Basin; 20) El Cuervo Bolsón; and 21) Laguna de Barbícora. Basin locations based on Brand (1937), Hawley (1969); Wilkins (1986), Ortega-Ramírez et al. (2000), Mace et al. (2001), Hibbs and Darling (2005), Land (2016).

From Fort Quitman through the Big Bend, the river falls about 700 m, mainly in the dramatic canyons of this reach, to the headwaters of International Amistad Reservoir. It is in this reach that the Río Conchos, the principal tributary of the Rio Grande (see Table 14.2), conflows with the river. The Conchos heads in the mountains of the Sierra Madre Occidental in Mexico and flows into the Rio Grande near Presidio (see Figure 14.1). Downstream from the confluence of the Rio Grande and Río Conchos, the river receives the flow of two important tributaries, the Pecos and the Devils, both from the United States, both of which flow into Amistad Reservoir. The Pecos Basin reaches north through west Texas into New Mexico almost to the latitudes of the Rio Grande itself, draining segments of the Rocky Mountains that form the eastern boundary of the main river system.

The reach of the Rio Grande below Amistad is primarily coastal plain with a gentle slope in elevation to the Gulf. Between Amistad and International Falcón Reservoirs, the river flows through the weathered and dissected inner coastal plain. In Mexico, this merges into the foothills of the Sierra Madre

Table 14.1 *Principal mainstem streamflow gauges on Rio Grande, with station identification (ID) used by agency maintaining records, river distance from mouth, elevation of gauge zero, contributing drainage area, and 1981–2010 average annual discharge*

Gauge	*ID*	*Records**	*Distance from mouth (km)*	*Elevation** of gauge (m)*	*Drainage area (km^2)*			*Mean flow 1981–2010 (Mm^3/yr)*
					Total	*US.*	*Mexico*	
Colorado								
Del Norte	8220000	USGS	2,910	2,433	3,419	3,419	0	784
Lobatos	8251500	USGS	2,777	2,264	12,328	12,328	0	422
New Mexico								
Cerro	8263500	USGS	2,728	2,167	14,245	14,245	0	452
Taos	8276500	USGS	2,675	1,844	17,586	17,586	0	695
Embudo	8279500	USGS	2,651	1,765	19,321	19,321	0	765
Otowi	8313000	USGS	2,605	1,673	29,422	29,422	0	1,350
Cochiti***	8317400	USGS	2,566	1,593	38,591	38,591	0	1,242
San Felipe	8319000	USGS	2,541	1,559	34,084	34,084	0	1,307
Albuquerque	8330000	USGS	2,489	1,508	37,555	37,555	0	1,171
San Marcial†	8358300 +8358400	USGS	2,306	1,435	49,806	49,806	0	1,050
Elephant Butte Dam***	8361000	USGS/ IBWC	2236	1,293	67,141	67,141	0	935
Caballo Dam***	08-3625.00	IBWC/ USBR	2,190	1,262	70,495	70,495	0	938
Texas								
El Paso	08-3640.00	IBWC	2,021	1,135	75,812	75,812	0	552
Texas/Chihuahua								
Fort Quitman/Colonia Luis Leon	08-3705.00	IBWC	1,888	1,052	82,734	79,269	3,465	238
Candelaria/San Antonio del Bravo††	08-3712.00	IBWC	1,672	871	87,500			247
Presidio/Ojinaga above Rio Conchos	08-3715.00	IBWC	1,551	783	90,649	83,532	7,117	218
Presidio/Ojinaga below Rio Conchos	08-3742.00	IBWC	1,529	772	164,047	88,308	75,739	952
Johnson Ranch/Santa Elena	08-3750.00	IBWC	1,388	623	175,497	93,910	81,587	1,001
Texas/Coahuila								
Foster Ranch/Rancho Santa Rosa	08-3772.00	IBWC	1,058	353	209,120	110,517	98,603	1,313
Amistad***	08-4509.00	IBWC	920	274	318,915	214,148	104,767	1,970
Del Rio/Ciudad Acuña	08-4518.00	IBWC	903	265	319,352	214,316	105,036	2,066
Quemado/Jimenez	08-4557.00	IBWC	853	234	325,265	216,813	108,452	1,414
Eagle Pass/Piedras Negras	08-4580.00	IBWC	800	208	329,735	218,188	111,547	2,488
El Indio/Villa Guerrero	08-4587.00	IBWC	741	177	334,692	218,802	115,890	2,572
Texas/Tamaulipas								
Laredo/Nuevo Laredo	08-4590.00	IBWC	580	107	343,374	222,003	121,371	2,672
Falcón Dam***	08-4613.00	IBWC	438	n/a	412,506	227,292	185,214	2,587
Rio Grande City/Camargo	08-4647.00	IBWC	378	30	451,595	227,867	223,728	2,911
Anzalduas Dam***	08-4692.00	IBWC	273	0	456,128	230,333	225,795	1,651
San Benito/Ramirez	08-4737.00	IBWC	156	0	456,626	230,385	226,241	655
Brownsville/Matamoros	08-4750.00	IBWC	78	0	456,701	230,421	226,280	514

* USGS – US Geological Survey
USBR – US Bureau of Reclamation
IBWC – International Boundary and Water Commission
** National Geodetic Vertical Datum relative to Ellipsoid of 1929 (NGVD29)
*** Gauge located downstream from dam
† Sum of floodway (400) and conveyance channel (300) gauges
†† Drainage area estimated

Oriental, marked by large anticlines rising from a desert plain. Below Falcón the river flows onto the coastal prairie, a largely featureless terrain with sandy soils and vegetated dunes. Irrigated agriculture below Amistad is concentrated in two areas. The first is the Winter Garden, mainly in the United States between Del Rio and Eagle Pass, above Falcón Reservoir. The second lies downstream from Falcón on both sides of the river and extends nearly to the river's mouth, known as the Lower Rio Grande Valley (LRGV).

14.1.2 Population

Albuquerque, El Paso–Juárez, and a string of cross-border cities in the LRGV are the main population centers on the mainstem of the Rio Grande. Population is highly variable along the river corridor. In the Colorado segment, basically the San Luis Valley, the present population (2010 census) is about 50,000, projected to grow to about 80,000 by 2050 (Dinatale Water Consultants, 2015). The New Mexico reach of the mainstem has a population (2010) of about 1,360,000, projected to reach 1,940,000 by 2050, mainly due to growth in the Albuquerque area. The Pecos reach in New Mexico has a population of 183,000 (2010), projected to grow to 219,000 by 2050. The total 2010 population in the New Mexico Rio Grande basin is therefore 1,540,000, projected to be 1,940,000 in 2050. (These population statistics were compiled from the 2016 Regional Water Plans for Regions 3, 7, 8, 10, 11, 12, 13, 14, and 15 from the website of the New Mexico Office of the State Engineer.) The Texas reach of the Pecos is sparsely populated, totaling 53,500 in 2010, projected to reach 64,000 by 2050. This includes the eighty-two noses making up the entirety of Loving County. (Data from the Texas Water Development Board.)

The main population centers in the border segment of the Rio Grande – Paso del Norte and LRGV – depend for their economic livelihood on their upstream reservoirs, viz. the Elephant Butte system, and the tandem International Reservoirs. The Paso del Norte (Las Cruces, El Paso, Cd. Juárez) has a current population of about 3 million. Las Cruces was included in the New Mexico statistics (Region 11). El Paso and Juárez have a total (2010) population of about 2.1 million, projected to reach 2.8 million by 2050. The Lower Rio Grande Valley, covering the last 270 river miles (434 km), has a current population of 2.3 million, about equally divided between both sides of the border. This is projected to reach 3.9 million by 2050. The total of all of the twin-city population centers along the Rio Grande border is about 5.4 million (2010), projected to be 8.4 million in 2050. (These data were compiled by the Texas Commission for Environmental Quality.) The population of the state of Chihuahua, which is dominated by the Rio Conchos, was 3.7 million in 2016, and is projected to reach 4.2 million in 2030 (National Water Commission, 2017).

Most of the borderlands reach of the river has seen rapid population growth since 1960, and this trend will continue into the future. Growth is driven by several factors: high birth rates, availability of cheap irrigation water, tax advantages for cross-border manufacturing (maquiladoras), and in-migration from interior Mexico to more prosperous border communities. In the border segment of the basin, population has doubled every twenty years since 1960.

14.1.3 Land Use

The majority of the basin is grassland and scrubland, characteristic of the arid to semi-arid climate. Extensive ranching is carried out in the basin in New Mexico and Mexico. Several parts of the basin enjoy fertile soil – the result of sediment accumulation in the days before the river was dammed. Rain-fed farming is impossible west of the 100th meridian, which crosses the basin close to the river's mouth. Therefore, irrigated farming is dominant. In the main agricultural regions, the Paso del Norte (Las Cruces, El Paso, and Cd. Juárez) and the LRGV, farmers can produce three to four crops a year. In Texas, urbanization and the right to sell or lease water rights to cities lead to a small decline in agricultural land use. New Mexico and Mexico laws do not allow creating water markets. As a result, they see no decline in agricultural land use.

The lower 250 km of the Rio Grande has rich alluvial soil. Taking advantage of good soil quality and Rio Grande water, the LRGV has become a center for irrigated agriculture on both sides of the international border.

14.1.4 Economy

Farming, services, commerce, and manufacturing (maquiladoras) are the main economic activities on the border. In the borderlands region, first in Texas, now in Mexico, drilling for oil and gas has re-emerged as an important sector of the economy as a result of horizontal drilling and fracking of shale resources. Paso del Norte and LRGV are important suppliers of fruits, vegetables, corn, and sorghum. The state of Tamaulipas, for example, is responsible for 17 percent of Mexico's agricultural production. Throughout the basin the amount of irrigated land varies greatly from year to year, primarily depending on water availability.

14.2 HYDROLOGY AND ENGINEERING

The basic geometry of the Rio Grande from headwaters to mouth has been abstracted in Figure 14.2, using data from Tables 14.1 and 14.2. The drainage area, of course, increases monotonically

Table 14.2 *Principal tributaries of the Rio Grande: Downstream-most gauges with station identification (ID) used by agency maintaining records, channel distance from confluence with Rio Grande, river distance of confluence from Gulf of Mexico, elevation of gauge zero, contributing drainage area, and 1981–2010 average annual discharge*

							Drainage area		
Tributary	*Gauge*	*ID*	*Records**	*Distance from confluence (km)*	*Elevation** of gauge (m)*	*Confluence distance upstream (km)*	*US (km^2)*	*Mexico (km^2)*	*Mean flow 1981–2010 (Mm^3/yr)*
Conejos River	Lasauses, CO	8249000	USGS	2.1	2,936	2,285	2,297	0	146
Red River	Questa, NM	8265000	USGS	14.4	2,717	2,272	293	0	42
Embudo Creek	Dixon, NM	8279000	USGS	0.8	2,655	1,786	790	0	77
Rio Chama	Chamita, NM	8290000	USGS	8.6	2,627	1,724	8,143	0	527
Jemez River	Jemez Canyon Dam***	8329000	USGS	3.8	2,528	1,554	2,688	0	58
Rio Puerco	Bernardo	8353000	USGS	5.9	2,399	1,440	16,110	0	26
Rio Conchos	Ojinaga	08-3730.00	IBWC	1.0	1,547	780	0	68,387	694
Pecos River	Langtry	08-4474.10	IBWC	24.1	991	345	91,114	0	211
Devils River	Pafford Crossing	08-4494.00	IBWC	41.0	925	345	10,259	0	330
Rio San Diego	Jimenez	08-4555.00	IBWC	7.0	856	254	0	2,209	184
Rio Escondido	Villa de Fuente	08-4581.50	IBWC	8.0	794	219	0	3,779	72
Rio Salado	Las Tortillas	08-4597.00	IBWC	39.8	482	0	0	59,971	348
Rio Alamo	Mier	08-4620.00	IBWC	8.0	422	0	0	4,339	72
Rio San Juan	Camargo	08-4642.00	IBWC	5.0	384	0	0	33,515	226

from headwaters to mouth, with large increments associated with the major tributaries. However, the flows carried by these tributaries do not correlate with the drainage areas because of the overriding importance of climatology. Substantial diversions, primarily agricultural, have now reduced upper Rio Grande flow on the average to practically nil at Fort Quitman. The river is rejuvenated farther downstream by the inflow from the Río Conchos and a few other major tributaries. (From Fort Quitman to Presidio, above the Conchos, there is generally little river flow. This reach is sometimes referred to as the "forgotten river.") For our present purposes, it is useful to divide the river system into two main reaches: the upper Rio Grande, from the headwaters in Colorado down to the streamflow gauge at Fort Quitman, Texas, below El Paso (Table 14.1); and the lower Rio Grande, from the Fort Quitman gauge to the river's mouth at the Gulf of Mexico. As will be seen, these are practically independent rivers.

By the early years of the twentieth century, the entire basin of the Rio Grande was already subjected to engineering practices to manipulate river flows and store water, mainly to support irrigated agriculture. In the upper Rio Grande, some of these practices extend back into the region's prehistory. Descendants of the Anasazi, the Pueblo culture in the upper Rio Grande and the upper Pecos, excavated ditches to intercept and re-direct the flow of tributaries, which then could be distributed to crops (Rivera, 1998, and citations therein). The strategy was to divert into a main canal, the *acequia madre* ("mother ditch"), from which a complex of sangrías (laterals) drained into parcels of farmland, the supply being controlled by headgates. Spanish colonists in the early seventeenth century brought essentially the same strategies of water management (acquired from the Moors), except implemented on a much larger scale than those of the Pueblos, and augmented by such structures as main-channel diversion dams to raise the level of the river, and canoas (wooden flumes) to facilitate overland transfer. The creation and operation of the acequia madre was a community project centered on the point of diversion. Collectively, acequias in Provincia del Nuevo México evolved a culture, administration, and jurisprudence that is

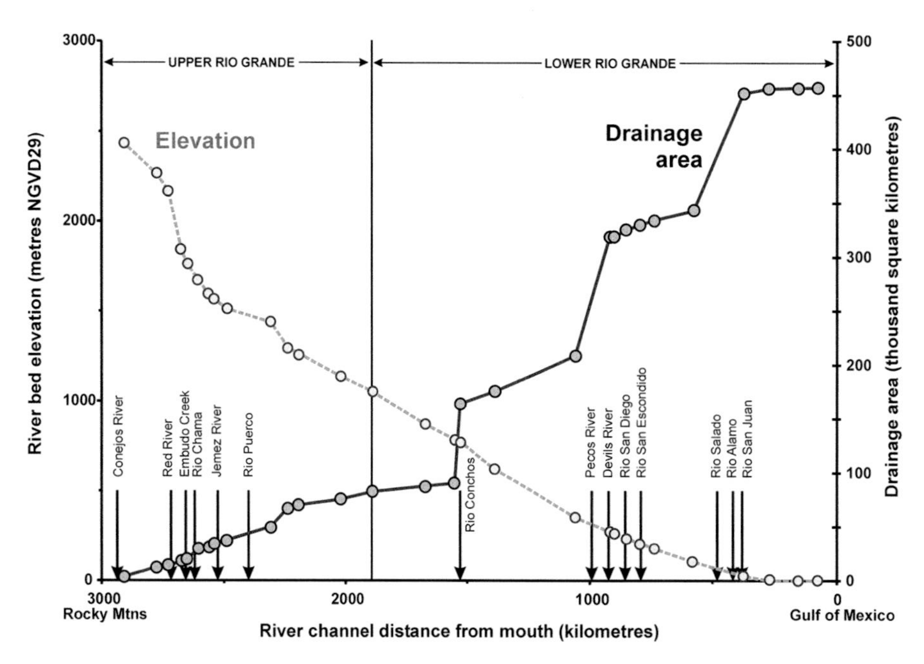

Figure 14.2 Longitudinal variation in elevation and drainage area along the Rio Grande mainstem.

retained in the management of water in New Mexico to the present (Hutchins, 1928; Clark, 1987; Rivera, 1998; Wozniak, 1998).

By the close of the eighteenth century, irrigation of the Rio Grande bottomlands was scattered from Santa Fe to El Paso del Norte. Climate variation, Indian raids, and governmental or financial instability created episodes of expansion and contraction of agriculture throughout the nineteenth century, with acequias being constructed as far north as the San Luis Valley and as far south as the Mesilla Valley. The *coup de theatre* was the mid-century annexation of New Mexico to the United States, with the associated installation of military posts and railroads. After 1880, there was accelerated development of irrigated agriculture throughout the basin. For example, in the headwater basin of the Rio Grande in the San Luis Valley, by the turn of the century the farmland area irrigated from the river totaled about 1,200 km^2 (3 x 10^5 ac), where before only a few acequias had been in operation (Wozniak, 1998). The nineteenth century closed with the construction of the first major dam in the Rio Grande system, Avalon on the Pecos River.

The hydrology of a basin is best characterized by measurements before substantial engineering alterations have been implemented. The historical sketch in this section demonstrates that this is impossible for the Rio Grande because major modifications to the watershed were already in place before routine stream gauging began (around 1890 at the gauge at Embudo). While much can be illustrated or inferred from the instrumental record of water and its movement in the Rio Grande basin, it must be qualified because of the effects of human actions.

14.2.1 Physical Controls on Hydrology

HYDROCLIMATOLOGY

Except for the high mountainous regions in the upper reaches of the Rio Grande basin, in both the United States and Mexico, the climate of the Rio Grande varies from arid (Köppen BWh) to semi-arid (Köppen BSh). Low cloudiness and high insolation, associated with large-scale atmospheric subsidence, conspire to create a warm, dry climate with high evapotranspiration. Large segments of the watershed lie in the Chihuahuan Desert. In general, throughout the basin, precipitation occurs in the form of intense storm events widely separated in both time and space (Schmidt, 1986; Ritchie et al., 2011). However, precipitation climatology is different in the upper and lower Rio Grande reaches with respect to the synoptic drivers of precipitation, the geographical distribution of precipitation (and therefore sources of runoff into the river channels), and its seasonality. An excellent summary of the regional climate is provided by Sheppard et al. (2002). Table 14.3 presents the 1981–2010 climate-normal monthly precipitation (millimeters, water equivalent) for the contributing drainage of the Rio Grande. Data are from the climate division averages of the US National Oceanic and Atmospheric Administration (NOAA), and from station averages in three ad hoc zones for Chihuahua (M = mountain, D = desert, E = elevated desert).

For the upper Rio Grande, one of the major producers of precipitation is midlatitude synoptic-scale storm systems (cyclones) of the westerlies (Bluestein, 1993; Barry and Carleton, 2001). The upper basin in Colorado and northern New Mexico becomes more exposed to these systems in winter with the seasonal descent to lower latitudes of the climatological equator. The associated precipitation is enhanced by the increase in elevation encountered by these systems as they track from west to east (or, more likely, northwest to southeast) across the North American Cordillera. Local orographic effects are exemplified by the annual precipitation at two stations 80 km (50 miles) apart in San Luis Valley in the Colorado part of the basin, Alamosa and Wolfe Creek Pass: though subject to essentially the same synoptic climatology, the former is 18 cm (7 ins), the latter 173 cm (68 ins). The superposition of a short-wave disturbance is particularly effective in enhancing precipitation. In the higher latitudes of the upper Rio Grande, this precipitation is generally frozen, forming over time a winter snowpack, rather than immediate runoff (Bryan, 1938; Schmidt, 1986; Wilson, 1999; Angersbach et al., 2002; Booker et al., 2005; Rango, 2006). This snowpack typically melts over several months starting in April, providing a stable flow in the river into the summer. These synoptic systems may persist into the spring, producing extreme rainfall events, which create immediate runoff and often flooding in the river channels. In the lower Rio Grande, while snow does occur in the higher elevations of the Pecos and, more rarely, the Conchos, it is insufficient to form a lasting snowpack. In the lower Rio Grande, therefore, rainfall is the main producer of runoff (Schmidt, 1986). The same midlatitude synoptic systems can be rainmakers for the lower Rio Grande as well, but, except for the Pecos, its southerly location makes this less likely. Only the most energetic midlatitude cyclones are able to reach the subtropical latitudes of the lower Rio Grande basin.

Table 14.3 *1981–2010 normal monthly precipitation (mm) in large segments of the Rio Grande basin, US climate divisions of the NOAA, and ad hoc zones for Chihuahua*

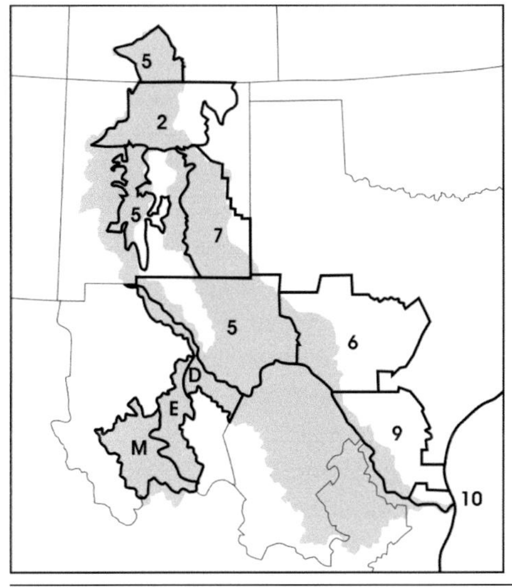

Division	*Jan*	*Feb*	*Mar*	*Apr*	*May*	*Jun*	*Jul*	*Aug*	*Sep*	*Oct*	*Nov*	*Dec*	*Annual*
Colorado													
5	29	30	42	40	32	24	50	62	42	39	33	32	455
New Mexico													
2	23	23	32	30	35	37	66	75	46	35	25	26	453
5	13	12	12	11	14	19	53	58	37	29	15	18	291
7	12	13	15	17	34	45	59	62	51	34	16	17	377
Chihuahua													
M	11	9	4	5	17	40	108	116	103	30	12	12	468
E	5	5	2	6	12	38	80	71	68	23	7	7	323
D	7	10	6	5	11	24	57	51	52	20	8	10	261
Texas													
5	13	14	11	15	30	43	53	53	47	35	13	14	342
6	28	37	45	44	76	80	50	55	66	71	39	29	621
9	30	34	36	41	70	69	58	50	83	62	35	28	595
10	29	34	28	36	65	63	55	51	119	71	34	31	616

A second major synoptic system affecting the upper Rio Grande is the Mexican monsoon (or North American monsoon), an essentially tropical system that results in increased July–September rainfall rates in Arizona, New Mexico, and peripheral regions of northwestern Mexico and the United States (Douglas et al., 1993; Adams and Comrie, 1997). Its intensity and geographical influence exhibit significant interannual variability. Nearly half of the annual rainfall in the upper Rio Grande basin, and more than half in the Conchos, is produced by the monsoon (Schmidt, 1986; Douglas et al., 1993; Ladwig and Stenrud, 2009). The potential importance of this seasonal circulation has only begun to be appreciated among meteorologists, and there are many aspects of its dynamics that remain conjectural or controversial. A consensus has emerged that the monsoonal circulation in fact extends a substantial distance down the Mexican Pacific coast, centered approximately at Mazatlan (latitude 23.3°N), a key component of which is a low-level jet directed to the north along the axis of the Gulf of California. Much of the precipitation is orographic, delivered as rainfall along the Mexican coast on the coastal foothills of the Sierra Madre Occidental. There is an apparent association between precipitation events in the monsoon with the transit of tropical easterly waves over the Gulf of California carried in the trade winds.

The enhanced rainfall in the upper Rio Grande in the Mexican monsoon is associated with depressed rainfall in the south-central United States, an "out of phase" relationship (Ladwig and Stensud, 2009). This is suggested to be forced – or reinforced – by the Madden–Julian oscillation (MJO), an intra-annual vacillation of the tropical atmosphere (Madden and Julian, 1994; Perdigon-Morales et al., 2019). Certainly, a well-known feature of Texas hydroclimatology is the "midsummer drought" (e.g., Ward, 2004), which occurs from Oklahoma into South Texas, and in the lower Rio Grande downstream from about Del Rio, and generally extends from July through September. This is a result of a stationary, upper-level ridge of high pressure that develops in late summer over northern Mexico and south-central United States, referred to by South Texas meteorologists (with tongue in cheek) as La Canicula. Whether this is connected to the Mexican monsoon, or to the MJO, or is simply a separate atmospheric circulation with annual seasonality, must await future research.

The hydroclimatology of the upper Rio Grande is affected by several large-scale oscillatory modes of the atmosphere, notably the El Niño-Southern Oscillation (ENSO), North Pacific Oscillation, Pacific Decadal Oscillation, and, apparently to a lesser extent, the Atlantic Multidecadal Oscillation (e.g., Angersbach et al., 2002; Weiss et al., 2009). Over a period of centuries, ENSO has displayed the predominant statistical association with moisture variability in northern Mexico and southwestern United States, most notably drought (Stahle et al., 2016). The interaction among these oscillation modes, possibly with the synoptic-scale factors noted earlier, may account for the high variability in climate of the region. Tropical and/or north Pacific sea-surface temperatures (implicitly invoking the above modes) are an important control on interannual variability (e.g., Higgins and Shi, 2000; Castro et al., 2001; Angersbach et al., 2002). Pacific sea-surface temperatures have been investigated as a potential causative factor in the occurrence of severe droughts in the southwestern United States and northern Mexico regions, with mixed success (Trenberth and Branstator, 1992; Seager et al., 2005; Cook et al., 2007; Seager, 2007).

One final important element of the hydroclimatology of the Rio Grande is the occurrence of tropical storms, including hurricanes. The primary data sources are the historical compilations of the US National Oceanic and Atmospheric Administration (NOAA) presented by McAdie et al. (2009) for the Atlantic and Blake et al. (2009) for the Pacific. These events are rarities. The long-term frequency of occurrence of Atlantic tropical cyclones (which includes hurricanes) on the Texas and Tamaulipas coastal zone is about one per year. The frequency of occurrence of Pacific tropical storms on the northern Mexican coastline is much greater because of the proximity to the main development region in the tropical eastern Pacific, but restricting these storms to only those that affect rainfall in the southwestern states (Corbosiero et al., 2009; Ritchie et al., 2011) yields a frequency of about one to three per year. (Martinez-Sanchos and Cavazos, 2014, found 1.9 hurricanes per year for the entire western coastline of Mexico.) Generally, Pacific cyclones affect the upper Rio Grande with greater frequency than Atlantic storms, while the reverse applies to the lower Rio Grande.

A relationship between tropical storms and the larger synoptic controls on the climate remains elusive, though there is some statistical evidence for association of cyclone frequency with the Atlantic warm pool and ENSO. For present purposes, we consider these as random events, superposed on the prevailing meteorological conditions. Their importance to the Rio Grande basin is as prolific rain producers. Tropical storms are central to the long-term hydrology of the lower Rio Grande because of their effectiveness in filling the international reservoirs.

GEOLOGY AND PALEOHYDROLOGY

The present-day terrain and soils of the Rio Grande watershed, the characteristics of its river channel and drainage network, even aspects of the river hydraulics and hydrology, are the result of the cumulative cataclysmic changes in geology and climate of the past, especially those wrought over the past 65 million years, i.e., the Cenozoic. During this era, western North America and Central America were subjected to three major tectonic episodes deriving from interactions of the Pacific Plate (and remnants of the Farallon plate) with the North American, Cocos, and

Caribbean Plates (Ewing, 2016; see also Bryan, 1938; Keller and Baldridge, 1999; Dickinson, 2004; Pazzaglia and Hawley, 2004; Mann, 2007; Dickerson, 2013; Blakey and Ranney, 2018)

1. The Laramide/Hidalgoan orogeny from 80 until 40 Mega-annus (Ma) before present (BP). Already underway at the beginning of the Cenozoic, this was a period of mountain building of the North American Cordillera.
2. Volcanism from 48 to 18 Ma BP, and early stages of rifting from 28 to 20 Ma BP. Volcanism, a feature of rifting, occurred throughout the western states, including Texas, and extended into Mexico and Central America. This early rift, extending south from Colorado, began forming in the early Oligocene and exhibited broad, shallow basins over a wide rift zone (Baldridge and Olsen, 1989; Kelly and Chamberlin, 2012). It preceded a "magma gap" up to 15 Ma in duration (Seager et al., 1984).
3. Uplift of the Colorado Plateau from 24 Ma to 10 Ma BP (Stearns, 1953; Baldridge et al., 1980; Graham, 1999), renewed volcanism, crustal extension and rifting around 10 Ma to recent (Baldridge et al., 1980; Adams and Keller, 1994; Ingersoll, 2001; Blakey and Ranney, 2018). The crustal extension created the Basin and Range Province, as well as the present Rio Grande rift.[1]

Volcanic zones during the late Cenozoic (3) were principally distributed on the southern periphery of the Colorado Plateau in Colorado extending into the northeast corner of New Mexico, and include magma dating as late as 0.1 Ma (Dickinson, 2004; Rasskazov et al., 2010). (There are no known volcanoes in Texas younger than 18 Ma, see Ewing, 2016.) The renewed development of the Rio Grande rift began about 10 Ma BP, and by 5 Ma BP its present morphology was more-or-less established (Seager et al., 1984; Ricketts et al., 2016). This modern Rio Grande rift, extending from Colorado, through New Mexico, and along the Texas–Chihuahua border corridor for a length of over 1,200 km, is an extended depression lying between mountainous headlands, situated on the crest of the Rocky Mountains, and made up of elongated basins separated by rises, barriers, or constrictions (Figure 14.1). It is now recognized as a major continental rift (Riecker, 1979; Keller and Cather, 1994; Keller and Baldridge, 1999).

Each individual basin or bolsón ("bag" in Spanish) is a surface expression of crustal extension, during which the resulting release of lateral support initiated faulting and subsidence of blocks of basement rocks. Over time sediment eroded from the surrounding highlands has collected in the basin. The floor of a bolsón is typically relatively level, with the possible exception of a shallow depression (playa) where water collects. Depths of basin-fill sediments in the bolsóns range from 1 to 6 km (Wilkins, 1986; Hawley and Kernodle, 1999; Bartolino and Cole, 2002). The basin-fill sediments are typically sandstone, mudstone, and conglomerates, with intermixed gravel and sands. The resulting formation is porous, excellent for groundwater retention, viz. an aquifer. The major basins are located on Figure 14.1. A diagram of a representative basin is shown in Figure 14.3.[2]

During the Cenozoic, climate also exhibited drastic changes, including an average decline of about 20°C over its 65 Ma duration, from "greenhouse to icehouse" (a turn of phrase of Fischer, 1981), and culminated in the excursions of the Pleistocene glaciations (Cronin, 2010). Most important to the evolution of the Rio Grande were the excursions in climate during the past 2 Ma characterized by polar glaciations, especially the past 0.5 Ma (= 500 kilo annus, ka).

The importance of the ancestral landscape of the Rio Grande rift is that it presented a trough for surface flow from the mountains of Colorado southward through New Mexico into Chihuahua and Texas (the "Great Trough," per Kelley, 1969).

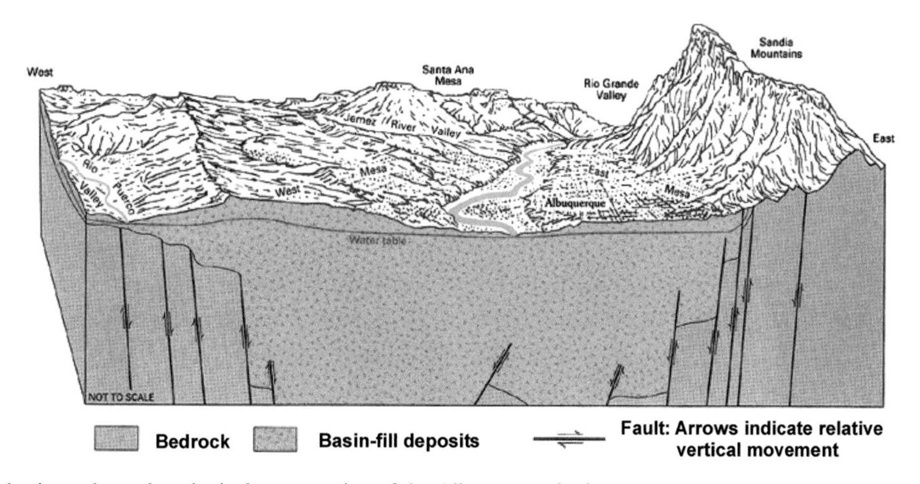

Figure 14.3 Surface physiography and geological cross section of the Albuquerque basin.
Figure source: Robson and Banta (1995)

Initially, the drainages into the bolsóns were internal, so that runoff water accumulated within the basins, evaporating to the atmosphere or percolating to the aquifer. Once the drainage divides between bolsóns were breached, establishing hydraulic continuity, at least part of this runoff could be transported downslope to the adjacent basin. The rift was a ready-made conduit for the flow of a river.

The ancestral upper Rio Grande ca. 5 Ma (i.e., at the beginning of the Pliocene Epoch) consisted of isolated riverine segments. For at least 3 Ma, the Rio Chama had been the primary channel of the stream network flowing from the west into the Española basin. The 5 Ma Rio Grande was a mere tributary rising in the San Juan mountains and flowing south to its confluence with the Chama. As summarized by Repasch et al. (2017), for the next five million years the rift was subjected to uplift, subsidence, faulting, and volcanism, which altered the paths available to the evolving river and its ability to flow between basins. This was a period of riverine incision, drainage integration, increased runoff, and acceleration of sediment transports. A potential additional causative factor was a shift in regional climate associated with the opening of the Gulf of California ca. 5 Ma BP, and the intensification of the Mexican monsoon (Chapin, 2008). During this period, the upper Rio Grande gradually became connected, proceeding downstream, typically ending in a lake created by a flooded bolsón. This ancestral segment of the Rio Grande extended southward through the New Mexico basins by about 3–4 Ma BP, and reached the Hueco bolsón by about 2–3 Ma, where it terminated in the ancestral Lake Cabeza de Vaca (Strain, 1971; Seager et al., 1984; Perez-Arlucea et al., 2000; Repasch et al., 2017).

Drainage to the Gulf of Mexico had occurred in the early and middle Cenozoic with substantial sediment discharge to the Gulf (Galloway et al., 2011), certainly involving the ancestral lower Rio Grande, but the connection of the entire basin from the Rocky Mountains to the sea seems to be fairly recent. The precipitant was the creation of Lake Alamosa in the San Luis basin ca. 4.5 Ma BP, and its subsequent breaching and draining at ca. 400 ka (Machette et al., 2007; Davis et al., 2017; Repasch et al., 2017; Ruleman et al., 2019). The breach not only delivered the contents of a 4,000 km^2 lake to the upper Rio Grande but also represented a permanent increase in flow due to the drainage area enlargement of about 24,000 km^2, an area that is also the most prolific source of runoff in the basin, deriving from the mountain winter snowpack. Seager et al. (1984) (which preceded the scientific discovery of the Lake Alamosa breach by about two decades) commented on the puzzling suddenness (on geological scale) of the transition from aggradation to incision of the Rio Grande in the southern section of the rift, and the upstream-directed entrenchment of the river from the Presidio bolsón to the Mesilla, all within 100–200 ka. Machette et al. (2007) and Ruleman et al. (2019) conclude that the causative agent for this occurrence was climatological, namely the Mid-Brunhes Event, the transition from glacial to interglacial of the first of five high-amplitude glacial cycles at 100 ka period that closed the Pleistocene (Cronin, 2010). The initial meltwater from the glaciers upstream triggered the breach of Lake Alamosa.

Meantime, early in the third tectonic episode enumerated earlier, regional uplifting along the central axis of the Rio Grande rift of some 2,000 m in elevation created a downward slope in the terrain to the east. The eastern highlands of the Rockies were exposed to fluvial erosion, with numerous streams carrying the resulting sediment load. This created a thick deposit of alluvium lying north–south, from the base of the Rockies eastward perhaps 400 km, which was to become the Ogallala formation. About 5 Ma BP, the Pecos, the upper reach of the ancestral lower Rio Grande, began eroding its valley northward through this alluvium into New Mexico, assisted by the dissolution and collapse of salt deposits in the Permian strata below (Ewing, 2016). The upper Rio Grande may have connected with the lower drainage at the modern Pecos confluence when Lake Cabeza de Vaca drained about 800 ka BP (Repasch et al., 2017), but the Alamosa event and the associated accelerated river incision are considered to have completed the drainage to the Gulf of Mexico in its approximately modern form.

While less is known about the ancestral Río Conchos, there is some indication of an established Pliocene (ca. 3 Ma BP) drainage from northern Mexico to the Gulf, perhaps approximately following its modern course to the lower Rio Grande to join with the Pecos (Gile et al., 1981; Galloway et al., 2011; Ewing, 2016). Like the advantage presented by the Rio Grande trough, the Conchos channel was probably fortuitous, following low points in the terrain to establish a course through the mountains, assisted by interglacial draining of lakes from the southern basins (Strain, 1971).

14.2.2 Aquifers and Groundwater

One fundamental difference between the two reaches of the Rio Grande is that in the upper Rio Grande groundwater and surface water are much more intimately connected than in the lower Rio Grande, a consequence of the differing physiography and geology of the two reaches. The basic component of groundwater storage in the upper Rio Grande is the bolsón, described previously. There are numerous such geological structures scattered throughout the Basin and Range province and the Rio Grande rift. The basin-fill deposits in these structures may range up to hundreds to thousands of meters in thickness, and many contain appreciable volumes of water. Those of primary importance to the Rio Grande region are shown in Figure 14.1. The river courses through several of the rift basins. These specific basins, Figure 14.1, are the key to the hydrology of the upper Rio Grande. In the words of Bryan (1938, p. 198), "it may be roughly likened to a stream flowing from one sand-filled tub to another through narrow trenches."

Recharge of the aquifer of a simple encircled bolsón is infiltration of rainfall and local runoff on its highland periphery, into recently deposited alluvial sediments on the slopes into the basin, and through the basin floor and central playa. When the basin is transected by a surface stream or tributary, an additional source of infiltration is through the channel of the stream, in which case the stream loses flow crossing the basin and may even terminate within the basin. Generally, the recharge from all of these sources is considered to be a small fraction of the annual rainfall, perhaps 5 percent (Hawley and Kernoble, 1999; Thiros et al., 2010): that is, the great majority of the rainfall is lost to evapotranspiration back to the atmosphere or is removed from the basin in surface streams. The stream networks in the upland regions (e.g., Jemez, San Juan, and Sangre de Cristo ranges) are considered to be the primary sources of recharge for these basin aquifers (Hawley and Kernoble, 1999). Often, the basin-fill sediments will be coupled between adjacent bolsóns, so that groundwater may flow between them, even if there is not a conveyance for surface flow. Though the flow magnitudes are much smaller than in surface streams, this transfer can become important over long periods of time.

Because of the great depths of basin-fill sediments in these bolsóns, the early assessment was that they provided a substantial source of groundwater, offering great opportunity for irrigation (e.g., Bjorklund and Maxwell, 1961; Cliet, 1969). Kelley (1969) enthused, "There is little or no chance of a water shortage, even with a substantial increase in population. Probably no other large city in the arid Southwest has such a bountiful supply of good water as Albuquerque has." This impression was reinforced by early wells fortuitously drilled in sediments of high hydraulic conductivity (Bartolino and Cole, 2002). Recent work has altered this view. Hawley and Haase (1992) showed that the productive zones of the Albuquerque aquifer comprise only a small part of the basin-fill aquifer. Plummer et al. (2004), using chemical and radiological properties of the groundwater as tracers, demonstrated that most of the basin aquifer water has been stored for thousands of years, dating from about 5,000 to 20,000 years (the latter being the last glacial advance). The bountiful water supply is, in fact, largely fossil water, which present recharge rates are inadequate to replace.

As a first approximation, the bolsón aquifer structure has been conceived as a two-layer system, with recent alluvial and fluvial deposits making up a shallow surficial layer, overlying a deeper and thicker volume of ancestral basin-fill sediment. While it is acknowledged that the geological structure of a basin is much more complex, and must be pieced together from a variety of observational strategies, including the mapping and interpretation of exposed sediments, analysis of drilling and well data, and application of geophysical techniques such as gravity surveys and radiological analyses, the two-layer conceptual model continues to retain value in analyzing and communicating the fundamental hydrologic behavior of the aquifer, especially for water budgeting (e.g., Bjorklund and Maxwell, 1961; Hearne and Dewey, 1988; ACMRGWA, 1999; Hawley and Kernodle, 1999; Wilson, 1999; Bartolino and Cole, 2002; Hawley and Kennedy, 2004; Eastoe et al., 2010).

The hydraulics of the upper Rio Grande has alternated between geological periods of aggradation of sediments (and filling of the channel and flood plain) and erosion (and incision of the river channel into older sediments). This is especially true of the late Pleistocene, as the river responded to the cycle of glacials and interglacials (Dethier, 1999; Passaglia, 2005; Connell et al., 2005; Lyle et al., 2012). One result of this vacillation is the morphology of the river, in which the active floodplain is generally constrained within an incised channel. The river cuts a channel into its previously deposited floodplain and then, in the later part of the cycle, aggrades this channel with fluvial deposits. From Otowi to Fort Quitman, this process has created a river corridor with steep channel boundaries whose level is some tens of metres below the basin floor, the net result of the cycle of erosion and subsequent aggradation. The river is free to meander across the flood plain, confined within the walls of the incised channel. This flood control has been reinforced by construction of levees. This incised floodplain is referred to as the Rio Grande Inner Valley (Kelley, 1969).

The recent deposits in the Inner Valley are unconsolidated rock fragments, sands, and silts, so are permeable, porous, and conductive. These sediments are, on the one hand, in contact with the river, and on the other hand, in contact with the surficial alluvium of the surrounding basin floor, allowing a free exchange between the water in the river and that in the surrounding shallow aquifer. Wells drawing water from the shallow aquifer will reduce the flow in the river, creating the apparently paradoxical circumstance that a groundwater withdrawal may deny a downstream water right on the river. This is a stark example of the intimate connection between surface water and groundwater in the upper Rio Grande, and the sorts of management considerations it necessitates.

In contrast, in the lower Rio Grande, the river does not interact to the same degree with the regional aquifers. Where interactions do occur, the exchange is almost entirely from aquifer to stream. Four major aquifers are crossed by the Rio Grande and its tributaries; proceeding downstream, the Pecos Valley (crossed by the Pecos), the Edwards-Trinity (Plateau), Carrizo-Wilcox, and Gulf-coast aquifers (George et al., 2011). An estimate of the contribution of these aquifers to the flow of the lower Rio Grande, based on the recent work of TWDB reported by Anaya et al. (2016), is an average annual inflow of about 350 Mm^3 (probably an overestimate since the spatial resolution is limited to counties through which the river passes, most of whom also contain the northern basin divide).

14.2.3 The Modern River System and Its Engineering

Reservoirs and canals have been a part of the drainage network of the Rio Grande for more than a century, as noted in the introduction to this section. While both are important engineering features of the basin, our emphasis in this chapter is on reservoirs. Canals are essentially tactical, delivering available water to the points in the basin where water is needed: fields, croplands, municipalities, industries, and ranches. Reservoirs, in contrast, impound and store water to be distributed as needed, so their role is essentially strategic.

The two central functions of reservoirs in the Rio Grande basin are flood control and water supply ("conservation" in the patois of engineering). The reader is reminded that flood-control reservoirs and water-supply reservoirs are operated in fundamentally different ways (see Chapter 3). For flood control, the objective is to maximize the capacity to store flood waters, so the reservoir is evacuated as quickly as possible after a flood event and stands empty most of the time. For water supply, the objective is to maintain a supply of water in storage, so inflow events are captured to the extent of available capacity and the reservoir is kept full as much of the time as possible. In the Rio Grande, some reservoirs are devoted solely to flood control and some solely to conservation. There are also multipurpose reservoirs, which accomplish both flood control and water supply, by subdividing the total capacity into separate "pools" for each function, the former being emptied after flood events, and the latter being maintained at capacity as inflows permit. (There may be a third pool for hydroelectric power generation. While historically some of the Rio Grande reservoirs – Boquilla, Elephant Butte, and Red Bluff, notably – had this provision, at present power generation is a by-product of releases for other purposes.) A list of the major reservoirs of the Rio Grande is tabulated in Table 14.4 chronologically by date of deliberate impoundment. The data in Table 14.4 reflect the most current information available to these writers. Over time, some of these reservoirs have experienced a re-allocation of their capacities to accommodate changing requirements for flood control, water supply, and power generation.

While a detailed map of the drainage system of the Rio Grande is presented in Figure 14.1, one problem of this sort of display is that, if everything is rendered to scale, hydraulic features such as reservoirs and tributaries can be very difficult to discern, some of them being smaller than the line width of the artwork. An alternative display is the stem diagram, in which the important drainage network features are deliberately distorted and simplified but presented in their relative configurations. The topography of the drainage network is lost, but its topology is clarified. Stem diagrams for the upper and lower Rio Grande reaches are shown in Figures 14.4 and 14.5, respectively. These emphasize only the mainstem features, showing the relative locations of gauges, tributaries, and reservoirs (see Tables 14.1 and 14.2 for details). Most of the reservoirs are in fact located on tributaries of the Rio Grande rather than its mainstem, see Table 14.4. It is apparent that the major branches of the Rio Grande, viz. the upper Rio Grande, the lower Rio Grande, the Rio Conchos, and the Pecos River, as well as many of the smaller tributaries, have been extensively engineered to store and control the waters of the basin.

To list Avalon on the Pecos as starting deliberate impoundment in 1894 in Table 14.4 requires some qualification. Avalon Dam and nearby McMillan Dam were built by private enterprise, the Pecos Irrigation and Investment Company, formed in 1888 (Bogener, 1993). Perhaps the best known of the original investors was Pat Garrett, former sheriff of Lincoln County, who gunned down Billy the Kid. The two dams were intended to work in tandem, McMillan providing storage, and Avalon handling distribution to the canal system. Work on both dams was underway by 1891, by which time Sheriff Garrett had left the company and returned to gunslinging, evidently finding it less risky. Avalon was essentially complete by mid-1893. Impoundment was deliberate but unsuccessful: floods in August took out the dam, its flume, and two bridges, and set back the work on McMillan. Apparently, Avalon was functional by 1894 since water deliveries were made in that year (Murphy, 1905; Schuyler, 1909; Hufstetler and Johnson, 1993; Bogener, 1997), hence the date in Table 14.4. McMillan was completed by September 1896 but was failing to hold water due to severe leakage into the bed (Bogener, 1993). In October 1904, an enormous flood on the Pecos again took out Avalon, along with three bridges, damage to the Pecos flume, and McMillan's west embankment. Two dam failures in eleven years, compounded with a major drought and financial panic, doomed the enterprise. In 1905, the Bureau of Reclamation took over the facilities, naming it the Carlsbad Project.

La Boquilla on the Rio Conchos, closed in 1913, and Elephant Butte on the Upper Rio Grande, in 1916, were among the largest impoundments in the world at the time of their completion. The original purpose of Boquilla was to provide hydroelectric power to the "American camp," the famous Parral mining operations in northern Chihuahua (Griggs, 1907; Duryea and Haehl, 1916), but its function has changed over the intervening century to water supply and irrigation. Elephant Butte is a multipurpose reservoir for water supply and power generation, with some limited flood control. The reservoir serves as the main hydrological structure of the US government Rio Grande Project that supports irrigated agriculture in southern New Mexico, West Texas, and Northern Mexico. The project served as a model for water engineering in the Colorado. A short distance downstream from Elephant Butte, Caballo dam was built in 1938. This allows for energy production at Elephant Butte without passing water out of the system, since Caballo can capture the power releases from Elephant Butte and release this water for irrigation. This works well for

Table 14.4 *Major reservoirs of the Rio Grande basin*

Name	Function	Capacity (Mm^3) Supply	Flood	Total	Location	Date* of closure
		Upper Rio Grande				
Colorado reach						
Sanchez	Irrigation San Luis Valley	127		127	Ventero Creek	1912
Rio Grande	Irrigation San Luis Valley	63		63	RG mainstem	1912
Terrace	Irrigation San Luis Valley	21		21	Alamosa River	1912
Mountain Home	Irrigation San Luis Valley	23		23	Trinchera Creek	1913
Smith	Irrigation San Luis Valley	6		6	Trinchera Creek	1913
Continental	Irrigation San Luis Valley	28		28	North Clear Creek	1928
Santa Maria	Irrigation San Luis Valley	56		56	North Clear Creek	1928
Platoro	Irrigation San Luis Valley**	66	7.5	74	Conejos River	1951
New Mexico reach						
Elephant Butte	Power & irrigation	2,433	62	2,495	RG mainstem	1916
Costilla	Supply	14		14	Costilla Creek	1922
El Vado	Irrigation	242		242	Rio Chama	1935
Caballo	Supply	280	403	683	RG mainstem	1938
Abiquiu	Flood & supply (SJ-C only)	247	1,442	1,689	Rio Chama	1963
Jemez Canyon	Flood		327	327	Jemez River	1965
Galisteo	Flood		188	188	Galisteo River	1970
Heron	Supply (SJ-C only)	495		495	Rio Chama	1971
Cochiti	Flood		886	886	RG mainstem	1973
		Lower Rio Grande				
Borderlands reach						
Falcón	Supply, irrigation, flood	3,273	6,181	9,454	RG mainstem	1953
Amistad	Supply, irrigation	4,174	3,883	8,057	RG mainstem	1968
Conchos basin						
La Boquilla	Supply & flood***	2,894	1,319	4,213	Conchos	1913
Chihuahua	Supply & flood	23	38	61	off channel	1948
Francisco Madero	Supply & flood	355	478	833	San Pedro	1949
Luis L. Leon	Supply & flood	292	832	1,124	Conchos	1968
San Gabriel	Supply & flood	245	317	562	Florida	1981
Pico del Aguila	Supply & flood	48	87	135	Florida	1993
Pecos basin						
Avalon	Irrigation & M&I	85		85	Pecos River	1896
McMillan†	Irrigation & flood	99		99	Pecos River	1908
Storrie	Irrigation & M&I	25		25	Gallinas Creek††	1921
Red Bluff	Irrigation	186		186	Pecos River	1936
Sumner†††	Irrigation & flood	50	116	166	Pecos River	1937
Santa Rosa	Irrigation M&I & flood	114	541	655	Pecos River	1980
Brantley	Irrigation & flood	49	511	560	Pecos River	1988
Mexico Rio Grande tributaries						
Venustiano Carranza§	Supply & flood	874	1,322	2,196		1930
Laguna de Salinillas	Supply			19	Rio Salada	1931
Centenario	Supply			47	Arroyo Las Vacas	1934
Marte R. Gomez	Supply & flood	782	2,304	3,086	Rio San Juan	1943
Rodrigo Gomez	Supply			40	Rio San Rodrigo	1963
La Fragua	Supply & flood	47	85	132		1991

Table 14.4 (*cont.*)

Name	*Function*	*Capacity (Mm³)* *Supply*	*Flood*	*Total*	*Location*	*Date* of closure*
El Cuchillo	Supply & flood	1,123	1,784	2,907		1994
Las Blancas	Supply & flood	84	134	218	Rio San Juan	2002

* Deliberate impoundment
** Original function flood control, non-flood pool now used jointly for both flood and supply
*** Original function power and limited supply
† Breached upon completion of Brantley
†† Tributary of Gallinas River
††† Née Alamagorda
§ Née Don Martín

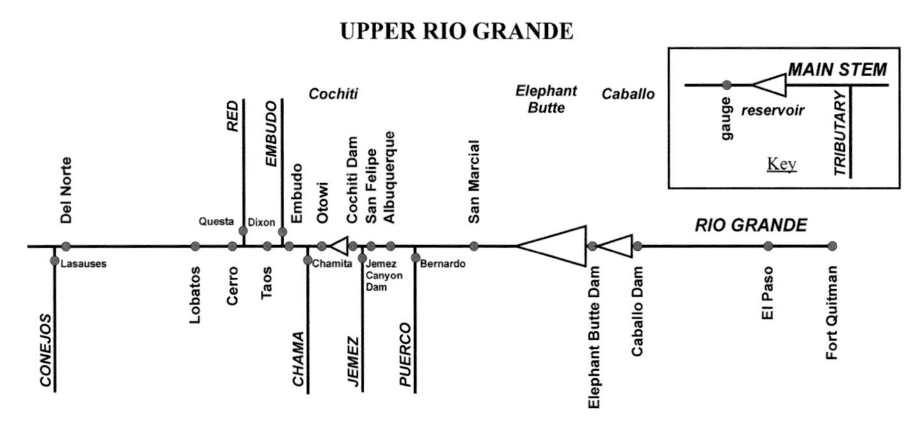

Figure 14.4 Stem diagram of the upper Rio Grande, showing tributaries, principal gauging stations, and major mainstem reservoirs.

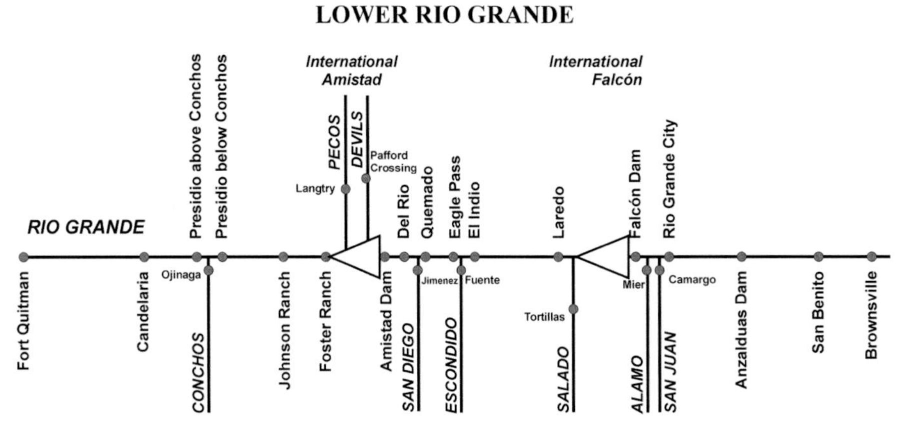

Figure 14.5 Stem diagram of the lower Rio Grande, showing tributaries, principal gauging stations, and major mainstem reservoirs.

the farmers, but severely reduces environmental flows during the months when no irrigation release is called for.

One of the most important projects in recent years has been the San Juan–Rio Chama (SJRC) Project of the US Bureau of Reclamation. This is a large-scale interbasin transfer and storage project, to move part of New Mexico's share under the Colorado River Compact from the Colorado basin to the Rio Grande. Annual diversions of 118.7 Mm^3 (96,200 ac-ft), the estimated firm yield (see Chapter 3), are made from three tributaries of the San Juan River, and transported through a series of tunnels and diversion dams to Azotea Tunnel, which carries the flow beneath the continental divide into Heron Reservoir. This reservoir was

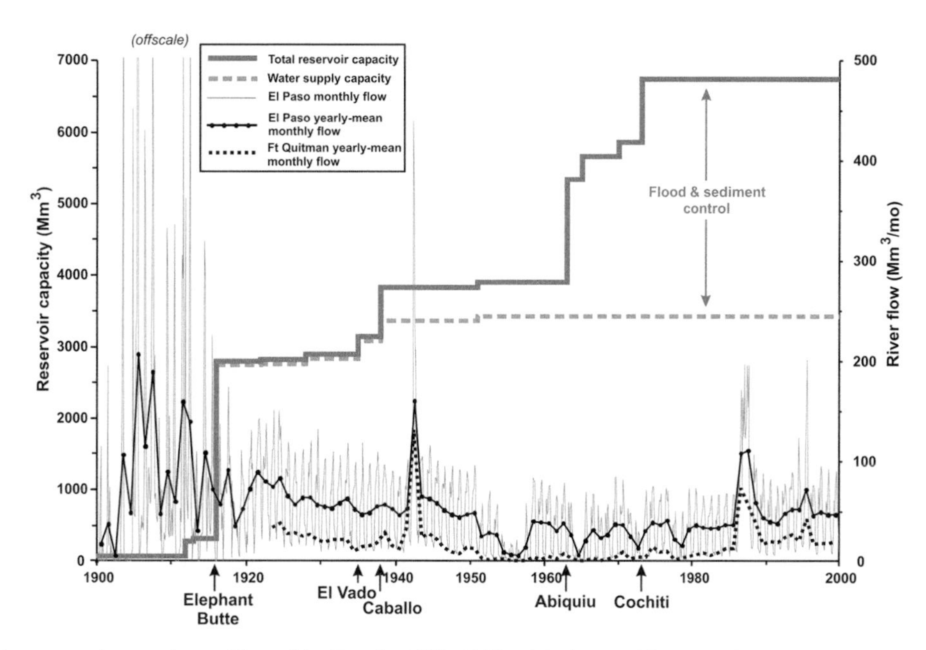

Figure 14.6 Cumulative reservoir capacity on Upper Rio Grande, 1900–2000, with observed flows at El Paso and Ft. Quitman.

built for the sole purpose of receiving and storing SJRC water. Additional temporary storage for SJRC water is available in El Vado, Abiquiu, and Cochiti. Abiquiu, Cochiti, and Jemez were originally intended for flood control and sediment retention only. Abiquiu is now also used to temporarily store water from the SJRC Project. Cochiti now includes a small permanent pool (50,000 ac-ft) for recreation. Jemez is operated as a dry reservoir.

The evolution of the upper Rio Grande over the twentieth century is illustrated by the time history plot of Figure 14.6. This graph shows the cumulative reservoir storage installed in the upper basin, with the principal reservoirs marked on the time axis. The monthly flows measured at the downstream gauge at El Paso are plotted as the light grey trace. Under "pre-development" conditions, i.e., before the institution of large reservoirs, the gauge at El Paso registered huge excursions in flow, several orders of magnitude in monthly volumes (and an even wider range in dailies). For 1900–1915, before closure of Elephant Butte, the standard deviation of monthly flows was about 150 Mm^3 (a coefficient of variation of about 1.7). In 1916–1930, this standard deviation dropped to about 50 Mm^3, and by the end of the century it had been reduced to about 30 Mm^3, a five-fold decrease in variation. This diminishment in the high-magnitude spikes is clear in Figure 14.6 as more reservoirs come online through the century. A more subtle, but equally important, change in the El Paso time signal with the 1916 closure of Elephant Butte is the regularity in river flows. A clear annual periodicity replaces the apparently random variation of 1900–1915. This is the regular midsummer releases from Elephant Butte. The amplitude of this nearly sinusoidal annual signal diminishes with time, as does the annual-mean flow at El Paso.

On the lower Rio Grande, i.e., downstream from Fort Quitman, the International Reservoirs Amistad and Falcón provide water for irrigation and for municipal and industrial use by cities between the two reservoirs, and in the Lower Rio Grande Valley (LRGV). The two reservoirs are operated as a tandem system by the IBWC. Falcón was the first to be constructed, in response to the devastating floods of the 1940s, and was completed in 1953, early in the Drought of the Fifties. Amistad followed later, deliberate impoundment beginning in 1968. These are the only two reservoirs on the mainstem (though there are several diversion dams). But, like the upper Rio Grande, much of the reservoir development on the lower Rio Grande took place on the tributaries. Avalon and McMillan on the Pecos were first. Boquilla on the Conchos, noted earlier, was the first large reservoir in the lower Rio Grande basin, built during the Mexican revolution. These were followed in the 1930s by Don Martín on the Salado, and Red Bluff and Alamagorda on the Pecos.

The evolution of the lower Rio Grande over the twentieth century is illustrated by the time history plot of Figure 14.7, following the same conventions (except a lagniappe in the form of an additional ten years). Like Figure 14.6, this displays the cumulative reservoir storage installed in the lower basin, with the principal reservoirs marked on the time axis. Unfortunately, there is not a gauging station on the river analogous to El Paso, located below the major reservoirs with a continuous long record of measurements predating reservoir development. The gauge at Eagle Pass, below Amistad but above Falcón, is our best

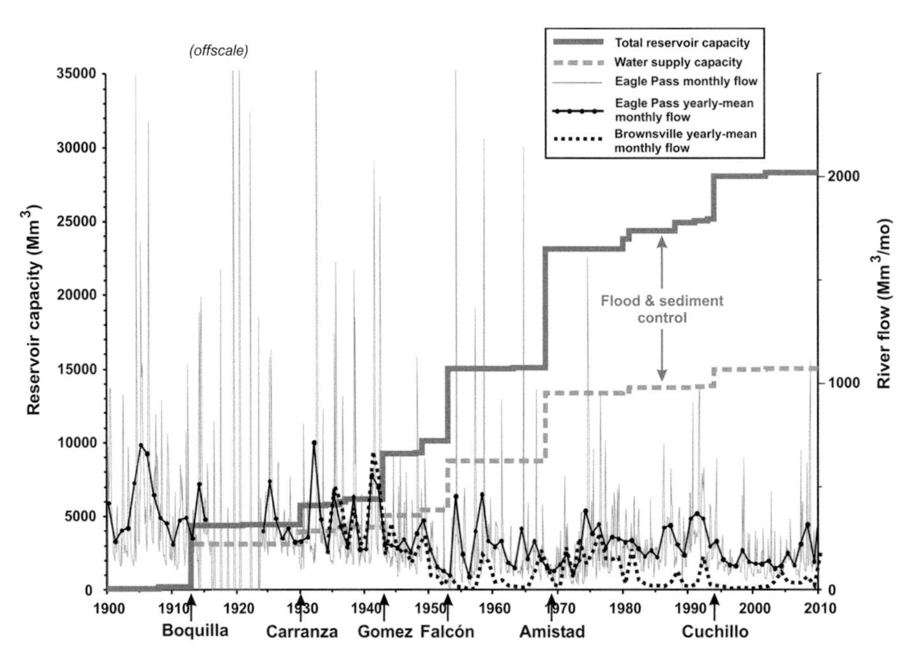

Figure 14.7 Cumulative reservoir capacity on Lower Rio Grande, 1900–2010, with observed flows at Eagle Pass and Brownsville. Figure by the authors

approximation. The monthly flows measured at Eagle Pass are plotted as the light grey trace on Figure 14.7. Under "pre-development" conditions, i.e., before the institution of large reservoirs around 1915, there is no substantive modification in the time signal at Eagle Pass, other than a decline in frequency and magnitude of the flood "spikes" in the monthly flows. For 1900–1915, before closure of Boquilla, the standard deviation of monthly flows was about 356 Mm^3 (a coefficient of variation of about 0.9). In 1916–1945, this standard deviation increased to about 458 Mm^3, in 1946–1965 it dropped to 326 Mm^3, and in 1966–2010, after Amistad, further decreased to 203 Mm^3. While there is a diminishment in the magnitudes of the flood spikes, there is no clear annual signal in the releases, but instead a noisy pattern with a weak, barely discernable annual component. In general, this agrees with the hydroclimatology of south Texas, viz. no distinct seasonality but largely random occurrences of floods in response to occasional diluvial storms from westerly synoptic systems, and occasional tropical storms from the Pacific or Atlantic.

The time history of inflows and storage in Elephant Butte reservoir, Falcón reservoir, and Amistad reservoir are shown in Figures 14.8, 14.9, and 14.10. Inflows for Elephant Butte are estimated by the Otowi gauge until 1949, then by the San Marcial gauge. For Falcón and Amistad, these are the deduced inflows reported by IBWC, with upstream gauges of Foster Ranch and Laredo, respectively, when deduced values are not available. For the period 1948–1980, Elephant Butte was chronically inflow-deficient, with the Drought of the Fifties embedded in this time period. Falcón was nearly empty for the first year after closure, but then was brought to 91 percent of capacity by a single event late in 1954, Tropical Storm Alice landfalling on the Rio Grande and following the river inland (McAdie et al., 2009). Its contents then diminished due to severe drought conditions until 1957. System (i.e., tandem) operation of the International Reservoirs began when both reservoirs achieved full storage, in late 1973. Typically, in tandem operation Falcón is drawn down first, being replenished by releases from Amistad, thereby maximizing the ability to capture inflow events.

The hydrology of the Rio Grande may be summarized in the variation of flow with downstream position on the mainstem, depicted in Figure 14.11. Since Figure 14.11 is based on the 1981–2010 means, a natural inquiry is whether there are time variations in the past such 30-year means. Table 14.5 is a compilation of selected mainstem and tributary gauges for the past four 30-year periods.

The upper Rio Grande receives substantial flow from upstream, the Rio Chama being the most important tributary, see Figure 14.11. This flow diminishes downstream from the Chama confluence almost monotonically to Elephant Butte. This is the combined effect of diversions, primarily for irrigation, and channel losses (including evaporation). Whether the diversions are from the river or from groundwater pumping, the effect on the river is basically equivalent because of the intimate connection of the river and the alluvial aquifer.

A one-sentence summary of the hydrology of the upper Rio Grande depicted in Figure 14.11 is all inflows are consumed by the end of the reach. There is some evidence in the data of Table 14.5 of increasing trends in the average flows. The flow

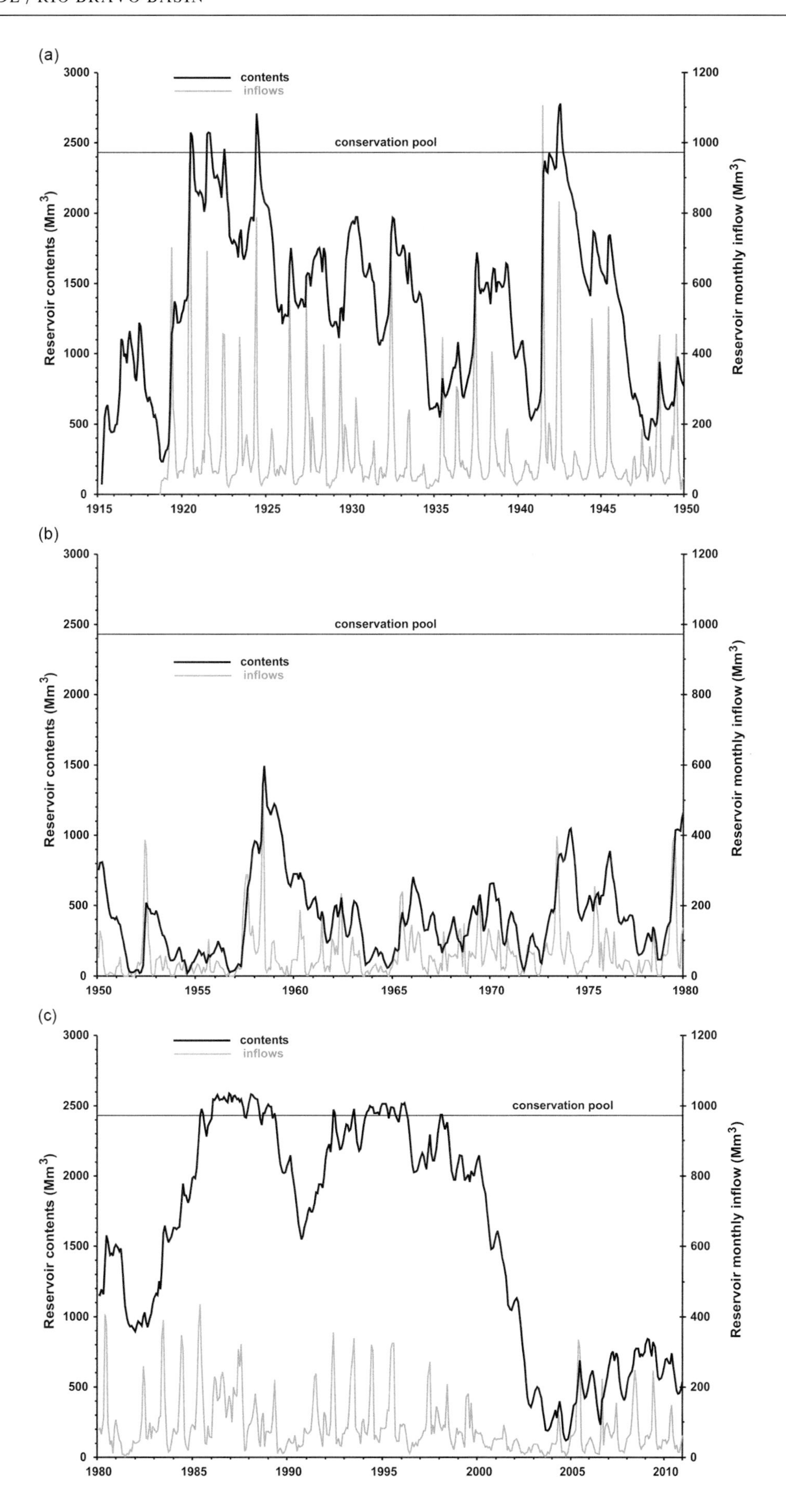

Figure 14.8 Elephant Butte contents and inflows (a) 1915–1949, (b) 1950–1979, and (c) 1980–2010.

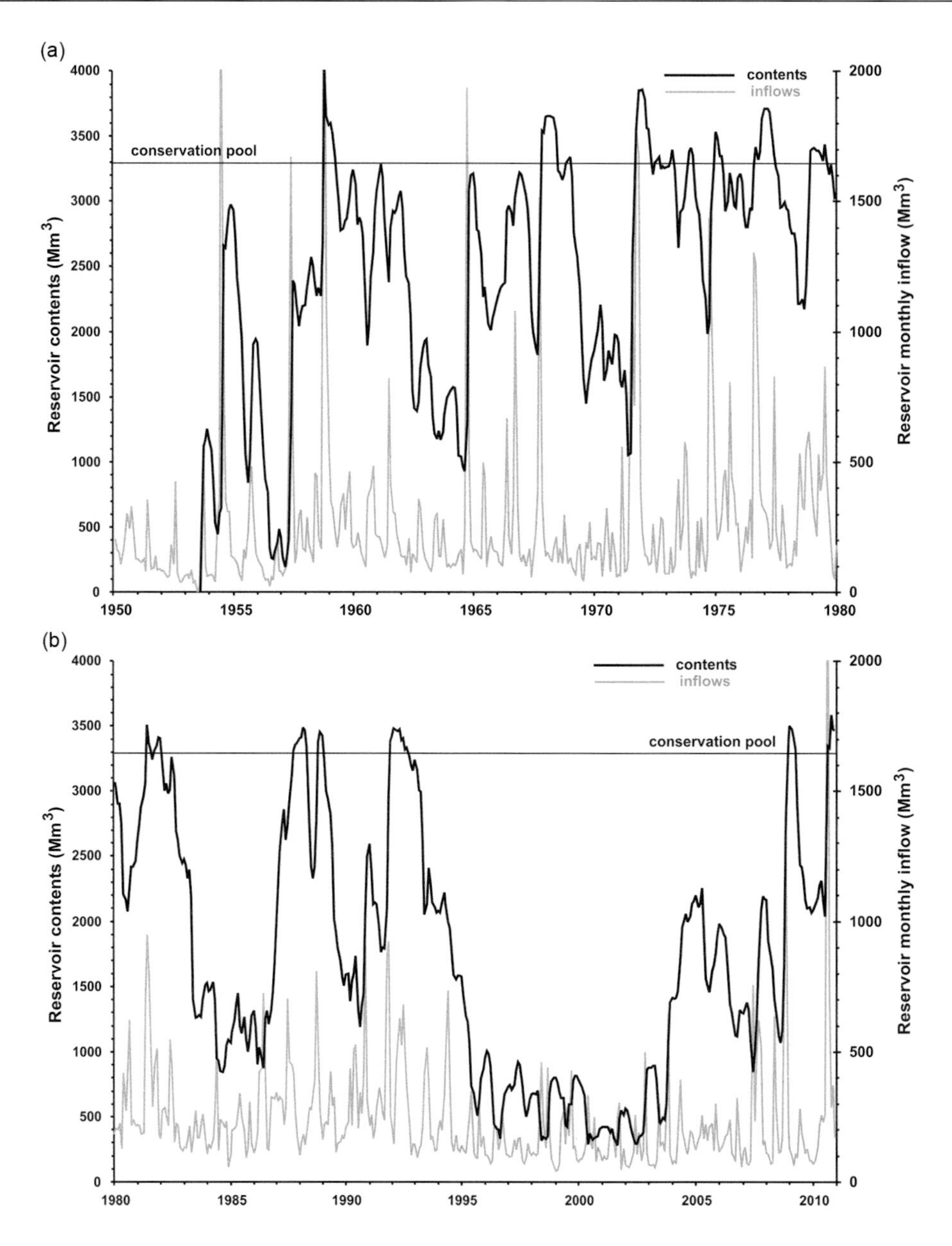

Figure 14.9 Falcón contents and inflows, (a) 1950–1979 and (b) 1980–2010.

at Otowi has increased about 30 percent from the 1951–1980 mean to the 1981–2010 mean, and similar trends are evident at San Marcial, Elephant Butte, and El Paso. These do not seem to be driven by tributary flows. The flow into the "forgotten reach" at Fort Quitman has increased by a factor of five.

After a streamflow minimum at Fort Quitman, the Rio Grande is rejuvenated by the Conchos, and to a lesser extent by the Pecos and Devils. With 694 Mm^3 average annual discharge, the Conchos is the most important tributary, providing about two-thirds of year-round streamflow in the lower Rio Grande. The Pecos, in contrast, delivers 211 Mm^3, and the Devils, 330 Mm^3. One impression drawn from Figures 14.8 to 14.10 is that both Amistad and Falcón can be described as being in a continuous state of drawdown, except for sporadic diluvial events that at least partially refill the reservoirs. While Brownsville is the last gauge station on the river, there is one more diversion dam and several withdrawals, so the flow reaching the Gulf is less than indicated by Figure 14.11. In the lowermost reach of the lower Rio Grande, below Brownsville, the combination of diversions for irrigation and municipal–industrial water use, and the diversion of floodwaters around the Lower Rio Grande Valley directly to the Gulf, has reduced the river flow to substantially zero. The only river flow in the channel below the last pumping-pool dam is the small amount of runoff deriving from rainfall on the coastal watershed, and fugitive releases of water due to inefficiencies in pumping the allocated withdrawals. There is no longer a true

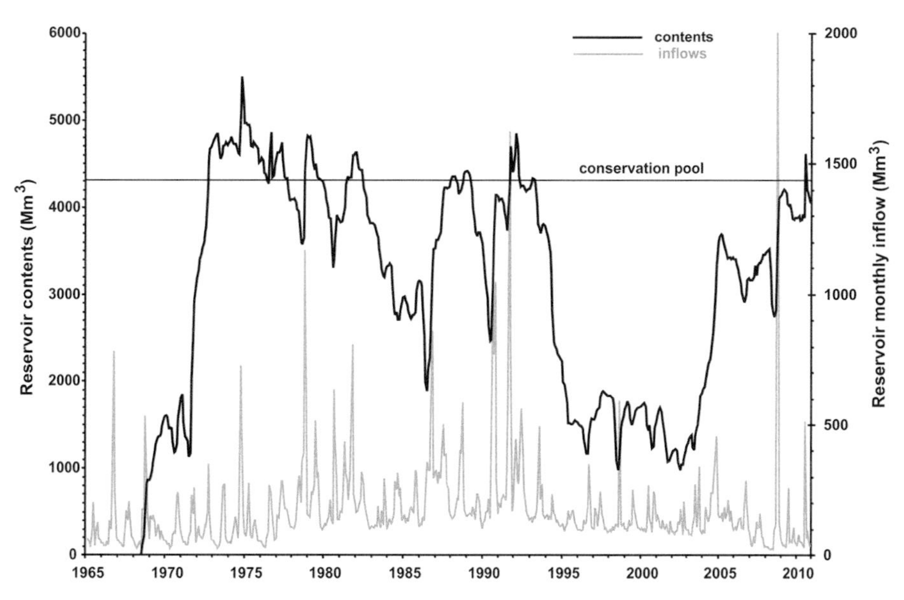

Figure 14.10 Amistad contents and inflows, 1980–2010.

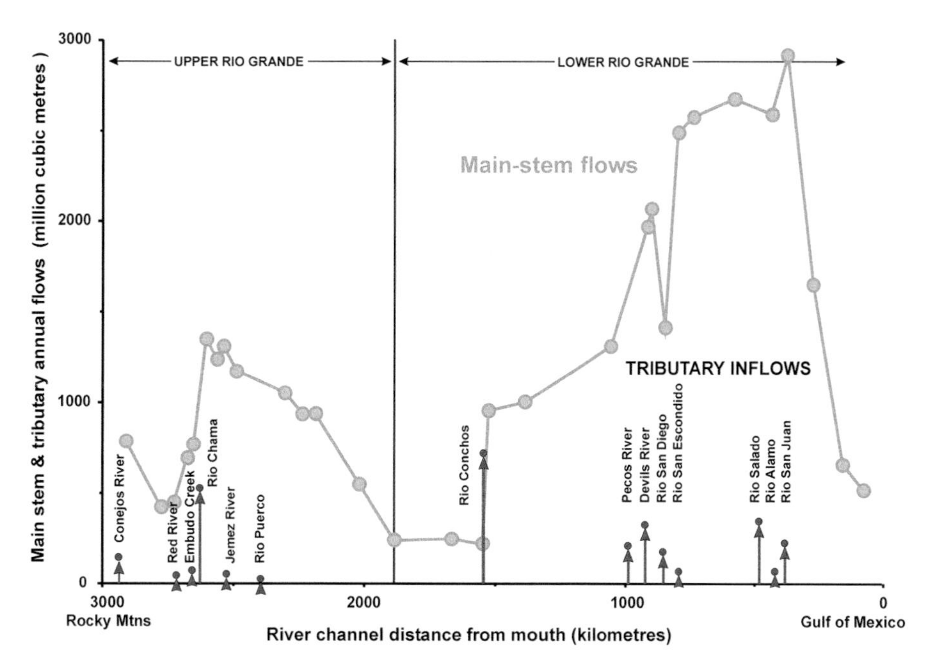

Figure 14.11 1981–2010 averaged flows at principal gauging stations on mainstem and tributaries (see Tables 14.1 and 14.2).

estuary, and much of the time, the mouth of the Rio Grande is blocked by a sandbar.

The natural flow in the lower Rio Grande is a low baseflow on which fluvials are randomly superposed. The lower Rio Grande exhibits a similar behavior to the upper Rio Grande, in that all of the inflows to the reach are consumed by its end, so the same one-sentence summary applies. The flow magnitudes are roughly a factor of two larger. There is little optimism to be drawn from the time trends of Table 14.5, however. Those above Falcón show no significant trend, while those below Falcón are definitely declining with time.

14.3 GOVERNANCE

14.3.1 Institutions

Water management in the basin is a complex maze of international, national, state, regional, and local institutions. Even so, the system works to the satisfaction of both agricultural and urban users, except during drought years. (See Conflicts section.)

The US Federal government, in 1902, established the Bureau of Reclamation, a part of the US Department of the Interior. The Bureau's mission was to assist with development of the

Table 14.5 *30-year average flows at mainstem and tributary stations along Rio Grande*

Gauge station		*1951–1980*	*1961–1990*	*1971–2000*	*1981–2010*
Conejos Lasauses	trib*	121	149	155	146
Del Norte		694	772	793	784
Lobatos		312	415	443	422
Embudo Dixon	trib	55	68	80	77
Chama Chamita	trib	403	463	541	527
Otowi		1,050	1,245	1,391	1,350
Cochiti		n/a†	1,229	1,271	1,242
Jemez Canyon Dam	trib	50	60	65	58
Albuquerque		n/a	1,144	1,237	1,171
Puerco Bernardo	trib	39	32	30	26
San Marcial		767	971	1,125	1,050
ElephantButte Dam		649	792	897	935
Caballo Dam		633	776	887	938
El Paso		326	445	532	552
Ft Quitman		44	138	214	238
Conchos Ojinaga	trib	800	892	844	694
Presidio/Ojinaga		843	1,075	1,081	952
Johnson Rch/Santa Elena		920	1,122	1,126	1,001
Foster Rch/Santa Rosa		n/a	1,458	1,442	1,313
Pecos Langtry	trib	n/a	274	252	211
Devils Pafford Xng	trib	n/a	329	366	330
Amistad Dam		2,160	2,133	2,158	1,970
San Diego Jimenez	trib	174	196	205	184
Eagle Pass/Piedras Negras		2,450	2,505	2,615	2,488
Escondido Villa de Fuente	trib	51	65	69	72
Laredo/Nuevo Laredo		2,778	2,848	2,917	2,672
Salado Las Tortillas	trib	435	409	359	348
Falcón Dam		n/a	2,900	2,915	2,587
Alamo Mier	trib	155	131	90	72
San Juan Camargo	trib	511	449	294	226
Rio Grande City/Camargo		3,657	3,469	3,263	2,911
Anzalduas Dam		2,142	2,118	1,937	1,651
San Benito/Ramirez		1,218	1,173	1,040	655
Brownsville/Matamoros		1,057	1,039	915	514

* mainstem station unless specifically identified as a tributary, as in this case

† data record too short to allow a valid average

American West. Control of Rio Grande waters at Elephant Butte reservoir was one of the earliest projects of the Bureau. As mentioned, it served as a dry run for larger projects on the Colorado.

In the US part of the basin, Colorado, New Mexico, and Texas share water under the Rio Grande Compact (53 Stat. 785). The compact was agreed in 1939 and has not been changed since. Colorado and New Mexico are represented on the compact commission by their state engineers, while the Texas commissioner is appointed by the governor of Texas. The chairman of the commission is appointed by the president of the United States. The compact provides for the apportionment of Rio Grande water above Fort Quitman, Texas, and provides for the operation and maintenance of about twenty gauging stations above Fort Quitman. The amount of water Colorado has to deliver to New Mexico, and New Mexico must deliver to Texas, is fixed as a function of the water flowing at defined river gauges.

Created by the 1906 convention between Mexico and the United States (see next section), the International Boundary and Water Commission (IBWC) is responsible for the construction, maintenance and operation of international dams and other water infrastructure, water accounting between the two countries, diplomatic problem solving (the minute system), flood control, and technical investigations. IBWC is organized in Mexican and US sections, headquartered in Cd. Juárez and across the river in El Paso. Both sections report to their respective ministries of foreign affairs. Conchos waters are overseen by the Mexican Comisión Nacional del Agua (Conagua).

Intense competition for river water in the LRGV has found a unique solution on the Texas side of the border. The courts appointed a Rio Grande Water Master to settle differences and allocate water received from the IBWC to irrigation districts and cities, both under normal and drought conditions. The water master is now an agent of the state of Texas.

At the local level, farmers are organized in irrigation districts. These are nonprofit organizations with an elected board and appointed managers responsible for allocating water to individual farmers and maintaining water distribution systems. City water utilities are responsible for management of urban water and treatment of wastewater.

14.3.2 Laws, Agreements, Compacts, Treaties

Competition for surface and groundwater prevails throughout much of the basin. In the Upper Rio Grande water rights are contested between irrigation, growing cities (specifically Albuquerque, New Mexico), and Indian tribes. Colorado, New Mexico, and Texas share water under the Rio Grande Compact (see earlier). Mexico and the United States share Rio Grande water under two international treaties.

Bi-national management of the Rio Grande began with the Convention of 1906. Under this agreement the United States allocates 60,000 ac-ft/yr (74 Mm^3/year) to Mexico to be used for irrigation in the Cd. Juárez area. In exchange, Mexico withdrew its longstanding complaints about limited access to Upper Rio Grande water. An existing International Boundary Commission, responsible for marking the border between the two countries, saw its authority expanded to manage river waters on the border with Mexico – from the Gulf of Mexico to California. The water mandate became the principal function of the reformed agency, as shown by the requirement that the two commissioners, one from each country, must be professional engineers. To this day the International Boundary and Water Commission (IBWC) functions as the most important bi-national institution on the US–Mexico border.

The Treaty of 1944 stipulates that Mexico, in exchange for receiving United States water in the Colorado basin, transfers to the US portion of Rio Grande water no less than 350,000 ac-ft/yr (432 Mm^3/yr, 36 Mm^3/mo), averaged over 5-year cycles. The purpose of the 5-year cycle averaging in the Treaty is to allow years of deficit delivery to be made up in years of excess flow. There are also complicated provisions to allow Mexico to carry over deficits into additional 5-year cycles.

While the 1944 treaty does not grant the IBWC authority to control water quality, this has changed gradually. The treaty allows for rulemaking by the IBWC, called the minute process. Under this system the Mexican and US sections of the IBWC study pressing water issues and submit recommendations for action. Over time, 324 Minutes were issued, covering a variety of issues, including construction of new reservoirs, salinity, flood control, and, in at least two dozen cases, sanitation (water quality). Each minute, at the end of sometimes year-long study and discussion, is signed by the American and Mexican commissioner of the IBWC, submitted to the two governments, and becomes part of the bi-national water management system if neither government raises objections within thirty days after execution. The vast majority of IBWC Minutes were approved by the two governments. This rule-making power works slowly but efficiently to address emerging issues. The system bypasses the difficulty of formal confirmation of international agreements by the US Senate.

14.3.3 Water Pricing

Agricultural water is heavily subsidized. Mexico does not put a price on water. In the United States, irrigation districts must repay a share of government subsidies for the construction of storage reservoirs. Once this debt has been paid, farmers are charged small amounts for the administrative costs of irrigation districts.

14.3.4 Conservation and Increased Efficiency

Education, water pricing, and model projects to advance these goals are in their infancy. The current use of more than 80 percent of river water for irrigation can be reduced substantially by improvements in water distribution and irrigation techniques. Water metering, changes in crop patterns, more realistic water pricing, and storage of flood waters need to be implemented on a large scale. Cities need to reduce losses from aging water pipes.

The potential for doing more with less is great. An irrigation district in the LRGV convinced their members to replace open air distribution ditches with underground pipes. Farmers reduced their water use by 40 percent, but still harvested the same value of crops. The same district did better than its neighbors when water supply was curtailed during a drought period.

14.3.5 Water Planning

In the United States, water planning is a function of the states. In Mexico, it is a responsibility of the Comisión Nacional del Agua (Conagua). In their water planning both countries consider changes from reservoir sedimentation, population growth, and changes in land use. Climate change, on the other hand is hardly considered. For example, the 2012 State Water Plan for Texas acknowledges that climate change and climate variability both pose challenges to water planning, because they add uncertainty. To meet the challenge, the Texas Water Development Board monitors climate science for applicability to the Texas water planning process (TWDB, 2012). In Texas, legislation passed in 1996 created regional water planning groups with membership of all stakeholders. One group each covers the Texas portions of the Paso del Norte and the LRGV.

14.4 WATER USE

14.4.1 Water Budgeting

The combination of natural hydrology and the influences of man can be usefully analyzed by constructing a water budget for a river. This is basically a quantitative accounting of the movement of water through a region of space over some period of time. A complete water budget includes both "transfers" and "stocks" of water in the system, that is, the movement, withdrawal, and discharge of water and the volumes accumulated in various storage units of the water system, e.g., the river, reservoirs, soil, and the atmosphere. For present purposes, a focus on the transfers alone is sufficient to depict the factors relevant to the disposition of water without considering storage. It is analogous to analyzing the income and outgoings in a bank account to determine whether it is being managed in a sustainable way, which does not require knowledge of the actual balance. In the case of the Rio Grande, separate water budgets are necessary for the upper and lower reaches of the river, as divided at Fort Quitman, see Section 14.2. This defines the regions of the water budget. For specificity, we chose the 1981–2010 averaging period as the time dimension of the budget. The procedure then is to identify the forms, locations, and physical states of the occurrence of water in the accounting region and the processes that transfer water from one to another, finally acquiring estimates of the magnitudes and directions of these transfers based on measurements or model analyses. A number of water budgets have been formulated for regions of the Rio Grande basin by past workers, and these were exploited for data. Not all such data are consistent with the space–time delineation sought in the present budget, which introduces a degree of error, or uncertainty, in the analysis. This is a common liability of a budget analysis of a distributed reactive physical variable, like water.

A summary water budget for the upper Rio Grande is given in Table 14.6, and for the lower Rio Grande in Table 14.7. The broad categories at the top of the water balance are "accounts," which are physical components of water "residence." For Table 14.6, the surface water account, the groundwater account, or the M&I account are the categories. Anything that provides water to a category is an "inflow" to that category, and any discharge or removal of water is an "outflow." For the upper reach, these are surface water, groundwater, and M&I usage. The lines of the table correspond to processes effecting the transfers. There is a category of transfer called M&I, the fourth line entry for each of the subreaches in the budget, which is the diversion or wastewater return associated with M&I usage. (So why is M&I also an account category? Because of the lack of imagination of the writers, who probably should have spent time considering account names, like "city tanks," or "brokers and authorities," but did not.) Each of these tables is an abridgement of a more detailed water budget, including citations for sources of data, which are given in the appendix to this chapter.

Because each water budget begins at the start of the reach with a flow measurement, and ends at the bottom of the reach with a flow measurement, it is desirable to close the internal surface water balances by accounting for all identifiable gains and losses. For the upper Rio Grande, the intimate exchanges between river and aquifer are not measured directly. From computational models of these aquifers, it is possible to infer the values of these exchanges. An example of this is the entry "Channel losses (seepage and recharge)" in the second and third subreaches. This represents the net exchange between river and aquifer. For the second subreach, it is equal between the two (i.e., balanced), for the third subreach it is not. The sums at the bottom of the water budget ("Totals") indicate the degree of balance in the budget. The surface component is always balanced because there is an unmeasured transfer determined by the difference among the others, the "unaccounted" term. For the lower Rio Grande, we note that this agrees satisfactorily with the estimated flux from aquifer to river given in Anaya et al. (2016).

14.4.2 Categories of Water Use

Agriculture uses 80–88 percent of river water. Irrigation districts in the Paso del Norte use groundwater during drought periods. The same strategy does not work in the LRGV where groundwater is of poor quality. Groundwater resources have been studied extensively, jointly by the IBWC and state water agencies. Yet conjunctive water planning and management of groundwater does not exist.

As recently as 2002, the sole source of water for municipal and industrial use in the middle Rio Grande valley was the bolsón aquifers of the Rio Grande (Bartolino and Cole, 2002). In particular, the entire water supply for the city of Albuquerque was groundwater. Cities and industry use 12 percent of river water. El Paso and Cd. Juárez meet urban demand almost entirely from groundwater. The main aquifer – Hueco Bolsón – has been over pumped and is expected to run dry by 2025. A secondary aquifer – Mesilla Bolsón – has mostly brackish water. El Paso Water Utilities, with help from the federal government, has built the largest land-based desalinization plant to use Mesilla water for drinking water. Both Hueco and Mesilla aquifers underlie New

Table 14.6 *Water budget for the Upper Rio Grande*

	Surface water		*Groundwater*		*M & I*	
Transfers & processes	*Inflows*	*Outflows*	*Inflows*	*Outflows*	*Inflows*	*Outflows*
Colorado reach (to Lobatos)						
Rio Grande at Del Norte (gauged)	784					
Tributaries & ungauged inflows	146					
Irrigation diversion and pumpage		543		285		
M & I water use				25	25	
Return flows	29		67			12
Unaccounted inflows in reach	5					
New Mexico reach (to Caballo Dam)						
Tributaries & ungauged inflows	1,523					
Irrigation diversion		321				
Irrigation pumpage						
M & I water use		76		650	226	
Return flows to river	93					93
Recharge & return flows			259			
Channel losses (seepage and recharge)		364	364			
Lake evaporation		173				
Unaccounted losses in reach		167				
Texas/Chihuahua reach						
Channel losses Caballo to El Paso		383				
Irrigation diversions		449				
Drains	110			110		
M & I water use (US & Mexico)		57		210	307	
Return flows to river	18					18
Channel losses (seepage and recharge)		183	287			
Unaccounted inflows in reach	245					
Flow at Fort Quitman (gauged)		238				
Totals	2,954	2,954	977	1,279	558	123

Mexico, West Texas, and parts of the Mexican state of Chihuahua. Mexico has started an ambitious program to tap additional aquifers.

Ever since the beginning of the century, horizontal drilling and fracking of shale oil and gas have caused sharply increased water demand on the US side of the LRGV.

Environmental flow has been substantially reduced by the heavily engineered river system. The floods and high flow pulses reaching the Paso del Norte and the LRGV are diminished. Increased water demand in the Rio Conchos has dramatically reduced the late summer/early fall high flow pulses entering the Rio Grande. As a result, significant ecological damage occurred in both upper and lower Rio Grande.

14.5 CHALLENGES

14.5.1 Siltation

The Rio Grande basin, particularly the higher elevations with their young mountains, are a prolific source of sediments, and siltation in the basin, particularly in the New Mexico reach, has been a historic problem. Several reservoirs, e.g., Chochiti, have been constructed for the express purpose of flood control and sediment retention.

The US Bureau of Reclamation has studied Elephant Butte sedimentation since 1915, a year before the dam was completed. By now a dataset covering almost a century is available. The 1999 survey reported a 23.3 percent loss of storage capacity since closure. The report projected additional 5 percent losses for future 14-year intervals. The 2007 survey corrected these findings as follows: "Since the last reservoir survey in 1999, the reservoir volume has increased 1,228 acre-feet (1.5 million m^3) due to the dewatering and resulting compaction of the previous measured sediments that have been exposed during the extended drought conditions. The average annual rate of sediment accumulation since 1915 is 6,575.6 acre-feet (8.1 million m^3) compared to the 1999 study computation of 7,253.2 acre-feet (8.9 million m^3)." Thus, between 1915 and 2007, 23.16 percent of storage capacity was lost (Bureau of Reclamation, 2008).

Table 14.7 *Water budget for the Lower Rio Grande*

	Surface water		*Irrigation*		*M & I*	
Transfers & processes	*Inflow*	*Outflow*	*Diversion*	*Return*	*Intake*	*Discharge*
Fort Quitman to Foster Ranch						
Rio Grande at Fort Quitman	238					
Tributary inflows	694					
Net channel gains in reach	364					
Unaccounted inflows	17					
Foster Ranch to Amistad Dam						
Tributary inflows	541					
Reservoir evaporation		537				
Storage trend		39				
Unaccounted inflows	691					
Amistad Dam to Falcón Dam						
Tributary inflows	604					
Net channel gains in reach	462					
Diversions & returns for irrigation	982	1,068	1,068	982		
M & I withdrawals and returns	31	123			123	31
Reservoir evaporation		479				
Storage trend		28				
Unaccounted inflows in Falcón	320					
Falcón Dam to Gulf of Mexico						
Tributary inflows	298					
Net channel losses in reach		314				
Diversions & returns for irrigation	76	2,171	2,171	76		
M & I withdrawals and returns	8	55			55	8
Rio Grande at Brownsville (gauged)		514				
Totals	*5,328*	*5,328*	*3,239*	*1,059*	*178*	*39*
US diversions Brownsville to Gulf of Mexico		124	124			
Rio Grande at Gulf of Mexico, estimated		391				

In the Lower Rio Grande, the greatest siltation occurs in the upstream reservoir – Amistad. From the closure of Amistad in 1968 through 1992, when the lake was extensively surveyed by the IBWC, 760,800 acre-feet (938 M m^3) had been lost to storage in the combined Falcon–Amistad system, about 12.5 percent of conservation capacity, of which 95 percent is in Amistad. Projected to the present, the loss of conservation capacity due to siltation is about 22 percent.

Available data document an annual storage volume loss of 0.25 percent in Elephant Butte. The Amistad loss is in the range of 0.5 to 0.6 percent (IBWC and Texas Water Development Board data). The measured plus projected loss for Elephant Butte (1915–2060) amounts to 36.5 percent. For Amistad (1968–2060) the loss will amount to 55.2 percent. It is possible that Amistad will lose somewhat less in future years because some siltation from the main tributary, the Río Conchos, is caught by recently built reservoirs in Mexico. Even so, reservoir losses in the 40 percent range are highly significant for both economic impact regions – the PdN and the LRGV rates of siltation after their construction, though this has ameliorated in recent years primarily due to interception in reservoirs upstream or in the Conchos tributary.

14.5.2 Climate Change Impacts

The Rio Grande basin has been plagued with multi-year droughts once or twice each century. The region has been identified in the IPCC model runs as one in which human-induced climate change will be exhibited as both increases in air temperature and decreases in rainfall. According to the IPCC, storm tracks reaching the American Southwest are moving northward. This will bring less precipitation to the Rio Grande basin. On the other hand, occasional storms from the Gulf of Mexico or the Pacific may become more frequent. Throughout the basin evaporation losses will increase as the temperature rises. Timing and volume of upstream snowfall have already changed over the last twenty years. Taking these factors together, there will be less river water in most years.

According to a 2013 study by the US Bureau of Reclamation and Sandia National Laboratory, climate change will make it difficult for New Mexico to implement Rio Grande Compact rules that guarantee a prescribed flow of water in the Rio Grande Basin to Texas. Climate change will cause reduced flows due to higher rates of evaporation and smaller expected snowpacks (Bureau of Reclamation, 2013). Texas water officials are suing New Mexico over water deliveries related to the Rio Grande Compact.

14.5.3 Water Quality

Natural water quality in the basin is variable. Water deriving from the snowpack in the upper basin and from seasonal runoff is typically of good quality, as is water from the Rio Conchos. Pecos water, particularly at low flows, is saline, due to interflow from a briny aquifer cut by the river in eastern New Mexico. The development of irrigated agriculture in the upper Rio Grande and in the Lower Rio Grande Valley has impacted the water quality in these respective reaches.

Mexican communities on the Rio Grande for many years released untreated wastewater. The single largest number of IBWC minutes is devoted to improving sanitation. Under side agreements to the North American Free Trade Agreement (NAFTA), the United States helped Mexico to build primary and secondary treatment plants. Tertiary treatment is still a way off in Mexico, as is sewage treatment in small Mexican communities.

14.5.4 Groundwater

In the Middle Rio Grande, Albuquerque has reduced its use of groundwater to 41 percent. The cities of Las Cruces, El Paso, and Cd. Juárez depend entirely on groundwater. In 1998 the IBWC published a study of the aquifers underlying the Paso del Norte (IBWC, 1998). The study predicted "the depletion of recoverable freshwater reserves of the bi-nationally shared aquifers by the middle half of the 21st century." Many years earlier, in the 1960s, Minute 242 had addressed United States concerns about Mexican groundwater pumping. The Minute was primarily concerned with shared groundwater in Arizona, but briefly addressed the groundwater future of the entire border region. It was agreed that the two governments shall consult with each other prior to developing new surface or groundwater projects that might adversely affect the other country. This is as far as things have come. Even today, there is no comprehensive agreement on groundwater. No IBWC Minute has ever focused on groundwater in the Paso del Norte. In 2006, the United States passed the United States–Mexico Transboundary Aquifer Assessment Act. The lead agencies are the US Bureau of Reclamation and the Water Resource Research Center at the University of Arizona. What little has been accomplished to date under the act is focused on aquifers in Arizona and Sonora.

14.5.5 Conflicts

Texas is suing New Mexico before the US Supreme Court, alleging that New Mexico interferes with water deliveries to Texas that are prescribed under the Rio Grande Compact. Specifically, Texas asks that New Mexico stop pumping groundwater along the border of the two states so that more of the river could flow south to farmers and residents in the Paso del Norte. Texas alleges that New Mexico water users are improperly intercepting (i.e., diverting or pumping) surface water and groundwater within New Mexico that reduces Rio Grande Project allocations to the Texas project beneficiary, El Paso County Water Improvement District No. 1. The federal government has also weighed in, arguing in its motion to intervene in the case that groundwater pumping in New Mexico is tapping the shallow aquifer that would otherwise drain back into the Rio Grande and flow to Texas.

New Mexico has appealed the case, arguing that the US District Courts, not the Supreme Court, should settle the case. In January 2014, the Supreme Court rejected the New Mexico appeal and the case continues before the Supreme Court.

In the Lower Rio Grande, the arid lower reach of the basin, recent droughts have resulted in a deficit of the treaty-mandated minimum delivery from Mexico to the United States of 432 Mm^3/yr (350,000 ac-ft/year). On average, the Mexican tributaries contribute about 80 percent of the flow into this reach of the Rio Grande, of which the Conchos is the most important.

For many years Mexico delivered more Conchos water to the Rio Grande than required by the 1944 water treaty. Exceptional drought and increased water demand by Mexican farmers and cities caused Mexico to fall behind in Conchos deliveries during two five-year delivery cycles from 1992 to 2002. As mentioned earlier, the treaty allows water debts to be repaid during five-year cycles. The 1990s drought led to a severe water conflict between the two countries

In 1995, IBWC Minute 293 stated that "the current storage of waters belonging to Mexico [in Conchos reservoirs] would be just sufficient to cover needs for Mexico through June 1996" (IBWC, 1995). Recognizing the threat to Mexican communities, Minute 293, in accordance with Article 9(f) of the Treaty, authorized Mexico to use some of the waters belonging to the United States that were stored in Amistad and Falcón reservoirs. Minute 308, agreed to in 2002, proposed new ways to deal with drought management. The two governments followed up on the IBWC recommendation and Mexico paid back its water debt in 2005.

14.6 CONCLUSION

How can the Rio Grande / Bravo cope with increasing water scarcity? To be on the safe side, water managers must plan for a Rio Grande by mid-century with 40–50 percent less water and doubled population. These are some of the measures they should consider:

- Dredge reservoirs to recover lost storage volume. This is, however, an expensive option, perhaps more so than building new reservoirs.
- Build new reservoirs. Unfortunately, the best places have been already been dammed. The Brownsville Weir and the Laredo Lower Weir, both located in the Lower Rio Grande Valley, have been on the planning board for many years. Both are controversial and fairly small.
- Capture flood waters. This has already been practiced to some extent following heavy rains several years ago, but more can be done. In particular, raising the levels of existing dams may be the most promising strategy.
- Desalinize brackish groundwater and salty sea water. El Paso has started the Rio Grande region along this road, with strong support from the federal government. The process is energy-intensive and getting rid of the salty brine is expensive. It will mostly help to improve urban water supply. Large cities can afford it. Small cities and irrigation districts will probably balk at the cost.
- Reduce irrigated acreage. Urbanization and market forces will reduce irrigated land by about 10 percent by mid-century.
- Improve maintenance of urban water systems. City distribution systems are losing up to 40 percent of water due to leaks from aging pipes. Repairs and maintenance should be higher on the list of city priorities.
- Use water more efficiently. The single most promising response to future water scarcity is conservation. One irrigation district in the Lower Valley estimates that farmers will be able to maintain the value of their harvest with 40 percent less water. Lining of canals and irrigation ditches, water metering, drip irrigation, leveling of land, and choice of less water-intensive crops are promising approaches. More realistic water pricing would help but is fiercely resisted by farmers and irrigation districts.
- For the Lower Rio Grande, IBWC/CILA should develop a bi-national Rio Grande Sustainability Plan, review progress in five-year intervals, and update management strategies. The Rio Grande Compact states should develop a similar plan for the Upper Rio Grande.
- Encourage stakeholders in sub-basins to develop drought management plans for their stretch of the river.

A prudent mix of these measures will make it possible to maintain *economic sustainability*. Irrigated agriculture, the lynch pin of the Rio Grande economy, and the riparian cities can be sustained. *Social sustainability* can be improved by reducing poverty. Improving the *ecological health* of the river with its plants and animals will be most difficult. The environment has suffered since dams were built and the inner valley was converted to irrigated agriculture. In all likelihood, ecology will suffer more in the future.

REFERENCES

Action Committee of the Middle Rio Grande Water Assembly (ACMRGWA) (1999). *Middle Rio Grande Water Budget (Where Water Comes from, & Goes, & How Much), Averages for 1972–1997*, Albuquerque, NM: ACMRGWA.

Adams, D. and Comrie, A. (1997). The North American Monsoon. *Bulletin of the American Meteorological Society*, 78(10), pp. 2197–2213.

Adams, D. and Keller, G. (1994). Crustal Structure and Basin Geometry in South-Central New Mexico. In G. Keller and S. Cather, eds., *Basins of the Rio Grande Rift: Structure, Stratigraphy, and Tectonic Setting*. Special Paper 291, Boulder, CO: Geological Society of America, pp. 241–255.

Anaya, R., Boghici, R., and French, L. et al. (2016). *Texas Aquifers Study: Groundwater Quantity, Quality, Flow, and Contributions to Surface Water*. Special report to Legislature, Austin: Texas Water Development Board.

Atwood, W. W. (1918). *Relation of Landslides and Glacial Deposits to Reservoir Sites*. Bulletin 685, US Geological Survey. Washington, DC: Government Printing Office.

Baldridge, W. and Olsen, K. (1989). The Rio Grande Rift: What Happens When the Earth's Lithosphere Is Pulled Apart? *American Scientist*, 77 (3), pp. 240–247.

Baldridge, W., Damon, P., Shafiqullah, M., and Bridwell, R. (1980). Evolution of the Central Rio Grande Rift, New Mexico: New Potassium-Argon Ages. *Earth & Planetary Science Letters*, 51, pp. 309–321.

Baldwin, B. and Kottlowski, F. (1968). *Santa Fe: Scenic Trips to the Geologic Past No. 1*, 2nd ed., Socorro: State Bureau of Mines & Mineral Resources.

Barry, R. and Carleton, A. (2001). *Synoptic and Dynamic Climatology*. New York: Routledge.

Bartolino, J. and Cole, J. (2002). *Ground-Water Resources of the Middle Rio Grande Basin*. Circular 1222, Denver, CO: US Geological Survey.

Benke, A. and Cushin, C. (2005). *Rivers of North America*. Burlington, MA: Elsevier Academic Press.

Bexfield, L. M. (2010). Section 11: Middle Rio Grande Basin, New Mexico. In Thiros et al., eds., *Conceptual Understanding and Groundwater Quality of Selected Basin-Fill Aquifers in the Southwestern United States*.

National Water-Quality Assessment Program: USGS, pp. 189–218.

Bexfield, L. and Anderholm, S. (2010). Section 10: San Luis Valley, Colorado, and New Mexico. In Thiros et al., eds., *Conceptual Understanding and Groundwater Quality of Selected Basin-Fill Aquifers in the Southwestern United States*. National Water-Quality Assessment Program: USGS, pp. 165–187.

Bjorklund, L. and Maxwell, B. (1961). *Availability of Ground Water in the Albuquerque Area, Bernalillo and Sandoval Counties, New Mexico*. Technical Report 21, New Mexico State Engineer, Santa Fe.

Blake, E., Gibney, D., and Brown, M. et al. (2009). *Tropical Cyclones of the Eastern North Pacific Basin, 1949–2006*. Historical Climatology Series 6-5, Asheville, NC: US National Oceanic and Atmospheric Administration.

Blakey, R. and Ranney, W. (2018). *Ancient Landscapes of Western North America: A Geologic History with Paleogeographic Maps*. Cham, Switzerland: Springer International Publishing.

Bluestein, H. B. (1993). *Synoptic-Dynamic Meteorology in Midlatitudes. Vol II: Observations and Theory of Weather Systems*. New York: Oxford University Press.

Bogener, S. D. (1993). *Carlsbad Project*. Albuquerque: Bureau of Reclamation.

(1997). *Ditches across the Desert: A Story of Irrigation along New Mexico's Pecos River*. Ph.D. dissertation: Texas Tech University, Lubbock.

Booker, J., Michelsen, A., and Ward, F. (2005). Economic Impact of Alternative Policy Responses to Prolonged and Severe Drought in the Rio Grande Basin. *Water Resources Research*, 41, https://doi.org/10.1029/2004WR003486

Brand, D. D. (1937). *The Natural Landscape of Northwestern Chihuahua*. Albuquerque: The University of New Mexico Press.

Bryan, K. (1938). Geology and Ground-Water Conditions of the Rio Grande Depression in Colorado and New Mexico. In *US Natural Resources Planning Board, The Rio Grande Joint Investigation in the Upper Rio Grande Basin in Colorado, New Mexico, and Texas 1936–1937*. Washington, DC: Government Printing Office, pp. 197–225.

Bureau of Reclamation, US Department of Interior (2008). *Elephant Butte Reservoir: 2007 Reservoir Survey*. Washington, DC: Government Printing Office.

(2013). *Impacts of Climate Change on the Upper Rio Grande Basin: Adaptation and Mitigation Strategies*. Washington, DC: Government Printing Office.

Castro, C., McKee, T., and Pielke, R. (2001). The Relationship of the North American Monsoon to Tropical and North Pacific Sea Surface Temperatures as Revealed by Observational Analyses. *Journal of Climate*, 14, pp. 4449–4473.

Chapin, C. E. (2008). Interplay of Oceanographic and Paleoclimate Events with Tectonism during Middle to Late Miocene Sedimentation across the Southwestern USA. *Geosphere*, 4 (6), pp. 976–991. https://doi.org/10.1130/GES00171.1

Chapin, C. and Cather, S. (1994). Tectonic Setting of the Axial Basins of the Northern and Central Rio Grande Rift. In G. Keller and S. Cather, eds., *Basins of the Rio Grande Rift: Structure, Stratigraphy, and Tectonic Setting*. Special Paper 291, Boulder, CO: Geological Society of America, pp. 5–25.

Chapin, C. and Seager, W. (1975). *Evolution of the Rio Grande Rift in the Socorro and Las Cruces Areas*. Guidebook, 26th Field Conference, Socorro: New Mexico Geological Society.

Clark, I. G. (1987). *Water in New Mexico*. Albuquerque: University of New Mexico Press.

Cliet, T. (1969). Ground-Water Occurrence of the El Paso Area and Its Related Geology. In D. Cordoba, S. Wengerd, and J. Shomaker, eds., *The Border Region (Chihuahua, Mexico & USA)*. 20th Annual Fall Field Conference Guidebook, New Mexico Geological Society, pp. 209–214.

Connell, S., Hawley, J., and Love, D. (2005). Late Cenozoic Drainage Development in the Southeastern Basin and Range of New Mexico, Southeasternmost Arizona, and Western Texas. In S. Lucas, G. Morgan, and K. Zeigler, eds., *New Mexico's Ice Ages*. Science Bulletin No. 28, Socorro: New Mexico Museum of Natural History and Science, pp. 125–150.

Cook, E., Seager, R., Cane, M., and Stahle, D. (2007). North American Drought: Reconstructions, Causes, and Consequences. *Earth-Science Reviews*, *81*, 93–134.

Cooper, D., Sanderson, J., Stannard, D., and Groeneveld, D. (2006). Effects of Long-Term Water Table Drawdown on Evapotranspiration and Vegetation in an Arid Region Phreaphyte Community. *Journal of Hydrology*, 325, pp. 21–34.

Corbosiero, K., Dickinson, M., and Bosart, L. (2009). The Contribution of Eastern North Pacific Tropical Cyclones to the Rainfall Climatology of the Southwest United States. *Monthly Weather Review*, 137, pp. 2415–2435.

Cronin, T. M. (2010). *Paleoclimates*. New York: Columbia University Press.

Davis, J., Hudson, M., and Grauch, V. (2017). A Paleomagnetic Age Estimate for the Draining of Ancient Lake Alamosa, San Luis Valley, South-Central Colorado, U.S.A. *Rocky Mountain Geology*, 52(2), pp. 107–117. https://doi.org/10.24872/rmgjournal.52.2.107

Dethier, D. P. (1999). Quaternary Evolution of the Rio Grande near Cochiti Lake, Northern Santo Domingo Basin, New Mexico. In F. Passaglia and S. Lucas, eds., *Albuquerque Geology*. New Mexico Geological Society 50th Annual Fall Field Conference Guidebook, pp. 371–378.

Dickerson, P. E. (2013). Tascotal Mesa Transfer Zone: An Element of the Border Corridor Transform System, Rio Grande Rift of West Texas and Adjacent Mexico. In M. Hudson and V. Grauch, eds., *New Perspectives on Rio Grande Rift Basins: from Tectonics to Groundwater*. Special Paper 494, Geological Society of America, pp. 475–500.

Dickinson, W. R. (2004). Evolution of the North American Cordillera. *Annual Review of Earth and Planetary Sciences,* 32, pp. 13–45.

Dinatale Water Consultants (2015). *Rio Grande Implementation Plan*. Boulder, CO.

Douglas, M., Maddox, R., and Howard, K. (1993). The Mexican Monsoon. *Journal of Climate*, 6, pp. 1665–1677.

Duryea, E. and Haehl, H. (1916). *A Study of the Depth of Annual Evaporation from Lake Conchos, Mexico*. Paper No. 1376, Transactions ASCE 80, pp. 1829–2060.

Eastoe, C., Hawley, J., Hibbs, B., Hogan, J., and Hutchison, W. (2010). Interaction of a River with an Alluvial Basin Aquifer: Stable Isotopes, Salinity, and Water Budgets. *Journal of Hydrology*, 395, pp. 67–78.

Everitt, B. (1991). Channel Responses to Declining Flow on the Rio Grande between Ft. Quitman and Presidio, Texas. *Geomorphology*, 6, pp. 225–242.

Ewing, T. E. (2016). *Texas through Time: Lone Star Geology, Landscapes, and Resources*. Austin: Bureau of Economic Geology, University of Texas.

Fischer, A. F. (1981). Climatic Oscillations in the Biosphere. In M. H. Nitecki, ed., *Biotic Crises in Ecological and Evolutionary Time*. New York: Academic Press, pp. 103–131.

Fox, D. G., Jamison, R., Potter, D. U., Valett, H. M., and Watts, R. (1995). Geology, Climate, Land, and Water Quality. In D. Finch and J. Tainter, eds., *Ecology, Diversity, and Sustainability of the Middle Rio Grande Basin*. Report RM-GTR-268, Ft. Collins, CO: Forest Service, US Department of Agriculture, pp. 52–79.

Galloway, W., Whiteaker, T., and Ganey-Curry, P. (2011). History of Cenozoic North American Drainage Basin Evolution, Sediment Yield, and Accumulation in the Gulf of Mexico Basin. *Geosphere*, 7(4), pp. 938–973.

George, P., Mace, R., and Petrossian, R. (2011). *Aquifers of Texas*. Report 380, Austin: Texas Water Development Board.

Gile, L., Hawley, J., and Grossman, R. (1981). *Soils and Geomorphology in the Basin and Range Area of Southern New Mexico: Guidebook to the Desert Project*. Memoir 39, Socorro: New Mexico Bureau of Mines & Mineral Resources.

Graham, A. (1999). *Late Cretaceous and Cenozoic History of North American Vegetation North of Mexico*. New York: Oxford University Press.

Griggs, J. (1907). *Mines of Chihuahua, History, Geology, Statistics, Mining Companies*. Chihuahua City.

Hawley, J. W. (1969). Notes on the Geomorphology and Late Cenozoic Geology of Northwestern Chihuahua. In D. Cordoba, S. Wengerd, and J. Shomaker, eds., *Guidebook of the Border Region*. Twentieth Field Conference, Socorro: New Mexico Geological Society, pp. 131–142.

Hawley, J. and Haase, C. (1992). *Hydrogeologic Framework of the Northern Albuquerque Basin*. Open-file Report 387, Socorro: New Mexico Bureau of Mines and Mineral Resources.

Hawley, J. and Kennedy, J. (2004). *Creation of a Digital Hydrogeologic Framework Model of the Mesilla Basin and Southern Jornada del Muerto Basin*. Technical Completion Report 332, New Mexico State University, Las Cruces: New Mexico Water Resources Research Institute.

Hawley, J. and Kernodle, M. (1999). *Overview of the Hydrogeology and Geohydrology of the Northern Rio Grande Basin: Colorado, New Mexico, and Texas*. Forty-fourth Annual New Mexico Water Conference, Santa Fe, Las Cruces: New Mexico Water Resources Research Institute.

Hearne, G. and Dewey, J. (1988). *Hydrologic Analysis of the Rio Grande Basin North of Embudo, New Mexico, Colorado and New Mexico*. Water-Resources Investigations Report 86-4113, Denver, CO: US Geological Survey.

Herzog, L. A. (1990). *Where North Meets South: Cities, Space, and Politics on the U.S.–Mexico Border*. Austin: Center for Mexican-American Studies, University of Texas.

Hibbs, B. and Darling, B. (2005). Revisiting a Classification Scheme for U.S.-Mexico Alluvial Basin-Fill Aquifers. *Ground Water*, 43 (5), pp. 750–763.

Higgins, R. and Shi, W. (2000). Dominant Factors Responsible for Interannual Variability of the Summer Monsoon in the Southwestern United States. *Journal of Climate*, 13, pp. 759–776.

Hill, R. T. (1900). *Physical Geography of the Texas Region*. Folio 3, Topographic Atlas of the United States. Washington, DC: US Geological Survey.

Hufstetler, M. and Johnson, L. (1993). *Watering the Land: The Turbulent History of the Carlsbad Irrigation District*. Rocky Mountain Region, Denver, CO: National Park Service.

Hutchins, W. (1928). The Community Acequia: Its Origins and Development. *Southwestern Historical Quarterly*, 31(3), pp. 261–184.

IBWC (1998). *Transboundary Aquifers and Binational Ground Water Database for the City of El Paso/Ciudad Juárez Area*. Available at www.ibwc.state.gov/Water_Data/binational_waters.htm.

(1995). Minute 293: Emergency Cooperative Measures to Supply Municipal Needs of Mexican Communities Located Along the Rio Grande Downstream of Amistad Dam.

Ingersoll, R. V. (2001). Structural and Stratigraphic Evolution of the Rio Grande Rift, Northern New Mexico, and Southern Colorado. *International Geology Review*, 43, pp. 867–891.

Kammerer, J. C. (1990). *Largest rivers in the United States*. Open-file Report 87-242 (revised 1990), Reston, VA: US Geological Survey.

Keller, G. and Baldridge, W. (1999). The Rio Grande Rift: A Geological and Geophysical Overview. *Rocky Mountain Geology*, 34(1), pp. 121–130.

Keller, G. and Cather, S. (1994). Introduction. In G. Keller and S. Cather, eds., *Basins of the Rio Grande Rift: Structure, Stratigraphy, and Tectonic Setting*. Special Paper 291, Boulder, CO: Geological Society of America, pp. 1–3.

Kelley, P. (1986). *River of Lost Dreams: Navigation on the Rio Grande*. Lincoln: University of Nebraska Press.

Kelley, V. C. (1969). *Albuquerque: Its Mountains, Valley, Water, and Volcanoes*. Scenic Trips to the Geologic Past No. 9, Socorro: New Mexico Institute of Mining & Technology.

(1977). *Geology of Albuquerque Basin, New Mexico*. Memoir 33, Socorro: New Mexico Bureau of Mines & Mineral Resources.

Kelley, S. and Chamberlin, R. (2012). Our Growing Understanding of the Rio Grande Rift. *New Mexico Earth Matters*, 12(2), pp. 1–4. New Mexico Bureau of Geology and Mineral Resources.

Ladwig, W. and Stensrud, D. (2009). Relationship between Tropical Easterly Waves and Precipitation during the North American Monsoon. *Journal of Climate*, 22, pp. 258–271.

Land, L. (2016). *Overview of Fresh and Brackish Water Quality in New Mexico*. Open-file Report 583, Socorro: New Mexico Bureau of Geology and Mineral Resources.

Langman, J. and Anderholm, S. (2004). *Effects of Reservoir Installation, San Juan-Chama Project Water, and Reservoir Operation on Streamflow and Water Quality in the Rio Chama and Rio Grande, Northern and Central New Mexico, 1938–2000*. Scientific Investigations Report 2004-5188, Denver, CO: US Geological Survey.

Llewellyn, D. and Vaddey, S. (2013). *West-wide climate risk assessment: Upper Rio Grande impact assessment*. Upper Colorado Region, Albuquerque: Bureau of Reclamation.

Lyle, M., Barron, J., and Bralower, T. et al. (2008). Pacific Ocean and Cenozoic Evolution of Climate. *Reviews of Geophysics*, 46. https://doi.org/10.1029/2005RG000190

Lyle, M., Heusser, C., and Ravelo, M. et al. (2012). Out of the Tropics: The Pacific, Great Basin Lakes, and Late Pleistocene Water Cycle in the Western United States. *Science*, 337, pp. 1629–1633.

Mace, R., Mullican, W., and Angle, E. (2001). *Aquifers of West Texas*. Report 356, Austin: Texas Water Development Board.

Machette, M., Marchetti, D., and Thompson, R. (2007). Ancient Lake Alamosa and the Pliocene to Middle Pleistocene Evolution of the Rio Grande. In M. Machette, M. Coates, and M. Johnson, eds., *Rocky Mountain Section Friends of the Pleistocene Field Trip–Quaternary Geology of the San Luis Basin of Colorado and New Mexico*. Open File Report 2007-1193, Reston, VA: US Geological Survey, pp. 157–167.

Madden, R. and Julian, P. (1994). Observations of the 40-50-Day Tropical Oscillation: A Review. *Monthly Weather Review*, 122, pp. 814–837.

Mann, P. (2007). Overview of the Tectonic History of Northern Central America. In P. Mann, ed., *Geologic and Tectonic Development of the Caribbean Plate Boundary in Northern Central America*. Special Paper 428, Geological Society of America, pp. 1–19.

Martinez-Sanchez, J. and Cavazos, T. (2014) Eastern Tropical Pacific Hurricane Variability and Landfalls on Mexican Coasts. *Climate Research*, 58, pp. 221–234.

McAdie, C., Landsea, C., and Neumann, J. et al. (2009). *Tropical Cyclones of the North Atlantic Ocean, 1851–2006*. Historical Climatology Series 6-2, Asheville, NC: US National Oceanic and Atmospheric Administration.

Murphy, E. C. (1905). Failures of Lake Avalon Dam, near Carlsbad, N. M. *Engineering News*, 54(1), pp. 9–10.

Ortega-Ramírez, J. Urrutia-Fucugauchi, J. and Valiente-Banuet, A. (2000). The Laguna de Babícora Basin: A Late Quaternary Paleolake in Northwestern Mexico. In E. Gierlowski-Kordesch and K. Kelts, eds., *Lake Basins through Space and Time*. Studies in Geology 46, American Association of Professional Geologists, pp. 569–580.

Passaglia, F. J. (2005). River Responses to Ice Age (Quaternary) Climates in New Mexico. In S. Lucas, G. Morgan, and K. Zeigler, eds., *New Mexico's Ice Ages*. Science Bulletin No. 28, New Mexico Museum of Natural History and Science, pp. 115–124.

Pazzaglia, F. and Hawley, W. (2004). Neogene (Rift Flank) and Quaternary Geology and Geomorphology. In G. Mack and K. Giles, eds., *The Geology of New Mexico, a Geological History*. Special Publication 11, Socorro: New Mexico Geological Society, pp. 407–438.

Perdigón-Morales, J., Romero-Centeno, R., Barrett, B., and Ordoñez, P. (2019). Interseasonal Variability of Summer Precipitation in Mexico: MJO Influence on the Midsummer Drought. *Journal of Climate*, 32, pp. 2313–2327.

Perez-Arlucea, M., Mack, G., and Leeder, M. (2000). Reconstructing the Ancestral (Plio-Pleistocene) Rio Grande in Its Active Tectonic Setting, Southern Rio Grande Rift, New Mexico, USA. *Sedimentology*, 47, pp. 701–720.

Petrie, M., Collins, S., Gutzler, D., and Moore, D. (2014). Regional Trends and Local Variability in Monsoon Precipitation in the Northern Chihuahuan Desert, USA. *Journal of Arid Environments*, 103, pp. 63–70.

Plummer, L. L., Bexfield, S., Anderholm, S., Sanford, W., and Busenberg, E. (2004). Hydrochemical Tracers in the Middle Rio Grande Basin, USA: 1. *Conceptualization of Groundwater Flow. Hydrogeology*, 12, pp. 359–388. https://doi.org/10.1007/s10040-004-0324-6

Rango, A. (2006). Snow: The Real Water Supply for the Rio Grande Basin. *New Mexico Journal of Science*, 44, pp. 99–118.

Rasskazov, S., Yasnygina, T., Fefelov, N., and Saranina, E. (2010). Geochemical Evolution of Middle-Late Cenozoic Magmatism in the Northern Part of the Rio Grande Rift, Western United States. *Russian Journal of Pacific Geology*, 4(1), pp. 13–40.

Repasch, M., Karlstrom, K., Heizler, M., and Pecha, M. (2017). Birth and Evolution of the Rio Grande Fluvial System in the Past 8 Ma: Progressive Downward Integration and the Influence of Tectonics, Volcanism, and Climate. *Earth-Science Reviews*, 168, pp. 113–164.

Ricketts, J., Kelley, S., and Karlstrom, K. et al. (2016). *Synchronous Opening of the Rio Grande Rift along Its Entire Length at 25-10 Ma Supported by Apatite (U-Th)/He and Fission-Track Thermochronology, and Evaluation of Possible Driving Mechanisms*. GSA Bulletin 128 (3/4), pp. 397–424.

Riecker, R. E., ed. (1979). *Rio Grande Rift: Tectonics and Magmatism. Special Publication*. Washington, DC: American Geophysical Union.

Ritchie, E., Wood, K., Gutzler, D., and White, S. (2011). The Influence of Eastern Pacific Tropical Cyclone Remnants on the Southwestern United States. *Monthly Weather Review*, 139, pp. 192–210.

Rivera, J. A. (1998). *Acequia Culture: Water, Land, and Community in the Southwest*. Albuquerque: University of New Mexico Press.

Robson, S. and Banta, E. (1995). *Arizona, Colorado, New Mexico, Utah*. Section HA 730-C of Ground water atlas of the United States. Denver, CO: US Geological Survey. Available at https://water.usgs.gov/ogw/aquifer/atlas.html

Ruleman, C., Hudson, A., and Thompson, R. et al. (2019). Middle Pleistocene Formation of the Rio Grande Gorge, San Luis Valley, South-Central Colorado and North-Central New Mexico, USA: Process, Timing, and Downstream Implications. *Quaternary Science Reviews*, 223, https://doi.org/10.1016/j.quascirev.2019.07.028

Sandoval-Solis, S. (2015). *Effect of Extreme Storms on Treaty Obligations in the Rio Conchos*. Research Report, University of California, Davis: Water Management Research Laboratory.

Schmandt, J., North, G. R., and Ward, G. H. (2013). How Sustainable Are Engineered Rivers in Arid Lands? *Journal of Sustainable Development of Energy, Water and Environment Systems*, 1(2).

Schmidt, J., Everitt, B., and Richard, G. (2001). Hydrology and Geomorphology of the Rio Grande and Implications for River Rehabilitation. In G. Garrett and N. Allan, eds., *Aquatic Fauna of the Northern Chihuahuan Desert*. Special Publications 46, Lubbock: Museum of Texas Tech University, pp. 25–45.

Schmidt, R. H. (1986). Chihuahuan Climate. In J. Barlow, A. Powell, and B. Timmermann, eds., *Chihuahuan Desert: U.S. and Mexico II*. Alpine, Texas: Chichuahuan Desert Research Institute, Sul Ross State University, pp. 40–63.

Schuyler, J. D. (1909). *Reservoirs for Irrigation, Water-Power and Domestic Water-Supply*, 2nd ed. New York: John Wiley & Sons.

Scurlock, D. (1998). *From the Rio to the Sierra: An Environmental History of the Middle Rio Grande Basin*. General Technical Report RMRS-GTR-5. Fort Collins, CO: US Department of Agriculture, Forest Service, Rocky Mountain Research Station.

Seager, R. (2007). The Turn of the Century North American Drought: Global Context, Dynamics and Past Analogs. *Journal of Climate*, 20, pp. 5527–5552.

Seager, R., Kushnir, Y., Herweijer, C., Naik, N., and Velez, J. (2005). Modeling of Tropical Forcing of Persistent Droughts and Pluvials over Western North America: 1856–2000. *Journal of Climate*, 18, pp. 4065–4088.

Seager, W., Shafiqullah, M., Hawley, J., and Marvin, R. (1984). New K-Ar Dates from Basalts and the Evolution of the Southern Rio Grande Rift. *Geological Society of America Bulletin*, 95, pp. 87–99.

Senay, G., Schauer, M., and Velpuri, N. et al. (2019). Long-Term (1986–2015) Crop Water Use Characterization over the Upper Rio Grande Basin of United States and Mexico Using Landsat-Based Evapotranspiration. *Remote Sensing Journal*, 11, pp. 1587–1611. https://doi.org/10.3390/rs11131587

Sheppard, P., Comrie, A., Packin, G., Angersbach, K., and Hughes, K. (2002). The Climate of the US Southwest. *Climate Research 21*, 219–238.

Stahle, D. W., Cook, E. R., and Burnette, D. J. et al. (2016). The Mexican Drought Atlas: Tree-Ring Reconstructions of the Soil Moisture Balance during the Late Pre-Hispanic,

Colonial, and Modern Eras. *Quaternary Science Reviews*, 149, pp. 34–60.

Stearns, C. E. (1953). Tertiary Geology of the Galisteo-Tonque Area, New Mexico. *Bulletin of the Geological Society of America*, 64, pp. 459–508.

Strain, W. S. (1971). Late Cenozoic Bolson Integration in the Chihuahua Tectonic Belt. In J. Hoffer, ed., *The Geologic Framework of the Chihuahua Tectonic Belt*. Publication 71-59, West Texas Geological Society, pp. 167–173.

Texas Water Development Board (TWDB) (2012). 2012 State Water Plan. Available at www.twdb.texas.gov/waterplanning/swp/2012/ind

Thiros, S., Bexfield, L., Anning, D., and Huntington, J., eds. (2010). *Conceptual Understanding and Groundwater Quality of Selected Basin-Fill Aquifers in the Southwestern United States*. Professional Paper 1781, Reston, VA: US Geological Survey.

Tipton, R., Mills, E., and Woods, B. et. al. (1942). *Pecos River Joint Investigation–Summary, analysis, and findings*. Part X, Regional Planning, National Resources Planning Board. Washington, DC: Government Printing Office.

Titus, F. B. (1961). Ground-Water Geology of the Rio Grande Trough in North-Central New Mexico, with Sections on the Jemez Caldera and the Lucero Uplift. In S. Northrop, ed., *New Mexico Geological Society 12th Annual Fall Field Conference Guidebook*, pp. 186–192.

Trenberth, K. and Branstator, G. (1992). Issues in Establishing Causes of the 1988 Drought over North America. *Journal of Climate*, 5, 159–172.

Umoff, A. A. (2008). An Analysis of the 1944 U.S.–Mexico Water Treaty: Its Past, Present, and Future. *Environs*, 32, pp. 69–98.

Ward, G. H. (2004). Texas Water at the Century's Turn: Perspectives, Reflections, and a Comfort Bag. In J. Norwine, J. Giardino, and S. Krishnamurthy, eds., *Water for Texas*. College Station: Texas A&M Press, pp. 17–43.

Weiss, J., Castro, C., and Overpeck, J. (2009). Distinguishing Pronounced Droughts in the Southwestern United States: Seasonality and Effects of Warmer Temperatures. *Journal of Climate*, 22, pp. 5918–5932.

Wilkins, D. W. (1986). *Geohydrology of the Southwest Alluvial Basin's Regional Aquifer—Systems Analysis, Parts of Colorado, New Mexico, and Texas*. Water-Resources Investigations Report 84-4224, Albuquerque, NM: US Geological Survey.

Wilson, B., Lucero, A., Romero, J., and Romero, P. (2003). *Water Use by Categories in New Mexico Counties and River Basins, and Irrigated Acreage in 2000*. Technical Report 51, Santa Fe: New Mexico Office of the State Engineer.

Wilson, L. (1999). *Surface Water Hydrology of the Rio Grande Basin*. New Mexico Water Rights Course, Forty-fourth Annual New Mexico Water Conference, Santa Fe, Las Cruces: New Mexico Water Resources Research Institute.

Wozniak, F. E. (1998). *Irrigation in the Rio Grande Valley, New Mexico: A study and Annotated Bibliography of the Development of Irrigation Systems*. Report RMRS-P-2. Fort Collins, CO: Rocky Mountain Research Station, Forest Service, US Department of Agriculture.

Zachos, J., Pagani, M., Sloan, L., Thomas, E., and Billups, K. (2001). Trends, Rhythms, and Aberrations in Global Climate 65 Ma to Present. *Science*, 292(5517), pp. 686–693.

Appendix

Table 14.A1 *Water budget and related data for the upper Rio Grande, CY 1981–2010 means where available*

COLORADO REACH INCLUDING SAN LUIS CLOSED BASIN

Watershed areas etc.			*Climate*			*Comments*
Rio Grande Del Norte	3,420	km^2	Air temp	4.3	°C	NOAA CO Div 5 81-10 mean
San Luis (Alamosa) closed basin	7,600	km^2	Annual precip	45.5	cm	NOAA CO Div 5 81–10 mean
Rio Grande Lobatos	12,328	km^2	Annual lake evap	145.0	cm	Bexfield & Anderholm (2010)
Irrigated area	2,100	km^2				Dinatale (2015)
Evapotranspiration over all irrigated areas	810	$Mm^3\ yr^{-1}$				Senay et al. (2019), averaged 1986-2015. Cf 1694 from Hearne & Dewey (1988).
Reservoirs total conservation storage	390	Mm^3				
Population	75,000					

	Surface water		*Surface- ground mix*		*M&I*		
Transfers in $Mm^3\ yr^{-1}$	*Inflows*	*Outflows*	*Inflows*	*Outflows*	*Intake*	*Return*	*Comments*
Rio Grande at Del Norte (gauged)	784						
Inflow Conejos	146						
Drain from "sump"	17			17			Closed Basin Project recent yields Dinatale (2015)
M&I use	12			25	25	12	250 gal per cap
River diversion for irrigation		543					Split 2:1, same proportion as found by Hearne & Dewey (1988)
Channel losses (evaporation)		7					Hearne & Dewey (1988)
Estimated irrigation entire basin				267			See River diversion
Irrigation return flow to aquifers			67				Assumed 25%. Hearne & Dewey (1988) estimate 95% returned, seems implausible.
Unaccounted inflow	12						
Rio Grande at Lobatos (gauged)		422					
Totals	971	971	67	310	25	12	
Depleted over reach				mined 243	consumed 13		

LOBATOS TO OTOWI

Watershed areas etc.			*Climate*			*Comments*
Rio Grande Lobatos	12,328	km^2	Air temp	8.3	°C	NOAA NM Div 2 81-10 mean
Otowi	29,422	km^2	Annual precip	45.3	cm	NOAA NM Div 2 81–10 mean
Reservoirs total conservation storage	242	Mm^3	Annual lake evap			

Table 14.A1 (*cont.*)

Transfers in Mm3/y	*Surface water* *Inflows*	*Surface water* *Outflows*
Rio Grande at Lobatos	422	
net inflow between Lobatos & Taos	273	
Red River inflow	42	
Embudo Creek inflow	77	
Rio Chama inflow	527	
unaccounted inflow	10	
Rio Grande at Otowi Bridge		1,350
Totals	1,350	1,350

OTOWI TO CABALLO

Watershed areas etc.:			*Climate*			*Comments*
Rio Grande Otowi	29,422	km^2	Air temp	13.7	°C	NOAA NM Div 5 81–10 mean
Caballo dam	70,495	km^2	Annual precip	23	cm	NOAA NM Div 5 81–10 mean
Irrigated area	1,190	km^2	Annual lake evap	220	cm	Evap Elephant Butte IBWC
Reservoirs conservation storage	2,955	Mm^3		330	cm	Wilson et al. (2003; evap Caballo, IBWC
Population	700,000					Bartolino & Cole (2002)

Transfers in $Mm^3\ yr^{-1}$	*Surface water* *Inflows*	*Surface water* *Outflows*	*Shallow aquifer* *Inflows*	*Shallow aquifer* *Outflows*	*Deep aquifer* *Inflows*	*Deep aquifer* *Outflows*	*M&I water uses* *Intake*	*M&I water uses* *Discharge*	*Comments*
Rio Grande at Otowi (gauge)	1,350								USGS
Inflow from San Juan-Chama project	76	76					76		Bartolino & Cole (2002) state 55 taf. BuRec (p A-18 in Llewelyn & Vaddey, 2013) states 62 taf avg.
Jemez d/s Jemez Dam inflow	58								USGS
Rio Puerco inflow	26								USGS
Other gauged tributaries	167								Bartolino & Cole (2002)
Albuquerque storm inflow	6								ACMREGWA (1999) for 72-97
Wastewater return flow	93							93	Bartolino & Cole (2002)
Surface seeps & springs	271			271					ACMREGWA (1999) for 72–97
Recharge to shallow aquifer		364	364						ditto
Open-water evaporation (incl ag irrgn)		74							ditto
Irrigated agriculture & valley floor turf ET		123							ditto
Riparian evapotranspiration				167					ditto

Transfers in $Mm^3\ yr^{-1}$	Surface water		Shallow aquifer		Deep aquifer		M&I water		Comments
	Inflows	*Outflows*	*Inflows*	*Outflows*	*Inflows*	*Outflows*	*Intake*	*Discharge*	
Mountain-front recharge					136				ditto
Groundwater underflow into basin					49				ACMREGWA (1999) for 72–97
Flux deep to shallow aquifer			62			62			ditto
Pumping from deep aquifer						150	150		According to p A-25 in App A of Llewelyn & Vaddey (2013), ABCWUA has reduced its pumping from 110 taf/y to "near half." So est reducing 170 to 120 Taf
Drainage from septic fields			12					12	ACMREGWA (1999) for 72–97
Open water, riparian & irrigation ET d/s Puerco		123							ditto
Lake evaporation, Elephant Butte & Caballo		173							ditto
Unaccounted losses over reach		177							
Rio Grande at Caballo dam (gauge)		938							IBWC
Totals	2,048	2,048	438	438	185	212	226	105	
Depleted over reach		159				mined 27		consumed 121	

CABALLO DAM TO FORT QUITMAN

Watershed areas etc.			*Climate*			*Comments*
Caballo dam	70,495	km^2	Air temp	17.8	°C	NOAA TX Div 5 81-10 mean
Fort Quitman	82,734	km^2	Annual precip	34.2	cm	81–10 mean
Irrigated area U.S. & Mexico	310	km^2				IBWC. Heywood & Yager (2003) report 227.
Irrigation water applied to U.S. per year	272	Mm^3				1.2 m in Heywood & Yager (2003)
U.S. Population	600,000					IBWC
Mexico population	1,500,000					estimated

Transfers in $Mm^3\ yr^{-1}$	Surface water		Shallow aquifer		Deep aquifer		M&I water uses		Comments
	Inflows	*Outflows*	*Inflows*	*Outflows*	*Inflows*	*Outflows*	*Intake*	*Discharge*	
Elephant Butte/Caballo (gauge)	938								IBWC
Drains to Rio Grande	110			110					Heywood and Yager (2003)
Return El Paso treated wastewater	18	57					57	18	IBWC; cf. Heywood & Yager, p. 21
Return flow Juarez wastewater									n/a
Losses from Caballo to El Paso		383							Gauge difference
Diversion American Canal (gauge)		383							IBWC
Diversion Acequia Madre Cd. Juarez		67							IBWC

Table 14.A1 (*cont.*)

	Surface water		Shallow aquifer		Deep aquifer		M&I water uses		
Transfers	*Inflows*	*Outflows*	*Inflows*	*Outflows*	*Inflows*	*Outflows*	*Intake*	*Discharge*	*Comments*
Underflow from Tularosa to Hueco					14				Heywood and Yager, p. 13
Underflow from Mesilla to Hueco					0.1				Heywood and Yager, p. 13
Rio Grande seepage		183	183						Heywood and Yager, Fig 18
Infiltration of irrigation return water			68		14				Shallow = 25% of applied irrigation water, Heywood & Yager
Mountain-front recharge of aquifer					8				Heywood and Yager, p. 12
U.S. pumping (El Paso & military)						70	70		ca 1997 in Heywood and Yager Fig. 5
Mexico pumping						140	140		ca 1997 in Heywood and Yager Fig. 5
El Paso from Mesilla basin	& treated RG water							40	Heywood and Yager
Unaccounted inflow	245								Includes seepage from shallow aquifer into river S of El Paso
Rio Grande at Fort Quitman		238							IBWC
Totals	1311	1311	251	110	36	210	307	18	
Depleted over reach						Mined 174		289	Return flows not accounted

Table 14.A2 *Water budget and related data for the Lower Rio Grande, CY 1981–2010 means where available*

FORT QUITMAN TO FOSTER RANCH (BIG BEND)

Watershed areas etc.			*Climate*			*Comments*
Ft Quitman	82,734	km^2	Air temp	17.8	°C	NOAA TX Div 5 81-10 mean
Rio Conchos	68,387	km^2	Precip	30.1	cm	NOAA TX Div 5 + Chihuahua D 81–10 mean
Foster Ranch	209,120	km^2	Evaporation	242	cm	Ojinaga, IBWC (2006)
Irrigated area	501	km^2				IBWC (2006), 407 upstream on Conchos basin
Reservoirs, conservation storage	3,857	Mm^3				All on Conchos

	River channel		*Irrigation*		*M&I usage*		
Transfers in Mm^3/y	*Inflow*	*Outflow*	*Diversion*	*Return*	*Intake*	*Discharge*	*Comments*
Rio Grande Ft Quitman (gauged)	238						IBWC
Channel losses to Presidio		37					
Rio Conchos inflow (gauged)	694						IBWC
Channel gains Presidio to Foster Ranch	401						
unaccounted inflow	17						
Rio Grande Foster Ranch (gauged)		1,313					IBWC
Totals	*1,350*	*1,350*					

AMISTAD RESERVOIR

Watershed areas etc.			*Climate*			*Comments*
Pecos + Devils Rivers	82,734	km^2	Air temp	18.2	°C	NOAA TX Div 5 + 6 81–10 mean
Amistad dam	209,120	km^2	Precip	48.1	cm	NOAA TX Div 5 + 6 81–10 mean
Irrigated area	501	km^2	Evaporation	267	cm	IBWC (2006)
Reservoirs conservation storage	4,922	Mm^3				Amistad + reservoirs on Pecos

	River channel		*Irrigation*		*M&I usage*		
Transfers in Mm^3/y	*Inflow*	*Outflow*	*Diversion*	*Return*	*Intake*	*Discharge*	*Comments*
Rio Grande Foster Ranch (gauged)	1,313						IBWC
Pecos River (gauged)	211						IBWC
Devils River (gauged)	330						IBWC
Reservoir evaporation		537					IBWC
Unaccounted inflows in Amistad	691						
Storage trend in reservoir (1981-2010)		39					
Rio Grande below Amistad Dam (gauged)		1,970					IBWC
Totals	*2,545*	*2,545*					

AMISTAD DAM TO FALCÓN DAM (WINTER GARDEN AND FALCÓN RESERVOIR)

Watershed areas etc.			*Climate*			*Comments*
Pecos + Devils Rivers	82,734	km^2	Air temp	18.2	°C	NOAA TX Div 5 + 6 81-10 mean
Falcón Dam	412,506	km^2	Air temp	22.2	°C	NOAA TX Div 9 81–10 mean
Mexican tributaries	65,969	km^2	Precip	59.5	cm	NOAA TX Div 9 81–10 mean
Irrigated area	426	km^2	Evaporation	274	cm	IBWC (2006), about 2/3 U.S. Winter Garden; 1/3 on Rio Salado
Reservoirs conservation storage	4,164	Mm^3				Falcón + Carranza

Table 14.A2 (*cont.*)

	River channel		*Irrigation*		*M&I usage*		
Transfers in Mm³/y	*Inflow*	*Outflow*	*Diversion*	*Return*	*Intake*	*Discharge*	*Comments*
Rio Grande below Amistad dam (gauged)	1,970						IBWC
Channel gains to Del Rio	96						
Cd. Acuna M&I + Del Rio return	5	9			9	5	IBWC
Maverick Canal diversions (gauged)		1,068	1,068				IBWC
Rio San Diego inflow (gauged)	184						IBWC
Channel gains Del Rio to Quemado	232						
Maverick Power Plant return	936			936			IBWC
Maverick ID misc drains above Eagle Pass	31			31			IBWC
Channel gains Quemado to Eagle Pass	108						
Piedras Negras M&I + Eagle Pass return	15	18			18	15	IBWC
Rio Escondido inflow (gauged)	72						IBWC
Maverick ID misc drains below Eagle Pass	16			16			IBWC
Channel losses Eagle Pass to Indio		24					
Channel gains Indio to Laredo	51						
Nuevo Laredo M&I + Laredo return	11	97			97	11	IBWC
Rio Salado inflow (gauged)	348						IBWC
Reservoir evaporation		479					IBWC
Storage trend in Falcón (1981-2010)		28					
Unaccounted inflows in Falcón	320						
Rio Grande below Falcón Dam (gauged)		2,672					IBWC
Totals	*4,395*	*4,395*	*1,068*	*982*	*123*	*31*	

FALCÓN DAM TO GULF OF MEXICO (LOWER RIO GRANDE VALLEY)

Watershed areas etc.			*Climate*			*Comments*
Pecos + Devils Rivers	82,734	km²	Air temp	18.2	°C	NOAA TX Div 5 + 6 81-10 mean
Gulf of Mexico	460,000	km²	Air temp	23.5	°C	Watershed estimated; NOAA TX Div 10 81–10 mean
Mexican tributaries	37,854	km²	Precip	61.6	cm	Alamo + San Juan; NOAA TX Div 10 81–10 mean
Irrigated area	5,186	km²	Evaporation	217	cm	IBWC (2006), about evenly split between U.S. & Mexico
Reservoirs conservation storage	1,905	Mm³				Gomez + Cuchillo

	River channel		*Irrigation*		*M&I usage*		
Transfers in Mm³/y	*Inflow*	*Outflow*	*Diversion*	*Return*	*Intake*	*Discharge*	*Comments:*
Rio Grande below Falcón Dam (gauged)	2,672						IBWC
Rio Alamo inflow (gauged)	72						IBWC
Lower Rio San Juan ID returns	7			7			IBWC
Rio San Juan inflow (gauged)	226						IBWC
US diversions above Rio Grande City		13	13				IBWC
Channel losses to Rio Grande City		53					
Lower Rio San Juan ID returns	70			70			IBWC
US diversions RG City to Anzalduas		251	251				IBWC
Diversions to Mexico at Anzalduas		964	964				IBWC
Channel losses RG City to Anzalduas		114					
US diversions Anzald to Progreso		252	252				IBWC
US diversions Progreso to San Benito		560	560				IBWC
Channel losses Anzalduas to San Benito		184					
US diversions San Benito to Brownsville		131	131				IBWC
M&I: Brownsville return + Matamoros diversion	8	55			55	8	IBWC
Channel gains San Benito to Brownsville	37						
Rio Grande at Brownsville (gauged)		514					IBWC
Totals	*3,091*	*3,091*	*2,171*	*76*	*55*	*8*	

Table 14.A2 (*cont.*)

US diversions Brownsville to Gulf of Mexico	124	124	IBWC
Rio Grande at Gulf of Mexico, estimated	391		

Notes

1 This physiographic feature is also referred to as the Rio Grande valley (e.g., Hill, 1900; Kelley, 1977; Gile et al., 1981), the Rio Grande depression (e.g., Bryan, 1938; Stearns, 1953), and the Rio Grande trough (e.g., Titus, 1961; Baldwin and Kottlowski, 1968; Baldridge and Olsen, 1989).

2 A distinction is often made between a bolsón, whose drainage is strictly internal, and a basin, which is intercepted by a through-flowing stream. In this chapter, following Hill (1900) and Hawley (1969), the terms are synonymous. On another matter of terminology, "basin" describes the watershed of an entire river or major component of its drainage network. It also refers to a lowered plain enclosed by highlands, created by extension of the earth's crust. The former is a feature of surface-water drainage. The latter is a geological structure. The two uses of the term are different, and the intended meaning should be clear from context.

15 The Jucar River Basin*

Jose Albiac, Taher Kahil, and Encarna Esteban

15.1 INTRODUCTION

Pressures on water resources have been mounting worldwide with water scarcity becoming a widespread problem in arid and semi-arid regions around the world. Global water extractions have climbed from 600 to 3,800 km^3 per year in the last century, which is above the rate of population growth (WWC, 2000). The degradation of water resources is a common threat to human water security and environmental biodiversity across the world. Large investments in developed countries ensure human security, but the threats to natural ecosystems are hardly accounted for (Vörösmarty et al., 2010). The water governance problem gravitates around the agricultural sector because most water resources are used for irrigation.

Irrigation is a key component of agricultural production, covering 20 percent of cultivated land and generating 40 percent of global food production (CAWMA, 2007). Irrigation covers 310 million hectares of land with the largest acreage located in Asia and the Americas (Siebert et al., 2013). Irrigation demand for water is close to 2,750 km^3 per year, of which 1,900 km^3 are surface water diversions and 850 km^3 are groundwater extractions. The construction of dams for irrigation has been reduced during recent decades, and most dams at present are being built for hydropower (Winemiller et al., 2016). The development of irrigation in recent decades has been based on the enormous expansion of groundwater extractions. Between 1960 and 2010, groundwater extractions from all sectors climbed from 300m^3 to 1,000 km^3 per year pushing depletion up to 150 km^3 (IGRAC, 2010; Wada et al., 2010; Konikow, 2011).

Water reforms are needed because groundwater extractions and surface water diversions are causing severe water scarcity and water quality problems, with substantial damages to human activities and natural ecosystems. Also, water quality impairment results from urban, industrial, and agricultural pollution loads of organic matter, heavy metals, nutrients, pesticides, and salinity.

Massive ecosystem damage in basins such as the Ganges, Indus, Nile, Yellow, Yangtze, Amu and Syr Darya, Tigris, Euphrates, Murray–Darling, Colorado, and Rio Grande (WWAP, 2006) call for a reconsideration of current water management, institutions, and policies, leading to far-reaching water reforms. The scale of the global growing water depletion indicates that water mismanagement is quite common, and that sustainable management of basins is a complex and difficult task. The upcoming water governance problem will be especially acute in arid and semi-arid regions. In these regions the combined effects of human-induced permanent water scarcity and climate change-induced water scarcity and droughts portend unprecedented levels of water resources degradation in the absence of remediating water reforms.

In the coming decades, climate change is going to be an important challenge for agricultural production. This challenge will be especially difficult to harness because global food demand will almost double by 2050 (Alexandratos and Bruisma, 2012), driven by the growth of world population and income. Climate change will increase temperatures and modify the pattern of precipitations, reducing crop yields in both irrigated and rainfed cropland and livestock productivity because of prolonged or extreme changes in temperature. The biological processes underlying the productivity of plants and animals will be negatively affected by increasing numbers of weeds, diseases, and pests along with changes in the development and pollination periods (USDA, 2012; OECD, 2014).

Water resources projections using coupled global hydrological and crop models indicate that crop losses from climate change could be in the range of 20–30 percent by the end of the century (Elliott et al., 2014). Further losses may occur from water resources scarcity in some regions, which will force the reversion of irrigation to rainfed cropland. Changes in precipitation regimes and extreme precipitation will have negative effects on water availability. Precipitation will decrease in midlatitude and subtropical dry regions, reducing renewable surface water and groundwater resources and escalating the competition for water among sectors (IPCC, 2014a).

* This study has been financed by the project INIA RTA2017–00082-00-00 from the Spanish Ministry of Science and Innovation, which includes partial funding from the European Regional Development Fund, and by support funding to the research group ECONATURA from the Government of Aragon. Additional support has been given by the Rockefeller Foundation and the Volkswagen Foundation. Regarding individuals, special assistance has been provided by Llorenç Avellà, Marta García-Mollà, and Carles Sanchis (UPV).

The sustainable management of water resources is quite challenging because of the different types of goods and services provided by water. These goods and services can be classified as private goods, common pool resources, or public goods depending on the degree of exclusion and rivalry in consumption. Treated drinkable water in urban networks is close to a private good (rivalry and exclusion), water in surface watercourses and aquifers is close to a common pool resource (rivalry and nonexclusion), while water sustaining ecosystems comes close to a public good (nonrivalry and nonexclusion) (Booker et al., 2012). The management of water resources is characterized by collective action processes since pure competitive markets cannot account for the common pool and public good characteristics of water. Collective action is needed to account for the externalities linked to the use of water resources, such as ecosystem damage or the depletion of groundwater systems. Finally, collective action is also driven by the technologies employed in water utilization, which involve economies of scale and indivisibilities resulting in natural monopolies (Rausser et al., 2011).

15.2 WATER RESOURCES IN SPAIN AND THE JUCAR RIVER BASIN

Water management has always been an important issue in Spain. Irrigation projects have been undertaken since Roman times, but it was during the twentieth century that the expansion of irrigation acreage accelerated, coupled with a strong process of economic development and industrialization. The consequence has been growing pressure on water resources and the ensuing problems of water scarcity and quality degradation.

In Spain, the average annual rainfall is 680 mm or 340,000 Mm^3, but with large variability across time and space. The average flow in rivers amounts to 110,000 Mm^3 and the groundwater renewable resources are around 30,000 Mm^3. The large time and space variability of water regimes in rivers has led to the construction of dams with a storage capacity close to 50,000 Mm^3 (MIMAM, 2000).

15.2.1 The Mounting Pressures on Water Resources in Spain

Water scarcity in Spain is mainly linked to the enormous development of irrigation, while quality degradation is mainly linked to pollution from urban, industrial, and agricultural sources. Water withdrawals for consumptive uses were 15,500 Mm^3 per year in 1960, at the beginning of a period of strong economic growth (MOPT, 1993). Five decades later, water withdrawals have doubled to 30,100 Mm^3 per year, mainly driven by the expansion of irrigation acreage from 1.8 to 3.5 million hectares and the growth in urban and industrial demands. The distribution of water withdrawals by sector in Spain is presented in Table 15.1. Withdrawals are close to 30,100 Mm^3, covering the demand from irrigation, water supplying companies, and other industrial and service sectors. Household demand is 2,400 Mm^3 with an average price of €1.90/m^3, and industrial and service demand is 4,100 Mm^3 with an average price of €2.50/m^3 for network supplied water. Net irrigation demand is 18,100 Mm^3 and prices are related to the type of agriculture; prices rank between €0.07/m^3 in less profitable irrigation areas and 0.12-€0.30/m^3 in high profitable areas (INE, 2014 and 2016).

Table 15.1 *Water withdrawals and utilization by sector in Spain (2010, Mm^3)*

	Total	Agriculture	Water supply companies	Other sectors
Withdrawals	30,100	22,100	5,400	2,600
Surface	23,900	17,900	3,800	2,200
Groundwater	6,200	4,200	1,600	400
Network losses	5,500	4,000	1,000	500
Utilization	18,100	18,100	2,400	2,100
Agriculture	2,400		2,000	
Households	4,100			
Other sectors				

Source: INE (2014) and Martínez and Hernández (2003). Figures do not include energy production cooling, hydropower, and aquaculture.

Figure 15.1 River basins in Spain.
Source: MITECO (2020)

There are severe water scarcity and quality degradation problems in all basins, except in the northern Ebro and Duero basins, which have lower pressures from human activities (Figure 15.1). The expanding water demands in Segura, Jucar, Guadiana, Guadalquivir, and Tajo required building a substantial dam

Table 15.2 *Streamflows, storage capacity, and environmental flows in basins*

Basin	Natural streamflows (Mm^3/y)	Dam storage capacity (Mm^3)	Minimum environmental flows[a] (m^3/s)
Segura	740	1,300	1
Jucar	1,700	3,000	0,5
Guadiana	4,430	9,600	3,5
Guadalquivir	5,750	8,800	7
Tajo	8,200	11,140	10
Duero	12,200	7,550	116
Ebro	14,600	7,600	107

[a] Environmental flows at river mouth, or at border with Portugal for Tajo and Duero.

storage capacity, around twice the annual streamflows of basins. This has resulted in the conversion of these rivers in almost closed basins, as indicated by the small environmental flows being enforced (Table 15.2).

The main pressures come from agriculture under intensive production systems, urban sprawling, and from tourism along the Mediterranean coast. A surge in aquifer pumping by farmers, which has occurred in recent decades in the southern and eastern basins, has reduced water flows considerably in the Segura, Jucar, Guadiana, and Guadalquivir basins. Competition for water is keen among economic sectors and territories (Albiac et al., 2013). Large surface diversions and groundwater extractions are degrading water resources and creating significant environmental damage. The resulting threats to human water security have been compensated in Spain with important investments in water technologies.

15.2.2 Water Governance Institutions

The institutional organization of water management is based on public administration bodies, the legal system, and water policies. These institutional components are integrated through the interplay between Spanish and European legislation, and the Spanish federal and state governments.

Spanish legislation grants an important role to public administrations at national, basin, regional, and subregional levels. Water management is based on water authorities in each basin, which elaborate and implement the river basin plans. The Ministry of Agriculture and Environment is the federal authority on water resources, and its main tasks are the design and enforcement of water policies and managing federal basin authorities. The state governments oversee urban supply and wastewater treatment, agriculture, land planning, environment protection, and state river basins (Varela and Hernandez, 2010).

The participation of water users is especially strong in agriculture. Agricultural users relying on water from a single common concession must create a water users' association, with their own equity and legal status. Water users' associations are responsible for the management of secondary infrastructure and allocation of water among members.

The approach to water management is institutional and relies on the river basin authorities, taking advantage of the strong tradition of cooperation among stakeholders in Spain dating back centuries. The common pool and public good characteristics of water are the reasons behind this institutional approach based on basin authorities enabling the collective action of stakeholders.

The responses and adaptation to water scarcity and water quality degradation in Spain have been shaped by national water policies of the last twenty years, with large investment proposals costing billions of euros. The main water policies have been the National Irrigation Plan of 2002, the AGUA Project of 2005 to build desalination plants, and the Sanitation Plans of 1995 and 2007 dealing with urban wastewater treatment. The National Irrigation Plan provided subsidies (70 percent) for investments of €4 billion in irrigation technologies to upgrade primary and secondary irrigation networks and parcel irrigation systems. The AGUA project involved investments of €2.4 billion to build desalination plants and expand water supply by 600 Mm^3, of which 300 Mm^3 is for irrigation purposes in the coastal fringe (MIMAM, 2004). The Sanitation Plans investments in tertiary and secondary wastewater treatment plants have amounted to around €12 billion since 1995.

15.2.3 Conversion of the Jucar River into a Closed Basin

The Jucar River Basin (JRB) is in the regions of Valencia and Castilla La Mancha in southeastern Spain. It extends over 22,300 km^2 and covers the area drained by the Jucar River and its tributaries. The basin has an irregular Mediterranean hydrology, characterized by recurrent drought spells and normal years with dry summers.

At present, renewable water resources in the JRB are nearly 1,700 Mm^3/year, of which 930 are surface water and 770 are groundwater. Water extractions are 1,680 Mm^3, close to renewable resources, making the JRB an almost closed water system. Extractions for irrigated agriculture are nearly 1,400 Mm^3. Urban and industrial extractions total 270 Mm^3, and supply households, industries, and services of more than one million inhabitants, located mostly in the cities of Valencia, Sagunto, and Albacete.

The basin includes thirteen reservoirs, the most important of which are the Alarcon, Contreras, and Tous dams. There are two major water distribution canals in the lower Jucar: the Acequia Real canal, which conveys water from the Tous dam to the traditional irrigation districts, and the Jucar–Turia canal, which transfers water from the Tous dam to modern irrigation districts. The irrigated area extends over 190,000 ha, and the main crops grown are rice, wheat, barley, garlic, grapes, and citrus. There are three major irrigation

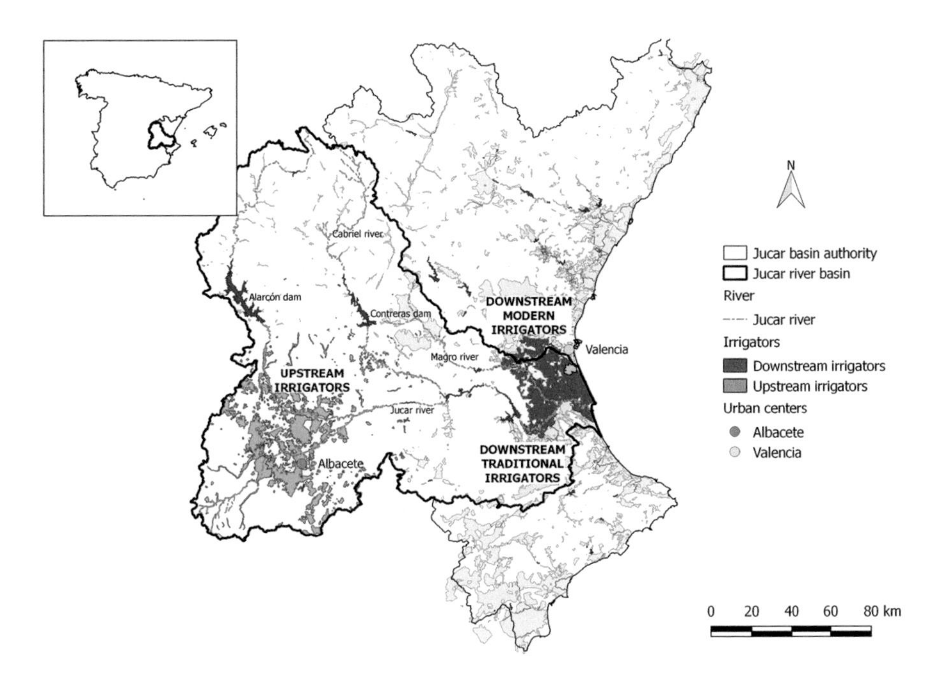

Figure 15.2 The Jucar River Basin.
Figure created by the authors

areas: the irrigation area in the upper Jucar based on groundwater pumping from the Eastern La Mancha aquifer, the downstream traditional irrigation area fed by the Acequia Real canal, and the downstream modern irrigation area fed by the Jucar–Turia canal and located north of traditional irrigators (Figure 15.2).

The expansion of water extractions and the severe drought spells in recent decades have triggered considerable negative environmental and economic impacts in the basin (CHJ, 2015). The growth of water extractions has been driven especially by groundwater irrigation from the Eastern La Mancha aquifer, the largest aquifer system in Spain (Estaban and Albiac, 2012). Intensive groundwater pumping since the 1980s has caused a significant drop in the water table, reaching 80 m depth in some areas, resulting in large storage depletion, fluctuating around 2,500 Mm^3. The Eastern La Mancha aquifer is linked to the Jucar River stream, and it used to feed the river with more than 250 Mm^3/year prior to the 1980s. Due to the depletion, aquifer discharges to the river have declined considerably over the past 30 years, and they are below 50 Mm^3/year at present (Figure 15.3). The consequence is that the lower Jucar is undergoing severe problems of low flows and water-quality degradation, with the riverbed in the middle Jucar being completely dry during recent droughts.

Environmental flows are dwindling in many parts of the basin, resulting in serious damage to water-dependent ecosystems but also significant costs to economic sectors. The minimum environmental flow in the final tract of the Jucar River is 0.5 m^3/s, which shows that the River is becoming a closed basin. Downstream water users are sustaining negative impacts, especially downstream irrigators. For instance, the water available to the Acequia Real irrigation district has been reduced from 700 to 200 Mm^3 in the last 40 years, with damage not only to irrigators but also to the environment because of the fall in irrigation return flows. The Albufera wetland, which is the main aquatic ecosystem in the Jucar River, is mostly fed by these return flows, which are dwindling (Garcia-Molla et al., 2013).

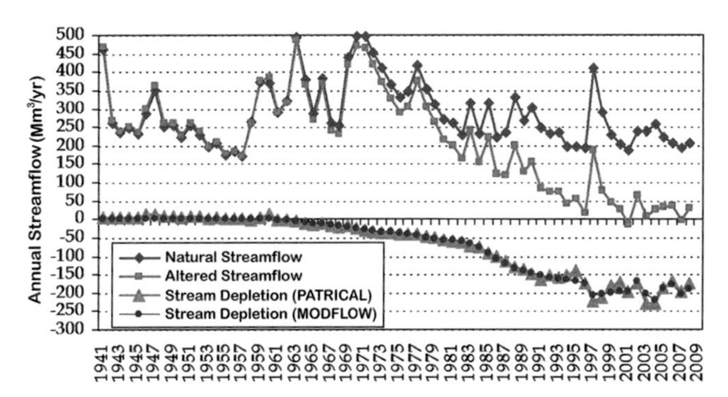

Figure 15.3 Annual streamflows in the Jucar River along La Mancha aquifer. The figure shows the natural and altered annual streamflows in the Jucar River stretch along La Mancha aquifer, and the pumping-induced stream depletion.
Figure adapted from Perez-Martin et al. (2014)

15.3 MANAGEMENT AND POLICIES TO ADDRESS WATER SCARCITY IN THE JUCAR BASIN

To confront progressive water scarcity and water quality degradation in the Jucar Basin and other Spanish basins, a large number of management and policy initiatives have been introduced in recent decades. The list entails massive investment in water technologies, such as modernization of irrigation technologies, urban wastewater treatment plants, seawater desalination plants, and interbasin water transfers. National legislation to promote water markets has also been introduced together with the creation of public water banks in basins, and regulation to facilitate the use of recycled urban wastewater in irrigation and the environment.

15.3.1 The Institutional Approach in the Jucar Basin Authority

The Jucar Basin Authority is the main administrative body responsible for water management not only over the Jucar Basin but also over the Turia Basin and other minor coastal basins (see Figure 15.2). The Basin Authority is organized around governing boards, stakeholder boards, and management services (Figure 15.4). The special characteristic of this institutional approach is the key role played by stakeholders in the Basin Authority. Stakeholders are inside the Basin Authority and include water users from each sector (irrigation, urban, industrial, hydropower), state and federal public administrations, farmers' unions, and environmental groups. The stakeholders' representatives are present in all governing and participation bodies at basin scale and run the watershed boards at local scale. Therefore, the stakeholders are involved in every level of decision making: planning, financing, water allocation and water public domain, waterworks, design and enforcement of measures, and water management at basin and watershed levels (MARM, 2008). The management of water is decentralized, with the Basin Authority in charge of water allocation, and water user associations in charge of secondary infrastructure, water usage, operation and maintenance, investments, and cost recovery. The main advantage of this institutional setting is that stakeholders

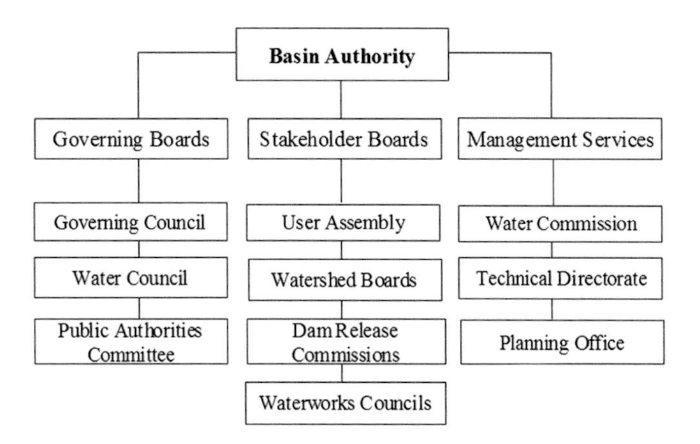

Figure 15.4 Organization of the Jucar Basin Authority.
Figure by the authors

cooperate in decisions, rules, and regulations, and therefore the implementation and enforcement processes are carried out smoothly.

During periods of water scarcity, the watershed boards reduce the level of surface diversions and groundwater extractions assigned to each water user association in the watershed. There is also a drought basin plan based on hydrological indicators to declare a state of alert or full drought, and the measures to be taken to minimize the environmental, economic, and social impacts of drought spells.

15.3.2 Engineering Solutions: Investment in Water Technologies and Public Subsidies

Investment in water technologies during the last two decades in the district covered by the Jucar Basin Authority corresponds mostly to investment in the Valencia region, where disaggregated data are available. The main components are an investment of €1,300 M in urban wastewater treatment plants, €1,000 M in irrigation modernization technologies covering 170,000 ha, €250 M in eight desalination plants for urban supply with a total capacity of 100 Mm^3/year, and €400 M in the Jucar–Vinalopo interbasin water transfer of 80 Mm^3/year. Public subsidies have covered the bulk of financing in wastewater treatment plants, desalination plants, and the Jucar–Vinalopo water transfer. Public subsidies have covered 65 percent of investment in irrigation modernization technologies and the rest has been financed by farmers.

Investment in wastewater treatment plants has considerably reduced the point pollution loads into water media from urban centers. Treated wastewater could be an important resource to confront water scarcity; however, less than one quarter (120 Mm^3) of the treated water is reused in the environment or irrigation because farmers are reluctant to substitute freshwater for urban recycled water.

The large investment in advanced irrigation technologies has improved the efficiency of irrigation at parcel level, but total water consumption at basin level has not been reduced with investment in advanced irrigation technologies. Efficiency gains could increase water consumption because efficient technologies convert a larger share of applied water to consumed water lowering the cost of consumption, so farmers respond by increasing water consumption and irrigated acreage, and changing the crop mix to more water-demanding crops (Scheierling and Treguer, 2016). Even if irrigation withdrawals are maintained, falling irrigation returns reduce basin streamflows. This could be remediated by reducing the water allocation of modernizing districts to avoid the expansion of water consumption, but this type of measure will be opposed by farmers.

Seawater desalination has been an important addition to urban water supply in Alicante, Sagunto, and other urban centers and tourist hubs. But these municipalities are not willing to pay the higher costs of desalination, and they prefer to continue overdrafting surface streams and groundwater, which are obtained at lower costs. These costs are lower because the environmental and opportunity costs of water are disregarded. The desalination facilities are being used at exceptionally low capacity and these public investments are being wasted. The alternative being considered is to subsidize desalinated water with public funds until desalination prices become competitive with current water sources.

Another engineering solution has been interbasin transfer of water between the Jucar and the Vinalopo rivers, completed in 2014. There was an important conflict over the water transfer point of diversion involving the Jucar downstream stakeholders. The initial point of diversion was the middle Jucar, but this would have reduced downstream flows. Finally, the point of diversion was moved from the middle Jucar River to the Jucar mouth. This interbasin water transfer adds further pressure on the almost closed Jucar basin.

15.3.3 Confronting Droughts and the Impending Climate Change with Different Policy Instruments

Severe droughts could have large impacts on agriculture, domestic and industrial users, tourism, and ecosystems. The average cost of drought damage have been estimated at 0.1 percent of gross domestic product (GDP) in the United States and the European Union during normal years (EC, 2007; NOAA, 2008), although the cost of droughts could be exceptionally higher during years of severe drought, up to 1 percent of GDP (Kirby et al., 2014).

Climate change is a major challenge for sustainable agricultural production in the coming decades in arid and semi-arid regions of Spain and other southern European countries. In those regions, climate change will likely increase temperature and evapotranspiration, reduce precipitation and snowmelt, and modify precipitation patterns, impacting negatively on water resources, irrigated and dryland agriculture, and water-dependent ecosystems. This challenge will be difficult to manage in a context of rising world food demand and growing competition between consumptive and environmental water uses (IPCC, 2014b).

The issue of irrigated agriculture adaptation to droughts and climate change has been addressed by many studies in the literature. One section of the literature calls for a reconsideration of water institutions and policies. These new approaches should implement incentive-based policies for an effective uptake of adaptation to more frequent and longer drought events driven by climate change (Zilberman et al., 2002; Booker et al., 2005). Three popular incentive-based policies to address irrigation adaptation to climate change that are widely considered in the literature are water markets, water pricing, and public subsidies for investment in efficient irrigation systems.

Water markets are a good policy alternative to address the economic impacts of climate change (Calatrava and Garrido, 2005; Gohar and Ward, 2010; Montilla-López et al., 2016). Water market benefits during the last drought in the Murray–Darling Basin of Australia, the most active water market in the world, have been around 1 billion US dollars per year (Connor and Kaczan, 2013). Also, the potential water market benefits in California during the recent drought are estimated at 1 billion US dollars per year, assuming that water markets would have been in full operation (Medellin et al., 2013). A challenge to water markets is the third-party effects such as environmental impacts. The reasons are that water markets reduce streamflows because previously unused water allocations are traded, and because gains in irrigation efficiency at parcel level from water trading reduce return flows (Qureshi et al., 2010), as indicated earlier.

Water pricing in irrigation, to achieve water conservation, has been a subject of debate since the 1990s. One section of the literature finds that irrigation water pricing has limited effects on water conservation (Moore, 1991; Scheierling et al., 2004), and some authors indicate that water markets seem far more effective than water pricing for allocating irrigation water (Cornish et al., 2004). Some studies in Spain support those previous findings but also find that water pricing involves disproportionate costs to farmers (Garrido and Calatrava, 2009; Calatrava et al., 2011).

Subsidizing investments in efficient irrigation systems is another important policy for climate change adaptation. The reason is that modernization reduces land abandonment, facilitates the adoption of diversified and high-value cropping patterns, and improves crop yields, leading to an increase in the value of agricultural production. In addition, modernization supports rural development and improves water quality (Playan et al., 2013). However, contrary to widespread expectations, modernization increases water depletion through enhanced crop evapotranspiration and reduction of return flows (Grafton et al., 2018).

Another policy option to be considered is the institutional cooperation approach, where affected stakeholders design the rules and enforcement mechanisms for the allocation of scarce water. This is the policy approach of basin authorities in Spain, although this approach has not received widespread attention in either research or policy circles.

The empirical analysis of these policy options is based on a modeling framework developed for the Jucar River Basin. The hydro-economic river basin model integrates hydrological, economic, institutional, and environmental components, and includes the irrigation, urban, and environmental sectors in the basin. Details of the modeling framework can be found in Esteban et al. (2016) and Kahil et al. (2015a, 2015b, 2016a).

A direct comparison of water markets, water pricing, and institutional policy instruments is made using the hydro-economic model of the Jucar. Table 15.3 shows the economic

Table 15.3 *Policies under drought: institutional, water markets, and water pricing.*

Water availability	Normal year	Drought		
Water policy	Current policy (Institutional cooperation)	Institutional cooperation	Water markets	Water pricing
Water use (Mm^3)				
Irrigation districts	1,030	683	683	683
EM	399	304	316	316
CJT	155	107	146	146
ARJ	200	131	185	185
ESC	33	18	31	31
RB	243	123	4	4
Urban use	119	74	74	74
Environmental flows to Albufera	60	34	29	29
Private and environmental benefits (million Euros)				
Private benefits				
Irrigation districts	190	136	148	54
EM	80	61	62	31
CJT	45	36	39	17
ARJ	34	23	25	4
ESC	7	4	5	2
RB	24	12	17	0
Urban use	283	241	241	241
Total	473	377	389	295
Environmental benefits	75	22	19	19
Social benefits	548	399	408	314

(Top) Water allocation by sector in million cubic meters. (Bottom) Private and environmental benefits by sector in million Euros. EM: Eastern La Mancha, CJT: Canal Jucar–Turia, ARJ: Acequia Real del Jucar, ESC: Escalona, RB: Ribera Baja. Source: Kahil et al. (2016a).

and environmental effects of drought under each policy instrument. The empirical results highlight that both water markets and institutional policies are economically efficient instruments to limit the damage costs of droughts, achieving similar social benefits in terms of private and environmental benefits. This finding is important because it shows that the status quo institutional policy can attain almost the same private benefits as water markets.

Water markets minimize the losses of private benefits from drought but disregard the environmental benefits. Results show that water markets entail a larger reduction of water for the environment than institutional policy, and this is because of the public good characteristic of environmental flows that are external to markets.

Water pricing is the policy advocated by the European Water Framework Directive. This policy involves important implementation challenges in arid and semi-arid regions such as the Jucar Basin in Spain, where irrigation is the main water use. Water pricing is quite detrimental to farmers because implementing water pricing instead of water markets or institutional policies triples farmers' drought losses from 50 to almost 150 million euros. Under the water pricing policy, drought reduces farmers' benefits by 72 percent, when the drought loss could be limited to 26 percent of benefits under water markets or institutional policies. These empirical results demonstrate that water markets and institutional policies are much more economically efficient and equitable than water pricing, and water pricing results in disproportionate costs to farmers. Enforcing water pricing seems a quite unfeasible task facing tough political and technical hurdles.

The protection of environmental flows from the water markets and institutional policies could be enhanced with additional measures. In the case of water markets, public water buyback programs to increase streamflows is a measure that allows reaping of the benefits of water markets while protecting ecosystems (Kahil et al., 2015a). In the case of institutional policy, the measure is greening institutional cooperation by including the environment as a full stakeholder in the process of water allocation among sectors and spatial locations (Esteban et al., 2016).

The policy of subsidizing investments in efficient irrigation systems has been compared to the water markets policy in the lower Jucar Basin (Kahil et al., 2015b). The results show the advantage of water markets over irrigation subsidies in terms of private and social benefits. Both water markets and irrigation subsidies reduce environmental flows compared to a drought scenario without any policy intervention, but with larger flow reductions from irrigation subsidies than water markets. The empirical results indicate that the benefits of the irrigation subsidies policy are small, especially when public subsidies and social costs of replacing lost environmental flows are accounted for. In contrast, the benefits of water markets are large, even though well-functioning water markets involve sizable monitoring and transaction costs that are not considered in the analysis but require evaluation.

Summing up, the water markets and institutional policies seem to be much more suitable than water pricing and irrigation subsidies policies to confront droughts and the forthcoming climate change. The main drawback of water pricing is the enormous burden placed on farmers' private benefits, making water pricing politically unfeasible. The main drawback of irrigation subsidies is the large fall in streamflows, above any other instrument, and the subsequent damages to ecosystems. As indicated previously, this fall in streamflows could be remediated by reducing water allocations in modernizing districts.

15.4 PROSPECTS FOR THE SUSTAINABLE MANAGEMENT OF THE JUCAR BASIN

The Jucar Basin is almost a closed basin, with groundwater extractions and surface diversions for human uses exhausting most basin resources. The consequence is very strong pressure on environmental assets, with minimum environmental flows at the Jucar mouth being only 0.5 m^3/s. Natural stream flows in the basin fell by 200 Mm^3 in the 1980–2012 period, and the water scarcity problem is going to worsen with climate change. The estimates are that water available in the basin will fall by 12 percent in 2040 and by 25 percent in 2100, with a much higher frequency of major drought events: up to three times higher in 2040 and almost ten times in 2100 (Perez-Martin et al., 2015).

15.4.1 The Jucar Basin in the Coming Decades

The change in the balance of water resources in the Jucar, estimated by the Jucar Basin Authority for 2040, is presented in Table 15.4. The impact of climate change reduces available water resources in the basin. The water balance deteriorates in 2040, since the planned reduction in water demand for human activities does not cover the fall of available water resources in the basin. Between 2015 and 2040, the reductions are 200 Mm^3 in water availability and 105 Mm^3 in net water demand, and the balance difference shrinks from 390 to 295 Mm^3 (last row in Table 15.4).

Table 15.4 *Water resources balance in the Jucar Basin (Mm^3/y)*

Period	Current (2015)	Future (2040)
Available resources	1,700	1,500
Water demand	1,670	1,530
Irrigation	1,400	1,260
Industrial	50	70
Urban	220	200
Return flows	360	325
Net water demand	1,310	1,205
Difference	390	295

Source: CHJ (2015).

However, this shortfall in the water balance could be underestimated. One reason is that the planned reduction in irrigation demand is based on investments in irrigation technologies after 2015. As indicated in the previous section, the modernization of irrigation will likely lead to higher evapotranspiration by crops, reduced irrigation returns flows, and falling basin streamflows. If irrigation modernization could not reduce irrigation demand, then the current net water demand will not decrease and the balance difference in 2040 will further deteriorate below 190 Mm^3 (1,500 availability – 1,310 net demand). Another reason for the shortfall in the water balance is the additional 80 Mm^3 of water diversions for the new Jucar–Vinalopo interbasin water transfer. Taken together, irrigation modernization and the Jucar–Vinalopo water transfer would cut down the current positive water balance and turn the Jucar basin into a fully closed hydrological basin.

Considering the entire territory covered by the Jucar Basin Authority, which includes the Jucar and Turia basins and other minor basins (Figure 15.2), the sustainability prospects are also dim. The current water demand of 3,300 Mm^3 is above the 3,100 Mm^3 of renewable water resources, and the projected trends for 2040 are water demand at 3,000 Mm^3 and water resources availability at 2,900 Mm^3 (CHJ, 2015). However, the projected reduction in demand is based on irrigation modernization, which could expand water consumption rather than reduce it. Water scarcity problems are likely to worsen in the coming decades, especially in the already vey stressed southern Alicante area. The engineering solutions provided in the last decade are 500 Mm^3 of recycled water from urban treated wastewater, of which only 120 Mm^3 are used at present, and the already built 100 Mm^3 capacity of seawater desalination plants being used at 10 percent of capacity.

The improvement of sustainability in the Jucar Basin Authority district depends on solving the problem of water governance, as the case of Alicante shows. Urban use in Alicante is 40 Mm^3, supplied with water from the Tajo–Segura interbasin transfer and overdrafted aquifers located in the Vinalopo basin.

The desalination plants built in Alicante could cover the full supply of urban water in Alicante, but the stakeholders do not want to pay the costs of desalinated water. They keep using water from the interbasin transfer and groundwater from overdrafted aquifers, which are much cheaper. This demonstrates that the engineering solutions based on investments in water technologies could not deliver sustainable outcomes, and the key issue is to solve social conflicts. The upcoming task is to promote the cooperation of stakeholders through suitable institutions, to advance the sustainable management of water resources and the protection of dependent ecosystems.

15.4.2 Potential Sustainable Outcomes and Stakeholders' Perceptions of Policy Reforms

One option to improve the sustainability in the Jucar basin is the substitution of freshwater for recycled water in irrigation. There is 200 Mm^3 of recycled water from urban wastewater treatment plants available in Valencia and other towns, but only 20 Mm^3 is being used, mostly to supplement environmental inflows to the Albufera wetland. Farmers are opposed to using recycled water at present.

A major contribution to sustainable management would be to curtail irrigation surface diversions and groundwater extractions, making sure that the water consumed by irrigation is reduced. The reduction of groundwater extractions in the upstream Eastern La Mancha irrigation district would recover the water table of the aquifer and the aquifer discharges feeding the Jucar River, which have almost disappeared during the 2000s (Figure 15.3). The reduction of diversions and extractions should also be substantial in downstream irrigation districts where irrigation demand is concentrated. The reduction of irrigation allocations will not be possible without the cooperation of farmers. Farmers can be compensated by public subsidies for irrigation technologies in exchange for reduced allocations to modernizing districts. These allocation reductions must be large enough to guarantee the decrease in water consumed by crops.

The current institutional approach to water management of the Basin Authority based on stakeholders' cooperation should be maintained and improved. Water allocation decisions under scarcity are taken with the involvement and support of stakeholders. Economic instruments such as water markets and water pricing can be ancillary instruments to this institutional approach, rather than instruments to substitute the institutional approach. The reason is that sustainable water management cannot be achieved without the collective action of stakeholders.

The improvement of water management requires detailed information of the water flowing through the basin. The Basin Authority already has access to measuring devices controlling the flows of water diversions and groundwater extractions in downstream irrigation districts. Remote sensing coupled with farmers' crop plans are used by the Basin Authority to control groundwater extractions in the upper irrigation district, although control will be enhanced with measuring devices in all wells.

Progressive water scarcity in the Jucar Basin has intensified conflicts between the interest groups, with keen competition over water allocation among sectors and spatial locations. The success of water policy reforms under such acute water scarcity is quite challenging and depends on accommodating the opposite interests of water users having different political power.

The perceptions of the different interest groups in Jucar regarding policy outcomes have been analyzed by Esteban et al. (2016). Basin upstream irrigators are a small group of large landholders that are very well organized, while downstream irrigators are large and heterogeneous groups of small landholders that are weakly organized. The upstream irrigators have more influence over water authorities and policy makers than downstream irrigators, because of their strong coordination and lobbying effort.

Both upstream and downstream irrigators consider irrigation modernization a good policy, but the policy of limiting extractions is mostly supported by downstream irrigators but not so much by upstream irrigators. Also, downstream irrigators consider that the policy of limiting extractions has been very unfair to them. The reason is that the upstream irrigators have increased water extractions fivefold since 1980, while downstream irrigators have seen their water extractions fall strongly (e.g., Acequia Real canal from 700 down to 200 Mm^3).

There are also different policy perceptions by sectors in the same location: Urban water utilities upstream prefer limits of extractions, and utilities downstream prefer irrigation modernization, just the opposite of irrigators' perceptions in each location.

The implication for achieving sustainable management in the basin is that irrigation modernization is supported by all stakeholders, and this is the policy choice of the Basin Authority for water planning in the coming decades (CHJ, 2015). As indicated previously, this policy will not result in the reduction of irrigation extractions. For sustainable outcomes addressing acute water scarcity, a policy of limiting water extractions is also needed. This policy would have the strongest support from downstream irrigators and upstream urban utilities but less support from the other stakeholders.

15.5 CONCLUSIONS AND POLICY RECOMMENDATIONS

Water policy reforms are needed in many river basins around the world facing water scarcity from excessive water abstractions and deteriorating water quality from large pollution loads. Water scarcity is common in arid and semi-arid regions with substantial irrigated agriculture, resulting in mounting competition among human uses and considerable environmental damage. Water quality degradation is driven by pollution coming from human activities and affects basins in all regions.

The scale of this global water depletion shows that correcting water mismanagement to achieve more sustainable outcomes is a challenging task. The challenges for successful water reforms are the sound design of reforms, and the support of key groups of stakeholders. Water reforms should be based on rigorous analysis based on economic and biophysical information that could support the appropriate measures and instruments for reform. Water reforms change the power and benefits of groups of stakeholders, so the active support of those groups that gain from the reform is needed, while the losing groups must be compensated to avoid the failure of reforms.

Water resources in Spain are under mounting pressures, with water scarcity linked to the large development of irrigation, and water quality degradation linked to urban, industrial, and agricultural pollution. There are severe water scarcity and quality problems in the Jucar River and the other southern rivers in Spain. Expanding water demands have resulted in the conversion of these rivers into almost closed basins with dwindling streamflows at river mouths. The ensuing threats to human water security have been compensated with important investment in water technologies.

The water management approach in Spain is institutional and relies on the river basin authorities enabling the collective action of stakeholders. Stakeholders are involved at all levels of decision making such as planning, financing, water allocation, design and enforcement of measures, and water management at basin, watershed, and district levels. The advantage of this institutional setting is the legitimacy gain in implementation and enforcement processes.

In the Jucar Basin, water demand is close to renewable water resources and the basin is becoming a closed water system. The growth of water extractions in recent decades has been driven mostly by groundwater irrigation in the upper Jucar, which has reduced streamflows in the lower Jucar and caused the desiccation of the middle Jucar during recent droughts.

To confront the progressive water depletion in the Jucar Basin Authority district, there has been a large set of management initiatives mostly based on engineering solutions supported with public subsidies. The list includes massive investment in water technologies such as modernization of irrigation systems (€1,000 M), urban wastewater treatment plants (€1,300 M), seawater desalination plants (€250 M), and interbasin water transfers (€400 M).

Three prevailing policies to address irrigation adaptation to water scarcity and climate change being considered in the literature are water markets, water pricing, and public subsidies for irrigation modernization. Another policy option receiving less attention is the institutional cooperation of stakeholders, which is the approach of basin authorities in Spain.

A direct comparison of water markets, water pricing, and institutional cooperation policies has been made in the Jucar Basin. The results show that both water markets and institutional policies are economically efficient instruments to deal with water scarcity, achieving similar social benefits, although environmental benefits are higher under the institutional policy since water markets disregard environmental outcomes. Water pricing is the worst policy option in terms of social benefits, and it is also very inequitable because farmers sustain disproportionate benefit losses. The implication is that water pricing in irrigation will face tough political and technical hurdles.

Water markets have also been compared in the Jucar with the policy of subsidizing irrigation modernization. There is a clear advantage of water markets over irrigation subsidies, with water markets attaining higher social benefits and larger river streamflows. This empirical evidence in Jucar indicates that water markets and institutional policies seem to be more suitable than water pricing and irrigation subsidies policies to confront water scarcity.

The prospects for achieving sustainable management of the Jucar Basin in the coming decades seem to be dim. Natural streamflows have been falling in the last three decades, and water scarcity is going to worsen with the impact of climate change. The predictions of falling natural streamflows are 12 percent in 2040 and 25 percent in 2100. The Jucar Basin Authority estimates that the water balance between available resources and net water demand is going to shrink by 25 percent in 2040. However, this balance shortfall could double to 50 percent because it is assumed that irrigation modernization would reduce the water consumed by irrigation and this reduction could not materialize.

For the whole territory covered by the Jucar Basin Authority that includes the Jucar, Turia, and other minor basins, the outlook for sustainable management is also dubious. The improvement of sustainability involves solving the problem of water governance, rather than pure engineering solutions. Some pressing water governance hotspots are first to convince farmers of substituting freshwater for the available urban recycled water, and second to make the arrangements for seawater desalination plants to work at full capacity. More long-term governance endeavours are to curtail irrigation surface diversions and groundwater extractions, and reallocating water to urban, industrial, and environmental uses. Therefore, the viability of reforms requires getting the support and cooperation of farmers by compensating them for the reallocation of water from agriculture to other sectors.

REFERENCES

Albiac, J., Esteban, E., Tapia, J., and Rivas, E. (2013). Water Scarcity and Droughts in Spain: Impacts and Policy Measures. In K. Schwabe, J. Albiac, J. Connor, R. Hassan, and L. Meza, eds., *Drought in Arid and Semi-Arid*

Environments: A Multi-Disciplinary and Cross-Country Perspective. Dordrecht: Springer.

Alexandratos, N. and Bruinsma, J. (2012). World Agriculture towards 2030/2050: The 2012 Revision, Global Perspective Studies Team, FAO Agricultural Development Economics Division, ESA Working Paper No. 12-03. Rome: FAO.

Booker J., Michelsen A., and Ward F. (2005). Economic Impacts of Alternative Policy Responses to Prolonged and Severe Drought in the Rio Grande Basin. *Water Resources Research*, 41, pp. 1–15.

Booker, J., Howitt, R., Michelsen, A., and Young, R. (2012). Economics and the Modeling of Water Resources and Policies. *Natural Resource Modeling*, 25(1), pp. 168–218.

Calatrava, J. and Garrido, A. (2005). Modelling Water Markets under Uncertain Water Supply. *European Review of Agricultural Economics*, 32(2), pp. 119–142.

Calatrava, J., Guillem, A., and Martinez-Granados, D. (2011). Análisis de alternativas para la eliminación de la sobreexplotación de acuíferos en el Valle del Guadalentín [Analysis of Alternatives for the Elimination of Overexploitation of Aquifers in the Guadalentín Valley]. *Journal of Agricultural Economics*, 11, pp. 33–62.

Comprehensive Assessment of Water Management in Agriculture (CAWMA). (2007). *Water for Food Water for Life: A Comprehensive Assessment of Water Management in Agriculture*. London: Earthscan-International Water Management Institute.

Confederacion Hidrografica del Jucar [Hydrographic Confederation of the Jucar] (CHJ) (2015). *Plan Hidrologico de la Demarcacion Hidrografica del Jucar [Hydrological Plan of the Jucar Hydrographic Demarcation]*. Memoria. Ciclo de planificación hidrológica [Memory. Hydrological Planning Cycle] 2015–2021. Valencia: CHJ.

Connor, J. and Kaczan, D., (2013). Principles for Economically Efficient and Environmentally Sustainable Water Markets: The Australian Experience. In K. Schwabe, J. Albiac, J. Connor, R. Hassan, and L. Meza, eds., *Drought in Arid and Semi-arid Environments: A Multidisciplinary and Cross-country Perspective*. Dordrecht: Springer.

Cornish, G., Bosworth, B., Perry, C., and Burke, J. (2004). *Water Charging in Irrigated Agriculture: An Analysis of international Experience*. FAO Water Report No. 28. Rome: FAO.

Elliott, J., Deryng, D., Muller, C. et al. (2014). Constraints and Potentials of Future Irrigation Water Availability on Agricultural Production under Climate Change. *PNAS*, 9, pp. 3239–3244. https://doi.org/10.1073/pnas.1222474110

Esteban, E. and Albiac, J. (2012). The Problem of Sustainable Groundwater Management: The Case of La Mancha Aquifers, Spain. *Hydrogeology Journal*, 20, pp. 851–863. https://doi.org/10.1007/s10040-012-0853-3

Esteban, E., Dinar, J., Albiac, J. et al. (2016). *The Political Economy of Water Policy Design and Implementation in the Jucar Basin, Spain*. UCR SPP Working Paper Series WP#16-04. Riverside: University of California.

European Commission (EC) (2007). *Communication from the Commission to the European Parliament and the Council, Addressing the Challenge of Water Scarcity and Droughts in the European Union*, COM 414/2007. Brussels: European Commission.

Garcia-Molla, M., Sanchis-Ibor, C., Ortega-Reig, M. V., and Avella-Reus, L. (2013). Irrigation Associations Coping with Drought: The Case of Four Irrigation Districts in Eastern Spain. In K. Schwabe, J. Albiac, J. D. Connor, R. M. Hassan, and L. M. Gonzalez, eds., *Drought in Arid and Semi-Arid Regions*. Dordrecht: Springer.

Garrido, A. and Calatrava, J. (2009). Trends in Water Pricing and Markets. In A. Garrido and M. Llamas, eds., *Water Policy in Spain*. Leiden: CRC Press.

Gohar, A. and Ward, F. (2010). Gains from Expanded Irrigation Water Trading in Egypt: An Integrated Basin Approach. *Ecological Economics*, 69, pp. 2535–2548.

Grafton, R., Williams, J., Perry, C., Molle, F., Ringler, C. et al. (2018). The Paradox of Irrigation Efficiency. *Science*, 381 (6404), pp. 748–750.

Instituto Nacional de Estadística [National Institute of Statistics] (INE) (2014). *Cuentas satélites del agua en España [Satellite Water Accounts in Spain]*. Madrid: INE.

(2016). *Encuesta sobre el suministro y saneamiento del agua [Survey of Water Supply and Sanitation]*. Madrid: INE.

International Groundwater Resources Assessment Center (IGRAC) (2010). *Global Groundwater Information System*. Delft: IGRAC.

Intergovernmental Panel on Climate Change (IPCC) (2014a). Summary for Policymakers. In Climate Change 2014. In R. Pachauri and core team, eds., *Synthesis Report*. Geneva: IPCC.

(2014b). Summary for Policy Makers. In C. Field, V. Barros, D. Dokken et al., eds., *Climate Change 2014: Impacts, Adaptation, and Vulnerability*. Part A: Global and Sectoral Aspects. Contribution of Working Group II to the Fifth Assessment Report of the Intergovernmental Panel on Climate Change. Cambridge: Cambridge University Press.

Kahil, M. T., Connor, J. D., and Albiac, J. (2015b). Efficient Water Management Policies for Irrigation Adaptation to Climate Change in Southern Europe. *Ecological Economics*, 120, pp. 226–233.

Kahil, M. T., Dinar, A., and Albiac, J. (2015a). Modeling Water Scarcity and Droughts for Policy Adaptation to Climate

Change in Arid and Semiarid Regions. *Journal of Hydrology*, 522, pp. 95–109.

Kahil, M. T., Albiac, J., Dinar, A. et al. (2016b). Improving the Performance of Water Policies: Evidence from Drought in Spain. *Water*, 8(2), p. 34.

Kahil, M. T., Dinar, A., and Albiac, J. (2016a). Cooperative Water Management and Ecosystem Protection under Scarcity and Drought in Arid and Semiarid Regions. *Water Resources and Economics*, 13, pp. 60–74.

Kirby, M., Bark, R., Connor, J., Qureshi, E., and Keyworth, S. (2014). Sustainable Irrigation: How Did Irrigated Agriculture in Australia's Murray–Darling Basin Adapt in the Millennium Drought? *Agricultural Water Management*, 145, pp. 154–162.

Konikow, L. (2011). Contribution of Global Groundwater Depletion since 1900 to Sea-Level Rise. *Geophysical Research Letters*, 38. https://doi.org/10.1029/2011GL048604

MARM (2008). *Instrucción de Planificación Hidrologica [Hydrological Planning Instruction]*. Orden ARM 2656, 38472–38582.

Medellin, J., Howitt, R., and Lund, J. (2013). Modeling Economic-Engineering Responses to Drought: The California Case. In K. Schwabe, J. Albiac, J. Connor, R. Hassan, L. Meza, eds., *Drought in Arid and Semi-arid Environments: A Multi-disciplinary and Cross-country Perspective*. Dordrecht: Springer.

Ministerio de Medio Ambiente [Ministry of the Environment] (MIMAM) (2000). *Libro blanco del agua en España, Dirección General de Obras Hidráulicas y Calidad de las Aguas [White Paper on Water in Spain, General Directorate of Hydraulic Works and Water Quality]*. Secretaría de Estado de Aguas y Costas [Secretary of State for Waters and Coasts]. Madrid: MIMAM.

(2004). *Programa A.G.U.A.: Actuaciones para la Gestión y Utilización del Agua [A.G.U.A. Program: Actions for the Management and Use of Water]*. Madrid: MIMAM.

Ministerio de Obras Públicas y Transportes [Ministry of Public Works and Transport] (MOPT) (1993). *Plan Hidrológico Nacional [National Hydrological Plan]*. Memoria y Anteproyecto de Ley [Memorandum and Draft Law]. Madrid: MOPT.

Ministerio para la Transición Ecológica [Ministry for Ecological Transition] (MITECO) (2020). *Delimitación de las demarcaciones hidrográficas [Delimitation of Hydrographic Demarcations]*. Dirección General del Agua [General Water Management]. MITECO. Madrid.

Montilla-López, N. M., Gutiérrez-Martín, C., and Gómez-Limón, J. A. (2016). Water Banks: What Have We Learnt from the International Experience? *Water*, 8(10), p. 466.

Moore M. (1991). The Bureau of Reclamations New Mandate for Irrigation Water Conservation: Purposes and Policy Alternatives. *Water Resources Research*, 27, pp. 145–155.

National Oceanic and Atmospheric Administration (NOAA) (2008). *Summary of National Hazard Statistics for 2008 in the United States*, National Weather Service. Washington, DC: NOAA.

Organisation for Economic Co-operation and Development (OECD) (2014). *Climate Change, Water and Agriculture: Towards Resilient Systems, OECD Studies on Water*. Paris: OECD Publishing.

Perez-Martin, M., Batan, A., Del-Amo, P., and Moll, S. (2015). Climate Change Impact on Water Resources and Droughts of AR5 Scenarios in the Jucar River, Spain. In J. Andreu, A. Solera, J. Paredes, D. Haro, and H. Van Lanen, eds., *Drought: Research and Science-Policy Interfacing*. Leiden: CRC Press/Bakelma.

Perez-Martin, M., Estrela, T., Andreu, J., and Ferrer, J. (2014). Modeling Water Resources and River-Aquifer Interaction in the Júcar River Basin, Spain. *Water Resources Management*, 28, pp. 4337–4358.

Playan, E., Lecinia, S., Isidoro, D. et al. (2013). Living with Drought in the Irrigated Agriculture of the Ebro Basin (Spain): Structural and Water Management Actions. In K. Schwabe, J. Albiac, J. D. Connor, R. M. Hassan, and L. M. Gonzalez, eds., *Drought in Arid and Semi-Arid Regions*. Dordrecht: Springer.

Qureshi, M., Schwabe, K., Connor, J., and Kirby, M. (2010). Environmental Water Incentive Policy and Return Flows. *Water Resources Research*, 46, pp. 1–12.

Rausser, G., Swinnen, J., and Zusman, P. (2011). *Political Power and Economic Policy. Theory, Analysis and Empirical Applications*. New York: Cambridge University Press.

Scheierling, S. and Treguer, D. (2016). *Investing in Adaptation: The Challenge of Responding to Water Scarcity in Irrigated Agriculture*. Federal Reserve Bank of Kansas City Economic Review, Special Issue 2016.

Scheierling, S., Young, R., and Cardon, G. (2004). Determining the Price Responsiveness of Demands for Irrigation Water Deliveries versus Consumptive Use. *Journal of Agricultural and Resource Economics*, 29, pp. 328–345.

Siebert, S., Henrich, V., Frenken, K., and Burke, J. (2013). *Update of the Digital Global Map of Irrigation Areas (GMIA) to Version 5*. Institute of Crop Science and Resource Conservation, Bonn: University of Bonn.

United States Department of Agriculture (USDA) (2012). *Climate Change and Agriculture in the United States: Effects and Adaptation, Agriculture Research Service*, USDA Technical Bulletin 1935. Washington, DC: USDA.

Varela, C. and Hernández, N. (2010). Institutions and Institutional Reform in the Spanish Water Sector:

A Historical Perspective. In A. Garrido and R. Llamas, eds., *Water Policy in Spain*. Abingdon: CRC Press.

Vörösmarty, C. et al. (2010). Global Threats to Human Water Security and River Biodiversity. *Nature*, 467, pp. 555–561.

Wada, Y. et al. (2010). Global Depletion of Groundwater Resources. *Geophysical Research Letters*, 37, pp. 1–5.

Winemiller, K. et al. (2016). Balancing Hydropower and Biodiversity in the Amazon, Congo, and Mekong. *Science*, 351(6269), pp. 128–129.

World Water Assessment Programme (WWAP) (2006). *Water: A Shared Responsibility*. New York: UNESCO-Berghahn Books.

World Water Council (WWC) (2000). *World Water Vision*. London: Earthscan.

Zilberman, D., Dunarm, A., MacDougal, N. et al. (2002). Individual and Institutional Responses to the Drought: The Case of California Agriculture. *Journal of Contemporary Water Research and Education*, 121, pp. 17–23.

Part IV Response

16 River Basin Management and Irrigation

François Molle

16.1 INTRODUCTION

It is often stated, in rounded-up numbers, that the shares of global water withdrawals for domestic, industrial and agricultural uses are 10 per cent, 20 per cent and 70 per cent, respectively.[1] Although these numbers usually serve to stress agriculture's 'lion's share' and to emphasise the need to achieve water savings in this sector in order to 'free' water for other competing uses, it is not well understood that agricultural water consumption actually amounts to more than 90 percent of global water consumption (Döll and Siebert, 2002). This is merely a reflection of the fact that the process of water consumption[2] occurs through vaporisation, i.e., the evaporation of water from the soil surface, water bodies and industrial processes, and transpiration from humans, animals and plants (needless to say the first two are dwarfed by the last).

In other words, nature dictates that most water consumption is due to vegetation, whether natural or irrigated crops. Thus, the 'closure' of river basins, or the process through which overall water consumption gradually comes close to the average amount available (and sometimes even exceeds it) (see Molle and Wester, 2009), is closely related to the unchecked (and, frequently, encouraged) growth in the infrastructural capacity to tap surface and groundwater for irrigation. This implies that increasingly frequent water shortages should chiefly be responded to by adjustments in crop evapotranspiration, whether intended (through mulching, changes in cropping patterns, fallowing, etc.) or otherwise (destruction of crops and farms), although this is often achieved by depleting aquifers or using more resources.

Irrigation is therefore the main cause of basin closure and the associated pathologies, as well as the principal locus of solutions to overexploitation and basin 'overbuilding' (see Molle, 2008). But just as resource overexploitation is typically caused by a sociopolitical process that distributes benefits attached to water use, the social and political implications of limiting access to water are unpalatable to managers or politicians. Where augmenting supply (e.g., through desalination or interbasin transfers) is physically or economically unfeasible, technological fixes purporting to avoid such a tough decision are understandably in much demand. 'Water saving' technologies and policies promise to solve the problem by reducing demand and, more frequently and more precisely, by reducing 'losses' in the system. Micro-irrigation has all the features of a good policy option, combining efficiency of water application (reducing 'losses') and increased yields without curtailing use by farmers.

This chapter discusses both the role of irrigation in river basin development and closure and how its share in total water use can be reduced. I first briefly outline the growing share of unchecked irrigation in water consumption and the closure of the basins under consideration in this volume. Indeed, understanding how irrigation came to play a peculiar role in river basin development is important when considering how its share can be reduced. I then recall the diversity of policy options available to respond to imbalances between supply and demand, noting that supply augmentation is generally favoured. Finally, I focus on the issue of 'water savings' by documenting how the irrigated sector has dealt with shortages and exploring how policies aimed at modernising irrigation technology may inadvertently lead to increased evapotranspiration and thereby counteract their purported conservation objectives.

16.2 IRRIGATION AND RIVER BASIN DEVELOPMENT

16.2.1 The Closure of River Basins

All nine river basins analysed in this volume are closed, either fully or to a large extent. A brief reminder of the unchecked development of irrigation in each of them illustrates the significance of irrigation demand and its key role in basin closure.

Total water use in the Murray–Darling Basin (MDB), of which irrigation accounts for 80 per cent, grew steadily over the past century (Figure 16.1), leading to overallocation of water and a neglect of environmental flows. The expansion of irrigation has now virtually ceased, although groundwater development is still possible in many areas (limited by water quality and abstraction costs). The frequency of zero or low-flow years at the mouth of

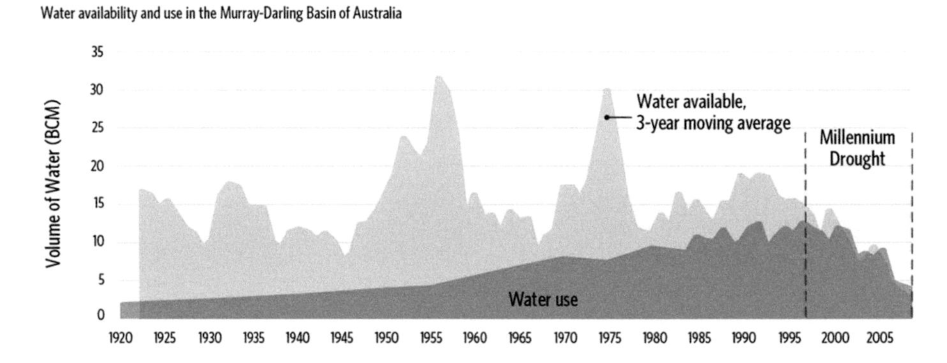

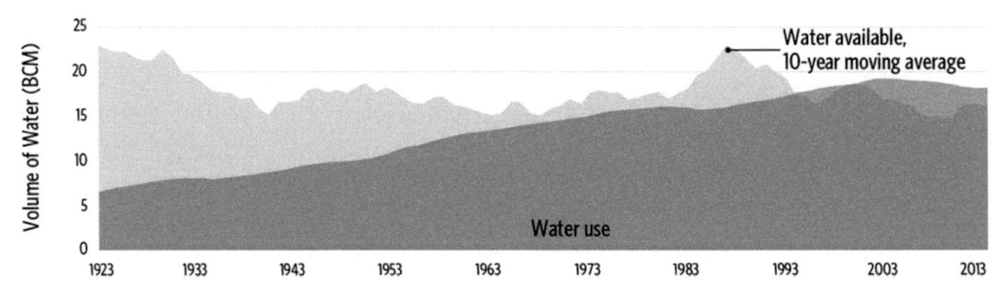

Figure 16.1 Historical water use in the Murray–Darling and Colorado River basins.
Source: Richter et al. 2017

the Murray River increased from about 1 per cent prior to European settlement to over 40 per cent (Hancock and Pietsch, 2008). During the 2002–2010 Millennium drought flows at the Murray mouth were close to zero (see Figure 16.2). Some climate change modelling scenarios predict that by 2030 such low-flow events may be expected 70 per cent of the time (Young and Chiew, 2011). The long-term average basin inflow (8.9 Bm^3) obscures large variability. Total withdrawals of 13.75 Bm^3[33], including 10 per cent from aquifers and a total of 10.9 Bm^3 by irrigation, are planned to be reduced by 30 per cent to restore environmental flows at the river mouth (MDBA, 2016).

Irrigation water use in the Colorado River Basin increased dramatically until the 1970s and has since levelled off, currently accounting for around 70 per cent of total consumptive use (Figure 16.1). The Colorado River has not reached the sea since the late 1990s. The basin storage capacity is around four times the average annual runoff (21 Bm^3), and only 9.1 per cent of the average annual flow is passed on from the United States to Mexico, where this share is used up in the Mexicali Valley. The Colorado Delta has reduced to 5 per cent of its historical area, with highly degraded wetlands (Grafton et al., 2012).

Irrigation in the Rio Grande Valley outstripped resources as early as the 1970s. Virtually all the renewable water available in the upper Rio Grande Basin has been consumed, with a near-zero flow at Fort Quitman, in the middle of the basin. The next tributary, the Conchos that flows from Mexico, has also seen all its surface water appropriated. The river has an average

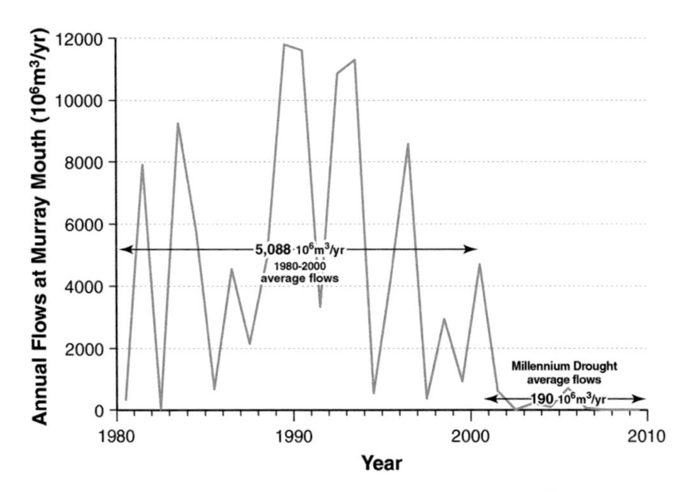

Figure 16.2 Flows at the Murray River mouth (million m^3/year).
Source: Grafton (2016)

outflow to the sea of 0.8 Bm^3 (of a total runoff of 48 Bm^3) and occasionally stops flowing in the delta. Flash floods that make it to the sea are now channelled to lateral lagunas to protect urban areas that grew along the estuary.

In the Tigris–Euphrates River Basin (TEB) the 68–84 Bm^3 of average runoff are largely depleted by irrigation and evaporation of water bodies, with only 45 m/s reaching the Shatt al-Arab on average (i.e., 1.5 Bm^3/year). Kucukmehmetoglu and Geyman (2014) identified a 144 Bm^3 decrease in water storage during the 2003–2009 period, of which 91 Bm^3 were from groundwater

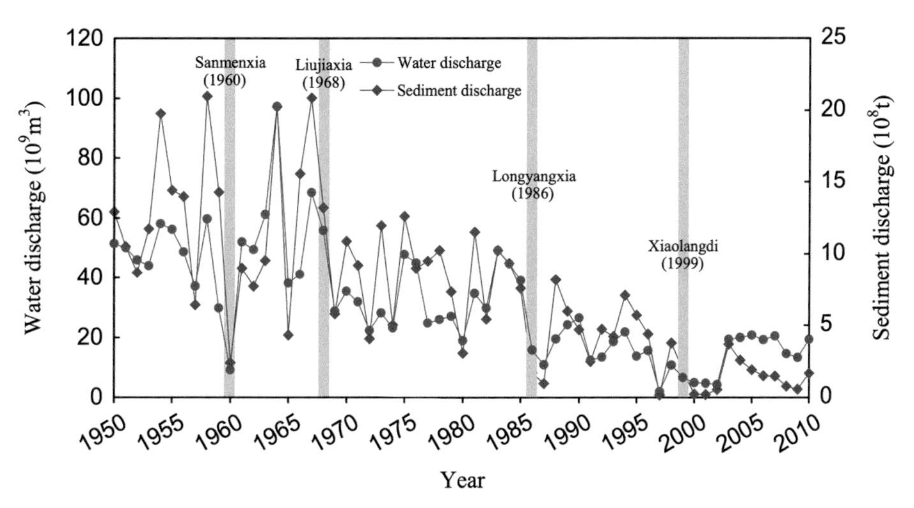

Figure 16.3 Average annual water flow and sediment discharge, Yellow River at Lijin (1950–2010). Figure is from Yu et al. (2013)

depletion, in the north-central Middle East, including portions of the Tigris and Euphrates River Basins and western Iran. The total water lost in the TEB alone was later assessed at 131 Bm^3 in the 12-year period from 2003 to 2015 (Chao et al., 2017). Water is increasingly stored and consumed in Iran and Turkey, while Syria and Iraq lack storage sites. In Turkey, the GAP project plans to develop as much as 1.7 million hectares of land, of which only 474,528 hectares are presently operational (Chapter 8).

In the Yellow River Basin, the area irrigated with water from the river increased enormously from some 0.8 million hectares before 1949 to 7.51 million hectares in 1997 (Grafton et al., 2012) but has now levelled off. Although in the late 1990s the outflow to the sea was zero for the majority of the year, the construction of the Xiaolangdi Dam made it possible to create artificial flood pulses with discharges of around 4,000 m^3/s in summer and a total outflow of around 15 Bm^3, i.e., roughly a third of the available resources (Figure 16.3). This allowed the river channel to be scoured and sediment to be removed, but only to around 10 per cent of the 1950s levels (Yu et al., 2013). The total runoff is reduced by changes in rainfall and ET (Meng et al., 2016), as well as by the revegetation of 1.6 million ha on the Loess Plateau that has reduced runoff by 8 per cent (Feng et al., 2016). These 'uses' contribute to the basin's sustainability and thus cannot be discounted as 'losses'.

Irrigation is still in full expansion in the Rio São Francisco Basin, where irrigation water demand doubled in the 2006–2016 period[4] and reached 800,000 ha of private irrigation and 180,000 ha of public irrigation,[5] accounting for 77 per cent of total consumptive use (Chapter 11). The São Francisco Integration Program plans for a total of 600,000 ha, of which 269,000 ha are to be fed by the Eixo Sul pipe/canal that branches off the Sobradinho Dam.[6] This is to take place in a region which the RSF Basin Committee (CBHSF, 2015) already describes as 'very critical' in terms of surface water balance, with 'existing over-exploitation of available water resources and water use conflicts'. The 'expansion of irrigation schemes' is held as one of the 'main threats' to the basin's water resources.

The basin still has a substantial outflow to the sea (half its total runoff). The average natural flow is 2,846 m^3/s (or 90 Bm^3/year), while the minimum flow after the dam furthest downstream (Xingó) is fixed at 1,300 m^3/s. However, in 2017 nothing could be done to prevent the flow from decreasing to 550 m^3/s, which caused saline intrusion inland with a severe impact on fisheries and the water supply to riparian zones and coastal cities. Basin managers try to keep a flow of at least 2,000 m^3/s to feed the series of hydropower dams, which puts pressure on consumptive uses in the basin. The Water Resources Plan for 2016–2025 estimates withdrawal for consumptive uses in 2015 at 786 m^3/s with a maximum of 1,073 m^3/s expected in 2035 (Chapter 11). This increase will necessitate reducing benefits from hydropower.

Irrigation has steadily increased along the Nile River in Sudan and Egypt and in the Nile Delta, with governments pursuing expansionist policies. This irrigation now forms the bulk of a total irrigated area of 5.4 million ha in the basin (FAO, 2011), and upstream countries, particularly Ethiopia and Sudan, are trying to develop their irrigation sectors. The Nile Basin appears to still have a lot of water flowing to the sea (12 Bm^3 of the ~56 Bm^3 released annually on average by the Aswan Dam). But closer examination of the water and salt balance shows that this outflow is necessary to flush polluted/saline water and keep the water level in the drains of the northern part of the delta low enough to allow cultivation (Molle et al., 2018). Considering the water 'consumed' by both productive and unavoidable processes, the current efficiency of the delta is 93 per cent of its potential, with possible 'savings' applying to only around 4 per cent of dam releases (Molle et al., 2018).

The Jucar Basin is also closed, with total water withdrawals estimated at 1.69 Bm^3 of a total renewable amount of 1.7 Bm^3 (85 per cent of which goes to irrigation) and an outflow to the sea of around 16 Mm^3 (1 per cent of total). The growth of surface and groundwater abstraction in the upper basin has both reduced the share to the ancient downstream systems and depleted groundwater stocks. The total irrigated area is now around 190,000 ha.

Three additional points can be made to complement the description of the remarkably similar dynamics and 'basin trajectories' observed in our sample basins.

- First, these basins undergo a double basin-closure, whereby groundwater resources are also overexploited, and stocks depleted. This is particularly salient in the ET, Yellow, Rio Grande, Jucar, and now Nile (Delta) basins. The depletion of groundwater stocks in the Jucar Basin currently stands at 2.5 Bm^3 – more than the annual available resource.
- Second, basin closure is not necessarily tantamount to a physical outflow slopping down towards zero. Some outflow may be vital to the very sustainability of the basin, such as in the Yellow River (flushing out sediment) or Nile Delta (flushing salt/pollutants and ensuring drainage). Non-consumptive use can also put pressure on water use, as in the São Francisco Basin, where the reserve for downstream hydropower dams limits upstream users, despite substantial average outflow to the sea.
- Third, by dramatically reducing the outflow discharge, basin closure means that sediment tends to be trapped within the basin, most commonly in reservoirs. The sediment load of the outflow of the São Francisco River is now less than 5 per cent of that in 1970 (Sabadini-Santos, Knoppers, Padua Oliveira, and Leipe, 2009). In the Colorado and Rio Grande River basins the historically large sediment load continues to be captured by reservoirs, but the huge storage capacity will delay any negative consequences in terms of regulation (but not for the scouring of channels downstream of the dams or decreased fluvial sandbars and beaches).

16.2.2 Regulatory Failure

It is a puzzling observation that river basins invariably see their water resources overallocated, even where the strict control of hydrological conditions and allocation of water rights and entitlements should avoid this trap (Molle and Wester, 2009). Basin 'overbuilding' – resulting from large-scale state infrastructure, unchecked private investment and, where such a system exists, the overallocation of water entitlements – creates similar patterns of scarcity.

In the Colorado Basin the allocation of water among riparian states was based on optimistic average hydrological data, considering neither evaporation losses from reservoirs to be built years later (now totalling 2 Bm^3) nor Native American rights. In the Murray–Darling Basin, and notably in the state of New South Wales, licences were granted despite recognising the ticking time bomb represented by large contingents of 'dozers and sleepers' who only use their rights occasionally or pay their fees without using water.

In the Limarí Basin, the allocation of water without a sound and meaningful knowledge of both supply and use have led to a system of private water rights that is not anchored in hydrological reality and invites overallocation, especially of groundwater.

The overallocation of water entitlements is an obvious political expedient to reduce tension by not denying access to resources and satisfying the maximum number of existing (or would-be) users, particularly agricultural constituencies (Molle and Wester, 2009).

16.2.3 Additional Pressure on Water Supply

Imbalances between supply and (potential) demand are compounded by climate change, which, tellingly, is rarely considered in basin master plans. Alexandra (2017) highlights the failure of the MDB Plan to explicitly include climate change in establishing sustainable diversion limits (SDLs): 'Possible reasons for not formally reducing water deemed available in the future include the complexity and uncertainty of climate science, the cultural construction of "climate normal" based on long-term averages, and institutional settings that reinforce dominant "hydro-logical" approaches and rationalities. Minimizing the political, legal and financial consequences of attributing reductions in water allocations to climate change are also potential reasons.'

Climate change scenarios for the São Francisco River yield widely varying results, but reductions in the inflow to the three main dams (Três Marias, Sobradinho and Xingó) were estimated to be in the following brackets: 0–55 per cent, 10–70 per cent, and 0–70 per cent. Natural runoff in the Jucar Basin dropped by 200 Mm^3 in the 1980–2012 period, and it is estimated that water availability in the basin will fall a further 12 per cent by 2040 and 25 per cent by 2100, with far more frequent major drought events (Perez-Martin et al., 2015).

The prospects for the Colorado are equally worrying (Chapter 13), with projected reductions in flows for different time periods, ranging from 4 to 18 per cent (Grafton et al., 2012). The projections are between 9 and 29 per cent for the Yellow and up to 69 per cent for the Murray–Darling. Climate change is also barely considered in the case of the Rio Grande River Basin. While the 2012 water plan for Texas acknowledges that 'climate change and climate variability both pose challenges

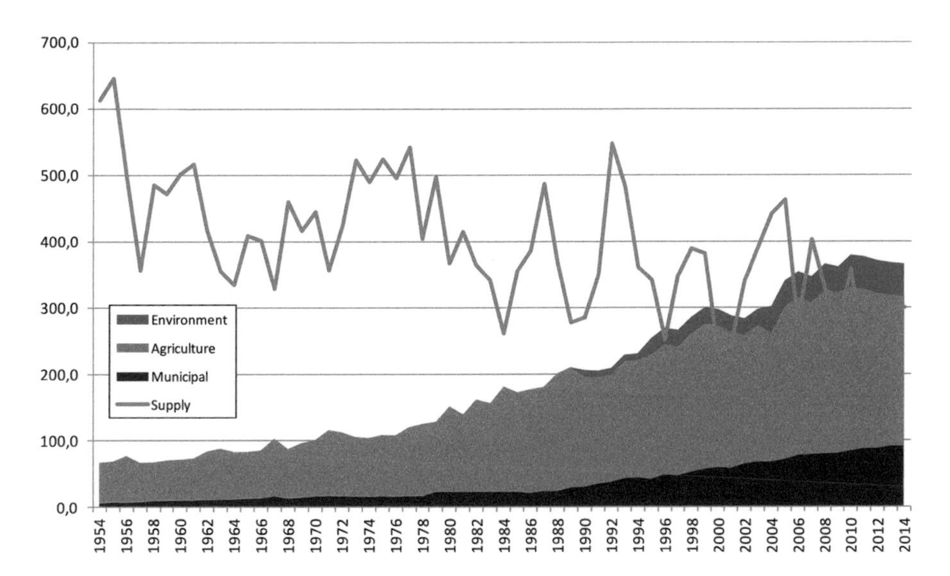

Figure 16.4 Irrigated agriculture and the 'double squeeze' (hypothetical case, no unit).
Source: Molle and Sanchis Ibor, 2019

to water planning, because they add uncertainty' (Chapter 14), no consideration is given to the reductions in withdrawals that this will entail.

The ubiquitous neglect of the impact of climate change is even more worrying in the Limarí Basin. As in most basins with snowmelt spring discharge, runoff is expected to increase over the next decade as the snowpack and glaciers are depleted. This will cause more water uses to face resource deprivation in the subsequent phase (Chapter 12).

Figure 16.4 shows a hypothetical but generic case of 'double squeeze', where the supply to agriculture falls increasingly short as supply decreases along with an ineluctable increase in non-agricultural uses (including the recognition of environmental flows).

In summary, consumptive use by irrigation varies between 50 per cent and 90 per cent of total consumption across our nine basins and its uncontrolled growth has driven most basins to closure. As a result, irrigation areas are now relatively stable, except for the São Francisco, (upper) TEB and (upper) Nile, where more irrigation development is planned despite anticipated negative impacts. The development of irrigation upstream will shift evapotranspiration spatially, affecting downstream countries or regions (although in the São Francisco this may also come from lower priority being given to hydropower). The over-expansion of agriculture, or more precisely the capacity to abstract water for use in irrigation (including surface water diversions as well as groundwater abstraction), mechanically generates recurrent shortages. This leads farmers to adaptation strategies that tend to overexploit groundwater resources, as mentioned previously. A growing proportion of irrigated agriculture is now at risk or unsustainable. This threatens local incomes but also broader economies as the issues affect more basins globally, since 40 per cent of the world's food production come from irrigated agriculture (Döll and Siebert, 2002).

16.3 WATER SHORTAGES AND POLICY RESPONSES IN THE IRRIGATION SECTOR

16.3.1 Irrigation Water Demand and Supply Augmentation

Basin 'overbuilding' and closure are commonplace processes globally (Molle et al., 2010) as represented by the Colorado, Rio Grande, Yellow, MDB and Jucar. Turkey (TEB) and Ethiopia/Sudan (Nile) illustrate the powerful geopolitical-cum-internal drivers of international river basin closure, while the São Francisco exemplifies the role of private investors and to some extent bureaucratic reproduction. Although state water policies generally include a mix of supply-and-demand management measures, supply augmentation is generally the favoured option as it minimises the political cost of reducing water deliveries to users or specific constituencies while spreading the financial cost to taxpayers at large. Capital-intensive infrastructure solutions appeal to hydrocracies, politicians, construction companies and development banks and are therefore systematically promoted (Molle, 2008). Increased supply begets increased use and claims to water, which inevitably cause repeated water shortages that in turn generate calls for further supply augmentation. Overallocation can lead basins to a lock-in situation so critical that demand management alone cannot generally reverse or remedy it. This allows the next supply augmentation project to be justified irrespective of cost or economic rationality, especially when big cities or large rural economies are threatened.

This is clearly demonstrated by the Yellow River's South–North transfer and, perhaps tomorrow, by the São Francisco River (with a possible transfer from the Tocantins River).

Typical supply-side responses to water shortages – usually produced by the overdevelopment of irrigation facilities and compounded by (greater) climate variability – thus include:

- new dams (e.g., Brazil), but this option is marginal in most basins except TEB and (upper) Nile
- interbasin transfers, as in China (South–North transfer), São Francisco (possible transfer from the Tocantins River), Jucar (planned but abandoned transfer from the Ebro River)
- desalination plants (Rio Grande [El Paso, Texas], Colorado, Jucar, Nile [under construction in Egypt])
- wastewater treatment and reuse (Yellow, Jucar, Nile [Egypt], MDB, etc.)

16.3.2 Irrigation Water Demand Management

The impacts of excessive supply augmentation projects have prompted a fashionable interest in 'soft approaches' and demand management. In the field of irrigation this has spurred calls for the use of water pricing, better user coordination/organisation, improved farming practices and technological change (adoption of micro-irrigation, canal lining and automation). I focus here on the last two.

FARM-LEVEL ADJUSTMENTS

Water scarcity elicits behavioural and technical changes in irrigation that are well documented by the literature and illustrated in our basins. During the Millennium drought in the Murray–Darling Basin farmers switched from surface water to groundwater and drilled additional or deeper wells. They also invested in more efficient irrigation techniques (drip irrigation) and the reuse of drainage water, and generally switched crops to horticulture during droughts.

In the area irrigated by the lower Yellow River (and more broadly in the northern plains[7]) various measures were found to have the potential to reduce ET. These included zero tillage, deficit irrigation, revised cropping patterns, improved cultivars and mulching (the last reducing consumption by up to 25 per cent) (Yan et al., 2015). Zhong et al. (2017) demonstrated the benefit of adjusting cropping patterns regionally, by adopting winter fallowing and the early sowing of summer maize as a monocrop in water-scarce areas of the region, while replacing winter wheat-summer maize sequential cropping systems by a wheat-maize relay intercropping system.

Most commonly, however, farmers who use surface water resources make up for shortfalls in supply by turning to groundwater. This involves the drilling of (often illegal) wells, thereby shifting the pressure to aquifers, which are quick to show signs of decline.

THE PROMISES AND FALLACIES OF IRRIGATION 'MODERNISATION' AND WATER SAVINGS

Due to its overwhelming responsibility for water diversions and consumption, irrigation is often depicted as the 'villain' that 'guzzles' water in a process where 'only half of the water distributed reaches the crop' (see Postel, 1997; Gleick, 2001; and Molle et al., 2010). Indeed, extremely low efficiency is often observed in countries/seasons where rainfall is plentiful and continuous flow a rational way to manage water and reduce labour costs. But even in water-scarce regions, gravity-irrigation systems typically have an overall efficiency of around 0.5 (e.g., the Nile Delta). However, two adjustments to water scarcity are glossed over (see Molle et al., 2010): first, when confronted with insufficient and/or unreliable supply, farmers in collective schemes often tap drainage water and aquifers. Second, in such water-scarce regions where demand is high and spatially widespread, any return-flow from irrigation schemes (and even from cities), whether to rivers, drains or aquifers, has in general been tapped by the same or other downstream farmers. 'Losses' at one point are someone else's resource (Grafton et al., 2018). Notable exceptions are situations where the return flow is degraded in quality and cannot be reused (saline aquifers, surface flows to sinks, pollution). Despite this (by now) well-recognised phenomenon, 'irrigation modernisation', 'canal lining' and 'water-saving interventions' are promoted based on a narrow view of plot-level efficiency/losses, without further screening of local specificity, and receive billions of dollars in investment.

In Australia, the Millennium drought prompted a A$10 billion 'rescue package', announced in 2007, which was increased to A$12.8 billion after a change in government under the banner 'Water for the Future' (Grafton, 2017; Grafton et al., 2020). The budget was to be spent (1) to purchase water entitlements from willing sellers; (2) to subsidise improvements to water infrastructure to reduce off-farm losses (primarily through state projects); and (3) to subsidise reductions in on-farm water losses, with at least 50 per cent of the 'savings' being transferred to the Australian government (in the form of water entitlements) (Grafton, 2017; Grafton et al., 2020). By May 2016 around 1,166 Mm^3/year had been acquired through the direct purchase of water entitlements, while 595 Mm^3/year had been 'saved' through infrastructure improvements and subsidies.

Yet a heated debate emerged around two major issues, the first being economic: the infrastructure improvement programme faced claims that buying out entitlements was far preferable financially (Wittwer and Dixon, 2013; Grafton, 2017). The second issue was whether the claimed 'savings' were accurate given the lack of information on how return-flows had been altered and downstream appropriators as well as the environment impacted (Crase and O'Keefe, 2009; Adamson and Loch, 2014; Horne, 2014). Uncertainty remains over the impact on river baseflows of both groundwater abstraction and 'infrastructure improvements'. Evans (2004, cited in Grafton, 2017) estimated that for every 100,000 m^3 of groundwater extracted in the MDB, surface water would be reduced by an average of 60,000 m^3. Wang et al. (2018) have assessed that the irrigation efficiency projects reduced base-flow by 121 Mm^3/yr, while the likely growth in groundwater use in the next 40 years will reduce it by 170 Mm^3/yr.

In the Yellow River Basin, the water-saving potential is obvious in irrigation schemes where insufficient drainage capacity causes waterlogging, as in the Qingtongxia irrigation district. Here the precision levelling of basins would reduce water application needs and increase yields (see Pereira et al., 2007). The past 15 years have seen much debate in the downstream section on how to save water and/or raise water productivity in irrigated areas. Kendy et al. (2003) emphasised that most of the 'losses' by infiltration in these plains reached the superficial quaternary aquifer and were readily taken up again through wells. They stressed the importance of looking at ET regionally and that 'net water savings' could only be achieved by reducing it (changes in crops, mulching, etc.).

In the Hai Basin Plain, canal lining, pipeline distribution technology (covering more than 2.7 Mha) and the introduction of drip and sprinkler irrigation (492,000 ha) have been widely promoted as 'water saving' since the 1980s. Their failure to restore the water balance was due to not having affected net consumption (ET) (Yan et al., 2015). Zhong et al. (2017) have attempted a regional balance to test the overall impact of water-saving measures in Hebei Province (where 80 per cent of the water used in irrigation is groundwater), including sprinklers, drip, low-pressure pipe irrigation, seepage-prevention measures and changes in farming practice. They found that groundwater overdraft had indeed been reduced. Although a third of the reduction was due to more abundant rainfall, they concluded that the combination of water-saving interventions (which unfortunately could not be disaggregated) had had a positive effect on groundwater overdraft.

The Upper Rio Grande is a typical basin with irrigation areas concentrated along the main river valley, a high level of water recycling and high connectivity between the aquifer and the river. It is also typical of a closed basin, where irrigation has outstripped resources and the shortfall is met by depleting the groundwater in falling aquifers (De Stefano et al., 2018). Ward and Pulido-Velazquez (2008) modelled various levels of subsidies to drip irrigation in the Elephant Butte Irrigation District (EBID) of southern New Mexico, where all return flows to the river or quaternary valley aquifer are reappropriated downstream. They found that better irrigation increases yields but also, predictably, ET, meaning that purported 'conservation measures' merely enhance local benefits from a finite resource to the detriment of downstream users. Fernald and Guldan (2006) found that seepage from earthen ditches could have multiple benefits, such as diluting agricultural chemicals or pollutants, providing groundwater recharge to shallow wells and delaying return flow to the stream, thereby evening out instream flow.

Spain has also designed an ambitious multi-billion euro programme to 'modernise' irrigation and 'save' water (1.16 Bm^3 at an estimated cost of €7.3 billion). In the Jucar Basin an ongoing water conservation programme (€790 million between 2009 and 2027) aims to save 300 Mm^3/year. It is acknowledged, however, that the net amount of water saved is much lower due to a decrease in aquifer recharge (which must be maintained at a high enough level to allow salt leaching) (CHJ, 2014). When these 'paper water' savings are not recognised as such they may be 'transferred' to other areas, as was the case when the water 'saved' through the modernisation of traditional irrigation systems in the Jucar Basin was planned to be transferred to the Vinalopó Basin (Sanchis-Ibor et al., 2019). This would have further desiccated the Jucar Basin.

Egypt's Ministry of Agriculture has floated proposals for massive technological shifts as part of a 640 billion pound[8] agricultural strategy until 2030. However, the aim of 10 Bm^3 of water savings is nonsensical considering the water and salt balance of the delta and the limitations mentioned earlier (Molle et al., 2018). More recently, the government has announced large investment in canal lining, also mistakenly expected to yield 5 Bm^3 of 'savings'.[9]

Last, the Colorado provides a textbook example of how investment in canal lining and farm-level efficiency in a closed basin are likely to amount to no more than the spatial and social reallocation of water. The All-American Canal diverting Colorado's waters to the Imperial Valley District has been lined with the aim of 'saving' 100 Mm^3 of seepage water and transferring the same volume to San Diego. But this water turned out to be the main source of recharge of the aquifer in the Mexicali Valley (in Mexico) (Maganda, 2005). This alleged win-win intervention ended up depriving Mexico of seepage water as well as reducing return flows to the Salton Sea, with major health hazards and environmental damage (Chapter 13).

Beyond the key question of the fate of return flows it appears that the volume of water effectively consumed at plot level is basically unchanged with the shift to drip, and even sometimes increased. This is due to substantially increased transpiration caused by the more frequent and better application of water that may offset gains in terms of soil evaporation (Perry and Steduto, 2017). In addition, the shift to drip irrigation is often accompanied by a number of key changes in farming: a) a shift to more water-consuming crops, b) a densification of tree plantations, c) an extension of the cultivated area made possible by reduced per-hectare application rates (the so-called rebound effect) (Willardson et al., 1994; Playan and Mateos, 2006; Ward and Pulido-Velazquez, 2008). Each of these factors results in greater evapotranspiration (i.e., water consumption). The expansion of irrigation by using 'saved water' has been observed in several places, including Spain (Mateos and Playán, 2006; Berbel et al., 2014; WWF/Adena, 2015), Morocco (Tanouti and Molle, 2013), China (World Bank, 2015; Zhang et al., 2017) and the United States (Huffaker et al., 2000; Huffaker and Whittlesey, 2003).

González-Cebollada (2015) showed that in seven schemes in Spain per-hectare water consumption had increased by between 4 per cent and 45 per cent due to intensification and a shift to crops with higher water requirements, notably trees. Likewise, Sese-Minguez et al. (2017) found that in the Cànyoles watershed drip irrigation was associated with intensification, a shift to trees and expansion into former rain-fed areas. This is not necessarily the case everywhere, however. Sanchis-Ibor et al. (2015) found that in the Mijares River Basin the main cropping patterns (citrus) were not affected by the shift to drip.

In the MDB, for example, on-farm infrastructure subsidies are likely to be taken up by farmers with perennial crops (vs. rice or cotton growers) or incentivise a shift to perennial crops. This will reduce flexibility in coping with water shortages by fallowing and increase the market price of water in times of scarcity (Grafton, 2017).

Modernisation means better control of water quantity and quality at the farm level but also of fertilisation, as well as improved technical services and market linkages (Mateos and Playán, 2006). Therefore, yields are in general substantially enhanced and diversification to cash crops generates higher incomes. In summary, drip and agricultural intensification generally achieve only half that expected or announced. They improve productivity and reduce labour costs and drudgery, and often the amount of water applied, but seldom do they reduce water consumption. At the system or river basin level they do not represent a solution to excessive water depletion.

16.4 CONCLUSION

Since irrigation is the main consumptive user of water (due to an increased evapotranspiration process), its unchecked development is the main driver of water resource overexploitation and basin closure. This fact is illustrated by our sample of basins, which are all overallocated and, as a result, prone to periodic shortages, conflicts between uses, instream and in-dam sediment accumulation and environmental degradation.

The historical reasons for the excessive expansion of irrigation in these basins and elsewhere include geopolitical considerations (or their equivalent in federal systems such as the United States and Australia) and the host of financial and political benefits that accrue to powerful players, including politicians, hydraulic bureaucracies, construction companies and bankers. This has been described in the case of the United States by the seminal works of Worster (1985) and Reisner (1986) and found to apply globally (see Molle et al., 2009). But this state-driven process has been paralleled by an equally unchecked growth in the use of groundwater resources, generally by individuals and dictated to a degree by local physical conditions. In some basins (for example, the Jucar), or countries such as India, groundwater abstraction now even exceeds surface water abstraction.

In our sample basins, irrigation – having outstripped the availability of water resources – is experiencing the 'double squeeze' found in the Mediterranean by Molle and Sanchis-Ibor (2019). On the demand side it will lose out to other sectors, such as domestic or industrial uses, as well as to the environment, whose 'needs' come to be recognised and factored in (as in the MDB). On the supply side it will be affected by the decrease in rainfall and runoff, as well as by more frequent and extreme events (in the case of drought, irrigation is the 'adjustment variable').

Irrigation is therefore forced to adjust. Governments and individual farmers alike tend to first favour supply augmentation. This can come from new reservoirs, the reuse of wastewater (whether treated or not), transfers, desalination or pumping from aquifers. Where these options are already fully exploited or unfeasible (physically or economically) there is no alternative but to reduce evapotranspiration (rather than 'use'). This is often forced by circumstances and may result in crop losses or short-term adjustments, such as deficit irrigation. In the medium term (seasonally or annually) farmers may adjust by fallowing land, changing crops or calendars, and improving on-farm techniques and practices (mulching, land-levelling, etc.), while long-term strategies may involve the sale of land or economic diversification.

This chapter did not attempt a detailed analysis of basin or national policy, but it must be noted that nowhere are these adjustments within the irrigation sector taking place smoothly. As pressure mounts on water resources, those available and the return flows tend to be tapped, since the abstraction capacity far exceeds the total supply. Reducing irrigation use without reducing evapotranspiration becomes increasingly difficult, and reducing evapotranspiration, whether through planning or enforcement, usually means less production and income. States therefore face two undesirable challenges: the need to implement measures that will likely dent the stream of benefits to irrigators, thereby fomenting discontent, and the sheer difficulty of implementing and enforcing such measures on so many spatially scattered, hostile users.

REFERENCES

Adamson, D. and Loch, A. (2014). Possible Negative Feedbacks from 'Gold-Plating' Irrigation Infrastructure. *Agricultural Water Management*, 145, pp. 134–144.

Alexandra, J. (2017). Risks, Uncertainty and Climate Confusion in the Murray–Darling Basin Reforms. *Water Economics and Policy*, 3(3).

Berbel, J., Gutierrez-Martin, C., Rodriguez-Diaz, J. A., Camacho, E., and Montesinos, P. (2014). Literature Review on Rebound Effect of Water Saving Measures and Analysis of a Spanish Case Study. *Water Resources Management*, 29, pp. 663–678. https://doi.org/10.1007/s11269-014-0839-0

CBHSF (2015). *Plano de recursos hídricos da bacia hidrográfica do rio São Francisco [Water Resources Plan for the São Francisco River Basin]*. Volume I.

Chao, N., Luo, Z., Wang, Z., and Jin, T. (2017). Retrieving Groundwater Depletion and Drought in the Tigris–Euphrates Basin between 2003 and 2015. *Groundwater*, 56(5).

Confederacion Hydrografica del Júcar [Hydrographic Confederation of the Jucar] (CHJ) (2014). *Memoria-Anejo 10*. Programa de medidas. Demarcación Hidrográfica Del Júcar. [*Memory-Annex 10*. Program of Measures. Júcar River Basin District.

Crase, L. and O'Keefe, S. (2009). The Paradox of National Water Savings: A Critique of 'Water for the Future'. *Agenda: A Journal of Policy Analysis and Reform*, 16(1), pp. 45–60.

De Stefano, L., Welch, C., Urquijo, J., and Garrick, D. E. (2018). Groundwater Governance in the Rio Grande: Co-evolution of Local and Intergovernmental Management. *Water Alternatives*, 11(3), pp. 824–846.

Döll, P. and Siebert, S. (2002) Global Modeling of Irrigation Water Requirements. *Water Resources Research*, 38(4), p. 1037.

Esteban, E., Dinar, A., Albiac, A. et al. (2016). *The Political Economy of Water Policy Design and Implementation in the Jucar Basin, Spain*. UCR SPP Working papers.

Evans, R. (2004). *River–Groundwater Interaction in the Murray–Darling Basin: Technical Status and Management Options*. In 9th Murray–Darling Basin Groundwater Workshop, Bendigo, 12–19 February.

FAO (Food and Agriculture Organization) (2011). *Water Balance in the Nile Basin*. Available at www.fao.org/nr/water/faonile

Feng, X. et al. (2016). Revegetation in China's Loess Plateau Is Approaching Sustainable Water Resource Limits. *Nature Climate Change*. https://doi.org/10.1038/NCLIMATE3092

Fernald, A. G. and Guldan, S. J. (2006). Surface Water–Groundwater Interactions between Irrigation Ditches, Alluvial Aquifers, and Streams. *Reviews in Fisheries Science*, 14(1–2), pp. 79–89. https://doi.org/10.1080/10641260500341320

Gleick, P. (2001). Making Every Drop Count. *Scientific American*, 284(2), pp. 40–45.

González-Cebollada, C. (2015). Water and Energy Consumption after the Modernization of Irrigation in Spain. *WIT Transactions on the Built Environment*, 168, pp. 457–465.

Grafton, R. Q. (2017). Water Reform and Planning in the Murray–Darling Basin, Australia. *Water Economics and Policy*, 3(3).

Grafton, R. Q. et al. (2012). Global Insights into Water Resources, Climate Change and Governance. *Nature Climate Change*. https://doi.org/10.1038/nclimate1746

Grafton, R. Q., Williams, J., Perry, C. J. et al. (2018). Paradox of Irrigation Efficiency and the Global Water Crisis. *Science*, 361(6404), pp. 748–750.

Grafton, R. Q., Colloff, M. J., Marshall, V., and Williams, J. (2020). Confronting a 'Post-Truth Water World' in the Murray–Darling Basin, Australia. *Water Alternatives*, 13 (1), pp. 1–28.

Hancock, G. and Pietsch, T. (2008). *Sediment Tracing and Dating Techniques Employed at CSIRO Land and Water*. CSIRO Land and Water Science Report, CSIRO.

Horne, J. (2014). The 2012 Murray–Darling Basin Plan: Issues to Watch. *International Journal of Water Resources Development*, 30(1), pp. 152–163.

Hu, Q., Yang, Y., Han, S. et al. (2017). Identifying Changes in Irrigation Return Flow with Gradually Intensified Water-Saving Technology Using HYDRUS for Regional Water Resources Management. *Agricultural Water Management*, 194, pp. 33–47.

Huffaker, R. and Whittlesey, N. (2003). A Theoretical Analysis of Economic Incentive Policies Encouraging Agricultural Water Conservation. *Water Resource Development*, 19(1), pp. 37–53.

Huffaker, R., Whittlesey, N., and Hamilton, J. R. (2000). The Role of Prior Appropriation in Allocating Water Resources into the 21st Century. *Water Resource Development*, 16(2), pp. 265–273.

Kendy, E., Molden, D. J., Steenhuis, T. S., Liu, C. M., and Wang, J. (2003). *Policies Drain the North China Plain: Agricultural Policy and Groundwater Depletion in Luancheng County, 1949–2000*. Research Report 71. Colombo: International Water Management Institute.

Kucukmehmetoglu, M. and Geymen, A. (2014). Transboundary Water Resources Allocation under Various Parametric Conditions: The Case of the Euphrates & Tigris River Basin. *Water Resources Management*, 28, pp. 3515–3538.

Maganda, C. (2005). Collateral Damage: How the San Diego-Imperial Valley Water Agreement Affects the Mexican Side of the Border. *The Journal of Environment & Development*, 14(4), pp. 486–506.

MDBA (2010). *Guide to the Proposed Basin Plan*.

(2016). *The Murray–Darling Basin at a Glance*.

Meng, F., Su, F., Yang, D., Tong, K., and Hao, Z. (2016). Impacts of Recent Climate Change on the Hydrology in the Source Region of the Yellow River Basin. *Journal of Hydrology: Regional Studies*, 6, pp. 66–81.

Molden, D., Oweis, T., Steduto, P. et al. (2010). Improving Agricultural Water Productivity: Between Optimism and Caution. *Agricultural Water Management*, 97, pp. 528–535.

Molle, F. (2008). Why Enough Is Never Enough: The Societal Determinants of River Basin Closure. *International Journal of Water Resource Development*, 24(2), pp. 247–256.

Molle, F. and Closas, A. (2020). Groundwater Management and the Pitfalls of Licensing. *Journal of Hydrogeology*.

Molle, F. and Sanchis-Ibor, C. (2019). Irrigation Policies in the Mediterranean: Trends and Challenges. In F. Molle, C. Sanchis-Ibor, and L. Avella, eds., *Irrigation in the Mediterranean: Technologies, Institutions and Policies*. Global Issues in Water Policy Series. Dordrecht: Springer.

Molle, F. and Wester, P. (2009). River Basin Trajectories: An Inquiry into Changing Waterscapes. In F. Molle and P. Wester, eds., *River Basins Trajectories: Societies, Environments and Development*. Wallingford and Cambridge, MA: CABI, pp. 1–19.

Molle, F., Wester, P., and Hirsch, P. (2010). River Basin Closure: Processes, Implications, and Responses. *Agricultural Water Management*, 97, pp. 569–577.

Molle, F., Wester, P., and Mollinga, P. P. (2009). Hydraulic Bureaucracies: Flows of Water, Flows of Power. *Water Alternatives*, 2(3), pp. 328–349. Available at www.water-alternatives.org

Molle, F., Gafaar, I., Al-Agha, D. E., and Rap, E. (2018). The Nile Delta's Water and Salt Balances and Implications for Management. *Agricultural Water Management*, 197(15), pp. 110–121.

Ortega-Reig, M., Sanchis-Ibor, C., and Garcia-Molla, M. (2017). Drip Irrigation in Eastern Spain. Diverging Goals in a Converging Process. In J. P. Venot, M. Kuper, and M. Zwareveen, eds., *Drip Irrigation for Agriculture: Untold Stories of Efficiency, Innovation and Development*. Abingdon: Earthscan from Routledge.

Pereira, L. S., Goncalves, J. M., Dong, B., Mao, Z., and Fang, S. X. (2007). Assessing Basin Irrigation and Scheduling Strategies for Saving Irrigation Water and Controlling Salinity in the Upper Yellow River Basin, China. *Agricultural Water Management*, 93(3), pp. 109–122.

Perez-Martin, M. A., Batan, A., del-Amo, P., and Moll, S. (2015). Climate Change Impact on Water Resources and Droughts of AR5 Scenarios in the Jucar River, Spain. In J. Andreu, A. Solera, J. Paredes-Arquiola, D. Haro-Monteagudo, and H. Van Lanen, eds., *Drought: Research and Science-Policy Interfacing*. London: CRC Press, pp. 189–196.

Perry, C. and Steduto, P. (2017). *Does Improved Irrigation Technology Save Water*. Cairo: FAO.

Pittock, J., Williams, J., and Grafton, R. Q. (2015). The Murray–Darling Basin Plan Fails to Deal Adequately with Climate Change. *Water: Journal of the Australian Water Association*, 42(6), p. 28.

Playán, E. and Mateos, L. (2006). Modernization and Optimization of Irrigation Systems to Increase Water Productivity. *Agricultural Water Management*, 80(1–3), pp. 100–116.

Postel, S. (1997). *Last Oasis: Facing Water Scarcity*. New York: World Watch Institute.

Rahi, K. A. (2018). Salinity Management in the Shatt Al-Arab River. *International Journal of Engineering & Technology*, 7(4.20), pp. 128–133.

Reisner, M. (1986). *Cadillac Desert: The American West and Its Disappearing Water*. New York: Penguin Books.

Richter, B. D. et al. (2017). Opportunities for Saving and Reallocating Agricultural Water to Alleviate Water Scarcity. *Water Policy*, 19, pp. 886–907.

Sabadini-Santos, E., Knoppers, B. A., Padua Oliveira, E., and Leipe, T. (2009). Regional Geochemical Baselines for Sedimentary Metals of the Tropical São Francisco Estuary, NE-Brazil. *Marine Pollution Bulletin*, 58, pp. 601–634. https://doi.org/10.1016/j.marpolbul.2009.01.011

Sanchis-Ibor, C., Garcia-Molla, M., and Avella-Reus, L. (2017a). Effects of Drip Irrigation Promotion Policies on Water Use and Irrigation Costs in Valencia, Spain. *Water Policy*, 19(1), pp. 165–180.

Sanchis-Ibor, C., Macian-Sorribes, H., Garcia-Molla, M., and Pulido-Velazquez, M. (2015). Effects of Drip Irrigation on Water Consumption at Basin Scale (Mijares River, Spain). 26th Euro-mediterranean Regional Conference and Workshops: Innovate to Improve Irrigation performances, Montpellier, France.

Sanchis-Ibor, C. et al. (2019). Spain. In F. Molle, C. Sanchis-Ibor, and L. Avella, eds., *Irrigation in the Mediterranean: Technologies, Institutions and Policies*. Global Issues in Water Policy Series. Dordrecht: Springer.

Sese-Minguez, S., Boesveld, H., Asisns-Velis, S., van der Kooij, S., and Maroulis, J. (2017). Transformations Accompanying a Shift from Surface to Drip Irrigation in the Semi-arid Cànyoles Watershed, Valencia, Spain. *Water Alternatives*, 10(1).

Shiklomanov, I. A., ed. (1997). *Assessment of Water Resources and Water Availability in the World, Comprehensive Assessment of the Freshwater Resources of the World*. Stockholm: Stockholm Environment Institute.

Soto-Garcia, M., Martinez-Alvarez, V., Garcia-Bastida, P. A., Alcon, F., and Martin-Gorriz, B. (2013). Effect of Water Scarcity and Modernisation on the Performance of Irrigation Districts in South-Eastern Spain. *Agricultural Water Management*, 124, pp. 11–19.

Tanouti, O. and Molle, F. (2013). Réappropriations de l'eau dans les bassins versants surexploités: le cas du bassin du Tensift

(Maroc) [Reappropriation of Water in Overexploited Watersheds: The Case of the Tensift Basin (Morocco)]. *Etudes Rurales [Rural Studies]*, 192(2), pp. 79–96.

Voss, K. A., Famiglietti, J. S., Lo, M. et al. (2013). Groundwater Depletion in the Middle East from GRACE with Implications for Transboundary Water Management in the Tigris–Euphrates–Western Iran Region. *Water Resources Research*, 49. https://doi.org/10.1002/wrcr.20078

Wang, Q. J., Walker, G., and Horne, A. (2018). *Potential Impacts of Groundwater Sustainable Diversion Limits and Irrigation Efficiency Projects on River Flow Volume under the Murray–Darling Basin Plan.* An Independent Review. Melbourne School of Engineering.

Ward, F. and Pulido-Velázquez, M. (2008). Water Conservation in Irrigation Can Increase Water Use. *PNAS*, 105, pp. 18215–18220.

Willardson, L. S., Allen, R. G., and Fredericksen, H. D. (1994). *Elimination of Irrigation Efficiencies.* 13th Tech. Conference USCID, Denver, CO.

Wittwer, G. and Dixon, J.(2013). Effective Use of Public Funding in the Murray–Darling Basin: A Comparison of Buybacks and Infrastructure Upgrades. *Australian Journal of Agricultural and Resource Economics*, 57(3), pp. 399–421.

World Bank (2015). *Project Performance Assessment Report People's Republic of China Irrigated Agriculture Intensification Project. Mainstreaming Climate Change Adaptation in Irrigated Agriculture Project. Hai Basin Integrated Water and Environment Management Project.* IEG Public Sector Evaluation, Independent Evaluation Group.

Worster, D. (1985). *Rivers of Empire. Water, Aridity, and the Growth of the American West.* New York: Panthenon Books.

WWF/Adena (2015). *Modernización de Regadíos: Un mal negocio para la naturaleza y la sociedad [Irrigation Modernization: A Bad Business for Nature and Society].* Madrid: WWF/Adena. Available at http://awsassets.wwf.es/downloads/modernizacion_regadios.pdf

Yan, N., Wu, B., Perry, C., and Zeng, H. (2015). Assessing Potential Water Savings in Agriculture on the Hai Basin Plain, China. *Agricultural Water Management*, 154, pp. 11–19.

Yang, H. and Jia, S. (2008). Meeting the Basin Closure of the Yellow River in China. *Water Resources Development*, 24 (2), pp. 265–274.

Young, W. J. and Chiew, F. H. S. (2011). Climate Change in the Murray–Darling Basin: Implications for Water Use and Environmental Consequences. In R. Q. Grafton and K. Hussey, eds., *Water Resources Planning and Management.* Cambridge: Cambridge University Press.

Yu, Y., Wang, H., Shi, X. et al. (2013). New Discharge Regime of the Huanghe (Yellow River): Causes and Implications. *Continental Shelf Research*, 69, pp. 62–72.

Zhang, H.. Singh, V. P., Sun, D., Yu, Q., and Cao, W. (2017). Has Water-Saving Irrigation Recovered Groundwater in the Hebei Province Plains of China? *International Journal of Water Resources Development*, 33(4), pp. 534–552.

Zhong, K., Sun, L., Fischer, G. et al. (2017). Mission Impossible? Maintaining Regional Grain Production Level and Recovering Local Groundwater Table by Cropping System Adaptation across the North China Plain. *Agricultural Water Management*, 193, pp. 1–12.

Notes

1 www.fao.org/nr/water/aquastat/tables/WorldData-Withdrawal_eng.pdf

2 Water lost to sinks or degraded in quality (to be unusable) is also often considered as an unproductive loss or 'consumed'.

3 During the last three years of the Millennium drought (2006–2009) withdrawals were down to just ~450 Mm^3 (MDBA, 2010).

4 http://cbhsaofrancisco.org.br/planoderecursoshidricos/aguas-do-sao-francisco-tem-maior-demanda-pelo-setor-de-irrigacao/

5 www.clubeklff.com.br/publicacao/cresce-area-irrigada-no-vale-do-sao-francisco-1227

6 www2.camara.leg.br/atividade-legislativa/comissoes/comissoes-temporarias/especiais/54a-legislatura/pec-368-09-recursos-destinados-a-irrigacao/audiencias-publicas/apresentacao-francisco-jacome-sarmento-28.08.13

7 Although the lower part of the YRB is very narrow, water (~10 Bm^3) is exported towards neighboring irrigated areas in North China Plain.

8 Around $120 billion at 2010 values. This is only half of the investment envisaged, with the other half expected to come from the private sector.

9 www.egypttoday.com/Article/1/85190/Canal-lining-project-saving-Egypt-5-bn-cubic-meters-of

17 "Intelligent" Water Transfers

Eduardo Sávio P. R. Martins, Antônio R. Magalhães, and James E. Nickum

17.1 INTRODUCTION

This chapter addresses issues related to water transfers between river basins, not only in hydrological terms but also, quite frequently, over political boundaries (transboundary water transfer). Interbasin water transfers have a long history in dealing with the water scarcity in a region by the transposition of water surpluses from other relatively more water-abundant areas. Naturally, water scarcity, water sharing, and water conflicts are aspects greatly present in interbasin water transfers, which are often promoted – not always successfully – as a way to support socioeconomic development of water-scarce regions (Gichuki and McCornick, 2008).

Generally, interbasin water transfers involve large hydraulic infrastructures (e.g., channels, dams, pumping stations, aqueducts, pipelines, and others), and because of this they are often criticized as the first thought of solution in water management to address water scarcity. When the water transfer involves different political regions, either countries or states, the debate tends to lose perspective of the important aspects to be analyzed in order to guarantee a rational water policy. These are two aspects of interbasin water transfers: inherent complexity and tendency to generate controversy; the latter could be attributed to the multitude of ways water interacts with society (UNESCO, 1999).

In this context, is the water transfer an issue, or are there other aspects that need to be further examined? New demands, or usual demands under drought conditions, must be met primarily through either the increase of efficiency in water use and in water transport or reallocation of existing supplies. Interbasin water transfers (IBWT) should be considered to deal with the mismatch between water supply and demand in the context of a sustainable development policy and not as an end in itself.

This chapter does not aim to make recommendations on how to implement water transfer projects; rather, it only discusses aspects related to these projects that need to be considered when engaging in such endeavors. Interest in interbasin water transfer systems increases as water availability decreases or water demand for different uses increases in a particular basin. Projects of this nature include ancient practice and there are records of water transfer to agriculture in Spain since the fifteenth century (Lund and Israel, 1995). Here, the authors present several case studies (see Table 17.1), chosen to provide a representative geographical view of IBWTs around the world and insights on the different aspects that involve IBWTs.

17.2 PRESSURE FOR WATER TRANSFERS

Interbasin water transfers have increased in developing countries during recent decades, generally guided by socioeconomic considerations. Water is a necessary condition for life and economic growth everywhere. So, if there is water deficit (actual or potential) in one basin and water surplus in another basin, it is only natural to transfer water from the water surplus basin to the water deficit basin. There are examples of interbasin water transfers all over the world, from the United States to Brazil to Australia to China to Africa to Europe.

Pressures for water transfers may come from the government and other interest groups with the aim of developing a region where there is lack of water for irrigation or for urban consumption, for example, or from irrigation districts that need more water, or urban systems that look for water security. Also, in some cases, the government may want to promote socioeconomic development in a region where there is much poverty and need for more employment and income. Pressures against interbasin transfers depend on the political economy of each country, but they usually come from politicians and other interest groups from the donor basin.

However, things are not so simple. There are many technical difficulties in assessing a situation of water deficit, present or future. Integrated water resources management and water accounting provide better tools to determine a situation of water deficit or water surplus. Advancement in water management is essential. Also, there are technical problems of civil construction, sometimes involving complex channels and tunnel systems. But these problems are all easily solvable as there are adequate technologies and companies prepared to enter construction bids worldwide.

What is more difficult is to identify and bring together a large body of different interests involved in the transfer. The groups of

Table 17.1 *Interbasin water transfers (IBWT) case studies*

IBWT project	Donor region	Recipient region	Purpose
Colorado Big Thompson Project	Colorado River Basin	South Platte River Basin	Irrigation, industry, hydropower generation
Chavimochic (Peru)	Santa River Basin	Inter-valleys of Chao, Virú, Moche and Chicama	Agriculture, agroindustry, water supply, industry, hydropower generation
São Francisco River Diversion Project (Brazil)	São Francisco River	Jaguaribe and Piranhas Açu River Basins, and East portion of Pernambuco and Paraíba States	Urban water supply, irrigation
Tagus - Segura Transfer	Tagus	Segura	Irrigation, urban water supply
El-Salam Channel in Sinai (Egypt)	Nile Delta and Nile River Basin	Deserts of north Sinai	Agriculture
Lesotho Highlands Water Transfer to South Africa (Southern Africa)	Orange/Senqu Basin	Vaal River System	Water supply, industry, hydropower generation
The South-to-North Water Transfer projects (China)	Yangtze River (Chang Jiang) basin	North and Northwest China	Agricultural, municipal, industrial, environmental, water transportation
Snowy Mountains Hydroelectric Scheme (Australia)	Snowy River and Murrumbidgee River Basins	Murray River Basin	Agriculture, hydropower generation

interest may involve governments (federal, provincial and local), water supply companies, irrigation districts, private sector institutions, and civil society. Two large groups of interest have prominence: those representing the donor basin and those representing the recipient basin. Each situation may be different from the other, but there are instances when political interests become intensely involved, as in the case of state governments situated in the donor basin versus state governments situated in the recipient basin. That is why a negotiation process is always necessary before construction of an IBWT is begun.

The negotiation process may take longer than the construction process itself. The "social engineering" is usually more complex than the "construction engineering." However, in some cases, construction starts before the negotiation process is concluded, and sometimes negotiations are still not finished when the water transfer is ready to start. Topics of negotiation may involve different types of compensation for the donor basin, as well as the volume and management of water transfer itself and the way the costs of transfer are paid by the final water users or not. In the case of the São Francisco interbasin transfer project, the federal government in Brazil committed itself to developing a project for revitalization of the river, including reforestation of river margins, sanitation of urban areas, control of pollution, and water quality. The government has committed to draw 1.4 percent of the river water flow, equivalent to 26.4 m^3/s. Total capacity of the water transfer system, which is 127 m^3/s, would be achieved only when there would be excess supply in the donor basin.

Each case is different. However, the need for negotiation is always present, and negotiation is more difficult the more decentralized decision processes are. In any circumstance, a good strategy is to start with the negotiations, to obtain agreement between donors and recipient stakeholders and, in many cases, to define precisely how the transfer water will be used and the costs of transportation will be covered.

17.3 WHAT IS "INTELLIGENT" WATER TRANSFER

The term "intelligent" water transfer is used to call attention to the point that water transfers shall not be the first alternative to address imbalances between water supply and demands, but rather one of the last. There is a set of alternatives, briefly discussed later, that should be explored before engaging in such costly and controversial projects. Also, before its implementation, a rigorous assessment of social and environmental impacts should be carried out, since this type of project may produce unintended adverse effects. During the project phase, one should explore different alternatives aiming to guarantee equitable sharing of interbasin water transfer benefits among donor and recipient basins (UNESCO, 1999). Preferably, such a solution should be considered within a sustainable development policy in order to promote socioeconomic development (Gichuki and McCornick, 2008):

1. increasing total water benefit (transfer of water surpluses from a region to a water-scarce basin/region);
2. facilitating reallocation of water from a low- to a high-value use;
3. reducing regional inequity by promoting socioeconomic development in water-scarce regions;
4. facilitating broader cooperation and promoting solidarity between donor and recipient basins as a possible response to water shortages;
5. restoring degraded freshwater ecosystems.

In other words, the IBWTs should be part of an integrated plan, combining the IBWT itself with both traditional water supply augmentation and demand management measures (Lund and Israel, 1995), the former being ideally the last implementation step in such a plan. The implementation of such an integration plan requires developing strong collaboration among different water use sectors, administrative regions, or countries, and addressing institutional issues.

One must recognize that (a) large-scale IBWTs may represent one of the most significant human impacts in natural processes, (b) such impacts may be addressed both in the donor and recipient basins, and (c) the decision to go ahead with an IBWT should always mean that overall benefits are larger than costs, including political costs of negotiation. An intelligent IBWT should then not only consider all positive and negative impacts but also, and above all, embody a comprehensive negotiation process with the participation of all key stakeholders in both donor and recipient regions. The construction process itself will be strongly facilitated if all agreements have been reached previously.

17.3.1 Is the IBWT Really Needed?

There is no doubt that to avoid controversies, policy makers and water resources managers of water-scarce basins should firstly work on all possible ways to better manage their own water. In this context, knowledge of the water balance in the recipient basin is necessary to establish how one could implement better management of one's own water before engaging in a water transfer project (Ghassemi and White, 2007). Once demonstrated that the recipient basin uses the available water optimally and significant water conservation measures are in place, one may now consider a water transfer solution. Of course, the water transfer solution assumes that the water balance of the donor basin allows such water transfer without compromising its current and long-term water requirements in the donor basin. Also, as already mentioned, the rights of the donor basin need to be respected and different alternatives need to be explored to guarantee equitable sharing of interbasin water transfer benefits between donor and recipient basins (UNESCO, 1999).

Discussions regarding large IBWTs have been accompanied by controversies and they have common elements but with their own specific features. One way to prevent such controversies is to address the water shortages in the recipient basin by first implementing better management practices available, as previously mentioned (Ghassemi and White, 2007). These practices, listed in Section 17.4, along with the prospective IBWT, should be preferably included in a technically sound sustainable development plan for the whole region. The donor basin may have enough water to guarantee current water requirements for its diverse uses (e.g., urban/domestic, agricultural, and industrial) but this assessment may not account for both the long-term trends of such uses and the potential impacts of climate change in the region.

It is reasonable to think that any possible IBWT should be properly evaluated to determine if it is justified or not. However, this seems to have not been always the case (Cox, 1999). In fact, as seems apparent in the literature, the simple need for water may have been enough to trigger the implementation of IBWT projects without a thorough assessment of the impacts on the donor basin (e.g., Cox, 1999). Although this is a reality, due to the nature and magnitude of its potential impacts, a thorough impact assessment before the actual implementation of IBWTs is necessary.

Cox (1999) suggests that such assessment should consider all effects of the IBWT, both tangible and intangible. The IBWT under consideration would then be justified if the produced benefits exceed losses, and the net benefits of the water transfer were equitably shared between donor and recipient basins. Such evaluation would include considerations of both efficiency and equity. Therefore, an IBWT must increase benefits from water use by promoting its efficiency and must address equity by taking into consideration benefit sharing between donor and recipient basins.

17.3.2 Criteria for Evaluation

A framework for determining if a given IBWT is justified or not was suggested by Cox (1999). This framework comprises benefit sharing assessment and impact assessments in three categories: (1) economic productivity; (2) environmental quality; and (3) sociocultural. Below, the five criteria within each of these categories are described.

Regarding economic productivity impacts, a proposed IBWT can be justified if the recipient basin faces significant deficits in meeting current and future demands, even after looking at all available alternatives to the proposed IBWT (Criterion 1). At the same time, the proposed IBWT must not limit significantly the development of the donor basin, which may be acceptable to some extent if the development of the recipient basin compensates the losses in the donor basin (Criterion 2).

The environmental quality and sociocultural assessments, for both donor and recipient basins, must indicate that environment and society are not substantially negatively impacted, and being so, the IBWT is justified when compensation to offset such impacts is provided by the recipient basin and the central government (Criteria 3 and 4, respectively).

Finally, as already mentioned, the net benefits must be shared equitably between donor and recipient basins (Criterion 5). Determination of the benefits and equitable distribution between the donor and recipient basins is not an easy task.

These criteria serve as a starting point for negotiations among parties. They should not be viewed as absolute criteria and should not, probably, be applied individually to a particular IBWT. Also, there may be trade-offs among the criteria which should be considered (Cox, 1999).

17.4 ALTERNATIVES TO INTERBASIN WATER TRANSFERS

A large scale IBWT may not the most cost-effective alternative or complementary option for meeting water demands in a basin. Moreover, the literature shows that IBWTs do not necessarily promote good practices in water management (See Section 17.5). This may be the case, for instance, when the IBWT is not considered within the scope of a sustainable development policy, as one of the available alternatives for addressing deficits between water supply and demand.

In this context, one could adopt a sequential implementation approach toward the consideration of an IBWT solution. The idea behind this approach is to guarantee that the most cost beneficial water management practices be implemented before considering other alternatives that are less cost effective, in particular, an IBWT. Therefore, policy makers and water managers should consider, at the basin level, the implementation of the following ordered alternatives or complementary actions within the context of an integrated planning process and not as an isolated solution:

(a) Reducing water demands;
(b) Reducing losses in the water supply network (urban and rural networks);
(c) Increasing water use efficiency for domestic/urban (water-saving practices) and agricultural water uses (increasing the share of the water taken up by plants, and producing more crop per unit of water);
(d) Implementing educational programs;
(e) Recycling wastewater;
(f) Supplementing water supplies locally (e.g., rainwater harvesting; water conservation; restoring traditional water management structures; desalinization);
(g) Reviewing policy and regulations;
(h) Using water management instruments to promote efficiency (e.g., improving monitoring of water uses and increasing water prices);
(i) Implementing conjunctive use of surface and groundwater resources where possible; and,
(j) INTER-BASIN WATER TRANSFER.

This order seems to be a natural one to follow; however, it does not consider the political capital in place for the water transfer decision, which may not be there in the future. In this case, either you take the decision in favor of the IBWT at the current moment, or simply hope for a forthcoming political opportunity.

17.5 SOME EXPERIENCES OF WATER TRANSFERS: ASSESSING WATER TRANSFER AND ITS EFFECTS

Here, we analyze transfers around the world, describing their characteristics in terms of infrastructure, diversion discharge, donor and recipient regions, main uses, environmental and social impacts, and other aspects. At the end of this section, we present, for all IBWT introduced here, a summary of their process weakness and their expected negative impacts (Table 17.2).

17.5.1 North America

COLORADO BIG THOMPSON PROJECT: C-BT (US)

The State of Colorado has specific geographical features that favored the construction of IBWT schemes from basins located west of the Rocky Mountains, a humid region, to the East, a region of low precipitation and higher demands, such as: (a) the Grand River Ditch Project, built in 1892, which transfers water from the Colorado River to the Cache La Poudre River through canals; and (b) the Colorado Big Thompson (C-BT), which transfers water from the headwaters of the Colorado River Basin, on the western slope of the Continental Divide, to the eastern slope of the Divide, toward the Big Thompson River, a tributary of the South Platte River. The transfer agreement with Western Slope water users guaranteed that the diversion would not compromise existing water rights.

The construction of the C-BT scheme started in 1930 within the scope of President Roosevelt's New Deal agreement for the United States' economic recovery. The agreement was based on big regional development projects such as the Tennessee Valley Authority, Hoover Dam, C-BT, and others. The first funding for

Table 17.2 *Process weakness or expected negative impacts of the interbasin water transfers (IBWT) case studies*

	North America	South America		Europe	Africa		Asia	Oceania
Process weakness or expected negative impacts of the IBWT	Colorado Big Thompson Project/ Colorado River Compact	Chavimochic (Peru)	São Francisco Basin Interlinking Project (Brazil)	Tagus-Segura Transfer, Spain	El-Salam Channel in Sinai (Egypt)	Lesotho Highlands Water Project, Lesotho and S. Afr.	The South-to-North Water Transfer projects in China (China)	Snowy Mountains Hydro-Electric Scheme (Australia)
Demand management in recipient basin was not considered as a pre-planning activity for IBWT		X	X	X	X	X	X	X
IBWT became (or may become) a driver for unsustainable water use in recipient basin (urban and irrigation)		X	X	X	X		X	X
Strong dependence on IBWT in recipient basins	X	X	X	X	X	X	X	X
IBWT now seen as inadequate and other water portfolio are required (groundwater, desalinization, recycling, etc.)				X				X
Proliferation of boreholes to access groundwater (over-exploitation of this resource too)				X				
Donor basin experienced serious environmental impacts through reduced flows especially	X	X		X	X	X	X	X
IBWT created or escalated threats to critically endangered, threatened species	X			X	X	X		
Scheme saw economic benefits in recipient basin at the cost of communities in the donor basin	X		X	X	X	X	X	X
Inadequate consultations with those likely to be affected either directly or indirectly		X			X		X	

Table 17.2 (*cont.*)

Process weakness or expected negative impacts of the IBWT	North America: Colorado Big Thompson Project/ Colorado River Compact	South America: Chavimochic (Peru)	South America: São Francisco Basin Interlinking Project (Brazil)	Europe: Tagus-Segura Transfer, Spain	Africa: El-Salam Channel in Sinai (Egypt)	Africa: Lesotho Highlands Water Project, Lesotho and S. Afr.	Asia: The South-to-North Water Transfer projects in China (China)	Oceania: Snowy Mountains Hydro-Electric Scheme (Australia)
IBWT catalyst for social conflict between donor and recipient basins (or among recipient basins) and/or with government		X	X	X	X		X	X
IBWT has not helped, as foreseen by the project, the situation of the poor affected or displaced by it		X	*		X	X	X	
Post-IBWT mitigation costs very high (environmentally and/or socially)		X	X	X	X	X	X	X
Governance arrangements for IBWT weak, resulting in budget blow-out and/or corruption		X	X		X	X		

* Potentially not as expected

** Modified from WWF (2007) and new IBWT cases included. The assessment here corresponds to the view of the authors based on several documents, papers, and reports analyzed, including the WWF (2007). Nonchecked process weakness or expected negative impacts may not reflect current conditions.

the construction of the project was sanctioned in 1937, and, in the same year, the management institution was created (Northern Colorado Water Conservancy District, NCWCD). Construction was initiated in 1938, the first water allocations started in 1947, and construction was finished in 1959 (USDI, 2004).

The design volume to be transferred is 0.38 km^3/yr, but the average transfer in the past 43 years has been about 72 percent of this value (Ghassemi and White, 2007). The main uses of the C-BT scheme are irrigation, industry, and hydropower generation. Its infrastructure consists of eighteen pumping stations (pumping capacity of 30.6 MW), canals (153 km), tunnels (56.3 km), twelve reservoirs (1.23 billion m^3), and eleven hydropower plants (total capacity 183.95 MW). This infrastructure delivers water to twenty-nine cities and towns and 607,000 hectares of irrigated land. The water delivered is computer-controlled and uses a communication system to provide realtime or near-realtime information to water users and system managers.

Ghassemi and White (2007) identified the following environmental and social impacts of this water transfer:

(a) Environmental impacts
1. introduction of renewable sources of hydroelectric power generation for the growth of municipalities, industry, and agriculture;
2. supply of drinking water to the rural areas of the eastern side;
3. increase of the threat level for critically endangered species;
4. altered/damaged indigenous archaeological sites dating from 4,000 to 8,000 years old.

(b) Social impacts
1. reduction of water availability to sustain high value irrigated fruit production in the Colorado River during the dry season. As a mitigation measure, a large compensation reservoir, the Green Mountain, was built in the Blue River;
2. losses in taxes due to land expropriation for the project;
3. dissatisfaction of the population on the west side, which was adversely affected due to the reduction of their water availability.

The water allocation, following the NCWCD policies and procedures, may be transferred between users annually or permanently (Ghassemi and White, 2007). This made possible the increase in water allocation necessary to the growth and expansion of municipal areas by transferring water from agricultural users to the water supply for supporting such growth. This transfer between uses sounds reasonable, since much of the land annexed by municipalities had been agricultural land (Zimbelman and Werner, 2001). These innovative water transfer arrangements were facilitated through a series of legal and institutional reforms (Lund and Israel, 1995). In this context, water market and water banking are two mechanisms that facilitate the efficient reallocation of water resources. The C-BT water project of the Northern Colorado Water Conservation District (NCWCD) is a good example of how water transfers are managed in the United States of America.

17.5.2 South America

CHAVIMOCHIC IBWT (PERU)

The lower portion of the Santa River divides the regions of Ancash and La Libertad, and a significant percentage of the Santa River Basin is within La Libertad's boundaries. The competition between the two regions, already marked, is aggravated by water competition, due to the presence of two large coastal irrigation and water transfer systems: Chinecas (Ancash) and Chavimochi (La Libertad). The Chavimochi Transfer in Peru derives waters, to a maximum amount of 3.31 km^3/yr, from the Santa River for the irrigation of the valleys and inter-valleys of **Cha**o, **Vi**rú, **Mo**che, and **Chi**cama, all located close to the north coast of Peru. In addition to the agricultural benefit obtained by the incorporation of new lands (65,000 ha) and the improvement of the irrigation system of the old lands (around 78,000 ha), the project will promote the installation of agroindustries destined for the export market. Complementarily, the project will supply water for domestic and industrial use in the city of Trujillo and hydropower generation through the construction of small plants. The infrastructure comprises tunnels (the longest one has 10 km); canals (the main canal has 270 km); pipelines; siphons; three hydropower plants (two at 300 MW and one at 7.5 MW); and a water treatment plant for water supply of the city of Trujillo (maximum capacity of 1 m^3/s). The total cost of the project was estimated at US$2.25 billion.

The identified impacts of this water transfer can be listed as follows (Azevedo et al., 2005; Casana, 2005; Lynch, 2013; Aguirre and Drinot, 2017):

(a) Environmental impacts
1. reduction of streamflow in the donor basin due to vast increase in demand, which may be even worse due to the accelerated glacial melt indicated by climate change studies (USAID, 2011)[1];
2. increase of water diversion from the Amazon basin and the highlands, where it is essential to the maintenance of natural systems, to the coast;
3. increase of the water table in Huanchaco, in the area of Boquerón, where there is underground water at a very shallow depth. It is surfacing with serious danger for the modest houses.

(b) Social impacts
1. reduction of water supply available to older and small coastal irrigation canals serving local food producers and to Chimbote, Peru's third largest city, also affected by competition between the two transfer schemes (Chavimochi and Chinecas);
2. increase of water consumption in the receiving basin due to all expectations generated by the IBWT. Casana (2005), general manager of the project, reports that in the Chao valley irrigators use twice as much water as needed;
3. increase of upstream–downstream imbalances in the agricultural sector, by bringing water from the Amazon basin and the highlands, where it is essential to the maintenance of natural and agroecosystems and to domestic food security, to the coast, where export agriculture predominates. This tends to be exacerbated once the latest phases of the two projects are completed;
4. generation of substantial profits, employment, and income for Peru, but also increased vulnerability of millions of people dependent on the glacier-fed Santa River. This may foster heated water conflicts in the future among different users, due to the already mentioned imbalance in the agriculture sector.

According to Lynch (2013), the water competition in Santa River Basin is more motivated by the incompatibility of its multiple uses than by water scarcity: (1) Peru Government policies have stimulated water demand increases for both hydropower and agroexports and favored the growth of mining, which has impacted water quality to a point that it cannot be used for human/animal consumption or irrigation and (2) the water scarcity discourse has been used to justify the water transfer to make feasible agroexport growth, or in its actual meaning, allow export of virtual water that could be used to improve food security. However, water scarcity can become a physical problem in the future due to climate change impacts in the Andean glaciers.

Unsustainable water use in the receiving basin was already reported by Casana (2005) due to all expectations generated by the IBWT. Measures aiming to both improve demand management and curb water demand in the receiving basins should be implemented urgently. Moreover, decision makers should pursue strengthening of the negotiation process among all players involved in the transfer.

SÃO FRANCISCO IBWT (BRAZIL)

The São Francisco Water Transfer in Brazil was initially thought of in the nineteenth century, but only taken up again at the end of the twentieth century and early 2000s. Construction started in 2007. The project is a large-scale diversion scheme to transfer water from the São Francisco River Basin to the north of the semi-arid Northeast. The volume transferred is 0.83 km^3/yr in normal conditions[2] from the donor basin and the infrastructure, interlinked to existing reservoirs, comprises the two following axes: Northern axis (mean discharge = 0.52 km^3/yr or 16.4 m^3/s; maximum discharge = 3.12 km^3/yr or 99.0 m^3/s): three pumping stations, thirty-six canals (161.37 km), eight aqueducts (2.48 km), three tunnels (21.48 km), fifteen reservoirs, and three hydropower plants (total capacity 123 MW); Eastern axis (mean discharge = 0.32 km^3/yr or 10.0 m^3/s; maximum discharge = 0.88 km^3/yr or 28.0 m^3/s): six pumping stations; twenty-three canals (167.66 km), one aqueduct (822.2 m), one tunnel (6.52 km), twelve reservoirs, and a pipeline (4 km).

The main purposes of this transfer are first, urban water supply and livestock water use, and second, irrigation, industry, and tourism use once the former purposes are guaranteed. In terms of impacts, one can list the following for this IBWT (Azevedo et al., 2005; WWF, 2007; Roman, 2017; Magalhães and Martins, 2019):

(a) Environmental impacts: The identified negative impacts refer mainly to the direct impacts of the implementation and operation of the project and can be, in general, mitigated and/or compensated by appropriate measures. The identified environmental impacts are:
  1. reduction in biodiversity of native aquatic communities in receiving basins;
  2. uncertainty regarding the adequacy of stream regimen determined;
  3. loss and fragmentation of areas with native vegetation.

(b) Social impacts: The impacts reported here can only be regarded as potential impacts, since the project is not fully operational – only the Eastern Axis is partially in operation. The potential social impacts are:
  1. increase of social conflicts between donor and recipient basins, among recipient basins and among users within the same receiving basin, in particular during extreme drought events;
  2. increase of water consumption in the receiving basin due to all expectations generated by the IBWT (the first signs are already reported in newspapers and workshops about the water use of this transfer).

The São Francisco IBWT may become a driver for unsustainable water use in the receiving basin and measures aiming to both improve demand management and curb the water demand in the receiving basins should be implemented urgently. Moreover, decision makers should continue to pursue the implementation of the São Francisco Basin and Semi-Arid Integrated Sustainable Development Program and strengthen the negotiation process among all players involved in the transfer.

The IBWT expected cost was US$4.5 billion, but the project faced several delays (as of December 2018, it was not completed yet) and substantive cost overruns (from US$4.5 to over US$10 billion). Also, some construction companies involved in the construction works are currently under scrutiny by the Federal Court of Accounts (Roman, 2017) and Justice System for contracting irregularities.

The legal and institutional bases needed to ensure the operation, maintenance, and economic-financial return of the project are yet to be properly defined. Likewise, although the construction works are about 95 percent finished, the IBWT management model is not yet fully defined, and neither is the project management.

Much debate, mobilization, conflict, and improvement in public policies occurred during the discussion of the project, during the negotiation of the water licenses, and now throughout its implementation. One might even think that this process is in its final phase, but solutions involving the "old chico" (as the São Francisco River is sometimes locally called) based solely on increasing supply continue to be thought out. These include a south axis for Bahia, a west axis for Piaui, and a water transfer from the Tocantins River to the São Francisco River. These may be early signs of a new heated federative discussion aiming at providing water security for economic development, independent of the cost.

TRASVASE DAULE-SANTA ELENA (ECUADOR)

The Santa Elena peninsula's economy is based on oil exports, tourism, aquiculture, and agriculture and one of the lowest illiteracy rates in the country. In contrast, the region has a high child mortality rate due to poor sanitation conditions (Azevedo et al., 2005). In this context, the IBWT project was implemented in the

Santa Elena peninsula, located close to Guayaquil in Ecuador, for multiple uses: irrigation, domestic, and industrial. The implementation included complementary infrastructure works to promote local development by the implementation of a water supply system, and collection and disposal of wastewater.

This IBWT has as donor basin the Daule Basin and the receiving region is the Santa Elena peninsula, which has a potential area of 42,000 ha for agriculture and industrial development.

The infrastructure comprises three sections: Section I comprises the capture of the waters of the Daule River and its conduction through a tunnel and channels to the Chongon dam; Section II comprises the Chongon dam, the Chongon–Cerecita canal, and the irrigation infrastructure of the Chongon, Daular, and Cerecita areas; and Section III comprises the Chongon pumping station and the Upper and Lower Chongon canal.

17.5.3 Europe

In Europe, in countries where the right for use of water is granted by the basin authority or public agency, transfer agreements have been implemented involving water fluxes at large scale, such as Rhone–Barcelona and Tagus–Segura, Spain, both for urban supply and irrigation (Ballestero, 2004).

TAGUS–SEGURA IBWT

The Tagus–Segura Transfer in Spain was completed in 1978, having as donor basin the Tagus Basin and as receiving basin the Segura Basin. The Tagus river is a bi-national river that has its longest stream in Spain but also crosses Portugal and traverses the city of Lisbon into the Atlantic, where it is called the Tejo river. The volume transferred is 0.6 km^3/yr and the infrastructure comprises five dams and 286 km of pipelines. The main purposes of this transfer were irrigation and urban water supply. In terms of impacts, one can list the following for this IBWT (Azevedo et al., 2005; WWF, 2007):

(a) Environmental impacts
   1. reduction of streamflow in the donor basin, which prevents maintenance of the minimum streamflow limits and imposes an increase in pollution;
   2. increase of the threat level for critically endangered species;
   3. proliferation of illegal boreholes, which results in overexploitation of the aquifers and an increase of the water deficit;
   4. destruction of thousands of hectares of native vegetation;
   5. serious degradation of the quality of the Tagus river due to effluent releases in the city of Madrid;
   6. increasing degradation of soil and pollution of streams due to increasing demand and excessive use of pesticides and fertilizers;
   7. deep alteration of the riparian vegetation by pollution, which has destroyed aquatic flora and fauna.

(b) Social impacts
   1. increase of social conflicts between donor and recipient basins;
   2. increase of water consumption in the receiving basin due to all expectations generated by the IBWT.

Some alternatives in order to mitigate or, at least, diminish some of the aforementioned impacts are related, mainly, to the improvement of demand management (WWF, 2007): closing down illegal wells, preventing the creation of new irrigated areas, promoting more sustainable urban land use, and recycling wastewater. The current knowledge and a Strategic Environmental Assessment would mitigate most of these negative impacts through a technically sound integrated plan and negotiation with the necessary stakeholders (Azevedo et al., 2005). As demand for the region's coastline was expected to continue to increase, the Spanish government then planned to promote desalination, especially for urban supply, and to treat and recycle wastewater (WWF, 2007).

The IBWT, as described in this section, was a driver for unsustainable water use in the receiving basin and should have been implemented together with measures aiming to limit the water demand in the receiving basin (Azevedo et al., 2005; WWF, 2007).

17.5.4 Africa

EL-SALAM CANAL IN SINAI (EGYPT)

The North of Sinai project is within the context of a regional development policy named Horizontal Expansion.[3] Annually, more than 12 km^3/yr of drainage water from the Nile Delta flows into the sea. The El-Salam canal brings a mixture of this drainage water and Nile water (1:1 ratio) eastward, crossing the Suez canal, to the deserts of north Sinai for irrigation of about 292,000 ha (Azevedo et al., 2005; Quosy, 2005) and a discharge of 4.25 km^3/yr (2.25 km^3/yr from two drains called Bahr Hadous and Lower Serw and about 2.2 km^3/yr of fresh water supplied from the Nile and transferred through the canal at its intake). The mix is to guarantee that the water salinity does not exceed the critical limit acceptable for agricultural crops (water salinity should not exceed 1,250 ppm in the canal; Donia, 2012). The purpose of the El-Salam canal project is not only to introduce the development of agriculture but it is also viewed by local authorities as an integrated project for the development of the territories of the Sinai Peninsula. It combines agriculture with agribusiness, mining, energy production, tourism, and other urban and industrial activities. Another positive aspect reported is the creation of new jobs for young people and the redistribution of the population, increasing the concentration of people in the uninhabited areas.

The intake of the El-Salam Canal is in the right bank of Damietta branch, located 3 Km upstream of Faraskur Dam. The El-Salam Canal crosses five states (Damietta, Dakahliya, Sharkiya, Port Said, and North Sinai) comprising a total length of 277 Km. It can be divided into two parts: (1) West of Suez Canal (86 Km), known simply as El-Salam Canal and (2) East of Suez Canal (191 Km), known as El-Sheikh Gaber Canal (Donia, 2012). The overall cost of the project is US$2.8 billion, and its infrastructure comprises 277 Km of canals, pumping stations, regulators, siphons, and bridges.

In terms of impacts, one can list the following for this IBWT (Abu-Zeid, 1983; Azevedo et al., 2005; Hafez, 2005):

(a) Environmental impacts: The identified environmental impacts are,
   1. loss of natural habitats and increase of pressure on the humid lands of the Delta;
   2. loss of known and unknown historical and archaeological monuments and sites;
   3. increased risk to human health[4] and the lives of wild animals;
   4. increased seepage of contaminated groundwater into Lake Bardawil;
   5. contamination of local wells used for drinking water;
   6. increase of bird harvesting;
   7. transfer of new human, animal, and plant diseases to Sinai.

(b) Social impacts: The impacts reported here can only be regarded as potential impacts, since the project is not fully operational. The potential social impacts are:
   1. displacement of both the population of the Sinai Peninsula (e.g. Bedouin) and their land use activities, which will result in the rupture of social and land use customs;
   2. loss of both traditional land rights and cultural heritage;
   3. emerging conflicts due to competing users of Nile Delta water, from which water is being withdrawn;
   4. local farmers negatively impacted by the project.

The project improved socioeconomic conditions in terms of land ownership and registration and dune fixation. Water availability is a fundamental condition for achieving the planned goals, and in order to do that, management measures shall be implemented in the context of the project: (a) restriction of cultivated area with high water consumption (e.g., sugar cane and rice); (b) a more efficient irrigation system (sprinkler and localized irrigation); (c) improvement of the efficiency of surface irrigation for old crops; (d) stimulus for nocturnal irrigation and soil leveling; (e) recycling of drainage water and treatment of sewage and industrial effluent; (f) change in the crop model and planning, as well as in the harvesting season; and (g) introduction of short cycle varieties (Quosy, 2005).

According to Quosy (2005), the major challenge facing current and future Egyptian generations is how to improve and develop their water resources, rationalize water use, and protect their sources against pollution and contamination.

LESOTHO /SOUTH AFRICA (SOUTHERN AFRICA)

The Lesotho Highlands Water Transfer (LHWT) was feasible due to a bi-national agreement celebrated between Lesotho and South Africa and the project was expected to be completed by 2020. This IBWT has as donor basin the Orange/Senqu Basin and the receiving basin is the Vaal River System, transferring at its end a volume of about 2.2 km^3/yr to Gauteng, South Africa's most industrial province, and generating about 90 MW of power for use in Lesotho. These, along with infrastructure for Lesotho and royalties paid by South Africa to Lesotho, are the main purposes of this transfer. The infrastructure comprises five dams, 200 km of tunnels, and a hydropower plant. The expected cost for the whole project is US$8 billion. In terms of impacts, one can list the following for this IBWT (Azevedo et al., 2005; WWF, 2007):

(a) Environmental impacts
   1. reduction of flow rates and less-frequent floods in the Lesotho donor basins; and
   2. increase in the threat level for critically endangered species of fish (e.g. Maloti minnow).

(b) Social impacts
   1. construction work affected 30,000 people and resulted in relocation of 325 households; and
   2. increased Lesotho's dependence on food imports due to the loss of more than 11,000 hectares of grazing or arable land, with reports of slow and inadequate compensation.

Some alternatives have been indicated by WWF (2007) and Azevedo et al. (2005) as a way to mitigate or, at least, diminish some of the aforementioned impacts by, mainly, improving demand management as outlined in South Africa's Water Act of 1998 and promoting water reuse and recycling among leading industry players in the basin.

The project was innovative in setting up a fund called the Lesotho Highlands Water Revenue Fund, based on royalties, dedicated to combating poverty in the affected areas. The fund has proved inefficient and opaque, leading to its termination and the creation of a new fund, now managed by a committee called the Lesotho Fund for Community Development. The irregularities verified have not yet been fully remedied, with the risk of cancellation of the fund.

The project failed to examine both environmental and social impacts from the outset, and the mitigation costs these would require. Also, demand management was not considered as a previous step before considering the IBWT, which may represent a much lower cost solution. In fact, as was the case here, the costs of these projects are generally higher than the proponents claim (World Commission on Dams, 2000).

17.5.5 Asia

THE SOUTH-TO-NORTH WATER TRANSFER PROJECTS (CHINA)

The South-to-North Water Transfer projects (SNWT: in Chinese, Nanshui Beidiao) constitute a set of giant interbasin diversions from the water-rich Yangtze River (Chang Jiang) basin into water-short north and northwest China. Two of these, the Eastern Route and the Middle Route, pass through tunnels dug under the elevated bed of the Yellow River. They are already delivering water to both the Huai River Basin south of the Yellow and to the Hai River basin to the north. The remaining transfers, collectively termed the Western Route, would deliver water from the headwaters of the Yangtze directly into the headwaters of the Yellow, but are technically difficult, extraordinarily costly, and still in the planning stage.

These diversions are huge and are to be built in stages over a period from 2002 to 2030 for the Eastern and Middle routes, with the Western Routes, if built, requiring up to another 20 years. According to the plan, all the SNWT projects together would be capable of transferring 44.8 km^3/yr by the year 2050 (Jia and Liu, 2014, p. 32), nearly ten times the aggregate capacity of all other diversion projects considered in this chapter. The smallest diversion, the 1,240 km long Middle Route, which since 2014 has been delivering a total of up to 9.5 km^3/yr, distributed among its terminus in Beijing (allocated 1.24 km^3/yr), the coastal city of Tianjin (1.02 km^3/yr, via an additional 142 km spur), and two provinces along the way (Henan [3.77 km^3/yr] and Hebei [3.47 km^3/yr]), has an ultimate design capacity of 13 km^3/yr.[5] Its primary uses are municipal (shenghuo – household and service industry), industrial, and then agricultural, as well as environmental (notably lake and river restoration). It draws from the Danjiangkou Reservoir on the Han River, a tributary of the Chang Jiang (Yangtze River), and required construction of a new canal. The Eastern Route makes use of the existing Grand Canal for its 1,156 km long primary canal but requires a significant number of pumping and water treatment facilities and involves a new 701 km lateral to serve portions of the Shandong peninsula. Its ultimate design capacity is 14.8 km^3/yr. Its primary uses are agricultural, municipal, industrial, and water transportation. The first stage of the Eastern and Middle routes alone cost at least US$25 billion.[6]

The government began serious consideration of SNWT toward the end of the 1970s. Early scoping efforts, such as that of the Chinese Academy of Sciences and United Nations University in 1980 (reported in Biswas et al., 1983), identified potential adverse environmental impacts, questioned its economic rationality, and suggested possibly more cost-effective alternatives. Deterioration of the surface water quality and flows and significant overdraft of groundwater in the north over the next two decades, together with a desire to enhance the image of Beijing, led to a go-ahead decision in 2002, in large part using ecosystem restoration as the decisive factor (Yang and Zehnder, 2005, p. 346). In terms of impacts, one can list the following for this IBWT:

(a) Environmental impacts
  1. Eastern Route
    (a) early concerns focused on increased soil salinization in transfer plain areas with high water tables, the spread of schistosomiasis northward, changes to the ecology of lakes along the route, and impacts on coastal and lacustrine fisheries (Wang and Liu, 1983; Yao and Chen, 1983)[7];
    (b) severe urban and industrial pollution along the transfer course requires heavy expenditure on wastewater treatment that is not readily forthcoming (Jia and Liu, 2014);
    (c) withdrawals during low flow periods could aggravate saltwater intrusion in the Yangtze delta.
  2. Middle Route
    (a) downstream impacts on Han River navigation and ecology are significant, requiring mitigation by a supplemental diversion from the mainstem of the Yangtze, in operation since 2014;
    (b) amount available for withdrawal during droughts may be restricted;
    (c) increasing the volume stored in the reservoir may aggravate earthquakes.

(b) Social impacts
  1. Eastern Route
    (a) "beneficiary" municipalities are unwilling to use the water due to high costs and concerns over quality;
  2. Middle Route
    (a) over 300,000 people were displaced to allow enlargement of Danjiangkou Reservoir in donor basin. Living standards fell, at least in the short run, for many of these "dam refugees";
    (b) polluting activities were banned upstream of the Danjiangkou Reservoir, affecting economic livelihoods of residents there, and necessitating several expensive compensatory projects.

The transfer projects have been controversial from their conception, on environmental, sociopolitical, and economic grounds (Biswas et al., 1983; Jia and Liu, 2014). Conflicts between donor and recipient basins have been relatively minor on the Eastern Route, which takes a small portion of the flow of the Yangtze near its estuary, and where the offtake is located in a province that is a major beneficiary of the diversion. Such conflicts are much more salient for the Middle Route, which takes a significant portion of the flow of the Han Jiang and impacts the economies and ecologies, lying both upstream and downstream, which lie in provinces that do not benefit from the diversion.

Compensation payments, often to build more diversion projects, comprise a significant share of the cost of the Middle Route. At least some of these payments are from beneficiaries such as Beijing Municipality to administrations in the donor basins.

Prior to receiving transfer water, Beijing, the "most enthusiastic and determined user" (Jia and Liu, 2014, p. 36), had modified its agricultural and industrial structures to lower total water demand; maintained household use at about 100 l/cap/day; recycled waste water; and universalized water meters to households. In 2017, 27% of Beijing's water supply was recycled waste water, compared to 22% originating from the SNWT. Despite these efforts, prior to receiving transfer water, China's capital relied on over pumping its groundwater and "emergency transfers" from nearby provinces.

Beneficiary administrations pay for fixed project costs based on their presumed benefit. They are also charged for variable costs on a usage basis and are responsible for constructing their own delivery systems. Those costs are presumably high for the Eastern Route, which has high pumping and water treatment outlays, and where desalination and nontransfer sources provide a cost-effective alternative for many, even at full-cost pricing (Beijing Qingnianbao, 2017).

17.5.6 Oceania

SNOWY MOUNTAINS HYDROELECTRIC SCHEME (AUSTRALIA)

This IBWT was built, between 1949 and 1974, by the Government of Australia and two state governments (Victoria and New South Wales). The scheme consists of two major developments (Ghassemi and White, 2007):

(a) the Snowy–Tumut Diversion, which transfers water from the upper Murrumbidgee, Eucumbene, upper Tooma, and upper Tumut rivers to the Murrumbidgee River; and
(b) the Snowy–Murray Diversion, which transfers water from the Snowy River waters at Jindabyne to the catchment of the upper River Murray.

The cost for the whole project was US$630 million (current prices), comprising the following infrastructure: sixteen dams (total capacity of 8.47 billion m^3), two pumping stations (capacities: 297.3 m^3/s and 25.5 m^3/s), seven hydropower plants (total capacity of 3756 MW), twelve tunnels (135 km), and twenty aqueducts (80 km).

This IBWT has as donor basins the Snowy River and Murrumbidgee River basins and as receiving basin the Murray River Basin, transferring a volume about 1.1 km^3/yr for hydropower and irrigation. In terms of impacts, one can list the following for this IBWT (Azevedo et al., 2005; Ghassemi and White, 2007):

(a) Environmental impacts: The environmental impacts on the Snowy River have been severe,
   1. reduction by 99 percent of the natural flow below the Jindabyne dam, which resulted in loss of wetland habitat, silting up of the river channel, invasion by exotic trees, saltwater intrusion in the estuary, and loss of migratory fish populations;
   2. restoration of flows to 21 percent for the Snowy River (0.21 km^3/yr) and diversion in the amount of 0.07 km^3/yr to the Murray, which helped to sustain the ecological values of Ramsar wetlands and the Murray River channel itself;
   3. significant increase in the amount of sediment and nutrients transported;
   4. modification/loss of riparian and aquatic vegetation, which results ultimately in loss of aquatic habitat;
   5. reduction of native fish population and probably invertebrates by the introduction of new species (brown and rainbow trout).
(b) Social impacts
   1. loss of income for the communities of the Snowy River, amenity values, and a natural asset;
   2. creation of significant employment locally in the recipient basin through recreation and tourism services;
   3. irrigators in the receiving basin were the most benefited by the transfer, since the Murray River is an over-allocated system that has 80 percent of its average annual flow diverted for irrigation.

The project did not consider properly the full costs and benefits and, because of this, caused conflicts for decades. The Snowy Mountains/Murray–Darling IBWT was not needed for both hydropower generation and expansion of agriculture in the Murray Basin. The latter could be attained by, for example, using more efficient irrigation practices in the recipient basin.

Some alternatives have been indicated by WWF (2007) and Azevedo et al. (2005) as a way to mitigate or, at least, diminish some of the aforementioned impacts by, mainly, improvement of demand management (improved water use efficiency) in the recipient basin. Also, upfront provision of environmental flows would have significantly reduced the costs.

17.6 RECOMMENDATIONS FROM EXPERIENCES

The problems of earlier implementations of IBWTs are quite well documented, but not always taken into account in either planning or management phases of an IBWT. Based on the analysis of previous IBWT implementations, a set of recommendations follows:

1. Prior to the implementation of any IBWT, policy makers and water managers should minimize the need for water transfer

by looking at good practices in water management for the recipient basin (such as those listed in Section 17.4). The implementation of such alternatives must be in order, within the context of an integrated planning process and not as an isolated solution.

2. Prior to the implementation of IBWT proposals, a thorough IBWT evaluation, comprising five criteria on economic productivity, environmental quality, and sociocultural impacts and benefit sharing assessments, should be carried out. These criteria serve as a starting point for negotiating among parties involved as there are trade-offs among them (Cox, 1999). It seems apparent in the literature that the simple need for water may be sufficient to trigger the implementation of such projects without a thorough assessment of the impacts on both basins.
3. Governance arrangements for some IBWTs and funding for sustaining the management and operation of the system should be thought through at the planning phase and well discussed with all parties involved.
4. In general, large scale IBWTs do not promote good practices in water management, in particular, if the IBWT is not considered, within the scope of a sustainable development policy, as one of the available alternatives for addressing deficits between water supply and demand in the recipient basin.
5. The implementation of an integrated plan, combining the IBWT itself with both traditional water supply augmentation and demand management measures, requires developing strong collaboration among different water use sectors, administrative regions, or countries, and addressing institutional issues.
6. The IBWTs should not represent limitations on the economic development of the donor basin and, if the IBWT imposes limitations on the economic development of the donor basin, compensation mechanisms should be implemented.
7. Measures should be implemented to minimize adverse impacts due to the IBWT both in the donor and receiving areas, including changes in the hydrological/ecological regimes, environmental pollution, and human interests.
8. Unbiased information through proper tools of analysis should be provided to decision makers and the general public to support participatory decisions within the context of a particular IBWT.

In summary, an intelligent IBWT should be implemented in the context of more general sustainable development strategies and Integrated Water Resources Management, and negotiation between donor and recipient stakeholders should be done and agreements reached prior to the construction of the projects. However, one should also consider the political momentum for the water transfer decision, which may not be favorable in the near future.

REFERENCES

Abu-Zeid, M. (1983). The River Nile: Main Water Transfer Projects in Egypt and Impacts on Egyptian Agriculture. In A. K. Biswas, D. Zuo, J. E. Nickum, and C. Liu, eds., *Long-Distance Water Transfer: A Chinese Case Study and International Experiences*. United Nations University.

Aguirre, C. and Drinot, P. (2017). *The Peculiar Revolution: Rethinking the Peruvian Experiment under Military Rule*. Paperback. Austin: University of Texas Press, p. 353.

Azevedo, L. G. T. de, Porto, R. L. L., Mello Jr, A. V. et al. (2005). Inter-Basin Water Transfers. *World Bank – Water Series*, 7, p. 93.

Ballestero, E. (2004). Inter-Basin Water Transfer Public Agreements: A Decision Approach to Quantity and Price. *Water Resources Management*, 18, pp. 75–88.

Beijing Qingnianbao [Beijing Youth News] (2017). *Caofeidian baiwan dun danhua haishui jiang jinjing [曹妃甸百万吨淡化海水将进京 (Caofeidian Will Provide a Million Tonnes of Desalinated Seawater to Beijing)]*. Available at www.bj.xinhuanet.com/bjyw/2017-12/08/c_1122079691.htm

Biswas, A. K., Liu, C., Zuo, D., and Nickum, J. E. (1983). *Long-Distance Water Transfer: A Chinese Case Study and International Experiences*. United Nations University, p. 432.

Casana, A. C. (2005). *Rationalize the Use of Water Demand: Chavimochic Project*. Available at www.laindustria. com/satellite

Cox, William E. (1999). Determining when Interbasin Water Transfer Is Justified: Criteria for Evaluation. In *Inter-basin Water Transfer: Looking for Solutions for the Future*, Technical Documents in Hydrology Vol. 28. Paris: UNESCO.

Donia, N. S. (2012). Development of El-Salam Canal Automation System. *Journal of Water Resource and Protection*, 4, pp. 597–604. https://doi.org/10.4236/jwarp.2012.48069

Ghassemi, F. and White, I. (2007). *Inter-Basin Water Transfer: Case Studies from Australia, United States, Canada, China, and India*. Cambridge: Cambridge University Press, pp. 1–463.

Gichuki, F. and McCornick, P. G. (2008). International Experiences of Water Tranfers: Relevance to India. In U. A. Amarasinghe and B. R. Sharma, eds., *Strategic Analyses of the National River Linking Project (NRLP) of India - Proceedings of the Workshop on Analyses of Hydrological, Social and Ecological Issues of the NRLP*, pp. 345–371.

Hafez, A. (2005). *Investigation of El-Salam Canal Project in Northern Sinai, Egypt – Phase I: Environmental Baseline, Soil and Water Quality Studies*. Proceedings of the Ninth International Water Technology Conference, Sharm El-Sheikh, Egypt: IWTC9 2005, pp. 953–970.

Jia, S. and Liu, J. (2014). *Daguo Shuiqing: Zhongguo Shui Wenti Baogao [The Water Regime of a Large Country: A Report On China's Water Problems]*. Wuhan: Huazhong Keji Daxue chubanshe.

Kuo, L. (2014). *China Is Moving More than a River Thames of Water across the Country to Deal with Water Scarcity*, Quartz. Available at qz.com/158815/chinas-so-bad-at-water-conservation-that-it-had-to-launch-the-most-impressive-water-pipeline-project-ever-built

Lund, J. R. and Israel, M. (1995). *Water Transfers in Water Resources Systems. 1–16*. Available at https://watershed.ucdavis.edu/shed/lund/ftp/Transfers.doc

Lynch, B. (2013). River of Contention: Scarcity Discourse and Water Competition in Highland Peru. *Georgia Journal of International and Comparative Law*, 42, pp. 69–92.

Magalhães, A. R. and Martins, E. S. P. R. (2019). The Case of the São Francisco River. In *Engineered Rivers in Arid Lands: Searching for Sustainability in Theory and Practice*. Philadelphia: CRC.

MWRI of Egypt (2003). *North Sinai Development Project: The Project Achievements*.

National Academy Press (2017). *Water Conservation, Reuse, and Recycling*, p. 293. https://doi.org/10.17226/11241

Quosy, D. E. (2005). *Agricultural Development in Egypt across Two Millenniums*. Virtual Global Super Project Conference, 2001: World Development Federation. Available at www.wdf.org/gspc

Roman, P. (2017). The São Francisco Interbasin Water Transfer in Brazil: Tribulations of a Megaproject through Constraints and Controversy. *Water Alternatives*, 10(2), pp. 395–419.

UNESCO (1999). *Interbasin Water Transfer: Looking for Solutions for the Future*. Proceedings of the International Workshop, Technical Documents in Hydrology Vol. 28, Paris.

USAID (2011). *Peru Climate Change Vulnerability and Adaptation Desktop Study*, p. 72.

USDI (2004). *Bureau of Reclamation. Colorado – Big Thompson Project*. Available at www.usbr.gov/dataweb.html

Wang, J. and Liu, Y. (1983). An Investigation of the Water Quality and Pollution in the Rivers of the Proposed Water Transfer Region. In Biswas et al., *Long-Distance Water Transfer*, pp. 361–371.

Wang, K. and Zhang, A. (2018). Climate Change, Natural Disasters, and Adaptation Investments: Inter- and Intra-port Competition and Cooperation. *Transportation Research Part B: Methodological*, 117, pp. 158–189. https://doi.org/10.1016/j.trb.2018.08.003

World Commission on Dams (2000). *Dams and Development, A New Framework for Decision Making*.

WWF (2007). *Pipedreams? Interbasin Water Transfers and Water Shortages*. Global Freshwater Programme, WWF, p. 49.

Yang, H. and Zehnder, A. J. B. (2005). The South-North Water Transfer Project in China: An Analysis of Water Demand Uncertainty and Environmental Objectives In Decision Making. *Water International*, 30(3), pp. 339–349.

Yao, B. and Chen, Q. (1983). South-North Water Transfer Project plans. In Biswas et al., *Long-Distance Water Transfer*, pp. 127–149.

Zimbelman, D. D. and Werner, B. R. (2001). Water Management in the Northern Colorado Water Conservation District. In J. Schaack and S. S. Anderson, eds., *Transbasin Water Transfers*. Proceedings of the 2001 USCID Water Management Conference, Denver, Colorado, 27–30 June 2001. Denver, CO: US Committee on Irrigation and Drainage, pp. 331–339.

Notes

1 Glaciers play an important role in the hydrology of the Andes and, in the Santa River, glaciers contribute to 80 percent of flows during the dry season and between 4 and 8 percent during the rainy season. USAID (2011) indicated that, over the last thirty years, Peru has lost 22 percent of its total glacier area. This same study refers to a prediction, using data from Peru's National Meteorological Service, that glaciers will lose up to 37 percent of their current area by 2030. Increase in temperatures due to climate change will accelerate glacial retreat, resulting in too much water during rainy season, which causes floods, and not enough during the dry season, which limits water availability for urban supply and irrigation.

2 Or 3.60 km^3/yr (114.16 m^3/s), when the level of the Sobradinho Reservoir is above the minimum value between: (a) the level corresponding to 94.0 percent of the useful volume; and (b) the level corresponding to the waiting volume for flood control. Under these conditions, it will be allowed to transfer a maximum daily flow of 114.3 m^3/s and an instantaneous discharge of 127 m^3/s. The option to relax the firm flow rate condition (26.4 m^3/s) and regard this value as a condition on the annual mean flow rate is under discussion, which would allow the transfer of greater volumes while observing the annual mean.

3 The Horizontal Expansion policy becomes especially important to the future of Egypt, a country with 1 million km^2, and only 4 percent of its territory occupied. These occupied areas correspond to the Nile Valley and Delta, which has 70 million habitants living in only 40,000 km^2 (1.700 hab/km^2).

4 Increase of human health risk due to the transfer of polluted irrigation water.

5 Allocations are from interview with Jiang Xuguang, Deputy Director of the South–North Water Transfer Office of the State Council (2013) http://news.ifeng.com/shendu/nfzm/detail_2013_12/13/32099046_0.shtml (accessed 3 December 2018).

6 Jia and Liu (2014) give 163 bn yuan to end Jan 2002 – about US$25 billion.

7 Other factors have probably rendered these concerns largely moot. The over pumping of the aquifer on the plain has reduced soil salinization risk, while pollution along the route and overfishing have no doubt impacted the ecologies in the absence of the transfer.

18 Better Basin Management with Stakeholder Participation

Jurgen Schmandt and Aysegül Kibaroglu

18.1 RIVER BASIN STAKEHOLDERS AND THEIR ROLE

Most SERIDAS rivers are overseen and managed by government-appointed or elected basinwide organizations, often called river agencies or commissions. One of their main functions is to allocate water to farmers, cities, and industries – the end users. End users want to have their voices heard in river management. For this purpose, they form intermediate or stakeholder organizations, such as irrigation districts and city water utilities. In some river basins there are additional stakeholders, such as native tribes, industry groups, and nongovernmental organizations working on environmental, economic, or social issues. In each case, stakeholders represent the interests of water users in specific river segments or sub-basins. In this chapter we discuss the role of stakeholder organizations in river basins, using the example of two SERIDAS rivers – the Rio Grande and Euphrates–Tigris.

In 2015, the Paris-based Organization for Economic Cooperation and Development (OECD) issued a massive report on Stakeholder Engagement for Inclusive Water Governance. Stakeholder participation is presented as the best way for achieving equitable and effective basinwide water management. The report, based on sixty-nine case studies of stakeholder participation, concludes,

> There are many economic, environmental, and social benefits to be gained from effectively engaging stakeholders in water policies and projects. Greater acceptance and trust are two of the main ones. Others include cost savings, greater policy coherence and more synergy among projects. There are also fewer tangible benefits from better co-operation: knowledge development, conflict avoided, and social cohesion.
>
> (OECD, 2015, p. 3).

Principle 8 of the OECD water governance recommendations also links improved water management to the work done by stakeholders: "Promote the adoption and implementation of innovative water governance practices across responsible authorities, levels of government and relevant stakeholders" (Neto, 2017, p. 8). In the same vein, the UN Sustainable Development Goal 6.5, which aims at implementing Integrated Water Resources Management (IWRM), emphasizes stakeholder participation as a major target and indicator for operationalizing sustainable river water management (UN Environment, 2018).

While the establishment of robust stakeholder institutions at all river basin levels should be the aim for sustainable, equitable, and integrated management of a river system, current stakeholder participation represents a mixed picture in the SERIDAS river basins. We use our work in the Rio Grande/Brávo to illustrate the important role assumed by stakeholders in sustainable basin management. In the case of the Euphrates–Tigris, we find that the lack of a basinwide management organization, has led to a much smaller role played by stakeholders. While irrigation districts/irrigation associations and city water utilities exist, their impact on basinwide management is minimal and mostly limited to ad hoc arrangements.

18.2 DEFINING AND MANAGING RIVER SUB-BASINS

Stakeholders work at the sub-basin level – a river segment with its unique hydrological and economic boundaries. For example, the Lower Rio Grande hydrological sub-basin reaches from south of Fort Quitman, Texas, to the Gulf of Mexico. (See map in chapter 14, above.) Two large reservoirs – Falcon and Amistad – store water for irrigation and cities. The sub-basin receives the bulk of its water from two tributaries, the Conchos, and the Pecos. Without them the Rio Grande would never reach the Gulf of Mexico because virtually all mainstem water above Fort Quitman has been used for irrigated agriculture and drinking water. To manage current and future water supply and demand in this segment of the Rio Grande it is necessary to carefully analyze conditions in the sub-basin and the role played by stakeholders.

Before discussing ways that this can be done, we make a general point: We propose that the managers of SERIDAS rivers carefully measure and evaluate water-related issues in each of their sub-basins. A sub-basin, in a reservoir-equipped river, encompasses the storage lake upstream from the first reservoir dam and the river segment below the reservoir outlet to the last drainage channel above the second reservoir (or the mouth of the river) downstream. It is important to know the hydrological, environmental, and economic issues that are specific to each sub-basin. This information is needed for efficient water management in the sub-basin. It is equally important to use this information for understanding the role of stakeholders as intermediate entities: How do they cooperate with the legal entity responsible for management of the river as a whole? How do they represent the interests of end users in the sub-basin?

18.2.1 Stakeholders as the Main Movers in Sub-basins

SERIDAS rivers provide water for irrigated agriculture, cities, and the environment. To understand and illustrate the role played by stakeholders in two SERIDAS river basins, we convened two faculty-graduate student research teams to conduct year-long studies of sub-basins in the Euphrates–Tigris (Kibaroglu and Schmandt, 2016) and the Rio Grande (Stolp and Schmandt, 2018). In this work we ranked stakeholder activities using the six-level engagement classification developed in the aforementioned study on the role of stakeholders in water management (OECD, 2015). They are,

1. Communication: water-related information is made available to stakeholders.
2. Consultation: stakeholders are identified, and their perception and goals are mapped.
3. Participation: stakeholders participate in projects and policy discussions.
4. Representation: stakeholders are represented in water management agencies.
5. Partnerships: agreed collaboration among stakeholders.
6. Co-decision: power among shareholders is shared.

18.2.2 Rio Grande / Rio Bravo del Norte

AGRICULTURAL WATER STAKEHOLDERS

Agricultural water users in New Mexico, Texas, and Chihuahua are organized in irrigation districts. Farmers use 90 percent of Rio Grande water. They can contribute to more efficient water use in a variety of ways: replacing field flooding by sprinkler systems, growing less water-intensive crops, rain harvesting, and maintaining distribution channels in good condition. The cost of improvements and market demand for water-intensive crops limit progress to date. While the amount of farmland has stayed constant in the Paso del Norte, the type of crops grown has changed substantially. Originally, traditional crops like cotton, alfalfa, and chiles were grown. More recently, pecan groves requiring much more intensive irrigation have been planted. Pecan trees are a long-term investment for farmers because they take longer to mature. This reduces the flexibility farmers used to have in the use of irrigation water.

MUNICIPAL WATER STAKEHOLDERS

Municipal water use is overseen by the city water utilities of Las Cruces, New Mexico; El Paso, Texas; and Ciudad Juárez, Chihuahua. The cities experience rapid population growth. Water use in El Paso and Ciudad Juárez has increased exponentially between 1974 and 2015 (TWDB, 2018). The demand for drinking water will increase 100 percent in the next 100 years (IBWC, 2018). In addition to conservation and repair of leaking distribution pipes, cities may consider adding to their drinking water supply by desalination of brackish ground water. El Paso has chosen to do this; the city is home to the world's largest inland desalination plant, capable of producing more than 27 million gallons of fresh water per day.

18.2.3 Stakeholder Engagement to Date

Since 1999 the US Section of the International Boundary and Water Commission (IBWC) has convened quarterly citizen meetings for the US part of the Paso Del Norte region (Percha Dam, New Mexico, to Fort Quitman, Texas). The Rio Grande Citizens' Forum comprises six appointed citizen board members bringing different areas of expertise to the table. The board members serve 2-year terms and are responsible for determining forum topics for each meeting. IBWC provides information about their activities. Other topics are developed by board members or stakeholders who want to provide further information regarding a specific topic. The presentations are coordinated by an IBWC Public Affairs Officer (Kuczmanski, 2018).

The forum meetings are open to the public and frequently attended by policymakers and their staff. Meeting minutes are taken, and the information is compiled into an email that is sent to an internal listserv that the public can sign up to receive. Forum meetings and email distribution listserv are the main procedures for information dissemination. The IBWC Forum serves OECD engagement levels 1 and 2.

In the same year as the Rio Grande Citizens' Forum began, the Houston Advanced Research Center[1] started the Paso del Norte Water Task Force. New Mexico State University, Universidad Autónoma de Ciudad Juárez, and the University of Texas at El Paso were actively involved in planning and administering the task force. The Commissioners of IBWC/CILA convened the first meeting of the task force and supported its activities throughout its lifetime. Task Force members – civic leaders, managers of municipal water utilities and irrigation districts, water users, and water experts from New Mexico, Texas, and Chihuahua – came together to create a more sustainable water future for the region. They selected water planning as the focus

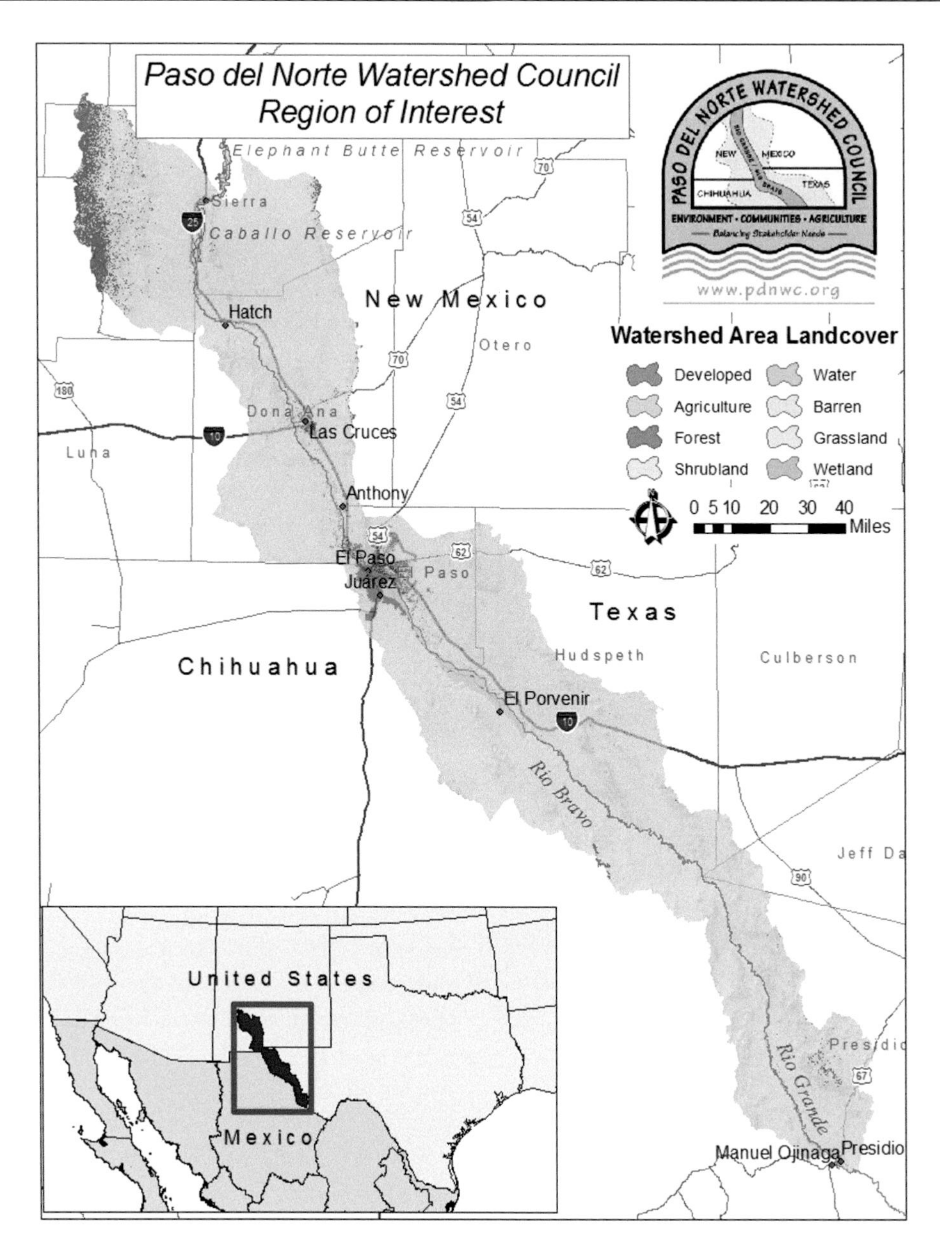

Figure 18.1 Map of Paso del Norte Watershed Council Region of Interest.
Map courtesy of Les Owen and the Paso del Norte Watershed Council

of their first project. Two research teams prepared case studies on water planning in the US (LBJ School of Public Affairs, 2000 and 2001) and Mexican (Centro de Estudios del Medio Ambiente, 2000) parts of the Paso del Norte region. The reports were submitted to water agencies in the Paso del Norte as well as federal and state agencies in the two countries. In the following years, the task force held bi-annual meetings that proved useful for information exchange, consultation, and policy advice (Figure 18.1).

Unfortunately, the Paso del Norte Water Task Force suffered a slow death when water conflicts, in this case between New Mexico and Texas, emerged and key participants no longer felt free to attend. Another weakness was the lack of stable financial support after successive grants from national foundations were exhausted. On the plus side, the task force enjoyed full participation from all subdivisions of the bi-national three-state region. With stable financial support and competent staff, the task force provides a good model for an NGO tasked with developing and updating a sustainable water management plan for the Paso del Norte. The task force served OECD engagement levels 1, 2, and 3.

A parallel group, the Paso del Norte Watershed Council, was convened in 2001 by New Mexico State University in Las Cruces. The Council worked to address issues related to the establishment and maintenance of a viable watershed. These included promoting projects to improve water quality and quantity, ecosystem integrity, quality of life, and economic sustainability in the Paso del Norte watershed. In addition to its role of assisting IBWC/CILA with mitigation and enhancement projects, the Paso del Norte Watershed Council also provided a forum for exchanging information about activities on the Rio Grande. The group met several times each year for discussion and exchange of information but closed in 2009. The Council served OECD engagement levels 1, 2, and 3.

The Texas Water Development Board prepares 5-year state water plans, designed to project future water supply and demand. The last report was published in 2016. Drafting of the plan is a decentralized effort entrusted to sixteen regional planning boards. One of these – Far West Texas – includes the El Paso region. Board members for the regional planning board include representatives of the public, counties, municipalities, industry, agriculture, environment, small business, electricity-generating utilities, river authorities, water districts, water utilities, and groundwater management areas. A similar effort, covering the entire Paso del Norte, should be undertaken for this sub-basin of the Rio Grande / Bravo. Preparation of a regional water plan allows for stakeholder engagement at levels 3 and 4 of the OECD scale.

18.2.4 Stakeholder Surveys

Several surveys were conducted to identify goals and aspirations of PdN water stakeholders. Luzma Nava (2020) found that stakeholders from the PdN region place great emphasis on the need to (a) strengthen communication and articulation among all of stakeholders in the Rio Grande Basin and related water agencies; (b) provide environmental education; (c) manage surface and groundwater jointly; and (d) renegotiate all river basin water agreements due to imbalances between water availability, supply, and demand.

At the request of the Paso del Norte Water Taskforce, graduate students at the LBJ School of Public Affairs, University of Texas at Austin, conducted a survey in 2016 on how to cope with regional water scarcity (Kibaroglu and Schmandt, 2016). Survey recipients included irrigation districts, water utilities, water governance and planning agencies, and NGOs. Stakeholder priorities for coping with water scarcity are listed in Table 18.1. Survey respondents ranked the importance of coping mechanisms on a scale from 10 (most important) to 1 (least important). Water conservation ranked highest.

Agricultural Sector: Farmers have different options for dealing with scarcity. These include: (1) crop choice, (2) irrigation techniques, and (3) abandonment of agriculture. The first two options are closely related. Crop choice is driven by strategies farmers use when deciding what to plant each year. Many farms convert to products that demand a high value on the market, like pecans (Turner et al., 2018). However, the tendency to rely primarily on one type of valuable crop makes farmers vulnerable to climatic change and water scarcity. Research suggests that diversifying the types of crops used and planting hardier crops can safeguard farmers from crop loss in the future (Draper, 2014). Farmers will need to carefully consider whether a sustainable water future outweighs possible economic losses.

Traditionally, farmers in the Rio Grande Valley relied on flood irrigation to get water to their crops. This method uses ditches to transport water and results in large amounts seeping back into the ground before the water ever benefits the farmer. As water prices rise and it becomes more difficult to find good-quality groundwater, farmers have shifted from flood irrigation to more sustainable practices. However, in many parts of El Paso del Norte, the practice of flood irrigation continues to this day.

New technologies have been introduced that can reduce the amount of water wasted while farming, but these methods can quickly become expensive for farmers, who often operate with exceedingly small profit margins. The main alternative to flood irrigation for farmers in the region is drip irrigation. This style of irrigation uses a series of plastic tubes to transport water in a more targeted manner. Drip irrigation significantly reduces water loss by preventing seepage and evaporation and results in more productive fields. The main drawback to this option is the higher labor costs associated with running the tubing to each individual row of crops.

The third and final option – abandoning agriculture completely in the region – is considerably less popular. While a reduction in agriculture may be feasible, even this possibility comes with many downsides for the community. If the region were to reduce or abandon agriculture, significant job losses would occur, and the cost of many foods would rise. Nonetheless, some sustainability advocates support abandoning agriculture in this hostile environment (Draper, 2014). A small reduction in agricultural production seems a likely scenario given future water availability projects. However, many people struggle to imagine a future where agriculture in El Paso del Norte is reduced.

Industrial Sector: With merely three responses to the LBJ survey, no reliable inferences could be made from the results. While industrial/business stakeholder interest in water use research may be inherently low, perhaps the most interesting response from this sector in the PdN is that all three respondents indicated they did not perceive an opportunity to participate in decisions regarding water use in their area. While the sample size is small, it may indicate that a gap exists between water management institutions and industrial/business stakeholders in the region.

Environmental Sector: Interestingly, the environmental stakeholders surveyed did not predict less water availability in the future. Most environmental stakeholders indicated they believed

Table 18.1 *Perceived importance of drought coping mechanisms*

	Coping measure	Level of importance	Responses
1	Greater emphasis on water conservation	7.73	22
2	Increased groundwater pumping	4.9	21
3	Letting fields lie fallow	4.45	20
4	More efficient irrigation technology	5.10	20
5	Putting distribution channels underground	3.70	20
6	Repair leaking pipes	5.19	21
7	Municipal water restrictions	5.29	21
8	Other (education, wastewater reuse, etc.)	Not ranked	5

Source: (LBJ School, 2016, pp. 164–168)

they could participate in regional decision making – more than any other sector surveyed. Respondents in the environmental sector also felt better equipped with the necessary tools and information to participate effectively in that process than other sectors. They were more supportive of economic sanctions that discourage water use, such as fees for excessive consumption, and more supportive of creating regional groups to engage the public in water management issues. Environmentalists were also more likely to see water use as a big picture, international issue rather than a local one.

18.2.5 Full Stakeholder Engagement: Preparing and Implementing a Regional Sustainability Plan

We find that PdN water stakeholders to date are not engaged at the two highest levels on the OECD scale: partnerships (agreed upon collaboration among stakeholders) and co-decision (share-holders share power with basin management agency). To reach these levels we propose that stakeholders take the lead in preparing and implementing a plan for sustainable water management in the PdN. To do this we propose alternative models for regional agencies to create, finance, and administratively support a revived PdN Water Taskforce. Under each model an important feature of the original PdN water taskforce would be preserved: full geographic participation – New Mexico, Texas, and Chihuahua – of the areas belonging to the PdN hydrological sub-basin.

PROPOSED MODEL 1: IBWC/CILA AS LEAD CONVENER

Under this model, the US and Mexican sections of the International Boundary and Water Commission establish water task forces for each of the bi-national sub-basins under their control. In the case of the Rio Grande / Brávo this means a task force for the PdN and a second taskforce for the Lower Rio Grande. Each group is organized as an administrative division of the IBWC/CILA, thus ensuring financial stability. Details on membership, organization, and tasks are spelled out in an IBWC Minute that is approved by the two national governments. The long-established Minute system, which allows for clarifications and additions to the 1944 water treaty between Mexico and the United States, is an important tool for updating the treaty. Specifically, Minute 308, passed in 2002, called for developing a bi-national sustainability plan for the Rio Grande Basin. Unfortunately, no action followed the adoption of Minute 308 (Kibaroglu and Schmandt, 2016). We suggest that the time has come to renew the effort.

PROPOSED MODEL 2: ACADEMIC INSTITUTIONS AS LEAD CONVENERS

Under this model, the three regional universities – New Mexico State University in Las Cruces, University of Texas at El Paso, and Universidad Autónoma de Ciudad Juárez – would create a nongovernmental organization for planning a sustainable water future in the Paso del Norte. To provide financial stability the initiative would be based on a multi-year grant provided by a consortium of international and national donors. The presidency of the PdN Water Taskforce would rotate between the three universities in three-year cycles.

PROPOSED MODEL 3: WATER AGENCIES AS LEAD CONVENERS

Here the convening and financing task would fall to the New Mexico State Engineer, Texas Water Development Board, and the Comisión Nacional del Agua (CONAGUA). Again, the presidency of the PdN Water Taskforce would rotate.

PROPOSED MODEL 4: NONGOVERNMENTAL ORGANIZATIONS (NGOS) AS LEAD CONVENERS

In this model, NGOs, in collaboration with other partners, including federal agencies, academic institutions, businesses, irrigation districts, and city water utilities, bring together stakeholders from across the basin to discuss challenges and solutions to water scarcity in the RGB basin. Next steps identified through this process could complement and be coordinated with results of the other proposed models, fostering much needed cross-basin, multisector traction on critical water issues.

Such a model was used when the World Wildlife Fund (along with the Desert Landscape Conservation Cooperative, The Coca-Cola Company of North America, Instituto Mexicano de Tecnología del Agua, Instituto Tecnológico de Monterrey, South Central Climate Science Center – US Geological Survey) organized and convened a bi-national forum on water management and scarcity in the Rio Grande / Bravo Basin in El Paso, Texas, on November 7–8, 2017. The RGB Water Forum brought together 189 invited participants from the United States and Mexico, representing multiple water stakeholder sectors, including farmers, ranchers, city officials, water managers, business leaders, conservationists, and scientists, to stimulate frank discussion between multiple sectors from both countries regarding the real, on-the-ground repercussions of water scarcity and to identify potential long-term solutions to those challenges.

Some of the solutions identified during the RGB Water Forum included the need to

- Maintain momentum following the RGB Forum with similar, though more targeted, water discourse and actions during 2018 and 2020 (in between the conclusion of the 2017 RGB Water Forum and the proposed next bi-national Forum, which is provisionally scheduled to take place in the Mexican portion of the basin in 2020)
- Summarize and distribute lessons learned from current efforts in the basin that are already addressing water-related issues (thus, addressing the question of how such efforts could potentially be expanded for greater impact)
- Develop a series of targeted, multi-media "splashes" on water challenges and solutions in the RGB basin that will be distributed in the public and political realm.

The task to be addressed by the stakeholders under each of the four models would be to prepare a 20-year sustainability plan for the management of PdN water resources. Sections of the plan might include current dependable yield, projected dependable yield, assessment of reservoir conditions and impact, changing priorities for water use, groundwater and surface water interactions, progress made in improving efficiency of water use, and actions needed to reconcile supply and demand. The plan would be updated in 5-year cycles. Task force members would take on the roles of data collectors, researchers, and policy advisers. Their recommendations would be submitted in draft form to the decision making agencies and the public. The final version, prepared with consideration of comments received, would be submitted for approval to the regional water agencies. The Water Task Force would monitor progress and start the process again for a new 5-year cycle.

18.3 EUPHRATES–TIGRIS

18.3.1 Stakeholder Involvement at Transboundary and National Levels

In the Euphrates–Tigris (ET) river basin no legal entity exists that is responsible for management of the entire river basin. Therefore, the basin lacks a joint body that represents the interests of all basin stakeholders at all levels. However, because the issues triggered by water development schemes along the Euphrates and Tigris became so complex and far-reaching in the late 1970s, the three riparians had to find ways to structure their discussions. To this end, Iraq took the initiative in the formation of a permanent joint technical body. The first meeting of the Joint Economic Commission between Turkey and Iraq in 1980 led to the establishment in 1983 of the Joint Technical Committee (JTC), whose members included participants from all three riparians assigned to lay down methods and procedures that would lead to the definition of a reasonable and adequate quantity of water for each country from both rivers. However, the JTC was not able to agree on any substantial resolution even after sixteen meetings (Kibaroglu, 2002). Negotiations were suspended in 1993. The JTC did not provide a platform for delineating the stakeholders' priorities and needs as a basis for addressing basin water issues. The meetings were attended by officials (technocrats and diplomats) from the concerned ministries in the absence of stakeholders from relevant sectors. In this respect, water use patterns and related legislation and institutional structures never had a chance of being discussed at the JTC meetings. National management and allocation policies were like "black boxes," and water management practices within the various countries simply could not be debated during those negotiations (Kibaroglu and Scheumann, 2013).

Efforts have been more meaningful at the national level. For instance, in Turkey efforts have been made to adopt river basin planning in accordance with the requirements of the European Union (EU) Water Framework Directive, which requires EU member states and candidate countries to produce and implement river basin management plans. The plans are to be designed and updated to include integrated and coordinated stakeholder participation (European Parliament and The Council of the European Union, 2000). In this context, River Basin Protection Action Plans have been completed in the designated twenty-five river basins of Turkey in the past decade. These plans are intended to be converted into EU-compliant River Basin Management Plans (RBMPs) by 2023. When completed, the RBMPs will enable Turkey to improve considerably the ecological, chemical, and quantitative status of surface and groundwater resources in river basins.

Basin organizations were established at the national, basin boundary, and provincial levels in 2015. They are created with the participation of related public and private institutions, NGOs, universities, irrigation associations, and so forth (Delipinar and Karpuzcu, 2017). Furthermore, Basin Management Committees (BMCs) have been established for each river basin and have responsibility for preparing the RBMPs with drought and flooding management reports, monitoring their implementation, and providing active public participation. Twenty-six BMCs have been established in twenty-five river basins across Turkey. The Turkish portion of the ET basin has two BMCs because of its large surface area. These new basin organizations have started to meet routinely according to their related legislation. However, public participation is partially achieved. NGOs, Organized Industrial Zones (OIZ), and irrigation associations have started to participate in the meetings of basin management at the basin and provincial levels in accordance with the related legislation. But there are still some deficiencies in stakeholder representation (Delipinar and Karpuzcu, 2017).

18.3.2 Agricultural Water Stakeholders

Agriculture is by far the dominant sector for water consumption in the ET basin. It accounts, on average, for more than 70 percent of water allocated and used in the riparian countries (FAO, 2008). Agriculture's contribution to Gross Domestic Product (GDP) has been declining in all riparian economies (FAO, 2008), yet a significant portion of the labor force is still employed in this sector.

Moreover, food security is still a prevailing priority, particularly during growing global food crisis. Accordingly, in political and economic contention, the achievement of food security depends on secure water supplies and the expansion of irrigated surfaces. In addition to these political, economic, and social reasons, there are physical facts on the ground that urge the expansion of irrigation: The evaporative demand is very high and crops require intensive irrigation because of low annual rainfall and hot and dry summers in the region. In this respect, the total area equipped for irrigation in the ET basin is estimated to be around 6.5–7 million hectares (ha), of which Iraq accounts for approximately 53 percent, Iran 18 percent, Turkey 15 percent, and Syria 14 percent. Agricultural water withdrawal is approximately 68 billion m^3 (FAO, 2008).

Traditionally, central water authorities have overseen building, operating, managing, and maintaining irrigation systems in the riparian countries. However, in Turkey, including the ET region, operation and management responsibility of 98 percent of the irrigated area equipped with irrigation facilities by the principal government water agency, namely the State Hydraulic Works (DSI, for its initials in Turkish), has been transferred to water users, namely the irrigation associations (IAs), since 1994. The Turkish experience of water user associations, which was supported by the World Bank, can be shared and exchanged with other riparian countries to increase water use efficiency and the water revenue collection rate, and to save water. However, both good and bad experiences should be shared. Following the transfer of irrigation schemes to the user organizations, some improvements are recorded in irrigation ratios, irrigation water fee collection rates, and financial cost reduction in irrigation systems operated by the IAs. However, system performance remained almost at the same level (Kibaroglu, 2020).

The participatory aspect of the transfers has been questioned due to the exclusion of irrigators from IA general assemblies and boards. The top-down approach, adopted rather than a grassroots approach generated by farmer interest and involvement, has caused fierce debate over the characterization of the associations as democratic. Critics also stress that maintenance, rehabilitation, and modernization of the irrigation canals, some of which are 50 years old, cannot be accomplished due to technical, administrative, and legal capacity deficiencies of the IAs.

18.3.3 Municipal Water Stakeholders

In Syria, the Euphrates supplies drinking (municipal) water to the governorates of Deir ez Zor and Raqqah as the major population centers. Aleppo also draws its water from the Euphrates, with a pipeline running west from Lake Assad and supplying cities and villages along its route (ESCWA, 2003). The combination of rapid urbanization and population growth has steadily increased the demand for municipal water in Iraq. Facing an increasing demand for domestic water in the ET basin provinces in Turkey, DSI has finalized a series of projects for drinking water supply providing 690 million m^3/year water to the nine provinces in the GAP[2] region, which constitute 10.8 percent of Turkey's population (Anatolian Agency, 2019).

Soon, it is envisaged that most, if not all, sectoral water allocations in the region will involve reducing water consumption in the agricultural sector and significant reallocation to other sectors. This reallocation will have a substantial impact on the domestic and industrial sectors, which could lead to increases in GDP, provide more water for rural communities, and create more jobs within the industrial sector, thereby helping to combat unemployment. Agriculture will therefore remain the focus of national water policies since it will be greatly affected by the reduction of water available to it. Better management of water in agriculture, including changing crop patterns and the adoption of modern irrigation technologies, will help the agricultural sector deal with the reduction in its water share.

18.3.4 Euphrates–Tigris Initiative for Cooperation: Stakeholder Involvement in Transboundary Water Relations

Regime change in Iraq, paradoxically, provided salient opportunities for Iraqi water professionals and scholars to interact more systematically with their colleagues in Syria and Turkey. As a result, 2005 saw the founding of the Euphrates–Tigris Initiative for Cooperation (ETIC), involving water professionals, former diplomats, technocrats, and academics from Iraq, Syria, and Turkey. The organization constituted a unique nongovernmental entity founded in the region, connecting government officials in a cooperative manner and transparent in all its actions.

The ETIC was crafted as a track-two diplomacy initiative and played a constructive role in transboundary water dialogue and scientific cooperation. Track-two diplomacy actors think outside the box. They address issues that may not yet be on governmental agendas, serving as a kind of early warning mechanism. Thus, even amid conflicts in Iraq and Syria and the deterioration of bilateral political relations between any pair of the three riparians, the ETIC managed to carry out research projects and training activities. The ETIC sought to lead dialogues not just about resolving narrow bilateral water disputes but also about creating a regional context through which important socioeconomic development issues affecting larger segments of the region could be discussed and addressed (ETIC, 2019).

18.3.5 The Role of NGOs and Civil Society in Water Policy and Management in Turkey and Syria

Since the 1990s, NGOs have steadily gathered more experience and knowledge regarding the practical implementation of sustainable water management initiatives in Turkey and have played a key role in demanding more participatory forms of river basin management, while further developing good practices on the ground (Divrak and Demirayak, 2011). Over recent decades, conservation activities have led to the emergence of many national and regional environmental NGOs, whose objectives comprise proposing efficient solutions and encouraging public participation in dealing with various environmental issues. Their strategies include boosting public awareness and providing a basis as a pressure group in decision making processes. The spirit of cooperation through international relations having gained momentum with Turkey's EU accession process, most of the Turkish NGOs have developed partnerships with international organizations such as the World Wide Fund for Nature (WWF), Greenpeace, Birdlife, and others, in order to benefit from their experience and increase fundraising opportunities. These partnerships have provided a good basis for the evolution of Turkish environmental NGOs, many of which have broadened their focus from specific species-oriented conservation efforts toward involving environmental concerns in the management of natural resources (Divrak and Demirayak, 2011).

To illustrate, WWF Turkey attended the Middle East Seminar: Cooperation Prospects in the Euphrates–Tigris Region, which was organized by the ETIC alongside a number of international development and aid agencies at the World Water Week in Stockholm in 2006 (SIWI, 2006). The seminar included a panel discussion on the role of civil society in transboundary water management, with the WWF Turkey chair sharing her experiences on the role of civil society in the peaceful settlement of water disputes, as well as sustainable development and protection of water resources in the Euphrates–Tigris region. The seminar was attended by senior diplomats and officials from the relevant Turkish, Iraqi, and Syrian ministries (SIWI, 2006).

Discussions on stakeholder involvement in the Euphrates–Tigris basin should also include one significant phenomenon: the emergence of a civil society in Syria during the civil war. Before the civil war, Syrian civil society was predominantly underdeveloped, fragile, and highly controlled. As the conflict became protracted and created an overwhelming amount of need, Syrian civil society organizations (CSOs) became important actors in the current crisis and have been largely focus on humanitarian relief activities. According to the United Nations Office for the Coordination of Humanitarian Affairs, "around 600 to 700 'local' groups were initiated between 2011 and 2015" (Svoboda and Pantuliano, 2015, p. iii).

Most of the Syrian CSOs were established after 2011; they have operated in regime- and nonregime-controlled areas. Generally, CSOs have had offices in humanitarian aid hubs, namely Turkey, Syria, and Jordan, as well as operational presences inside Syria. Their sizes differ: some have over 100 members and employees and some are rather small in terms of membership and employee capacity. They work in the sectors of health, food security, water, sanitation, and hygiene (WASH), education (for children, orphans, and so on), shelter, agricultural projects, livestock projects, relief, psychological support, assisting women inside Syria and in camps; protection for children, women, and orphans; and documenting detentions, massacres, and forced disappearances; creating database and reporting monthly (Dixon et al., 2015).

With the eruption of the uprising in Syria, there was a revival of civil society represented by youth groups, grassroots movements, local coordination committees, leaders, activists, religious groups, civil courts, religious courts, Local Councils, humanitarian support groups, media groups, and others. Their efforts and objectives include health; education; medical aid; civil disobedience; political, social, and economic empowerment; citizenship; election monitoring; service provision; law enforcement; conflict resolution; peace-building; human development; psychosocial support; and state and institution building. However, it is important to note that the nature and role of civil society during conflict is in continuous change and depends on the context in which it exists.

A bleak future demands concerted efforts by regional governments and the international community to extend humanitarian aid and economic support to the region in systematic and determined ways. A strategy should be adopted by the riparian states, as well as local and international funding agencies, which focuses on strengthening Syrian civil society, supporting their actions in the water sector, and enhancing their ability to obtain funds for rehabilitation and reconstruction. If one can overcome the challenge of identifying and working with local actors, their proximity, access, and contextual knowledge present opportunities to improve the overall humanitarian response and the prospects for sustainable relief and development. With the continuation of the civil war, there is a need to look at the conflict since actions during and after the conflict are linked. During the conflict, there is an immediate need to improve drinking water supply and to support agriculture in areas (less) affected by the fighting. From a post-conflict perspective, rehabilitation of the domestic and agricultural water infrastructure will be a priority to ensure the sustainable return of displaced populations. Beyond emergency relief interventions, the prioritization and allocation of resources for reconstruction will be determinant factors in the reconciliation process (Saade-Sbeih et al., 2016).

18.4 RECOMMENDATION FOR STAKEHOLDER INVOLVEMENT IN SERIDAS RIVER BASINS

All SERIDAS rivers can learn from the initiatives in the two river basins that we reviewed in this chapter. In particular, the four options for creating sub-basin water councils recommended for the Rio Grande provide useful guidance for other SERIDAS basins. Other models, reflecting different basin conditions, may emerge. But whichever model is selected, stakeholders should always organize to address the water agenda of their sub-basin. Doing so directly contributes to reaching and maintaining sustainability of the river as a whole.

- Assess current and future water supply and demand.
- Project future dependable water yield.
- Conduct reservoir impact assessments.
- Draft and implement a plan for sustainable water management.
- Cooperate closely with river agencies/commissions.

REFERENCES

Anatolian Agency (2019). *GAP kapsamındaki illerde içme suyu problemine son [An End to the Drinking Water Problem in the Provinces Covered by the GAP]*. Available at www.aa.com.tr/tr/turkiye/gap-kapsamindaki-illerde-icme-suyu-problemine-son/1549063

Carter, K. et al. (2018). Stakeholder Engagement. In *Sustainable River Management on the US–Mexico Border*. Policy Research Project Report, Number 202. LBJ School of Public Affairs, University of Texas at Austin.

Centro de Estudios del Medio Ambiente, Universidad Autónoma de Cd. Juárez [Center for Environmental Studies, Autonomous University of Cd. Juárez] (2000). *Planeación del Recurso Agua: Comparación de Planes de Administración del Agua en Cd. Juárez, Chihuahua [Water Resource Planning: Comparison of Water Management Plans in Cd. Juárez, Chihuahua]*.

Delipinar S. and Karpuzcu, M. (2017). Policy, Legislative and Institutional Assessments for Integrated River Basin Management in Turkey. *Environmental Science & Policy*, 72, pp. 20–29.

Dixon, S., Moreno, E. R., and Sadozai, A. (2015). *Syrian Civil Society and the Swiss Humanitarian Community*. Geneva: The Graduate Institute Geneva. Available at https://docs.water-security.org/document/syrian-civil-society-and-the-swiss-humanitarian-community

Divrak, B. B. and Demirayak, F. (2011). NGOs Promote Integrated River Basin Management in Turkey: A Case-Study of the Konya Closed Basin. In A. Kibaroglu, W. Scheumann, and A. Kramer, eds., *Turkey's Water Policy: National Frameworks and International*. Verlag Berlin Heidelberg: Springer.

Draper, M. (2014). Envisioning the Farms of the Future: In Pursuit of Sustainable Agriculture in the Rio Grande Valley. University of New Mexico Digital Repository Online. Available at http://digitalrepository.unm.edu/cgi/viewcontent.cgi?article =1006&context=ltam.etds

El Paso Water Utility (2018). *Conservation*. Available at www.epwater.org/conservation

ESCWA (2003). *Sectoral Water Allocation Policies in Selected ESCWA Member Countries, An Evaluation of the Economic, Social and Drought Related Impact*. New York: United Nations.

ETIC (2019). Available at https://euphratestigrisinitiativeforcooperation.wordpress.com/

European Commission (2000). Water Framework Directive (2000/60/EC). *Official Journal*, 22 December 2000.

European Parliament and The Council of the European Union (2000). Directive 2000/60/EC of the European Parliament and of the Council of 23 October 2000 Establishing a Framework for Community Action in the Field of Water Policy. *Official Journal of the European Communities*. L327, 22.12.2000.

FAO (2008). *Water Reports 34*, Irrigation in the Middle East Region in Figures, Aquastat Survey.

International Boundary and Water Commission (IBWC) (2018). *About the Rio Grande*. Available at https://ibwc/CRP/riogrande.htm

Khalaf, R. (2015) *Governance without Government in Syria: Civil Society and State Building during Conflict*. St Andrews University: Syria Studies Journal.

Kibaroglu, A. (2002). *Building a Regime for the Waters of the Euphrates–Tigris River Basin*. London, The Hague, New York: Kluwer Law International.

(2020). The Role of Irrigation Associations and Privatization Policies in Irrigation Management in Turkey. *Water International*, pp. 1–8.

Kibaroglu, A. and Scheumann, W. (2013). Evolution of Transboundary Politics in the Euphrates–Tigris River System: New Perspectives and Political Challenges. *Global Governance*, 19(2), pp. 279–307.

Kibaroglu, A. and Schmandt, J. (2016). *Sustainability of Engineered Rivers in Arid Lands: Euphrates–Tigris and Rio Grande/Brávo*. Policy Research Project Report, Number 190. LBJ School of Public Affairs, University of Texas at Austin.

Kuczmanski, L. (2018) Public Affairs Officer, IBWC/US Section. Phone interview. Available at www.ibwc.gov/Citizens_Forums/citizens_forums.html

LBJ School of Public Affairs (2000). *Water Planning Initiatives in the Paso del Norte: A Review of El Paso, Las Cruces, Doña Ana County and Far West Texas Planning Region*.

(2001). *Water Planning in the Paso del Norte: Toward Regional Coordination*.

Nava, L. (2020). The Transboundary Paso del Norte Region: Stakeholders' Preferences Allowing Water Resource Adaptation. Available at www.springerprofessional.de/en/the-transboundary-paso-del-norte-region/16802426

Neto, S. (2017). The OECD Principles on Water Governance. *Water International*, Special Issue, 43(1).

Organization for Economic Cooperation and Development (OECD) (2015). *Stakeholder Engagement for Inclusive Water Governance*. OECD Studies on Water. Paris: OECD Publishing

Saade-Sbeih, M. et al. (2016). Post Conflict Water Management: Learning from the Past for Recovery Planning in the Orontes River Basin. *Proceedings of the International Association of Hydrological Sciences*, 374, pp. 17–21.

Stolp, C. and Schmandt, J. (2018). *Sustainable River Management on the US/Mexico Border: Recommendations for the Paso del Norte*. Policy Research Project Report, Number 202. LBJ School of Public Affairs, University of Texas at Austin.

Stockholm International Water Management Institute-SIWI (2006). Beyond the River, Sharing Benefits and Responsibilities: Final Programme. World Water Week. Available at www.yumpu.com/en/document/read/24632793/final-programme-beyond-the-river-world-water-week

Svoboda, E. and Pantuliano, S. (2015). International and *Local/Diaspora Actors* in the Syria *Response:* A *Diverging Set* of *Sys*tems? Humanitarian Policy Group (HPG) Working Paper. Overseas Development Institute.

Texas Water Development Board (TWDB) (2018). *Historical Water Use Estimates*. Available at www.twdb.texas.gov/waterplanning/waterusesurvey/estimates/index.asp

Turner, C., Hamlyn, E., and Ibanez, O. (2018). *The Challenge of Balancing Water Supply and Demand in the Paso del Norte*.

UN Environment (2018). *Progress on Integrated Water Resources Management*. Global Baseline for SDG 6 Indicator 6.5.1: Degree of IWRM Implementation.

Notes

1 HARC is a research institute that seeks to provide independent scientific analyses on energy, air, and water issues with the goal of creating a sustainable future that helps people thrive and nature flourish. HARC was created by philanthropist and sustainable-development advocate George P. Mitchell, to whom this book is dedicated.

2 The Euphrates–Tigris basin includes almost one-third of the surface water supply in Turkey. An integrated water and land resources development program was designed with the specific aim of irrigating the fertile lands in south-eastern Anatolia, about one-fifth of the country's irrigable land. For this purpose, Turkey implemented the GAP, which included twenty-two large dams, nineteen hydropower plants, and several irrigation schemes. When fully developed, the GAP will provide irrigation for 1.7 million hectares of land. *See* www.gap.gov.tr/en.

Part V Conclusion

19 Conclusion

What We Found and What We Recommend

Aysegül Kibaroglu, Jurgen Schmandt, and George H. Ward

19.1 THE RIVERS WE STUDIED

Humans have long relied on river basins to grow and maintain civilizations. China, Egypt, and India stand out as ancient examples. In 1899, British engineers built the Lower Aswan Dam on the Nile. This was the beginning of modern water engineering, which has dramatically changed river systems in many parts of the world. Large dams store water in new lakes – sub-basins – with their own hydrological, economic, and ecological characteristics. Downstream from the dams, water is distributed through complex networks of canals and pipelines. The most important goal of engineered rivers is to provide water for irrigated agriculture and steadily growing cities. Reservoirs make it possible to support these goals during wet and dry seasons – a key advantage compared to natural systems, in particular in arid or semi-arid areas. Engineered rivers also produce electricity, enable water-intensive industries, and, last but not least, control environmental flows. In this volume we seek to enhance our understanding of how engineered rivers in arid or semi-arid regions have fared in the years since reservoir construction, what problems they face at present, and how they should deal with current and future challenges.

For our study we selected engineered river systems from six continents: the Nile (Burundi, Congo, Egypt, Eritrea, Ethiopia, Kenya, Rwanda, South Sudan, Sudan, Tanzania, and Uganda), Euphrates–Tigris (Turkey, Syria, and Iraq), Yellow (China), Murray–Darling (Australia), São Francisco (Brazil), Limarí (Chile), Colorado (United States and Mexico), Rio Grande (United States and Mexico), and Jucar (Spain) – a total of nine systems (or ten, when Euphrates and Tigris are looked at separately). These rivers share important characteristics. First, mountainous headwaters provide reliable river flow from glaciers, snowpack, or rainfall. Second, hundreds of kilometers downstream the river flows through areas with arid or semi-arid climates. Here the water flows calmly, and the fertile soil particles brought downstream by each year's spring flow are deposited along the riverbanks, creating ideal conditions for irrigated agriculture. Engineered reservoirs and distribution systems make it possible to plan the timing and volume of water release carefully and, thereby, significantly increase quantity and quality of food production. The river basin economies, as well as world food production, rely on this unique combination of engineering, soil, and water (Figures 19.1 and 19.2).

19.2 THE WAY WE ORGANIZED OUR WORK

The contributions of the SERIDAS rivers to basin economies and global food supply are large. Yet they must not be taken for granted. All of the SERIDAS rivers face serious challenges that are changing the quantity, quality, timing, and reliability of water supply. We studied these challenges, present in all SERIDAS rivers, though often with different intensity and timing: climate change (both global warming and climate variability), loss of storage volume due to reservoir siltation, declining groundwater levels and environmental flows, and threats to economic and food security. The various challenges are often interconnected with one another. Addressing one may necessitate addressing others – which complicates policy choices exponentially. We then examined the following response strategies: water-wise irrigated agriculture, intelligent interbasin water transfers, and active stakeholder participation in water management.

In the following pages we summarize the research discussed in detail in the main sections of the book: Challenge, Engineered Rivers, Response. We conclude with our recommendation on how to achieve river sustainability within the limits dictated by nature's dependable yield.

19.3 CHALLENGE

The elephant, or elephants, in the room for all our river basins are the still uncertain effects of climate change, which are likely to exacerbate many of the other challenges menacing the basins. Chapter 2 presents a thorough and accessible discussion of climate modeling to estimate future conditions in the

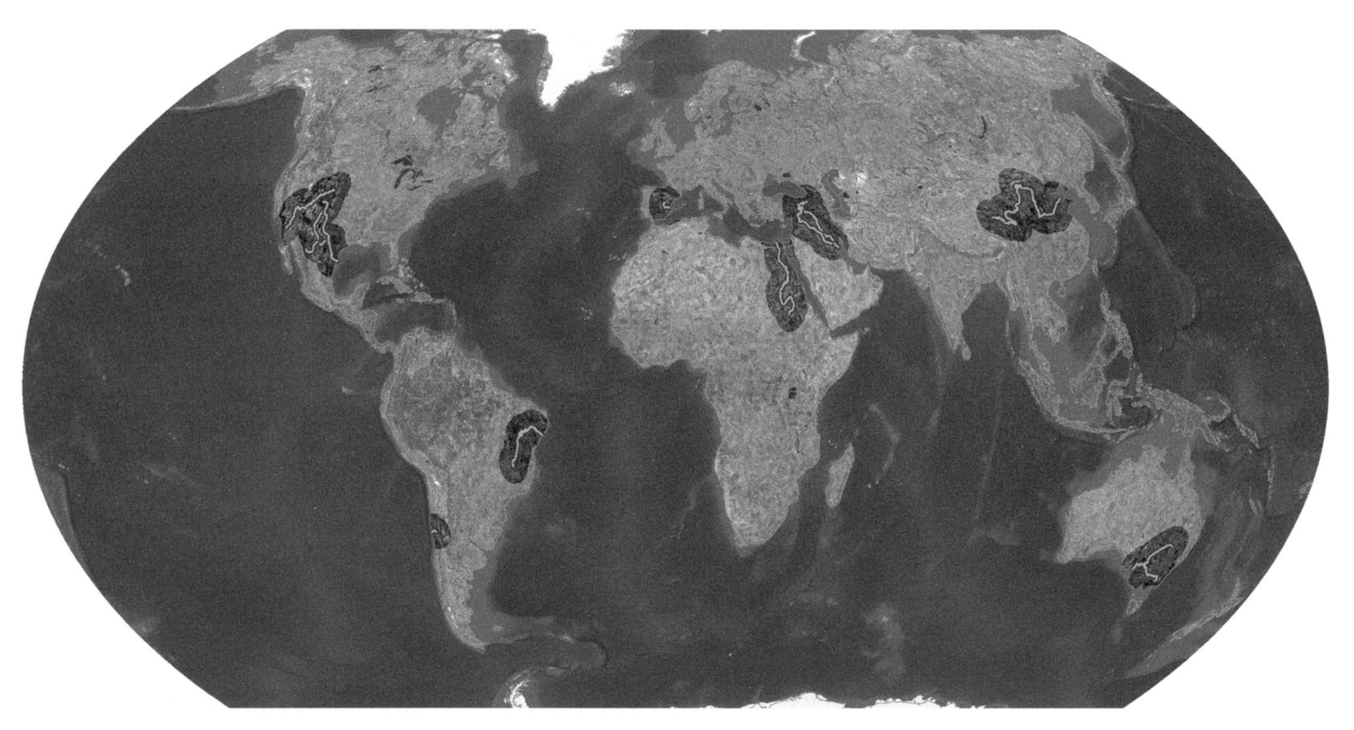

Figure 19.1 The SERIDAS Rivers.
Figure by Houston Advanced Research Center (HARC)

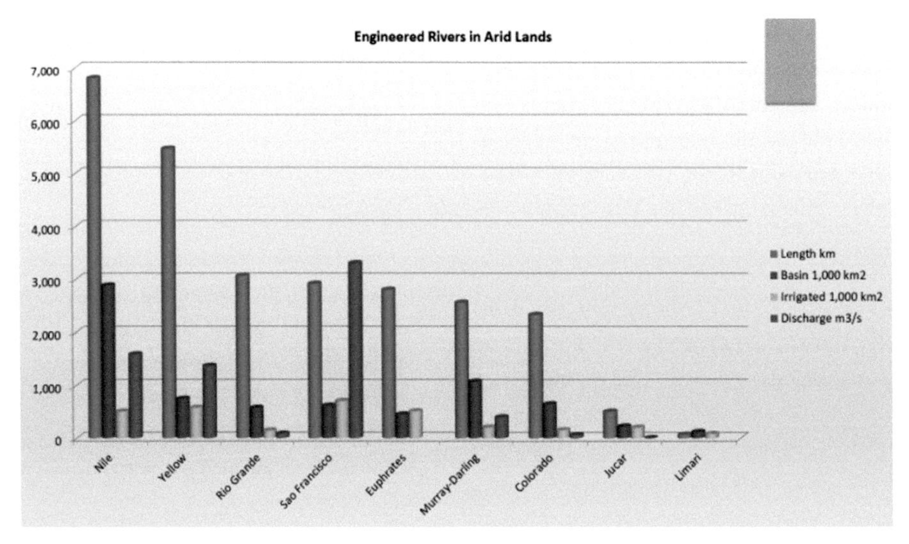

Figure 19.2 Physical characteristics of the SERIDAS rivers.
Figure by the Houston Advanced Research Center (HARC)

Earth-hydrological system. The author considers patterns of rainfall, potential evapotranspiration, actual evaporation, and runoff, among other factors, as well as the feedback mechanisms in the climate system that will impact water availability and use in the SERIDAS basins. Currently available climate models predict the planet will be warmer at the end of this century by about 3°C plus/minus 1.50°C, while certain regions (the polar regions and the interiors of larger landmasses) will warm slightly more and ocean surfaces will warm at a slower pace. Climate modeling also tells us that the midlatitude storm belts will recede gradually toward their respective poles, causing the Hadley Cells to widen and expanding the arid subtropical zones. Warming in the cloud-free subtropics will lead to more warming and hence more evaporation of soil water.

Chapter 3 describes design and functions of river reservoirs. Special attention is given to a reservoir's dependable yield.

The main challenges faced by SERIDAS reservoirs are highlighted. Humans have long augmented the water storage available in nature by constructing dams – barriers to the flow of water – in streambeds to create reservoirs. Broadly, the function of a reservoir determines whether storage of water is temporary or indefinite, e.g., flood-control reservoirs are kept empty while water-supply reservoirs are kept full. Most reservoirs are single-purpose, though multi-purpose reservoirs have been increasingly built in the twentieth century. For water-supply reservoirs, the key parameter is yield, whose evaluation requires a long period of flow records for the impounded river(s). Dependable yield is that draft which can be maintained constantly without failure throughout the time history of reservoir storage. Use of dependable yield as a basis for safe allocation therefore assumes the historical drought is the worst possible, an assumption that is flawed if the measurement record is too short, or if hydroclimate changes. A reservoir is a replacement of a segment of the river with a watercourse that is quite different, a larger, more quiescent water body with different water quality and capable of stratification. All reservoirs act as sediment traps and will eventually silt up unless special actions are taken to manage sediment. Reservoirs significantly alter the hydrology of the river downstream from the dam and can affect its water quality and its ecosystem. With increases in population and agriculture, water demands will increase, exacerbated by climate changes. Skillful management will be needed to temper conflicts that arise over how reservoirs will be operated under situations of incompatible objectives.

Chapter 4 explores the connection between groundwater and surface water. This is an important, but often overlooked, component of managing engineered rivers in arid lands. In order to make good management decisions, decision makers must understand the composition of water supplies and hydrological connections at work in their basin, including what portion of the overall water budget is supplied by groundwater, both in terms of availability and usage. Data gaps may impede this understanding in certain river basins. Directing policy and resources to meeting this information need should be a key priority.

Chapter 5 explores the challenge of maintaining food security in the context of increasing water scarcity. The authors encourage basin managers to closely cooperate with agriculture's main sub-sectors – farming and aquaculture – and actively contribute to planning and managing limited water, land, and soil resources. They highlight examples of holistic approaches to sustainable use of water, land, and soils. The best efforts focus not only on specific resources but serve wider objectives, such as conserving the landscape, developing ecosystem-based agriculture, and nature-based solutions. All of them include essential elements needed for coping with water scarcity.

In Chapter 6 we discuss the impact of river engineering on natural flow regimes and the attendant effects on ecosystems. Changes in flow regimes have resulted in huge ecological change, including the creation of exotic species that replace or compete with the native species. Many rivers in arid regions have growing populations and economic activity dependent on their water supplies. Restoring natural flows may not be possible.

19.4 ENGINEERED RIVERS: PAST, PRESENT, AND FUTURE

Chapters 7–15 examine challenges and response strategies of individual SERIDAS rivers. Placing the generic challenges listed earlier in the context of specific river conditions is critically important for guiding the development of good water management strategies, both at basin and sub-basin levels. Understanding how the broad, general challenges impact individual rivers and whether and how they are addressed should become an essential part of good river management.

The Nile must urgently address upstream–downstream issues. The projected large increase in future riparian population and the resulting push for development will increase demand of the Nile flow in upstream riparian countries where headwaters are sourced. Variability in temperature and precipitation due to climate change will impact the heavily populated river delta the most. This will have a profound economic impact as Egypt is the most prosperous country along the Nile. However, the increased development and engineering upstream will affect how much of the Nile is available at a specific time for Egypt. These upstream–downstream tensions have not been resolved. There is no basinwide management. The construction of the Grand Ethiopian Renaissance Dam is fraught by lack of an agreement on water sharing between riparian countries. If sufficient flow is not released to Egypt, saltwater intrusion in Egypt's river delta can endanger the quality of stored groundwater. The country would be well advised to invest more in desalination.

The water question emerged on the international agenda in the Euphrates–Tigris (ET) basin when the three riparian nations, namely Iraq, Syria, and Turkey, initiated major water and land resources development projects in the late 1960s. The political linkages established between transboundary water issues and nonriparian security issues also exacerbated the disagreements over water sharing and allocation. In 1987 and 1990, two bilateral Euphrates water sharing protocols were negotiated. They are acknowledged by all riparian states as being interim agreements. However, these bilateral accords failed to include basic components of sustainable water resources management, namely water quality management, environmental protection, and stakeholder engagement. In the early 1980s, the Euphrates–Tigris basin countries created an institutional framework, namely the Joint Technical Committee. However, they did not empower the committee with a clear and jointly agreed mandate. Instead, the riparian countries continued unilateral and uncoordinated water and land development ventures.

Impacts of climate change add to the already complex list of management shortcomings. The basin is one of the most affected regions. The findings of science project significant decreases in the Tigris and Euphrates flows. Examining the water–food–energy nexus in the ET basin is important because there are serious pressures on the river system due to population growth, agricultural practices, hydropower development, and ecosystem mismanagement. We recommend that transboundary institutions should apply the nexus approach, which helps to identify key development drivers as well as to unpack and clarify the development challenges and necessary tradeoffs in the basin. Sustainability of water resources requires stability, cooperation, and peace. The sub-state level conflicts and illegal control of water resources and water infrastructure in the basin deprive people of access to sufficient clean water, energy, and food resources in Syria and Iraq. The prerequisites for establishing or restoring sustainability in a river basin include stability as well as establishing participatory, transparent, inclusive, and accountable governance structures.

Among the SERIDAS rivers, the Yellow is most like the Nile in its length, relatively modest discharge to the sea, total area under irrigation, and historically important conveyance of silt. Like the São Francisco (reviewed later), it lies entirely within one large country, namely China, with a comparably large basin area, although with nearly twice the length. The Yellow can lay claim to being the most engineered river discussed in this volume. At the same time, the Yellow is by its nature not sustainable since it carried the world's heaviest silt load. This silt load (loess plateau) has fallen considerably in recent years, but perhaps at the cost of other forms of sustainability, such as streamflow. The reasons for this dramatic decline in runoff are not entirely clear. In addition to reducing silt load, terraces and vegetation have led to a marked reduction in runoff. The fall in natural runoff can also be attributed to groundwater and mining extractions, as well as reservoir filling. Population per se is not a major driver of water demand compared to irrigated agriculture and other sectors, notably mining and industry. With the population projected to peak by 2030, and decline thereafter, population is unlikely to be a major factor in the foreseeable future. While China is not a federal system, it is organized in a complex hierarchical system where provinces play an important role and can serve their own interests in negotiating usages and allocations of the river. Nine province-level administrations impose their specific goals on the Yellow River basin.

Sustainable river management in Australia's Murray–Darling Basin (MDB) has been the goal of governments in the country since at least the 1980s, when growing salinity problems and decreasing security of supply began to cause political conflict between and within the four states that share the catchment. In many years, the MDB is a closed basin with flows to the sea only maintained by dredging. In the future, under climate change conditions, the MDB is predicted to become drier and more variable. Since 1917, a catchment-wide organization, the Murray–Darling Basin Authority (MDBA), has been responsible for coordinating the management of storage reservoirs, major infrastructure, and cross-border flows. Over the last half century this body has adapted to changing conditions, successfully taking on new functions such as environmental management and sustainability planning. Large efforts have been made by the MDBA and state governments to reduce water salinity caused by natural and human-induced changes in native vegetation. In both dryland and irrigated farming, the focus was initially on the distribution of water to irrigators with only marginal concern for environmental values. An interstate water-sharing framework based essentially on proportional shares of available flow was introduced. It has proved successful through more than a century of intermittent droughts, some very severe. In recent years, water entitlements have been given stronger legal definition and there is now a well-developed water market in the MDB, particularly within states. The water needs of different activities are now mediated through the dynamics of the market rather than the decision of government officials. In times of crisis, however, it is still possible for governments to intervene and give priority to urban centers. As a response to severe drought in the 2000s, Water Act 2007 and the MD Basin Plan were approved in 2012. In many, if not most, regions in the MDB, however, water management is more complex now than it was only a few decades ago. The shortage of skilled personnel to manage Australia's highly modified hydrological systems, which is already making itself felt, could well prove the greatest risk in the medium and longer term.

The São Francisco is an entirely Brazilian river. The river crosses some of the driest parts of the Brazilian semi-arid region and brings life to ecosystems and to millions of people. Droughts are a recurrent problem in the basin. Since the second part of the twentieth century, several dams, irrigation projects, and water and sanitation systems have been built, and these developments have impacted river conditions, such as streamflows, sedimentation, silting, and water quality. Currently, four projects are planned, under construction, or functioning that will take waters from the São Francisco Basin to other basins in the northeastern part of Brazil. Water conflicts in the river involve mainly consumptive water uses, such as irrigation, and nonconsumptive water uses, such as hydroelectricity. Sedimentation is one of the big challenges presently facing the river. Others include changes in land use, deforestation, loss of biodiversity, desertification, and poverty. This is presently a big concern both at the governmental and societal levels. A program for the revitalization of the São Francisco Basin, with reforestation and sanitation measures, is a current priority. Scenarios indicate that, in the future, precipitation and streamflows may be reduced significantly due to climate change. At the same time, water demand will increase due to population growth and development. So far, the development of the river, and its engineering solutions, have prioritized the hydroelectric and agricultural sectors. This has started to change with the new water law of 1997 and the creation of the National Water Agency (ANA) in 2000. Both in the medium and long run, river water will become scarcer and will have to be managed more carefully. Water uses must be evaluated and a conclusion reached on what should be the best uses of the river waters.

The Limarí basin is the most engineered basin in Chile. The reservoirs were built by the national government between 1930 and 1972 with the objective of reducing climate vulnerabilities and enabling the development of intensive, export-oriented agriculture. Water rights and water market mechanisms are key characteristics to describe water management and allocation in the Limarí basin. The 1981 Water Code strengthens private water use rights and declares them freely tradable. The water resources access should be treated as a commodity delivered to a free-market regime. This extreme free-market approach makes Chile a unique case around the world. In other countries with water markets, such as Australia or the United States, markets are regulated by a wide framework of water use regulations and policies. In Chile, by contrast, water law and policy are dominated by the free market. Engineering infrastructure, climatic conditions, and institutional capacities in terms of tradable water rights and private water user associations allowed the economic development in the Limarí Valley. However, the lack of governmental regulation has led to overexploitation of water resources threatening water security, such as environmental and agricultural sustainability. In the face of climate change and decreasing water availability, the current infrastructural and management system requires reforms.

The Colorado River today is at a tipping point, with water storage at unprecedented low levels, forcing the unwelcome recognition by many water managers and water users that historic patterns of management and use are likely no longer sustainable. Though decades of often short-sighted management decisions have contributed to the crisis environment that currently surrounds the river, cooperative action has emerged on many levels. Over the past 15 years, stakeholders and managers have enacted significant changes to the rules and regulations governing the Colorado, and additional reforms remain close to enactment. Unlike the deal-making that characterized negotiations in the twentieth century, recent efforts focus primarily on strategies to promote coordinated management, reduce consumption, and restore a more holistic watershed perspective to a river that was legally apportioned long ago among a highly complex amalgam of jurisdictions and water users. The Colorado is a river in transition – a newly "closed" basin where demands have exhausted reliable supplies. In many respects, the river management system is a successful example of adaptation to changing social and hydrological conditions. Yet, it remains an open question whether the pace and scale of reform is sufficient to deal with the basin's future.

The Rio Grande, in Mexico called the Río Bravo (del Norte), is the fourth largest river in North America by both mainstem length and drainage area. The river marks the border between the United States and Mexico for 1,600 km, which makes its management particularly complex. Modern water engineering began a century ago when the US Bureau of Reclamation built the world's first large dam and reservoir on the Rio Grande River in New Mexico. Multiple reservoirs, diversion channels, and irrigation canals have since been added to the river.

In the border segment of the basin, population has doubled every twenty years since 1960. This trend will continue. Two large socioeconomic sub-basins have seen continued growth as a result of Rio Grande engineering – Paso del Norte (PdN) and Lower Rio Grande Valley (LRGV). The PdN is home to extensive agriculture and the cities of Las Cruces (New Mexico), El Paso (Texas), and Cd. Juárez (Mexico) with a current population of 3 million people. Population is expected to double by 2060. The Upper Rio Grande water is divided under interstate agreements between Colorado, New Mexico, and Texas and, under a 1906 treaty, between Mexico and the United States. Reliable projections to 2060 can be made for reservoir sedimentation and population growth. To date the three main reservoirs on the Rio Grande have lost about a quarter of their storage capacity. By 2060, total losses will have reached or surpassed 40 percent. Population in the economically important parts of the basin has increased rapidly since the 1950s. The current population of 6.5 million people in the bi-national economic sub-basins (PdN and LRGV) will reach 13 million by mid-century.

Climate change will have a significant impact on water supply by mid-century. To date, water planning in the Rio Grande basin – in Mexico as well as in the United States – considers changes resulting from reservoir sedimentation, population growth, and changes in land use. Climate change, on the other hand, is barely considered. There have been multiple disputes among states and countries. A sustainable 2060 scenario – limited climate change impacts on river hydrology, reduction of reservoir sedimentation, low population growth, more efficient agricultural and municipal use, and improved environmental flow – is not achievable. However, the basin will be able to supply drinking water to its projected 13 million people. In normal years, irrigated agriculture can continue to be the backbone of the basin economy and ensure food security, provided that farmers and managers begin now to learn how to do more with less. During drought years this will not be possible. The PdN may cope better than the LRGV, due to the availability of good or fair quality groundwater.

The Jucar basin faces the challenge of meeting an enormous demand for irrigation water while water quality degrades due to urban, industrial, and agricultural pollution. The expansion of water extractions and severe drought spells in recent decades have triggered considerable negative environmental and economic impacts in the basin. Within five decades, water withdrawals have doubled per year as irrigated areas almost tripled. This expansion of extraction along with intensive groundwater pumping and quality degradation has left a minimal flow. The management of water is decentralized, with the Basin Authority in charge of water allocation, and water user associations in charge of secondary infrastructure, water usage, operation and maintenance, investments, and cost recovery. The main advantage of this institutional setting is that stakeholders cooperate in decisions, rules, and regulations, and therefore the implementation and enforcement processes are carried out smoothly. To confront the progressive water depletion in the Jucar Basin Authority district, there has been a large set of management initiatives mostly based on engineering solutions supported with public subsidies. The list includes massive investments in water technologies, such as modernization of irrigation systems, urban wastewater treatment plants, seawater desalination plants, and interbasin water transfers. However, increasing river sustainability requires better water governance. Relying on engineering solutions is not enough. Empirical evidence in Jucar indicates that water markets and institutional policies seem to deal with water scarcity more successfully than water pricing and irrigation subsidies. A first water governance priority is to convince farmers of substituting freshwater for the available urban recycled water. Second, seawater desalination plants must be upgraded so they will work at full capacity. More long-term governance goals are to curtail surface irrigation diversions and groundwater extractions, and reallocating water to urban, industrial, and environmental uses. These reforms will only work if they get the support and cooperation of farmers by compensating them for the reallocation of water from agriculture to other sectors.

19.5 RESPONSE

We studied three response strategies that address our key interest – the need for improving water supply and use in irrigated agriculture. Many others exist. These will serve as examples.

Chapter 16, *River Basin Management and Irrigation*, addresses the need for irrigated agriculture, currently using 90 percent of river water, to proactively deal with increasing water scarcity. Governments and individual farmers alike tend to first favor supply augmentation. This can come from new reservoirs, the reuse of wastewater (whether treated or not), transfers, desalination, or pumping from aquifers. Where these options are already fully exploited or unfeasible (physically or economically) there is no alternative but to reduce evapotranspiration (rather than "use"). This is often forced by circumstances and may result in crop losses or short-term adjustments, such as deficit irrigation. In the medium term (seasonally or annually) farmers may adjust by fallowing land, changing crops or calendars, or improving on-farm techniques and practices (mulching, land-levelling, etc.), while long-term strategies may involve the sale of land or economic diversification.

Chapter 17, *Intelligent Interbasin Transfers*, should be taken in the context of general sustainable development strategies and Integrated Water Resources Management. This requires negotiations between donor and recipient stakeholders, so that agreements are reached prior to the construction of transfer projects. However, one must also consider the broader political momentum for the water transfer decision, which may not be favorable in the near future.

Chapter 18, *Better Water Management with Stakeholder Participation*, reviews the role of irrigation districts, city water utilities, and environmental groups in basin management. Examples from two river basins – Rio Grande and Euphrates–Tigris – illustrate vast differences in stakeholder participation. The authors recommend that all SERIDAS rivers pay increased attention to this option for better management. The four options for creating sub-basin water councils recommended for the Rio Grande provide useful guidance. Other models, reflecting different basin conditions, may emerge. Whatever model is selected, stakeholders should always organize to address the water agenda of their sub-basin. Doing so directly contributes to reaching and maintaining sustainability of the river as a whole. Each sub-basin organization should (1) assess current and future water supply and demand; (2) project future dependable water yield; (3) conduct reservoir impact assessments; (4) draft and implement a plan for sustainable water management; and (5) cooperate closely with river agencies/commissions.

19.6 SUSTAINABILITY OF ENGINEERED RIVERS IN ARID LANDS

How do some of the cornerstone documents on sustainable development address river sustainability? We begin with a look at The Club of Rome, perhaps the best known nongovernmental organization working on sustainability and growth. Over the course of five decades the organization has published widely read reports, drawing worldwide attention to sustainable development in the 1970s, then updating their findings several times, last in 2018 for the Club of Rome's fiftieth anniversary.

There is a second reason why we begin with the Club of Rome. In their work they also use the challenge-response sequence that is central to our book. River sustainability is not given by nature. It is always the result of natural or social change, conflict, debate, and finally action – *Challenge and Response*. Alexander King, co-founder and president of the Club of Rome, put it this way, "The danger spots of the problematique – population, environment, urban decay, global economy poverty, technological development, etc. (must be complemented) by The Resolutique, ... strategies for the resolution of problems, suggested lines of action" (King, 2006, p. 398).

Obviously, the Club of Rome reports cover a much broader agenda than the SERIDAS project. But we find it useful to see how our topic – sustainability of engineered rivers in arid lands – is addressed by a world leader in the sustainability debate. We then review a few additional sources discussing river sustainability. This prepares the ground for us asking the following question: Do we restate common knowledge, or do we cover new ground? We believe the latter is the case. To make this point we end with a new definition of sustainability of engineered rivers in arid lands.

The Limits to Growth report discusses the availability of arable land for food production. "There has been an overwhelming access of potentially arable land for all of history, and now, within 30 years ... there may be a sudden and serious shortage." The reason: "Arable land is removed for urban-industrial use as population grows" (Meadows, 1972, pp. 50–51). Irrigation, reservoirs, and water shortages are not discussed; food production is.

The 20-year update, *Beyond the Limits*, includes a four-page discussion of water as a resource for food production. It concludes: "Globally water is in great excess, but because of operational limits and pollution, it can in fact only support one more doubling of demand, which will occur in twenty to thirty years" (Meadows, 1992, p. 56). There is a reference to dependable yield, defined as follows: the "rates of use of renewable resources do not exceed their rates of regeneration" (Meadows, 1992, p. 209). Again, irrigation and water shortages are not discussed; food production is.

Limits to Growth: The 30-Year Update, devotes a ten-page chapter to water. It recognizes that water is a regional resource, available in specific watersheds. Dams have increased usable runoff by about 3,500 cubic kilometers per year, but they also flood prime agricultural land and increase evaporation from the river basin. "Sooner or later they silt up and become ineffective, so they are not a source of sustainable flow" (Meadows, 2004, p. 68). Ways to work toward water sustainability should include doing more with less, such as using drip irrigation instead of flooding fields. Most noteworthy: "Water sustainability is not possible without climate sustainability, which means energy sustainability" (Meadows, 2004, p. 74).

Come On!, the 50-year update, discusses water in the context of sustainable agriculture. The report finds that "the currently dominant global agriculture is in no way sustainable" and needs to be replaced by "agro-ecology (that) preserves soils and water supplies, regenerates and retains natural soil fertility and encourages biodiversity; its yields are sustainable in the longer term. To a large extent, it avoids agrochemicals by growing diverse crops together. ... It sequesters carbon rather than live on it" (Von Weizsacker and Wijkman, 2018, p. 123).

On balance, in the 50-year work of the Club of Rome, we find a continuing focus on the importance of water for food production, a growing recognition of ecological challenges, and hints that reservoirs are undermining warnings about the threat to river sustainability. Action recommendations are limited to changing farming practices; they do not address specifics on river governance, stakeholder participation, or sustainability planning. Other organizations have been more specific, and closer to the definition we submit.

The 2030 Agenda for Sustainable Development (Rio+20) defines seventeen Sustainable Development Goals (SDGs), including a dedicated goal for water and sanitation (SDG 6) that calls for "availability and sustainable management of water and sanitation for all" (United Nations, 2013).

In April 2016, the UN Secretary-General and the President of the World Bank appointed ten heads of states and governments and two special advisors to the High-Level Panel on Water. The group is charged with mobilizing action to accelerate the implementation of the aforementioned Sustainable Development Goal number 6 (United Nations, 2016). Previously, in 2015, a group of fifteen countries from all parts of the world launched the Global High-Level Panel on Water and Peace at a ministerial gathering in Geneva. The mandate of this panel is to propose global architecture to transform water from a source of potential crisis to an instrument of cooperation and peace (United Nations, 2016).

In 1998, the American Society of Civil Engineers published a report entitled *Sustainability Criteria for Water Resource Systems*. Their definition of a sustainable water resource was further refined by ASCE committee member D. P. Loucks in a UNESCO Hydrology Series publication, published in 1999 and also titled *Sustainability Criteria for Water Resource Systems*: "Sustainable water resource systems are those designed and managed to fully contribute to the objectives of society, now and in the future, while maintaining their ecological, environmental and hydrological integrity" (Loucks and Gladwell, 1999, p. 131). In 2015, ASCE concluded that their suggested definition had not had the desired impact: "Somewhat surprisingly, given the urgent stressors associated with rising global temperatures and competitive demands on water, definitions of sustainability [of water resources] have not been particularly evident in the intervening years ... " (McMahon, 2015). Brian Richter proposed specific sustainability principles based on the water management approach by the Australian Murray–Darling basin authority. The principles are: (1) build a shared vision for your watershed; (2) set limits on consumptive water use; (3) allocate a specific volume to each user, then monitor; (4) invest in water conservation; (5) enable trading of water entitlements; (6) subsidize reductions in consumption; (7) learn from mistakes or better ideas; and (8) adjust as you go (Richter, 2014, p. 140).

These are excellent principles. They are part of our recommendation: Use the calculated yield of water available in a reservoir under normal and historical drought conditions – the dependable yield – as the critical starting point for sustainable reservoir management. Without this information, river management operates without a firm foundation. Let us explain how to determine and use dependable yield by using the example of the Amistad–Falcon reservoirs in the Rio Grande / Bravo basin.

The fundamental problem for engineered basins in arid regions is the time variability in the natural water supply, specifically river flow. Reservoirs for storage of water are essential in engineered basins because of their capability to retain water during surfeit periods for distribution during drought. Design and operation of heavily engineered water systems require a long period of record (POR) of river flow, either from historical measurements or from long-term simulations, to meet minimal needs during periods of low river flow or drought. Mathematically, these are periods of negative trends in the measure of water availability. Drought for a reservoir is defined by a period of declining trend in stored water, called a drawdown period. The *dependable yield* of a reservoir is the constant volume of withdrawn water, *met without failure*, during the POR (Linsley and Franzini, 1964). This occurs during severe drought, specifically, the "drought of record," also called the "critical drawdown period." An example determination of dependable yield for the reservoirs in the Lower Rio Grande is shown in Figure 19.3.

For the period 1945 to 1960, the first two estimates of the constant withdrawal clearly bound the dependable yield since the reservoirs meet 200 and 250 Mm^3/month during the entire time period. Yet the latter withdrawal was no longer possible during the drought of record in the mid-fifties. Additional approximations within these bounds converge to 229.6 Mm^3/month as the dependable yield during the critical drawdown period.

While dependable yield is a theoretical construct, it is the single most important management indicator for assessing reservoir performance and is often used as a basis for allocation of water from a reservoir system. We argue that it needs to be carefully monitored and recalculated whenever natural or social conditions in the basin reduce water supply or increase water use. Climate change and population growth are change factors that need special attention. Based on the concept of dependable yield we propose this definition for sustainability of engineered rivers in arid lands:

A reservoir-dominated river in arid lands is sustainable when these conditions are met,

1. Nature's water supply, averaged over the period of the most severe drought experienced in the historical record, delivers a dependable yield sufficient to meet human and ecological needs in the basin.
2. To prepare for increased water scarcity from either natural or human causes, water managers and stakeholders proactively search for ways to use water more efficiently.
3. Whenever observed or projected changes in the natural system or human actions modify river flow, the dependable yield is recalculated, and water managers, after consultation

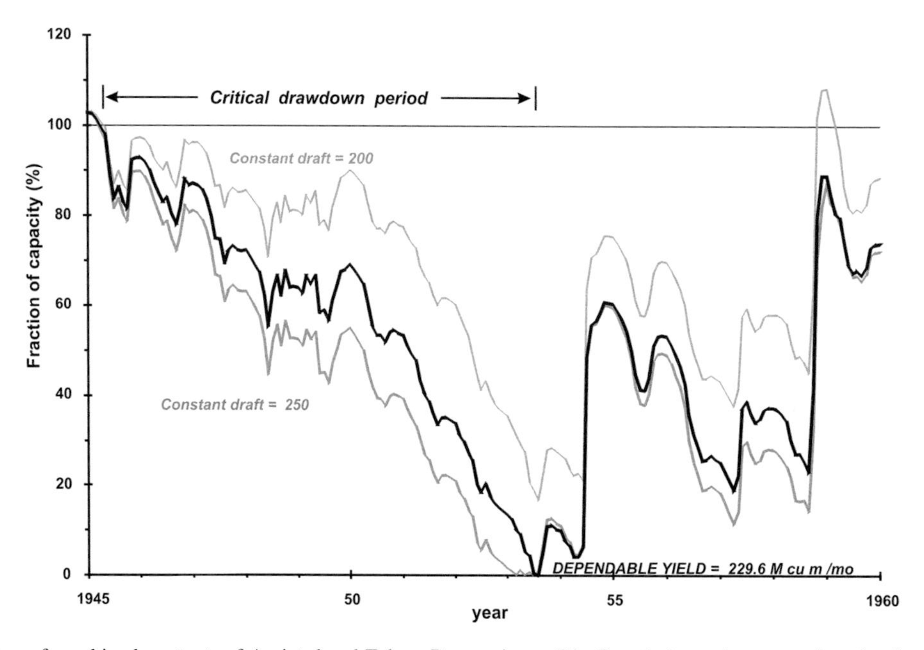

Figure 19.3 Simulation of combined contents of Amistad and Falcon Reservoirs on Rio Grande in tandem operation, for drought of record. Dependable yield (229.6 Mm3/month) calculated by iteration.
Figure by George Ward

with stakeholders, adjust existing rules for water allocation and water use to match the new level of dependable yield.
4. An ecologically prudent level of instream flow is maintained or restored.

We conclude with three sobering comments to remind us of the big obstacles that stand in the way of meeting these sustainability goals for engineered rivers in arid lands.

Management practices: Each of our four criteria calls for a departure from the usual management practices of most river basins. Their application requires willingness of the prevailing management culture to consider changes in their traditional operational practices. We invite river management agencies, as well as water stakeholders, to adjust their views and procedures so that maintaining or moving toward sustainability becomes the principal goal for river management.

Conjunctive management: Sustainable water management requires integrative consideration of connections between river water and groundwater. Existing law in some rivers, for example the Texas part of the Rio Grande basin, makes this goal difficult to reach. We urge movement toward conjunctive management of water resources.

Water conflicts: Where water becomes the core of, the trigger to, and a weapon in active conflict, the goal of sustainable water supply and demand is unattainable. Sustainability of water resources requires stability, cooperation, and peace. The sub-state level conflicts and illegal control of water resources and water infrastructure in the Euphrates–Tigris deprive people of access to sufficient clean water, energy, and food resources in Syria and Iraq. The lack of water undermines the search for sustainable development and causes agricultural, economic, and political decline. Hence, we argue that the prerequisites for establishing or restoring sustainability in a river basin include stability as well as participatory, transparent, inclusive, and accountable governance structures.

Only when these issues are addressed can river managers and stakeholders cope with the looming problems of water scarcity, climate change and variation, reservoir sedimentation, population growth, economic losses, food security, and ecological damage. It can be done. We repeat the Club of Rome's admonition: *Come On!*

REFERENCES

American Society of Civil Engineers (ASCE) (1998). *Sustainability Criteria for Water Resource Systems*. Reston, Virginia.

King, A. (2006). *Let the Cat Turn Round. One Man's Traverse of the Twentieth Century*. London: CPTM.

Linsley, R. and Franzini, J. (1964). *Water Resources Engineering*. New York: McGraw-Hill Book Co.

Loucks, D. P. and Gladwell, J. S. (eds.) (1999). *Sustainability Criteria for Water Resources Systems*. UNESCO International Hydrology Series, New York: Cambridge University Press.

McMahon, G. F. et al. (2015). Special Issue on Sustainability. *Journal of Water Resources Planning and Management*, 141(12).

Meadows, D. H. et al. (1972). *The Limits to Growth*. A Report to the Club of Rome's Project on the Predicament of Mankind. New York: Universe Books.

(1992). *Beyond the Limits: Confronting Global Collapse: Envisioning a Sustainable Future*. Post Mills, VT: Chelsea Green Publishing Company.

(2004). *Limits to Growth: The 30-Year Update*. White River Junction, VT: Chelsea Green Publishing Company.

Richter, B. (2014). *Chasing Water: A Guide for Moving from Scarcity to Sustainability*. Washington, DC: Island Press.

United Nations (2013). *Transforming Our World: The 2030 Agenda for Sustainable Development*. Available at https://sustainabledevelopment.un.org/post2015/transforming ourworld

(2016). *UN and World Bank Chiefs Announce Members of Joint High-Level Panel on Water*. UN News. Available at https://news.un.org/en/story/2016/04/527352-un-and-world-bank-chiefs-announce-members-joint-high-level-panel-water

Von Weizsacker, E. U. and Wijkman, A. (2018). *Come On! Capitalism, Short-terminism, Population and the Destruction of the Planet*. Report to the Club of Rome. New York: Springer.

Index